# 6th International Conference on Recent Trends in Structural Materials (COMAT 2020)



IOP Conference Series: Materials Science and Engineering
Volume 1178

Pilsen, Czech Republic
30 November - 4 December 2020

**Editor:**

**Jan Dzugan**

ISBN: 978-1-7138-4638-3
ISSN: 1757-8981

**Printed from e-media with permission by:**

Curran Associates, Inc.
57 Morehouse Lane
Red Hook, NY 12571

**Some format issues inherent in the e-media version may also appear in this print version.**



Institute of Physics
Dirac House, Temple Back
Bristol BS1 6BE UK

Phone: 44 1 17 929 7481
Fax: 44 1 17 920 0979

techtracking@iop.org

**Additional copies of this publication are available from:**

Curran Associates, Inc.
57 Morehouse Lane
Red Hook, NY 12571 USA
Phone: 845-758-0400
Fax: 845-758-2633
Email: curran@proceedings.com
Web: www.proceedings.com

# TABLE OF CONTENTS

IOP Conf. Series: Materials Science and Engineering 1178 (2021) 011001 doi:10.1088/1757-899X/1178/1/011001

# Preface

COMAT deals with recent trends in structural materials that cover production technologies (casting, forging, rolling, additive manufacturing…), heat and thermo-mechanical treatments, computer modelling and physical simulations and materials characterisation. COMAT started in 2010 so the last year we celebrated the 10th anniversary of the conference and also the 20th anniversary of COMTES FHT Inc. The conference was organized by traditional tandem COMTES FHT Inc. and TANGER Ltd. There were big plans to celebrate these anniversaries with adequate dignity. However, the COVID-19 has changed it, as well as many other things and we ended up with a significantly reduced event. Although the situation was not positive, we highly appreciated interest of participants from Europe and Asia. We received many papers for publication in proceedings out of which 67 were selected for the proceedings publication. Due to an uncertain situation and almost a global lock-down it took much longer to finalize the review process this time. We would like to thank very much to reviewers and authors for their comments and papers corrections. Anyway, we are finally here and we hope the papers presented in the proceedings will be interesting for all readers.

We are looking forward to seeing you all personally at COMAT 2022.

Yours faithfully,

Jan DŽUGAN

Proceedings editor

June 2021

IOP Conf. Series: Materials Science and Engineering 1178 (2021) 011001 doi:10.1088/1757-899X/1178/1/011001

## Scientific Commitee

Prof. Peter M. ANDERSON - The Ohio State University, Columbus, Ohio, U.S.A.

Prof. Tomasz BABUL - Institute of Precision Mechanics, Warsaw, Poland, EU

Assoc. Prof. RNDr. Maria BEHÚLOVÁ, CSc. - University of Technology in Bratislava, Slovakia, EU

Prof. Irene J. BEYERLEIN - University of California, Santa Barbara, CA, U.S.A.

Prof. Dr. Michael BRÜNIG - Bundeswehr University Munich, Germany, EU

Prof. Dr. Wenquan CAO - China Iron and Steel Research Institute, China

Dr. Daniel CASELLAS - Eurecat Technology Centre, Manresa, Spain, EU

Prof. Dr. Jose CESAR de SA - University of Porto, Portugal, EU

Prof. RNDr. Matěj DANIEL - Czech Technical University in Prague, Czech Republic, EU

Prof. Dr. Jan DŽUGAN - COMTES FHT a.s., Dobrany, Czech Republic, EU

Prof. Alexander GOKHMAN - South Ukrainian National Pedagogical University, Odessa, Ukraine

Prof. RNDr. Pavel LUKÁČ, DrSc. - Charles University, Prague, Czech Republic, EU

Prof. RNDr. Zuzanka TROJANOVÁ, DrSc. - Charles University, Prague, Czech Republic, EU

Dr. Mohsen SEIFI - ASTM International, Pennsylvania, U.S.A.

Prof. Ruslan Z. VALIEV - Ufa State Aviation Technical University, Ufa, Russian Federation

Prof. Dr. Ing. Dalibor VOJTĚCH - University of Chemistry and Technology, Prague, Czech Republic, EU

## Programme Committee

Prof. Dr. Jan DŽUGAN - COMTES FHT a.s., Dobrany, Czech Republic, EU

Ing. Libor KRAUS - COMTES FHT a.s., Dobrany, Czech Republic, EU

Dr. Ing. Zbyšek NOVÝ - COMTES FHT a.s., Dobrany, Czech Republic, EU

Ing. Tasilo PRNKA - TANGER Ltd., Ostrava, Czech Republic, EU

## Organizing Committee

Ing. Kateřina SEDLÁČKOVÁ - TANGER Ltd., Ostrava, Czech Republic, EU

Bc. Monika BAROŠOVÁ - TANGER Ltd., Ostrava, Czech Republic, EU

Mgr. Petra TUŠKOVÁ - COMTES FHT a.s., Dobrany, Czech Republic, EU

Eva HŮLOVÁ - TANGER Ltd., Ostrava, Czech Republic, EU

IOP Conf. Series: Materials Science and Engineering 1178 (2021) 011002 doi:10.1088/1757-899X/1178/1/011002

# Peer review declaration

All conference organisers/editors are required to declare details about their peer review. Therefore, please provide the following information:

- **Type of peer review: Single-blind / ~~Double-blind / Triple-blind / Open / Other~~**

Single-anonymous: authors' identities are known to the reviewers, reviewers' identities are hidden from authors

Double-anonymous: author and reviewer identities are hidden to each other

Triple-blind: author and reviewer identities are hidden to each other, and from the Editor(s)

Open: author and reviewer identities are known to each other

- **Describe criteria used by Reviewers when accepting/declining papers. Was there the opportunity to resubmit articles after revisions?**

  **Reviewers' criteria:**
  Do you think that the paper title corresponds with its contents?
  Do you think that the abstract expresses the paper contents well?
  Are the results or methods presented in the paper novel?
  Do the author(s)state clearly what they have achieved?
  Do you find the terminology employed proper?
  Do you find the bibliography representative and up-to-date?
  Do you find all necessary illustrations and tables?
  Do you think that the paper will be of interest to the journal readers?

  Based on reviewers comments the authors were asked for the papers corrections.

- **Conference submission management system:**
  System was provided by the conference organiser TANGER Ltd. The registration systems has been used by the company for many conferences they are organizing.

- **Number of submissions received:**
  84 abstracts were submitted, and of those 75 papers were finally accepted
  **Number of submissions sent for review:**
  75
- **Number of submissions accepted:**
  67
- **Acceptance Rate (Number of Submissions Accepted / Number of Submissions Received X 100):**
  89%
- **Average number of reviews per paper:**
  2.28 review per contribution

- **Total number of reviewers involved:**
  Altogether 121 reviewers were involved in the review process

- **Any additional info on review process (ie plagiarism check system):**
  There were up to three iterations until the paper was finally accepted for the publication.
- **Contact person for queries:**
  Prof. ing. Jan Džugan, Ph.D. – jan.dzugan@comtesfht.cz

Please submit this form along with the rest of your files on the submission date written in your publishing agreement.

The information you provide will be published as part of your proceedings.

# Microstructure evolution of Cu-Fe-based immiscible alloys prepared by powder metallurgy

**O Adam[1], V Jan[1]**

Institute of Materials Science and Engineering, Brno University of Technology, Technicka 2896/2, Brno 616 69, Czech Republic

E-mail: Ondrej.Adam@vutbr.cz

**Abstract**. The work is focused on the preparation of bulk immiscible Cu-Fe-based alloys by powder metallurgy. Three samples with chemical composition $Cu_{70}Fe_{30}$, $Cu_{70}Fe_{15}Co_{15}$, and $Cu_{50}Fe_{25}Co_{25}$ were prepared by mechanical alloying and spark plasma sintering (SPS). Microstructure evolution during sintering and the effect of Co on the resulting microstructure and hardness of the bulk samples were evaluated. Despite the immiscibility of Cu with Fe and Co, the FCC supersaturated solid solution was formed upon mechanical alloying. This supersaturated solid solution was decomposed during SPS and fine microstructure, consisting of separated BCC and FCC phases, was formed. The results showed that cobalt influenced the particle size of milled powders as the particle size of $Cu_{70}Fe_{30}$ alloy was about an order of magnitude higher compared to other alloys. Cobalt also affected the resulting microstructure of bulk samples, however, its effect on the hardness was negligible. It has been shown that powder metallurgy can be used for the preparation of bulk immiscible alloys with the fine microstructure consists of separate phases, which can be individually alloyed by the selected elements, and therefore, powder metallurgy can be considered as a suitable alternative to the more used casting.

## 1 Introduction

In the last years, much attention has been paid to heterogeneous materials due to their ability to overcome strength-ductility trade-off [1–4]. The good combination of strength and ductility is achieved by hetero-deformation induced strengthening (HDI). It is the result of the heterogeneous microstructure composed of two domains with different mechanical properties (typically hard and soft domains) that can be caused by microstructural, crystal structure, or/and compositional heterogeneities [5,6]. Soft domains are easily plastically deformed than hard domains, which creates gradients of plastic deformation in soft domains. These plastic gradients need to be accommodated by geometrically necessary dislocations (GNU) that contribute to the overall strengthening of the material.

Immiscible alloys can be noted as heterogeneous materials due to their crystal structure heterogeneity. These alloys are characterized by the positive enthalpy of mixing that significantly limited mutual solubility of present elements and the microstructure at room temperature is composed of separated phases of nearly pure elements [7]. Creating domains with different properties is an inherent feature of the immiscible alloys, therefore, these alloys seem to be a promising group of materials for creating advanced heterogeneous materials with a superior combination of strength and ductility. Cu-Fe system is one of the most studied immiscible alloys because it is assumed to be suitable material for electronic components [8,9] and recently has been used as a base for new advanced alloys [3,10]. At room temperature, the microstructure consists of two domains, BCC-Fe phase and FCC-Cu phase [7]. Each domain can be alloyed individually which allows controlling of its microstructure and properties that leads to the tailoring of mechanical properties. Currently, most bulk Cu-Fe-based alloys are produced by casting although in order to achieve fine and uniform microstructure (thus better mechanical properties) by this method, it is often necessary to subsequently perform homogenization, followed by mechanical and thermal treatment [3,11,12]. On the other hand, powder metallurgy (PM) provides a relatively easy way how to produce material with comparable mechanical properties, in some cases PM can even outperform cast materials [13]. Moreover, PM enables better control of microstructure during processing by setting various parameters during milling and sintering. It has been proven that it is possible to prepare a metastable homogeneous solid solution of Cu-Fe alloy by mechanical alloying (MA) [7,14]. However, there is a lack of studies focusing on sintering such a prepared powder and on the evolution of microstructure during sintering.

This study deals with the microstructural evolution of Cu-Fe alloys prepared by PM and with the effect of cobalt on the resulting microstructure and hardness. For this purpose, three alloys $Cu_{70}Fe_{30}$, $Cu_{70}Fe_{15}Co_{15}$, and $Cu_{50}Fe_{25}Co_{25}$ (at. %) were prepared using the same parameters of MA and sintering. Cobalt was chosen due to its immiscibility with copper, which should lead to alloying only BCC-Fe phase and therefore to increasing heterogeneity between Fe and Cu phases.

## 2 Materials and method

Immiscible alloy powders with the chemical composition of $Cu_{70}Fe_{30}$, $Cu_{70}Fe_{15}Co_{15}$, and $Cu_{50}Fe_{25}Co_{25}$ were prepared by a high-energy ball milling process. Commercially pure Cu, Co, Fe (purity 99.5 wt. %) elemental powders with particle size corresponds to 325 mesh were used as starting materials. The powder mixtures were milled together with hardened bearing steel balls of 10 mm and 15 mm in diameter, at a ball-to-powder weight ratio of 10:1. Mechanical alloying was performed in an argon atmosphere at the milling speed of 240 RPM for 16 hours using planetary ball mill Fritsch Pulverisette 6. The time scheme of MA was set to 60 min of milling and 30 min of idle to prevent overheating. Additionally, wet milling in ethanol for 15 min was done to fully remove the powder stuck to the milling equipment.

The milled powders were sintered by SPS (Dr. Sinter SPS625) in a 20 mm diameter graphite die in a vacuum atmosphere. Graphite foil was placed between the powder and die to ensure proper electrical contact. Sintering was performed from room temperature with a heating rate of 100 °C/min to the final temperature of 800 °C with 5 min dwell time and uniaxial pressure of 50 MPa. After sintering, the setup was left to spontaneous cool down, until the chamber was opened at ~ 200 °C. The amount of used powder was calculated to obtain 6 mm thick cylindrical samples.

The phase composition of powder and bulk samples was measured by X-ray diffraction (XRD). Philips X'Pert Pro diffractometer was used with Cu-Kα radiation (λ = 1.5418 Å), operated at 40 kV with a current of 30 mA. A continuous scanning was performed for 2θ from 20° to 105° by a speed of 0.011°/min and a step size of 0.0167°. The microstructural analysis of bulk samples was performed by field-emission scanning electron microscope (SEM; ZEISS Ultra Plus) using back-scattered electrons (BSE) for identifying present phases. Secondary electrons (SE) were used for observing milled powders. Chemical composition was measured by Energy-dispersive X-ray spectroscopy (EDS; Aztec, Oxford Instruments). Microhardness measurement was carried out by Qness Q10A hardness tester with a load of 200 g and 10 s dwell time. The presented values are an average of at least six measurements.

## 3 Results and discussion

### *3.1 Mechanically alloyed powders*

XRD analysis showed that despite the positive enthalpy of mixing, at least partially dissolution of Fe and Co into Cu occurred in all samples, resulting in the formation of FCC supersaturated solid solution (SSS). $Cu_{70}Fe_{30}$ and $Cu_{70}Fe_{15}Co_{15}$ powders showed complete dissolution and their microstructure consisted of only FCC SSS. The microstructure of $Cu_{50}Fe_{25}Co_{25}$ powder contained approximately

14% of BCC phase, which was caused by a higher amount of Fe and Co that had to be dissolved during mechanical alloying. To achieve fully alloyed powder, prolongation of milling time is required. Regardless of the chemical composition, the powder particles flattened to platelet shape during milling (Figure 1). The particle size of powders containing Co was very similar, the average value was in the order of tens micrometers (Figure 1 b), c)). The average particle size of $Cu_{70}Fe_{30}$ powder was one order of magnitude higher, suggesting its low hardness that leads to a reduction of powder fragmentation and an excessive cold-welding effect. On the other hand, greater platelet thickness of $Cu_{50}Fe_{25}Co_{25}$ powder indicates that it had the highest value of hardness.

**Figure 1.** Mechanically alloyed powders: a) $Cu_{70}Fe_{30}$; b) $Cu_{70}Fe_{15}Co_{15}$; c) $Cu_{50}Fe_{25}Co_{25}$. Note that the image of $Cu_{70}Fe_{30}$ was taken at a lower magnification due to the powder size.

### *3.2 Bulk samples*

XRD analysis revealed that FCC SSS decomposed during SPS and the microstructure of bulk samples was composed of BCC and FCC phase. FCC phase corresponds to Cu and BCC phase corresponds to Fe, in the case of $Cu_{70}Fe_{30}$ alloy [7] or to (Fe, Co) solid solution, in the case of $Cu_{70}Fe_{15}Co_{15}$ and $Cu_{50}Fe_{25}Co_{25}$ alloys [15]. The BCC (Fe, Co) solid solution should be ordered under 730°C (B2 structure), however XRD patterns of samples containing Co were similar to the pattern of $Cu_{70}Fe_{30}$ alloy and no superlattice peak was observed. Therefore it was assumed that the structure of BCC (Fe, Co) solid solution was disordered (A2 structure). The amount of BCC phase of all samples is presented in Table 1. It can be seen that amount of BCC phase of $Cu_{70}Fe_{30}$ and $Cu_{70}Fe_{15}Co_{15}$ alloys was equal to the equilibrium amount, which means that full decomposition of FCC SSS occurred. The equilibrium amount of BCC phase for $Cu_{50}Fe_{25}Co_{25}$ alloy is 48 wt. % indicating the decomposition of FCC SSS during SPS was incomplete and Fe and Co are still partially dissolved in the FCC phase. The separated phases formed a very fine mixture in the microstructures of bulk samples (Figure 2), but relatively large single-phase areas with the chemical composition of pure Cu were also observed. The shape of these areas indicates that they were formed on the surfaces of the milled powder particles, but the origin of their formation is not clear. However, it is evident that the presence of these areas was more pronounced in the samples containing Co, which forms a solid solution with Fe and is immiscible with Cu. It is therefore assumed that Co increased the driving force of phase separation that leads to the formation of large pure Cu areas. The effect of Co on the resulting microstructure can be also seen on the micrographs of the fine mixture of separated phases that are shown in Figure 2 d)–f). Microstructures of $Cu_{70}Fe_{30}$ and $Cu_{70}Fe_{15}Co_{15}$ alloys were very similar, composed of the submicron BCC phase embedded in the continuous FCC phase. However, $Cu_{70}Fe_{15}Co_{15}$ alloy contained coarser BCC phase suggesting that phase separation was more pronounced in this case, and thus that Co promotes phase separation. The microstructure of $Cu_{50}Fe_{25}Co_{25}$ alloy was relatively coarse with an obvious gradual merging of BCC phases. It has to be noted, that in this case, the phase separation was not completed and the continuous FCC phase was still formed by the SSS as was evaluated by XRD. The microstructures of all alloys contained also homogeneously dispersed round nano-sized particles, which were determined as oxides and carbides by EDS. The presence of carbides and oxides is common in materials prepared by PM. Moravcik et al. [16] have shown that the main oxides source is the oxidation of elemental powder surfaces and carbides are formed due to the carbon contamination during milling (wear of milling bowl and balls) and SPS (carbon-rich environment inside graphite die).

Table 1 also presents the average hardness of all samples. By comparing the results of $Cu_{70}Fe_{30}$ and $Cu_{70}Fe_{15}Co_{15}$ alloys it is clear that Co has no effect on the hardness of bulk samples. In this study, $Cu_{70}Fe_{30}$ alloy achieved even slightly higher average value of hardness (234±2 HV0.2) than $Cu_{70}Fe_{15}Co_{15}$ alloy (228±4 HV0.2), although, with respect to data scattering, the values can be considered as equal. $Cu_{50}Fe_{25}Co_{25}$ alloy showed the highest hardness (282±12 HV0.2). Since the amount of BCC phase is comparable to other samples, this increase in hardness is attributed to the solid solution strengthening of present FCC SSS.

**Table 1.** Amount of phases measured by XRD and average Vickers hardness.

| Alloy | Amount of phases [wt. %] | | Hardness [HV0.2] |
|---|---|---|---|
| | FCC | BCC | |
| **$Cu_{70}Fe_{30}$** | 72 | 28 | 234±2 |
| **$Cu_{70}Fe_{15}Co_{15}$** | 72 | 28 | 228±4 |
| **$Cu_{50}Fe_{25}Co_{25}$** | 76 | 24 | 282±12 |

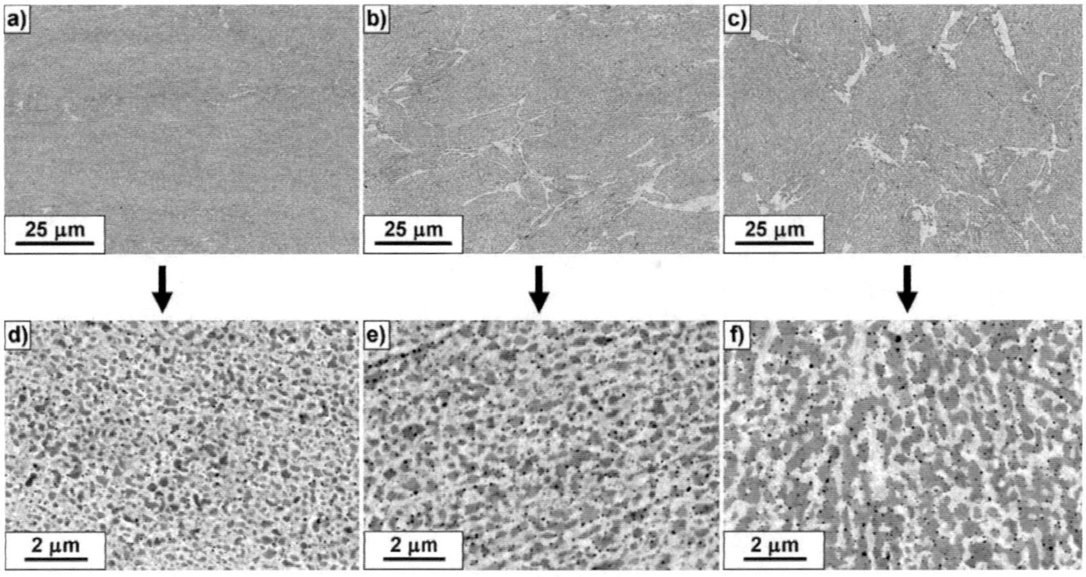


**Figure 2.** Microstructures of bulk samples. The bottom row of images shows a close-up view of a fine mixture of separated phases. a), d) $Cu_{70}Fe_{30}$; b), e) $Cu_{70}Fe_{15}Co_{15}$; c), f) $Cu_{50}Fe_{25}Co_{25}$. FCC phase is bright, BCC phase is dark. Black dots in the images in the bottom row are oxides and carbides.

## 3 Conclusion

In this paper, three immiscible alloys with the chemical composition of $Cu_{70}Fe_{30}$, $Cu_{70}Fe_{15}Co_{15}$, and $Cu_{50}Fe_{25}Co_{25}$ were prepared by powder metallurgy. Based on the results, the following conclusions can be drawn:

- combination of mechanical alloying and SPS provides a suitable method for the preparation of immiscible alloys with fine microstructure
- Co reduced excessive cold-welding effect during MA that led to the smaller particle size of milled powder
- the microstructure of bulk samples was composed of pure Cu areas, a mixture of separated BCC and FCC phases, and round nano-sized oxides and carbides
- it is assumed that Co promoted phase separation leading to the more pronounced formation of pure Cu areas and a coarser BCC phase
- Co had a negligible effect on hardness of bulk samples

**Acknowledgements**

This work was supported by the project no. FSI-S-20-6484 of BUT. The authors thank Ing. Vaclav Pouchly, Ph.D. in CEITEC for helping to perform consolidation of milled powders by SPS and Ing. Zdenek Spotz, Ph.D. for XRD analyses.

**References**

[1] Su J, Raabe D and Li Z 2019 Hierarchical microstructure design to tune the mechanical behavior of an interstitial TRIP-TWIP high-entropy alloy *Acta Mater.* **163** 40–54
[2] Park H K, Ameyama K, Yoo J, Hwang H and Kim H S 2018 Additional hardening in harmonic structured materials by strain partitioning and back stress
[3] Moon J, Park J M, Bae J W, Do H-S, Lee B-J and Kim H S 2020 A new strategy for designing immiscible medium-entropy alloys with excellent tensile properties *Acta Mater.*
[4] Zhu X, Cao G, Liu J, Zhao K and An L 2020 Enhanced strength and ductility synergy in aluminum composite with heterogeneous structure *Mater. Sci. Eng. A* **787** 139431
[5] Wu X and Zhu Y 2017 Heterogeneous materials: a new class of materials with unprecedented mechanical properties *Mater. Res. Lett.* **5** 527–32
[6] Ma E and Zhu T 2017 Towards strength–ductility synergy through the design of heterogeneous nanostructures in metals *Mater. Today* **20** 323–31
[7] Ma E 2005 Alloys created between immiscible elements *Prog. Mater. Sci.* **50** 413–509
[8] Stepanov N D, Kuznetsov A V., Salishchev G A, Khlebova N E and Pantsyrny V I 2013 Evolution of microstructure and mechanical properties in Cu-14%Fe alloy during severe cold rolling *Mater. Sci. Eng. A*
[9] Gao H, Wang J, Shu D and Sun B 2005 Effect of Ag on the microstructure and properties of Cu-Fe in situ composites *Scr. Mater.* **53** 1105–9
[10] Sarkar S, Srivastava C and Chattopadhyay K 2018 Development of a new class of high strength copper alloy using immiscibility route in Cu-Fe-Si system: Evolution of hierarchical multi-scale microstructure *Mater. Sci. Eng. A*
[11] Guo M X, Zhu J, Yi L, Wang F, Li G J and Lei R S 2017 Effects of precipitation and strain-induced martensitic transformation of Fe-C phases on the mechanical properties of Cu-Fe-C alloy
[12] Moravcik I, Hadraba H, Li L, Dlouhy I, Raabe D and Li Z 2020 Yield strength increase of a CoCrNi medium entropy alloy by interstitial nitrogen doping at maintained ductility *Scr. Mater.* **178** 391–7
[13] Moravcikova-Gouvea L, Moravcik I, Omasta M, Veselý J, Cizek J, Minárik P, Cupera J, Záděra A, Jan V and Dlouhy I 2020 High-strength Al0.2Co1.5CrFeNi1.5Ti high-entropy alloy produced by powder metallurgy and casting: A comparison of microstructures, mechanical and tribological properties *Mater. Charact.* **159**

IOP Conf. Series: Materials Science and Engineering 1178 (2021) 012001 doi:10.1088/1757-899X/1178/1/012001

[14] Suryanarayana C 2001 Mechanical alloying and milling *Prog. Mater. Sci.* **46** 1–184
[15] Dreval L A, Turchanin M A, Agraval P G and Du Y 2018 Cu–Fe–Co System: Verification of the High-Temperature Phase Equilibria and Thermodynamic Modeling of the Low-Temperature Phase Relations Involving Ordered Phase *Powder Metall. Met. Ceram.* **56** 546–55
[16] Moravcik I, Kubicek A, Moravcikova-Gouvea L, Adam O, Kana V, Pouchly V, Zadera A and Dlouhy I 2020 The Origins of High-Entropy Alloy Contamination Induced by Mechanical Alloying and Sintering *Metals (Basel).* **10** 1186

IOP Conf. Series: Materials Science and Engineering 1178 (2021) 012002 doi:10.1088/1757-899X/1178/1/012002

# Analysis of strength properties of 3D printed lightweight structures made of AlSi10Mg material

**V Andronov[1], J Šimota[1], L Pelikán[1], L Beránek[1], V Novák[1]**

Faculty of Mechanical Engineering, Czech Technical University in Prague

E-mail: vladyslav.andronov@fs.cvut.cz

**Abstract**. The main aim of this article is to compare two approaches to decreasing the overall weight of components by using additive manufacturing from AlSi10Mg alloy. This optimization is advantageous as it lessens both the production time of parts (components) and overall costs, in key industries such as aviation. The first method used to reduce the weight of a structure is the implementation of lattice structures, which replace filled parts with selected structures of different shapes, while maintaining key properties of the original structure. The other method to decrease weight is the topological optimization of prepared samples. In this article, these two options have been compared. The specially designed sample was first optimized topologically. Subsequently, it was lightened using selected structures (Octet truss and Rhombic dodecahedron) in SW Magics with different lattice parameters. These samples were then printed using DMLS technology from AlSi10Mg on the ConceptLaser M2 Cusing system. The material used was analysed by SEM, and a particle size distribution was also performed. Furthermore, a three-point bending test was performed on the bodies produced in this way. The results of this test were analysed and evaluated in terms of load (bending force), depending on the type of sample and the volume/weight saved.

## 1 Introduction

Today, industry is placing increasing emphasis on reducing the weight of manufactured components. One of the most important production areas concerned with lightening a structure's weight is aviation, although this trend can also be seen in other industries. Thanks to modern, metal 3D printing technology, two possibilities have opened up to decrease the weight of products: the use of lattice structure and topological optimization.

Other articles have dealt with the topic of lattice structures prepared by 3D printing and its mechanical properties. The SLM method for cell lattice structures was used by Hanzl P. [1] This article proved the suitability of using gyroid structures in the machinery industry. The conclusion of the same article was the ability to create stiffer and lighter samples. The same method of additive manufacturing as thc one used one in Hanzl's work (DMLS) was also used by Yan C [2]. His article proved that this technology can manufacture quality lattice structures, as well as that strength and modulus increase by increasing the volume fraction and decrease by increasing the size of the cell. Rashid R's article demonstrated the suitability of using a three-point bending test to determine mechanical properties, such as flexural strength and flexural modulus. [3] Tests made in this article were also key for proving a strong influence of lattice structure geometry on the deformation.

## 2 Experimental procedure

The experiment, which is described in more detail below, consists of creating sample designs and comparing lattice structures with topologically-optimized samples, and full, non-optimized samples. A comparison is made using a 3-point bending test. As mentioned in the introduction, the objective of this article is to reduce the weight of the sample and compare the strength of individual variants, not to strictly achieve better mechanical properties as it was in many articles [4-6].

### *2.1 Sample design and topology optimization*

Sample design was based on topological optimization of the rectangle shape of a 2D beam 30x135 mm supported at the distance of 70 mm, and loaded with variable force between supports (3-point bending test). Software used for sample design was a specially created script in Matlab. The goal of optimization was to create a geometry to reduce the amount of material to 50 %, and to maximize the stiffness of the sample. The resulting sample was adjusted on the sides. The material was added so that the sample could be supported. The original geometry had a significantly thinner sidewall. Unlike in other articles [1-7], where the structures are not enclosed in any shell, lattice structures in this article were created into the shell (**Figure 1**) which is much closer to real applications. The shell was created in Autodesk Inventor.

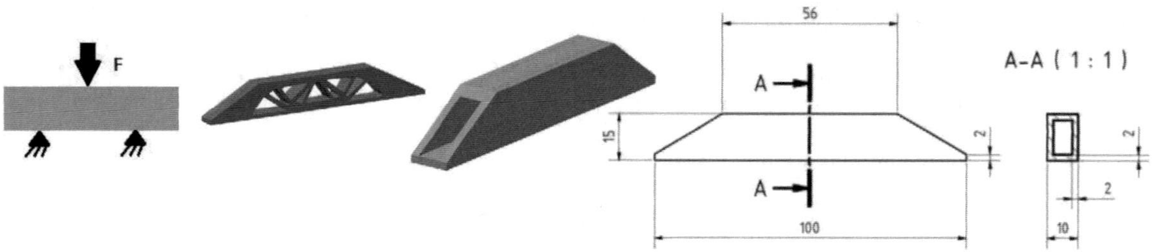


**Figure 1.** Design of topologically-optimized specimen and shell for lattice structure

### *2.2 Lattice structures*

Lattice structures were created in the software Materialise Magics 23. These structures were designed to be axially symmetrical and have similar volumes (lattice structure + shell) as the topological structure. The designed and used lattice structures (octet truss (OT) 3 and 6 mm and rhombic dodecahedron (RD) 3 and 6 mm) from Magics software are in **Figure 2**.

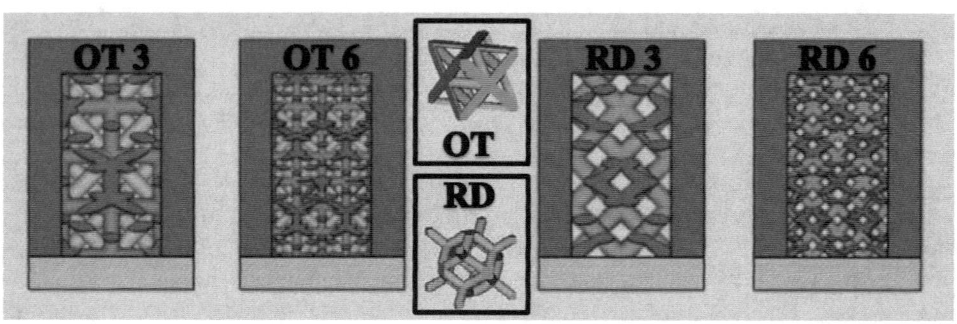


**Figure 2.** Lattice structures from left: octet truss 3 and 6 mm; rhombic dodecahedron 3 and 6 mm

### *2.3 Sample production and heat treatment*

Samples were printed using Direct Metal Laser Sintering (DMLS) technology, which is a very common technique for creating lattice structures [1-2]. The material of the samples is AlSi10Mg. The printer used was a ConceptLaser M2 Cusing system. DMLS technology is used only for alloys, and is one of the most used AM technologies in the industry. The process (printing) parameters used are listed in **Table 1**. The layer thickness used for the material was 25 µm.

**Table 1.** Printing parameters for AlSi10Mg on ConceptLaser M2 Cusing system

| Area | Power [W] | *Scan speed [mm/s]* | Spot size [mm] | *Offset to original contour [mm]* | Trace spacing |
|---|---|---|---|---|---|
| **Support** | 200 | 1600 | 0,05 | 0,1 | - |
| **Skin** | 200 | 1250 | 0,1 | 0,175 | - |
| **Core** | 370 | 1400 | 0,19 | - | 0,112 |

Each test sample was printed in 3 pieces. After printing, the building pallet with samples was heat-treated according to the recommendations of the powder manufacturer [8]. The process was stress-relief annealing: Heat up in 1 hour to 240°C, and then maintain the temperature for 6 hours. Then, cool down the oven to 100°C. Afterward, allow the parts to cool in the ambient atmosphere.
After the heat treatment, the samples were cut from the building pallet and treated by sandblasting. The samples modified in this way are shown in **Figure 3.**

**Figure 3.** Each type of sample after heat treatment and sandblasting

## 2.4 *Material (powder) analysis*

The default material for the DMLS technology is in powder form. For the powder analysis, the scanning electron microscope Tescan Vega 3 LMU with accelerating voltage 20 kV, detector SE +BSE was used following the particle distribution analysis by laser diffraction (ASTM B822).

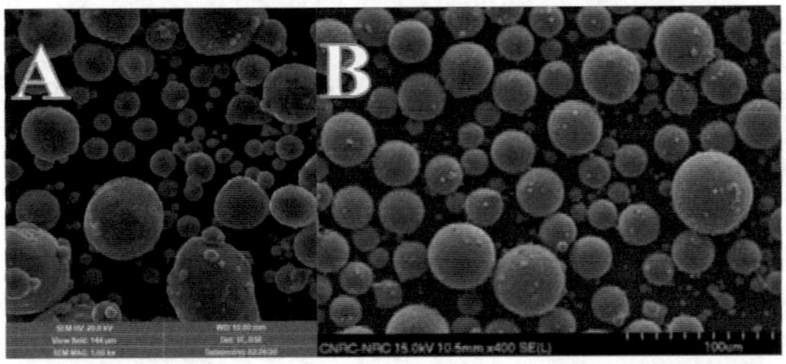


**Figure 4.** Powder quality comparison: powder used by us (A) x manufacturer's analysis (B) [7]

**Figure 4.** shows a comparison of the quality of the powder that was used with the powder analyzed by the manufacturer. At first glance, the powder from the manufacturer contains more regular and high sphericity particles. Furthermore, a smaller occurrence of satellites and imperfect particles can be seen.

Particle size distribution declared by the producer is 15-63 μm. [7] In this analyze the average mean diameter 26,29 μm was measured. In **Figure 5.** we can see particle size distribution graph of analyzed powder.

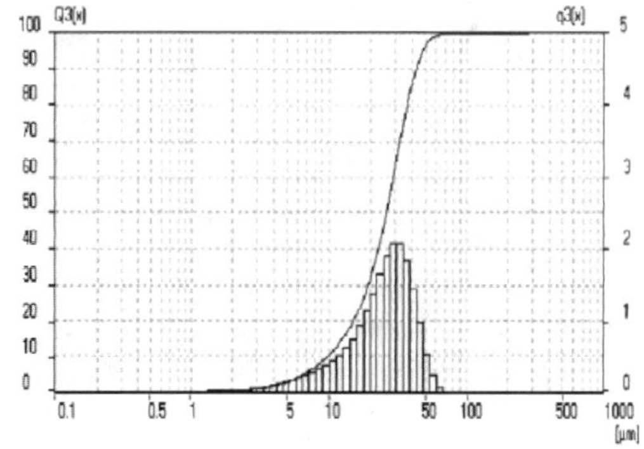


**Figure 5.** Particle size distribution graph of AlSi10Mg powder

*2.5 3-point bending test*

For the 3-point bending test, the 100 kN LabTest 5.100SP1 testing machine was used. The test progress is shown in **Figure 6.** and the loading rate was 2 mm/min.

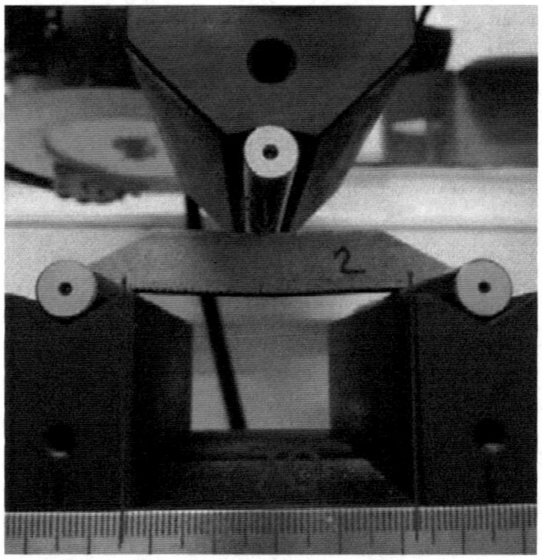

**Figure 6.** Three-point bending test

In **Figure 7.** we can see the results of the three-point bending test depending on the type of sample. Based on the experiment, the best result from the optimized samples was achieved by the sample after topological optimization.

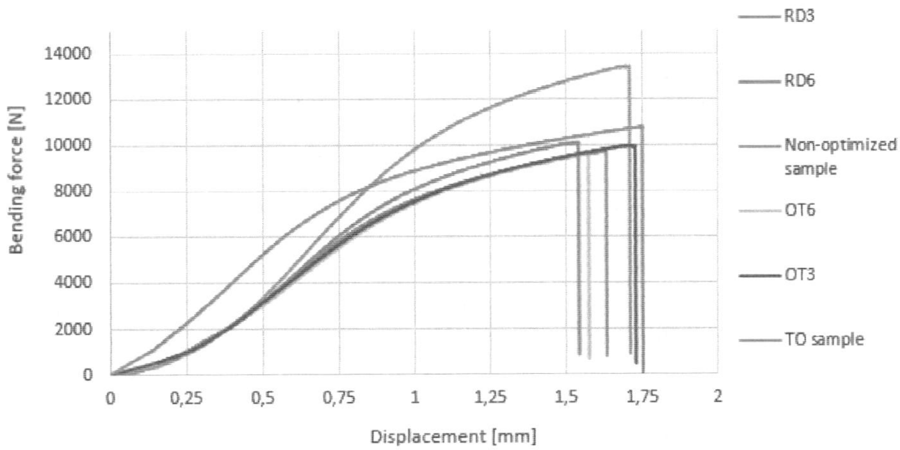


**Figure 7.** Bending force-displacement behavior of different samples

## 3 Results

Results of the 3-point bending test and sample weight are in **Table 2**. The best results were archived by topologically-optimized samples, which were the lightest and had the highest endurance.

If we evaluate the bending force / saved weight ratio, the RD6 sample turned out best from the lightweight structures. By comparing the topologically-optimized sample and the RD6 sample, we conclude that the RD6 sample was about 5% heavier than the TO sample and had about 10% lower load capacity. However, if we consider that specialized software is needed for TO and the lattice structure can be created in common software for print preparation, then this result is acceptable.

**Table 2.** Results of the experiment

| Structure | RD3 | RD6 | OT3 | OT6 | TO | Non-OPT |
|---|---|---|---|---|---|---|
| **Average value of bending force [N]** | 10 057 | 9 690 | 9 944 | 9 564 | 10 730 | 13 399 |
| **Percentage difference of maximum bending forces of samples and full body [%]** | 24,94 | 27,68 | 25,79 | 28,62 | 19,92 | - |
| **Average value of the weight of the sample [g]** | 28,01 | 24,43 | 26,96 | 24,68 | 23,22 | 31,85 |
| **Percentage difference in weight of samples and full body [%]** | 12,06 | 23,30 | 15,35 | 22,51 | 27,10 | - |

## 4 Conclusion

In this article, the two options for the optimization of 3D printing have been compared. The specially-designed sample was first optimized topologically. Subsequently, it was lightened using selected structures (Octet truss and Rhombic dodecahedron) in SW Magics with different lattice parameters. These samples were then printed using DMLS technology from AlSi10Mg on the ConceptLaser M2 Cusing system. Based on this experiment, the following was found:

- Topologically-optimized samples have the highest endurance and smallest average value of the sample weight, but can be disadvantageous as they require special software to make them
- Lattice structures had ~ 10 % less endurance than a topologically-optimized sample; the highest endurance was reached with a rhombic dodecahedron 3 mm lattice structure
- The finer the lattice structure, the better endurance, but also higher weight
- Issue of fine lattice structures is that powder might stay inside the structures, which may be dangerous for one's health

Based on the results of this experiment, it is necessary to verify the effect of the outside shell in the next step. The shell, which encloses the structures, could be so rigid that the influence of inner structure is almost insignificant.

**Acknowledgement**
The research was supported by the projects SGS19/164/OHK2/3T/12 "The development of 3D printing technology for the aerospace industry" and the ESIF, EU Operational Programme Research, Development and Education, and from the Center of Advanced Aerospace Technology (CZ.02.1.01/0.0/0.0/16_019/0000826), Faculty of Mechanical Engineering, Czech Technical University in Prague.

**References**
[1] Hanzl P, Zetek M and Zetkova I 2016 Cellular Lattice Structure Produced by Selective Laser Melting and its Mechanical Properties. *Proc. of the 26th Int. DAAAM Symp.* pp 748-752
[2] Yan Ch, Hao L, Hussein A, Bubb S L, Young P and Raymont D 2014 Evaluation of light-weight AlSi10Mg periodic cellular lattice structures fabricated via direct metal laser sintering *Journal of Materials Processing Technology* pp. 856-864.
[3] Rahman Rashid R A, Mallavarapu J, Palanisamy S and Masood S H. 2017 A comparative study of flexural properties of additively manufactured aluminium lattice structures *Materials Today: Proceedings* pp 8597-8604
[4] Alsalla H, Hao L and Smith C 2016 Fracture toughness and tensile strength of 316L stainless steel cellular lattice structures manufactured using the selective laser melting technique *Materials Science and Engineering: A* **669** pp 1-6
[5] Mahshid R, Hansen H N and Højbjerre K L 2016 Strength analysis and modeling of cellular lattice structures manufactured using selective laser melting for tooling applications *Materials & Design* **104** pp 276-283
[6] Takezawa A, Koizumi Y and Kobashi M 2017 High-stiffness and strength porous maraging steel via topology optimization and selective laser melting. *Additive Manufacturing* **18** pp 194-202
[7] AP&C material data sheet [https://www.advancedpowders.com/powders/aluminum/al-si10-mg]
[8] Conceptlaser and GE Additive material data sheet [https://www.ge.com/additive/sites/default/files/2018-12/CLMAT_30_31AL_DS_EN_US_2_v1.pdf]

IOP Conf. Series: Materials Science and Engineering 1178 (2021) 012003 doi:10.1088/1757-899X/1178/1/012003

# Influence of processing parameters and heat input level on the functional properties of atmospheric plasma sprayed $Al_2O_3$-$3TiO_2$ commercial coating

**Antoš J[1], Lencová K[1] and Keslová A[1]**

[1]Research and Testing Institute Pilsen, Surface Solution Laboratory, Jižní Předměstí 3005, 301 00 Pilsen, Czech

E-mail: antos@vzuplzen.cz

**Abstract.** This paper concerns the combined influence of spraying parameters and heat flux level during atmospheric plasma spraying process of commercial alumina-titania coating. One of the main industrial application of alumina-titania are wear and abrasion resistant protective thermally sprayed coatings. Commercial high throughput plasma torch with cascaded construction is used as the spraying equipment. Three sets of spraying parameters were used for the basic evaluation of their influence. Additionally, three levels of heat input during spraying process were added to some of those sets of parameters. Those heat input levels are called "cold", "hot" and "extra hot" depending on the relative amount of direct heat input and maximum achieved surface temperature. Comparison of surface hardness, microstructure and especially abrasion resistance is carried out. Results confirmed that for successful design of such ceramic coatings, appropriate and sufficient heat input during spraying process into the complex of layer-substrate is required.

## 1 Introduction

Atmospheric plasma spraying (APS) process, being one of thermal spraying technologies, is able to melt and subsequently deposit almost any material thanks to the temperature above 15 000 K in the plasma jet. Such high temperature is sufficient to deposit most materials, including refractory metals and ceramics and APS technology is commercially used mainly for deposition of ceramic coatings. Alumina-titania coatings were developed to produce wear resistant coatings, where titania addition to alumina significantly increases toughness of the coating compared to pure alumina. $Al_2O_3$ - 3 % $TiO_2$ commercial powder was used in this experiment as feedstock material, deposited by cascaded design plasma torch. Coatings sprayed on low carbon steel substrate underwent superficial Rockwell indentation, optical microscopy on polished cross section, adhesion strength testing and dry sand abrasion test. [1] [2] [3]

## 2 Spraying equipment, feedstock and substrate material

Atmospheric plasma spraying of all samples was done with cascaded plasma torch (SinplexPro by Oerlikon Metco) with argon - hydrogen mixture as plasma gas and 9 mm torch nozzle diameter. Samples were sprayed in department of thermal spraying of Research and Testing Institute Pilsen (VZÚ Plzeň s.r.o.).

Commercial powder Amperit 742 (fused) by Höganäs (Sweden) with nominal particle size distribution between 22 and 45 micrometers was chosen as the default material. Nominal chemical

composition is 97 % of $Al_2O_3$ and 3 % of $TiO_2$.. Trace amount of other oxides is presented ($SiO_2$, $Fe_2O_3$).
S235 construction steel served as the substrate material. Specimens with cross section of 25 mm x 5 mm and 75 mm in length were used for Dry sand rubber wheel test and specimens of same cross section and 50 mm length for metallography and indentation. In addition, cylindrical samples of 25 mm diameter and 6 mm height were used to evaluate adhesion strength of the coating according to ASTM C633.

## 3 Testing methods

Because good wear behavior and particles abrasion resistance are one of the commercially desired properties of alumina-titania coating, dry sand abrasion test was chosen in order to compare such resistance of specimens sprayed with different spraying parameters and heat input levels. Dry sand rubber wheel test according to ASTM G-65 were deployed on tree specimens of each parameter set. An output of this test was average material mass loss of tested specimens after five testing cycles. The ASTM testing standard however demands evaluation of volume loss of tested material, but due to the general difficulties of measuring the specific density of thermally sprayed coating (especially caused by closed porosity), mass loss was employed instead. It was assumed, that no relevant differences of density between samples of same material are presented and therefore mass loss can be compared with relevancy. Rockwell superficial indentation (HR 15N) were carried out on specimen from every set. Polished cross sections also underwent optical microscopy to compare microstructure, porosity and eventually presence of cracks developed in coating due to the possible overheating. Surface temperature was measured with laser pyrometer right after finishing every spraying cycle (at the start of the cooling cycle) on the coating surface. Pyrometer was calibrated to the emissivity of the coating material at ambient temperature.

## 4 Spraying parameters and heat input levels

Based on the former experience with deposition of both pure $Al_2O_3$ and $Al_2O_3$ based ceramic coatings and with utilization of results of papers concerning this topic [4, 5, 6], set of deposition parameters was designed. Main goal of spraying parameters optimization was to achieve crack-free, hard, wear resistant coating, especially with good abrasive resistance. As main variable process parameters, input torch power and argon flow rate were chosen. Those parameters have chief influence on the formation of plasma jet. Three levels of Critical Plasma Spraying Parameters (CPSP) [4, 5, 6] were deployed valuing 0.83, 1.01 and 1.17. Those three levels represent low, normal and high plasma conditions respectively. Constant spraying distance of 120 mm was used for all three sets of parameters. A fourth set of parameters was added, with the same spraying parameters and CPSP value as set No. 3 (1.17 CPSP) but with decreased spraying distance from 120 mm to 105 mm. This correction was done in order to deal with relatively short plasma jet of No. 3 set resulting from lowering argon flow from 60 l/min down to 50 l/min. For all variable parameters see **Table 1**. All other process parameters were constant during whole spraying, with 6 l/min of hydrogen flow rate as secondary plasma gas, 36 g/min of powder feed rate and argon also as a carrier gas. Torch linear speed was constant during spraying of all specimen.

**Table 1**. Parameter overview for all spraying sets.

| | CPSP | Power input *kW* | Argon flow *l/min* | Spraying distance *mm* |
|---|---|---|---|---|
| **No. 1** | 0.83 | 47 | 60 | 120 |
| **No. 2** | 1.01 | 57 | 60 | 120 |
| **No. 3** | 1.17 | 55 | 50 | 120 |
| **No. 4** | 1.17 | 55 | 50 | 105 |

With process parameter sets chosen, all four sets were sprayed several times with different heat load levels to the complex sample – coating. To achieve different heat loads with same spraying parameters, different system of spraying passes and cooling cycles were utilized. Three heat load levels were named “cold”, “hot” and “extra hot”. For spraying of “cold” sets, temperature of the substrate-coating complex did not exceeded 150 °C. After every cooling cycle, spraying was resumed after coating surface temperature dropped below 30 °C (this “ambient” temperature cooling threshold was same for all heat levels). “Hot” sets were sprayed with cooling cycles employed after 3 continuous torch passes. Maximal surface temperature after 3 continuous torch passes varied from 240 °C (set No.1) up to 300 °C (sets No. 2 and No.4). “Extra hot” was the highest heat load level, with cooling cycle after 5 continuous torch passes. Temperatures of the coating surface of those “extra hot” samples went up to 340 °C in set No. 3 and even up to 360 °C for sets No. 2 and No. 4. Not all three heat load levels were used in each spraying parameter sets. **Table 2** in the section 5 displays the distribution of heat load levels in individual spraying sets as well as the maximum temperatures during spraying mentioned above.

## 5 Results

Dry sand rubber wheel test was used as the principal test for the evaluation of the deposited coatings, with the search for the most abrasive resistant coating and thus revealing proper spraying parameters and also evaluation of the influence of heat load to the quality of the coating. Abrasive mass loss during individual cycles of the abrasive track test was plotted to a graph. Linear regression was employed to get the linear function of mass loss on the abrasive track length. Slope of this linear function was used to evaluate abrasive resistance of all coatings (results shown in **Table 2**), where lower the value of the slope, the lower mass loss during test and thus higher abrasive resistance of the specific coating. Rockwell superficial indentation results are also included in **Table 2** along with average thickness per torch pass. Rockwell superficial indentation results shown quite positive correlation with results of abrasion resistance.

Examples of microstructure images are shown below (**Figure 1** and **Figure 2**). Cracks at the interpass interfaces and some minor vertical cracks were observed in all extra hot heat load level sets (**Figure 1b** and **Figure 2b**). [7]

**Table 2**. Results of abrasive testing, hardness indentation, deposition efficiency (thickness per pass) and with maximum surface temperature during spraying with different heat loads. Colored values in the abrasive mass loss line slope goes with red representing the worst results, over tones of orange and yellowish (average results) to the green values representing the best results.

| Set nubmer | Plasma jet conditions (CPSP) | Heat input level | Maximum surface temperature °C | Thickness per pass *μm/pass* | Abrasive mass loss line slope (lower the better) | Superficial hardness (HR 15N) |
|---|---|---|---|---|---|---|
| **No. 1** | Low (0.83) | **Cold** | 150 | 32 | **0.330** | 85.8 ± 0.9 |
| | | **Hot** | 240 | 46 | **0.277** | 87.5 ± 0.4 |
| **No. 2** | Normal (1.01) | **Hot** | 300 | 45 | **0.163** | 88.3. ± 2.1 |
| | | **Extra hot** | 360 | 61 | **0.118** | 89.4 ± 1.6 |
| **No. 3** | High (1.17) | **Cold** | 150 | 40 | **0.380** | 86.7 ± 0.4 |
| | | **Hot** | 240 | 47 | **0.194** | 91.0 ± 1.4 |
| | | **Extra hot** | 340 | 48 | **0.183** | 91.3 ± 0.7 |
| **No. 4** | High (1.17) | **Hot** | 300 | 38 | **0.150** | 90.2 ± 0.5 |
| | | **Extra hot** | 340 | 47 | **0.089** | 90.9 ± 0.8 |

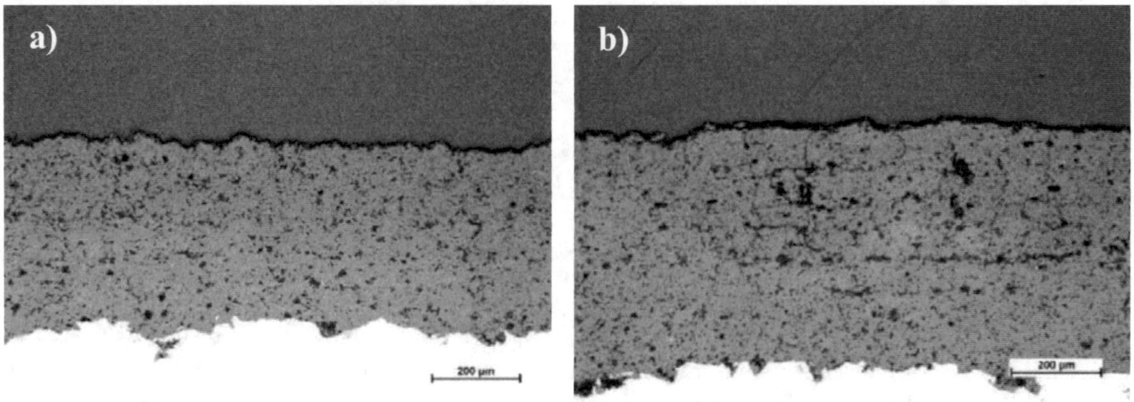


**Figure 1.** Set No. 2 samples microstruture images (100x); a) hot and b) extra hot

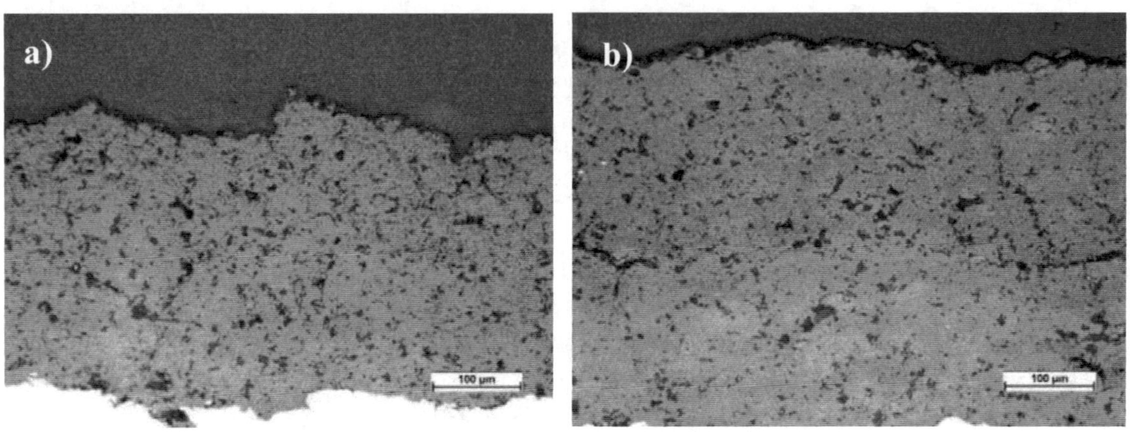


**Figure 2.** Set No. 4 samples microstruture images (200x); a) hot and b) extra hot

## 6 Discussion

Results of abrasive resistance test shown that there are two ways to increase abrasive resistance and thus presumably overall wear resistance of alumina-titania coatings. First way is increasing CPSP parameter of plasma jet above value of 1. The simplest way to increase CPSP is to increase input electrical current and thus increase power of the torch. A decrease in a primary argon flow also leads to increase in CPSP, but at a cost of somewhat reduction in plasma jet length and velocity, which may lead to inferior performance of plasma jet during spraying (see differences between spraying sets No. 2 and No. 3, where a decrease in argon flow had to be utilized in order to achieve high CPSP value). To balance plasma jet velocity reduction caused by lowering plasma gas flow, a decrease in spraying distance (i.e. distance between torch nozzle and substrate) may be employed. Effect of this can be observed in comparison between spraying sets No. 3 and No. 4., which were sprayed with exactly same parameters, except lowering spraying distance from 120 mm to 105 mm.

Second way to increase wear resistance of alumina-titania coating is to increase heat flux during spraying. This can be done simply by optimizing spraying passes and cooling cycling. Results proved that significant increase in heat input during cooling cycles leads to increasing abrasive resistance. This is probably caused by more intensive phase transformation (from unstable phases (mostly $\gamma$-$Al_2O_3$) to stable α-$Al_2O_3$ phase, that leads to improvement in mechanical properties) due to the higher temperature of the coating during spraying [8] [9].

Important fact is that both way of increasing wear resistance of alumina-based coating can be combined - increasing CPSP of plasma jet and optimizing spraying cycle to achieve higher temperatures leads to highly wear resistant ceramic coating. On the other hand, careful approach must be taken while increasing heat load and thus temperature of the coating during spraying, since subsequent increase in quenching stresses and thermal expansion mismatch between coating and substrate material may lead to development of cracks and also significant reduction in coating adhesion strength.

## Acknowledgement

The paper has originated in the framework of institutional support for the long-time conception development of the research institution provided by Ministry of Industry and Trade of the Czech Republic.

## References

[1] *Handbook of thermal spray technology*. Materials Park, OH: ASM International, 2004. ISBN 0-87170-795-0.

[2] Bolelli G, Cannillo V, Lusvarghi L and Manfredini T, *Wear behaviour of thermally sprayed ceramic oxide coatings*. Wear, 26, 2006, p 1298-1315

[3] Serkan A, *Effect of TiO2 doping on microstructural properties of Al2O3-based single crystal Ceramics*, Journal of Ceramic Processing Research, 2011, p 21-25

[4] Chidambaram Seshadri R, Sampath S, *Characteristics of Conventional and Cascaded Arc Plasma Spray-Deposited Ceramic Under Standard and High-Throughput Conditions*, Journal of Thermal Spray Technology, 2019, p 690-705

[5] Dong Zhao, Fa Luo, Wancheng Zhou, Dongmei Zhu, *Effect of critical plasma spraying parameter on complex permittivity and microstructure by plasma spraying Cr/Al2O3 coatings*, Applied Surface Science, 64, 2012, p 545-551

[6] Yugeswarana S, Selvarajana V, Vijaya M, Ananthapadmanabhanb P V, Sreekumar K P, *Influence of critical plasma spraying parameter (CPSP) on plasma sprayed Alumina–Titania composite coatings*, Ceramics International, Vol. 36, Issue 1, 2010, p 141-149

[7] Valarezo A, Dwivedi G, Sampath S, Musalek R, Matejicek J, *Elastic and Anelastic Behavior of TBCs Sprayed at High-Deposition Rates*, Journal of Thermal Spray Technology, 2014

[8] Sabiruddin K, Joardar J and Bandyopadhyay P P, *Analysis of phase transformation in plasma sprayed alumina coatings using Rietveld refinement*, Surface and Coatings Technology, 204(20), 2010, p 3248-3253

[9] Sánchez E, Bannier E, Cantavella V, Salvador M D, Klyatskina E, Morgiel J, Grzonka J and Boccaccini A R, *Deposition of Al2O3-TiO2 Nanostructured Powders by Atmospheric Plasma Spraying,* Journal of Thermal Spraying Technology, 17(3), 2008, p 329-337

IOP Conf. Series: Materials Science and Engineering 1178 (2021) 012004 doi:10.1088/1757-899X/1178/1/012004

# Microstructure and mechanical properties of Multimetal "Stahl 1018" Composite

**A Arustamian[1], D Kalisz[1]**

[1] AGH University of Science and Technology, Faculty of Foundry Engineering, 23 Reymonta Street, 30-059 Kraków, Poland

E-mail: aac0979560476@mail.ru

**Abstract.** The Stahl1018 composite is a multiphase material containing steel and ceramic particles placed in a polymer matrix. The scanning microscope and X-ray microanalysis techniques were using for chemical analysis, which allowed for the identification the content of elements in particular areas of tested samples. Then, a series of tests was performed to determine the mechanical strength depending on the temperature, hardness and impact strength. The results of the experiments show that the composite material Stahl1018 can be used for maintenance of machine parts in the temperature range 20-80 °C.

## 1. Introduction

Composites cover a numerous group of materials differing in their matrix and type of reinforcement [1]. The structure of composite materials depends on the production technology, material from which the matrix was made, properties, type and participation of reinforcement material. It is also conditioned by the matrix/reinforcement interface and distribution of the reinforcement phase. It is typical for the correct composite structures to have a uniformly distributed reinforcement [2-8].

The properties of composites can be shaped and designed to meet the actual needs. Composites have properties unattainable in conventional monolith materials. The former ones have increased parameters, i.e. strength, Young modulus, fatigue characteristic, wear resistance, tribological characteristic, resistance to corrosion both in room and elevated temperature [9-11]. The main applicability area of modern composites is space technology, military industry, communications and production, e.g. sport equipment. Multimetal Stahl 1018 composite was worked out in the early 1990s as a result of a close cooperation with a German bridge company and a company producing metal constructions. Since that time this material has been successfully applied in Germany and abroad for fixing and replacing pillars in railway, road or canal bridges. Multimetal Stahl 1018 is based on the best polymers, which almost do not shrink after they are hardened, and which have a good chemical resistance. Among the filler materials are stainless steel powder, ceramics as well as additives improving surface tension and chemical resistance parameters. This composite has maximal compressive strength equal to 60 N/mm$^2$, tensile strength 72 N/mm2, bending strength 89 N/mm2, elastic modulus 14000 N/mm$^2$ and thermal resistance up to 90 °C and is used for fixing construction objects and worn away machine parts [12-14].

## 2. Microstructural analysis of samples

The composite MM stahl1018 is a complex and multiphase material, therefore it is so important to analyze its content, filler fraction (steel 1018) as well as the distribution of that fraction in the composite. As the properties and application of the composite depend on intermolecular distances, any changes in these distances may cause high local tensions and formation of compressive stress zones.

IOP Conf. Series: Materials Science and Engineering 1178 (2021) 012004 doi:10.1088/1757-899X/1178/1/012004

Hence, it is vital to analyze both the filler phase distribution and the matrix/filler division surface. The chemical and phase composition of the filler and components' interface is important in the aspect of mechanical properties and may also be an element of a structure favoring premature decaying of the material. For this reason the microstructural analysis of composite MMStahl1018 is justified. The analyses were performed on 3 samples of the material compressed in plastic moulds. The microstructure of unetched surfaces has been presented in images. The basic SEM analyses were preceded with opticl microscopy analyses – Microscope Nicon Elipse LV100 (figs. 1-3). The microstructure of sample 1 was presented in fig. 4, with two areas marked: 1 – elongated and fusiform area, and 2- square-like area. The experimental part is very important for further analyses, including the analysis of mechanical properties of the composite.

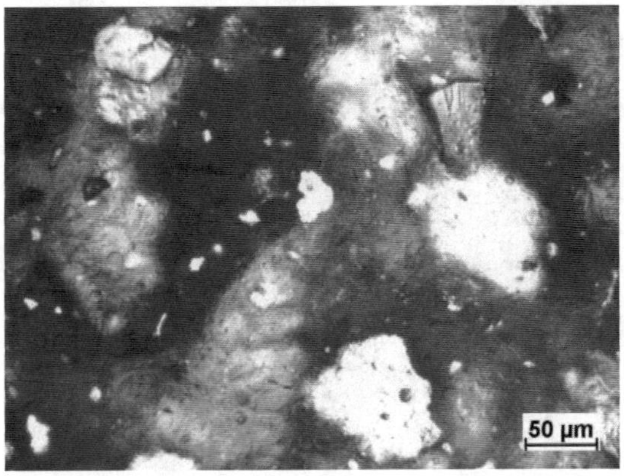


**Figure 1.** Microstructure of sample 1 performed of MM Stahl 1018 material.

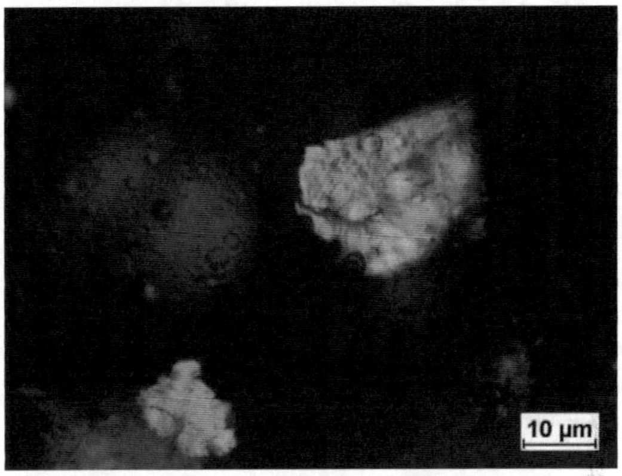


**Figure 2.** Microstructure of sample 2 performed of MM Stahl 1018 material.

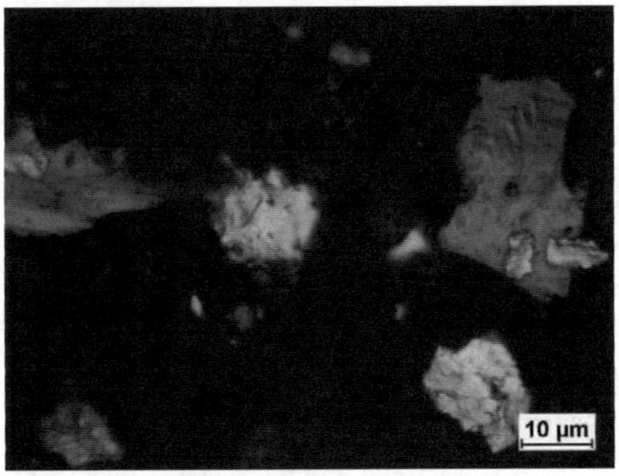


**Figure 3.** Microstructure of sample 3 performed of MM Stahl 1018 material.

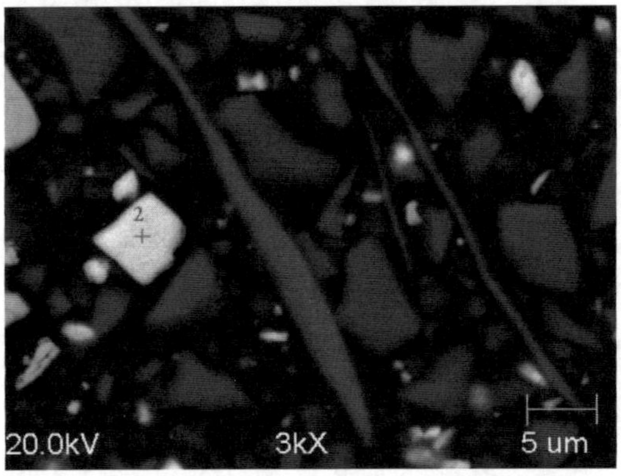


**Figure 4.** Microstructure of sample 1 performed of MM Stahl 1018 material, obtained with SEM.

The results of X-ray microanalysis of phases from the marked areas are presented in figure 5-6. Point 1 is located in an elongated tapering area. The analysis revealed that Al was the main component of the analyzed phase. Area 2 has a regular geometrical shape, where metals Ti, V, Mo, Cd and Cr as well as oxygen were identified [15].

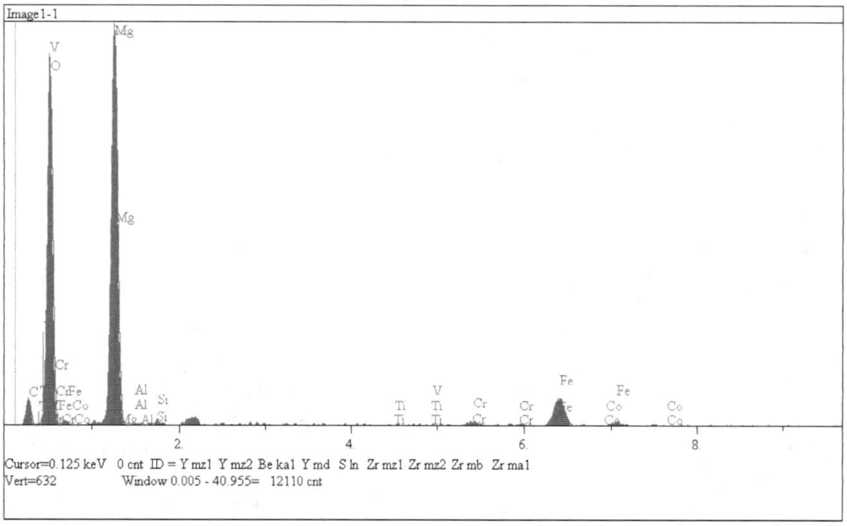


**Figure 5.** Result of X-ray microanalysis of sample1, area 1.

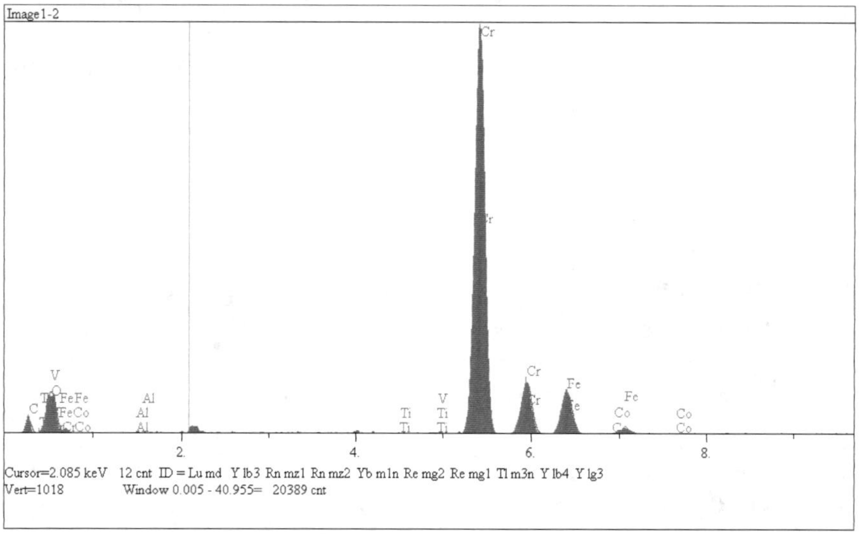


**Figure 6.** Result of X-ray microanalysis of sample 1, area 1.

## 3. Strength tests on MM Stahl 1018 composite samples

The compressive test was carried out using a testing machine Zwick Roell Amsler HB 100. This machine provides close to real conditions because it has a special heat chamber and ability to add the load in vibration mode. Using a special compressive testing device (mounted on a T-slotted plate), specimens of various sizes and shapes can be determined. The integrated device can be used in a heat chamber (temperature range -80 °C to 250 °C). The maximum load 100 kN in compression; Piston stroke 250 mm. During experimental strength tests performed with Zwick Roell Amsler HB 100, the composite samples were compressed at a specified load and temperature of 20-80 °C. The pressure force equaled to 15-30 kN and the samples were d = 10 mm., h = 4 mm in size. The results are presented in table 1. Exemplary results obtained for a sample subjected to a load of 30 kN at a temperature of 20-80 °C are visualized in figure 7.

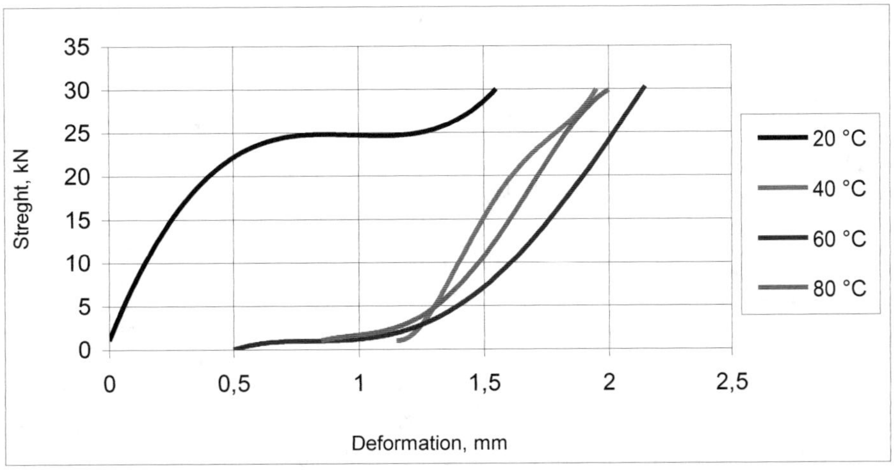


**Figure 7.** Thickness changes in a sample at a maximal load of 30 kN and temperatures 20-80 °C

It was observed for the analyzed samples that the maximal admissible load equaled to 30 kN, and the yield point of the material depended on temperature, i.e. the higher was the temperature, the lower was the yield point.

**Table 1.** Results of experiments – compressive test with MM Stahl 1018.

| | Maximum load, kN | Thickness change, mm | Temperature, ° C | Yeld point, MPa |
|---|---|---|---|---|
| **1** | 15 | 0.36 | 20 | 89.2 |
| **2** | 15 | 0.44 | 40 | 8.0 |
| **3** | 15 | 0.46 | 60 | 4.8 |
| **4** | 15 | 0.55 | 80 | 2.5 |
| **5** | 30 | 1.43 | 20 | 111.5 |
| **6** | 30 | 1.56 | 40 | 15.1 |
| **7** | 30 | 1.97 | 60 | 6.4 |
| **8** | 30 | 2.34 | 80 | 3.2 |
| **9** | 55 | 2.53 | 20 | 127.4 |
| **10** | 47 | 2.71 | 40 | 16.7 |
| **11** | 41 | 2.79 | 60 | 8.0 |
| **12** | 36 | 2.72 | 80 | 4.6 |

## 4. Hardness tests

Successively, the MM 'Stahl 1018' samples underwent hardness and impact strength tests were. Hardness was determined with the Brinell and Shore methods, whereas microhardness with the Hanemman method. The average hardness for three series of tests with the Brinell method was 192 HB, hardness measured with the Shoer method was similar, i.e. 97 HB.

Microhardness measured with the Hanemann method equaled to 33. The differences in the obtained results can be explained with considerable nonhomogeneity of phases in the composite. Further the efficiency of the composite as an agent for binding two surfaces was analyzed. The samples consisted of two steel rods 1 cm in diameter and 2.7 cm long, and they were combined with a binder made of MM "Stahl 1018". The following results of impact stress tests were obtained (table 2).

**Table 2.** Test results to be used for determining hardness and microhardness of the material "Stahl1080"by different methods.

| Sample no., № | 1 | 2 | 3 |
|---|---|---|---|
| Brinell hardness, HB | 180±2 | 195±2 | 200±2 |
| Shore hardness, HB | 195±2 | 195±2 | 200±2 |
| Microhardness (Hanemann method), HB | 35±1 | 27±1 | 29±1 |

## 5. Concluding remarks

We test this material because it is very easy to use and the damaged parts of machinery and metallurgical equipment can be easily restored. Much of the fixing work has been already done with the composite material MM "Stahl1018". The presented experiments are needed to extend the applicability of this material and prove that it can be used under high temperatures and vibration loads. The conducted series of experiments concentrated on the observation of microstructure with the optical and scanning microscopy methods, identifying chemical composition of selected areas in the analyzed sample and distribution of the filler in the matrix. It was also noted that an even distribution of the filler and its form (elongated and regular particles) cause that the obtained distribution is even, which consequently affects the mechanical properties and compressive strength of the composite.

The results of impact stress tests were over 100 times lower than in the analogous tests on uniform steel samples. This signifies that the material cannot be used for fixing devices exposed to strokes. The results of hardness and microhardness tests show a considerable discrepancy and this can be attributed to the presence of organic phases – polymeric matrix. Compressive tests performed at an elevated temperature showed that the material can be used for fixing wearing away machine parts at a room temperature.

## 6. Acknowledgements

The authors acknowledgement the AGH University of Science and Technology for its financial support within grant no POWR.03.05.00-00-z307/17

## 7. References

[1] Królikowski W 2020 Polimerowe kompozyty konstrukcyjne *Wydawnictwo Naukowe PWN* pp. 65-70

[2] Rabek J 2017 Współczesna wiedza o polimerach. Tom 2. Polimery naturalne i syntetyczne *Wydawnictwo Naukowe PWN* p 28

[3] Konopka K, Miazga A K 2017 Kompozyty ceramika – metal *Oficyjna Wydawnicza Politechniki Warszawskiej* pp. 71-79

[4] Śleziona J 1998 Podstawy technologii kompozytów *Wydawnictwo Politechniki Śląskiej* pp 81-83.

[5] Rabek J 2020 Polimery i ich zastosowania interdyscyplinarne. Tom 2 *Wydawnictwo Naukowe PWN* pp 119-124

[6] Romankiewicz F, Skocovsky P, Gorockiewicz R 2020 Niekonwencjonalne materiały kompozutowe *Wydawnictwo Politechniki Zielonogórskiej* pp 73-75

[7] Hyla I 1978 Wybrane zagadnienia z inżynierii materiałów kompozytowych *Wydawnictwo naukowe PWN* pp. 71-74

[8] Boczkowska A, Kapuściński J, Puciłkowski K, Wojciechowski S 2000 Kompozyty *Wydawnictwo Naukowe PWN* pp. 119-120

[9] Grabowska B, Bulwan M, Zapotoczny S, Grabowski G 2012 Biodegradation of new polymer foundry binders composition of poly(acrylic acid)/ dextrin *Polimery* vol **57** 529–534

[10] Lelito J, Żak P L, Gracz B, Szucki M, Kalisz D, Malinowski P, Suchy J S, Krajewski W K 2015 Determination of substrate log-normal distribution In the AZ91/SiCp composite *Metalurgija* vol **54** 204–206.

[11] Ischenko A A 2012 Technological bases of restoration of the industrial equipment by modern polymeric materials *PSTU* (Mariupol) pp 27-39
[12] Arustamian A, Kalisz D 2015 Doświadczalne wyznaczenie granicy plastyczności materiału kompozytowego *Arch. Found. Eng.* vol **15** pp 7–11
[13] Arustamian A, Sołek K, Kalisz D 2016 Identification of yield point of polymer-based composite material in the conditions of increased temperatures *Arch. Metall. Mater.* vol **61** (3) pp 1561–1566.
[14] Sołek K, Kalisz D, Arustamian A, Ishchenko A 2017 Analysis of strength characteristics of composite materials under vibration loads at higher temperatures *J. Mach. Constr. Maint.* vol **1** pp 93-97.
[15] Arustamian A, D Kalisz 2020 Multimetal Stahl 1018 composite – structure and strength properties *Arch.Found. Eng.* (in print)

# Microstructure of Al-Mg thin free-standing films

**L Bajtošová[1], R Králík[1], B Křivská[1], H Libenská[1], J Veselý[1], M Cieslar[1]**

[1] Charles University, Faculty of Mathematics and Physics, Ke Karlovu 3, 12116 Prague, Czech Republic

E-mail: lucibajtos@gmail.com

**Abstract**. Thin Al-Mg films were prepared by a DC magnetron sputtering on glass substrates covered with photoresist and subsequently free-standing samples were released from the substrate. The surface morphology, grain size and orientations were characterized by atomic force microscopy and transmission electron microscopy equipped with automated orientation and phase mapping software. The grain growth mechanism during sputtering is consistent with sputter deposition oblique incidence theory for growth. Strong preferred (110) orientation in direction perpendicular to the sample surface has been observed in all studied samples.

## 1 Introduction

Thin free-standing metallic films are commonly essential parts of microelectronic devices and micromechanical systems (MEMS) where they are exposed to a wide range of operating velocities and frequencies. Since the decrease of the sample size to micro and nanoscale results in properties considerably different from the ones observed in bulk materials, plasticity and mechanical behavior of thin films should be examined. In the case of thin films adhered to a substrate it is unclear how much the presence of the substrate influences measured properties. Therefore, a separation of the films from the substrate is unavoidable and experiments should be performed on free-standing films.

The relation between mechanical strength and grain size in bulk materials can be expressed by an inverse power law known as Hall-Petch relationship [1,2], which is a result of dislocation pileups formed in the vicinity of grain boundaries. In the case of very small grains, the validity of this relation ceases. On the contrary, a decrease of the yield stress and hardness with the decrease of grain size for sufficiently small grains has been observed in numerous studies [3,4]. Therefore an understanding of mechanical behavior of nanoscale materials is conditioned by the knowledge of their microstructure. Information about grain size and orientation distribution is crucial to determine their role in deformation processes.

Aluminum belongs to the most widely used lightweight and conductive metals in micro-mechanical structures. Aluminum films are often used for example as microswitches and membranes in microdevices or as interconnectors in integrated circuits. Adding Mg to pure Al could significantly increase the strength of the material [5].

In this work experimental evaluation of grain size and misorientations in Al-Mg free-standing thin films has been carried out using atomic force microscopy (AFM) and transmission electron microscopy (TEM) with automated orientation and phase mapping system (ASTAR) [6].

## 2 Experimental

50 nm, 100 nm and 150 nm thick films were prepared by a DC magnetron sputtering from a 2-inch target from Al – 3 wt.% Mg alloy with deposition rate 11.2 nm/min. The distance between the target and glass substrate coated in photoresist, a light sensitive polymer forming a smooth layer on the surface [7], was 7 cm. Layers were prepared using Ar gas under working pressure of 2 Pa.

Subsequently, free-standing film samples were prepared by a dissolution of photoresist in acetone and releasing of the film from the glass substrate. Afterwards, the pieces of thin film were fixed on a copper grid (diameter 3 mm, 300 mesh) and rinsed in methanol.

Surfaces of films on glass substrates were studied by AFM in tapping mode using Bruker Dimension Edge device. Free-standing samples on the copper grids were studied by a conventional TEM and ASTAR at JEOL 2200FS operated at 200 kV equipped with "Spinning Star" electron precession from NanoMEGAS.

## 3 Results and discussion

### *2.1 Atomic force microscopy*

AFM images of Al-Mg 50 nm, 100 nm and 150 nm thick films grown on the photoresist covered glass substrate are shown in Figure 1. The surface morphology shows grains with size 40-60 nm for 50 nm thick film, 60-90 nm for 100 nm thick film and 80-100 nm for 150 nm thick film. Features of sample surfaces are visible in Figure 1 d) – e). The surface profile suggests a formation of columnar structures which is characteristic for thin films grown by the sputtering deposition [8]. Columns of different heights are gradually formed during deposition and a screening of shorter columns by their higher neighbors occurs [8]. When a columnar grain is screened by another one, it stops growing. Similar surface symmetries have been observed in different materials prepared by magnetron sputtering such as Au [9], Cu [10] or ZnO [11].

**Figure 1.** AFM height sensor images of a) 50 nm, b) 100 nm and c) 150 nm Al-Mg films, d – f) corresponding topographic images.

### *2.2 Transmission electron microscopy*
Bright field images shown in Figure 2 reveal evenly distributed polyhedral grains 20 – 70 nm large. The grain size continuously increases with film thickness, the largest grains in respective thicknesses reaching size 40 nm, 50 nm and 70 nm. In 150 nm thick films, a noticeable overlapping of neighboring grains can be observed. Comparing with the AFM images and considering anticipated above described columnar growth of the grains, the smallest grains can be ascribed to the screened columns with suspended growth, whereas the larger grains are the ones reaching the surface and therefore observable in AFM scans. The grain sizes measured in AFM were slightly larger than the ones measured from TEM images. This may have been because the radius of curvature of the AFM tip approached the measured grain size and became a limiting factor in the scan resolution.

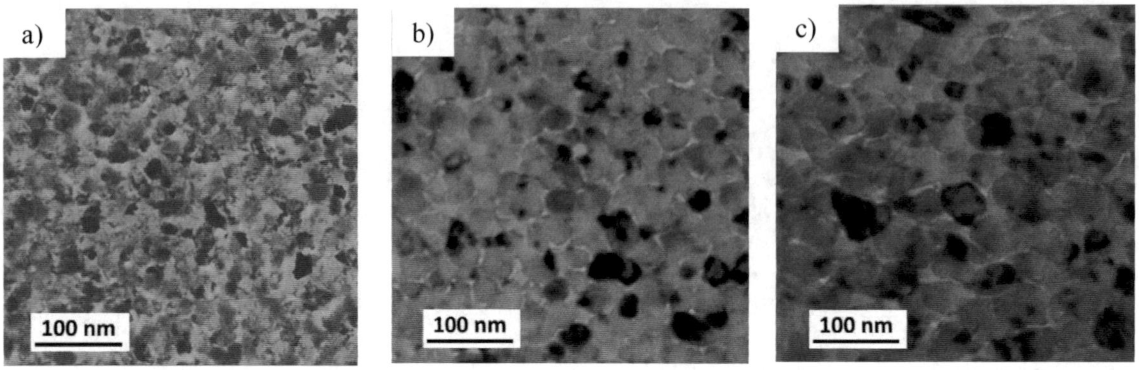


**Figure 2.** a) STEM BF image of 50 nm thick Al-Mg film, TEM BF images of b) 100 nm and c) 150 nm Al-Mg films.

### *2.3* ASTAR
Orientations maps of the samples combined with reliability maps obtained by automatic phase and orientation mapping in TEM are shown in Figure 3. The grain size could not be directly determined from the orientation mapping due to moderately large areas of a low reliability marked by a black color in the images (Figure 4). This is most probably caused by the overlapping of the neighboring grains, which leads to ambiguity in evaluated diffraction patterns and a loss of information about the crystal orientation in these areas as a consequence.

Nevertheless, random grain orientation distribution in x and y axes of the samples is clearly visible as well as a preferential Al-(101) orientation in the z-axis corresponding to the orientation perpendicular to the film surfaces. Generally, the texture is strongly dependent on deposition conditions e.g. substrate material and orientation or deposition rate, but (110) along with (111) orientation is commonly observed in aluminum thin films [12].

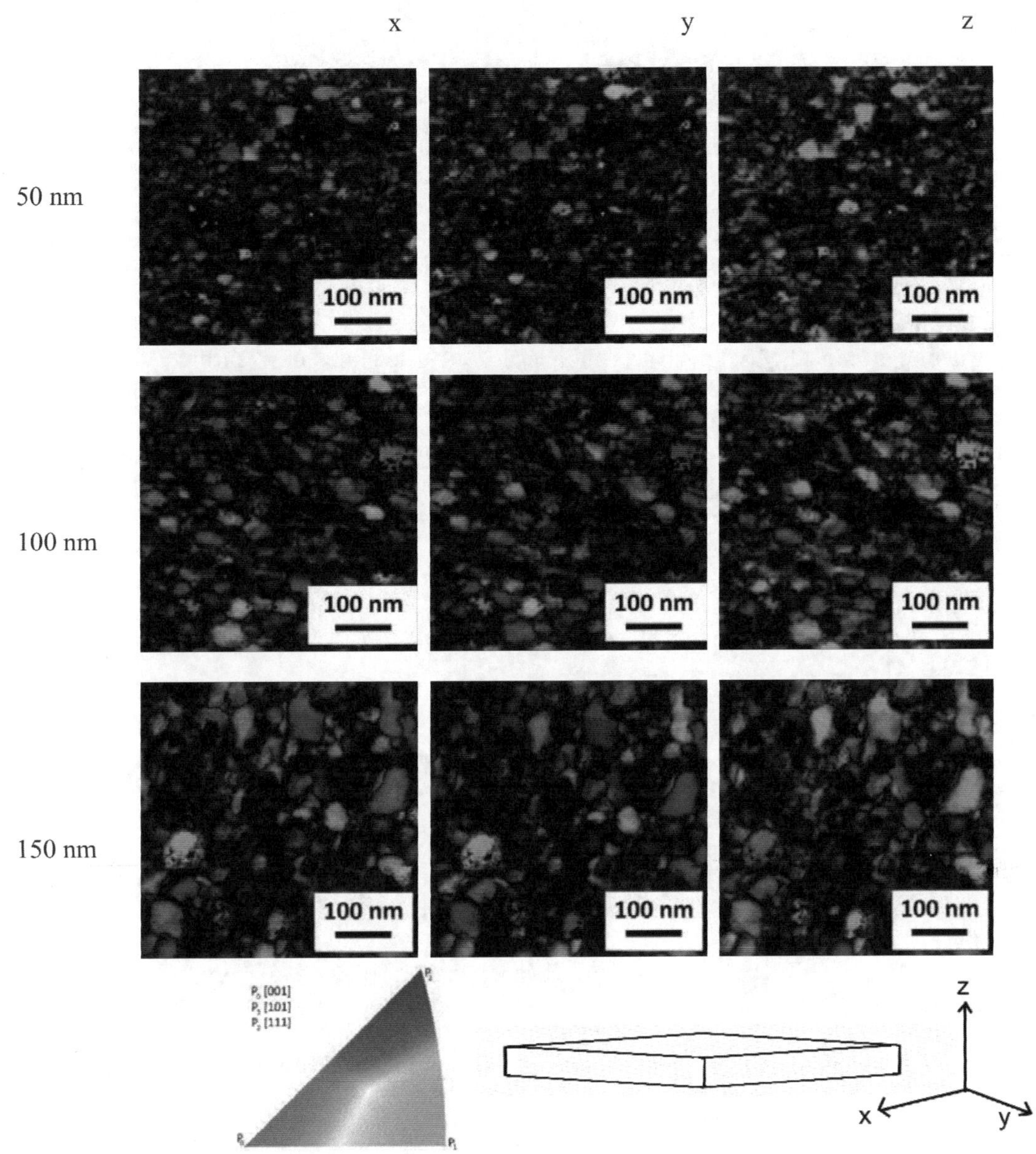


**Figure 3.** Orientation reliability maps of samples with different thickness.

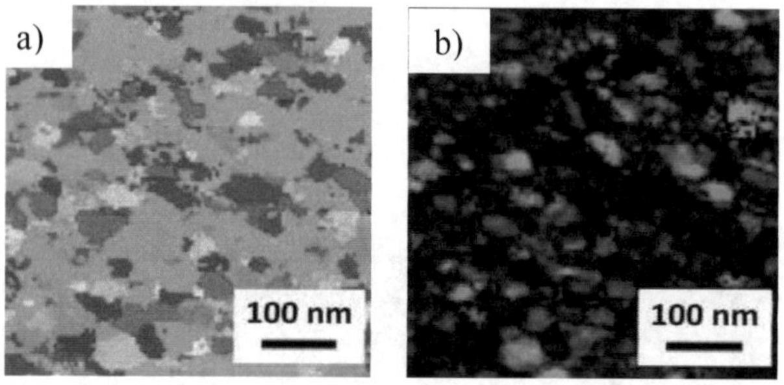


**Figure 4.** Original orientation map oriented in direction z (notation consistent with Fig. 3) of 100 nm thick film a) compared to the map overlapped with reliability index areas marked by black color b).

## 4 Conclusions

The microstructure of thin Al-Mg films has been investigated. The grains in size from 20 to 70 nm are evenly distributed in all 50 nm, 100 nm and 150 nm thick films. The average size of the grains increases proportionally to the film thickness. Performed measurements suggest a columnar mechanism of grain growth where several columns are growing in a rate corresponding to the rate of deposition. The smaller grains are gradually screened out and only the largest ones continue to grow. Orientation maps revealed moderately high texture in the material with (110) orientation perpendicular to the sample surface.

**Acknowledgements**
Authors would like to acknowledge the support from Grant Agency of Charles University under the project 574120.

**References**

[1] Hall E O 1951 The Deformation and Ageing of Mild Steel: III Discussion of Results *Proc. Phys. Soc. B 64* p 747-753
[2] N J Petch 1953 The Cleavage Strength of Polycrystals *J. Iron Steel Inst.* vol 174 p 25-28
[3] Swygenhoven H V, Weertman J R 2006 Deformation in nanocrystalline metals *Mater. Today 9* p 24-31
[4] Weertman J, Sanders P 1993 Plastic Deformation of Nanocrystalline Metals. *Solid State Phenomena 35-36* p 249-262
[5] Pun S C, Wang W, Khalajhedayati A, Schuler J D, Trelewicz J R, Rupert T J 2017 Nanocrystalline Al-Mg with extreme strength due to grain boundary doping *Mater. Sci. Eng. A 696* p 400-406
[6] Rauch E and Véron M 2014 Automated crystal orientation and phase mapping in TEM *Mater. Charact. 98* 1-9
[7] Wilson C G, Stewart M D 2001 Encyclopedia of Materials: Science and Technology p 6973-6977.
[8] Barabási A and Stanley H E 1995 *Fractal concepts in surface growth* Cambridge: Cambridge University Press
[9] Xin Z, Xiao-Hui S and Dian-Lin Z 2010 Thickness dependence of grain size and surface roughness for dc magnetron sputtered Au films *Chinese Phys. B, 19* 086802
[10] Kim H S 2018 Formation of Crystalline Copper Thin Films by a Sputtering-assisted Magnetic Field System at Room Temperature *Appl. Sci. Convergence Technol. 27* p 1-4
[11] Rosa A, Silva E, Amorim E, Chaves M, Catto A, Lisboa-Filho P, Bortoleto J 2012 Growth evolution of ZnO thin films deposited by RF magnetron sputtering *J. Phys. Conf. Ser.* 370
[12] Rodbell K P, Knorr D B and Mis J D 1993 The microstructure, mechanical stress, texture, and electromigration behavior of Al-Pd alloys *J. Electron. Mater 22* p 597-606

IOP Conf. Series: Materials Science and Engineering 1178 (2021) 012006 doi:10.1088/1757-899X/1178/1/012006

# The influence of quenching on fatigue life of Ti6Al4V alloy

**J Belan[1], O Bokůvka[1], M Uhríčik[1], L Kucharikvá[1] and A Vaško[1]**

[1]University of Žilina, Faculty of Mechanical Engineering, Department of Materials Science, Univerzitná 8215/1, 01026 Žilina, Slovakia

E-mail: juraj.belan@fstroj.uniza.sk

**Abstract**. Titanium Ti6Al4V alloy falls into the α+β category alloy system. It is widely used due to its unique mechanical and physical properties which can be modified by applied heat-treatment. The α-phase is considered to be the main phase influencing mechanical properties of an alloy such as ultimate tensile strength or ductility. However, it has also influence on the fatigue life where the form and presence of α-phase in prior β grains may increase or decrease fatigue life. The length and width of α-lamellae influence on fatigue life is well-known. A lot of authors have done their experiments on alloys with Widmanstätten or acicular α-phase structures. Our goal was to obtain an α′-martensitic structure in prior β grains and investigate its influence on fatigue life. To fulfil this goal the heat-treatment consisted of heating near above β-transus temperature 1050°C for a dwell time of 3 hours followed by water quenching was applied. After heat-treatment, the set of specimens with dimensions 10 mm x 9 mm x 55 mm were subjected to three-point bending fatigue test with the parameter of cycles asymmetry R < 1 and test frequency f = 69 Hz. The S-N curve of fatigue life was plotted after the fatigue test and SEM fractography analysis of fractured surfaces was done as well. Our results were compared to previous fatigue test done on alloy with α-lamellae structure.

## 1 Introduction

Titanium alloy Ti6Al4V is commonly used at constructions where high strength, low weight and good corrosion resistance are demanded [1, 2]. This alloys very often face not only static loading but more often a dynamic loading which results in fatigue damage of components (e.g. airfoils in an aero jet engine or turbocharger fan and so on). From that reason was alloy subjected to various fatigue tests at low (LCF) or high cycle (HCF) fatigue loading and at low (LFFL), moderate and even high frequencies (HFFL), but mostly with a parameter of cycle asymmetry R = -1. In experimental work of various authors is discussed the influence of microstructure on fatigue life of Ti6Al4V alloy with bimodal and Widmanstätten microstructures. Shital et al [3] shows the effect of volume fraction of α and transformed β on HCF properties of bimodal Ti6Al4V alloy. The ultimate tensile strength (UTS) and hardness were decreased with increasing of α percentage, as well as fatigue strength, reduces with increasing of α phase volume. Another comparison of bimodal and lamellar microstructure and its influence on HCF with R = 0.1 and 0.5 was done by Nalla et al [4]. They concluded that differences between bimodal and lamellar structures of Ti6Al4V in HCF life are not so dramatically large. However, lamellar microstructure shows slightly higher fatigue strength at higher ($10^5$ to $10^8$) number of cycles due to its higher UTS and also resistance to fatigue crack propagation was higher compared to bimodal structures. At a lower number of cycles, bimodal structures provide better results due to higher ductility of such microstructure. With relationship microstructure – mechanical properties – fatigue life in thin Ti6Al4V

IOP Conf. Series: Materials Science and Engineering 1178 (2021) 012006 doi:10.1088/1757-899X/1178/1/012006

alloy deals authors Fan et al [5]. They report that 0.2% Yield stress increase with $\alpha'$ martensite presented in microstructure, however, ductility and fatigue life decrease compared to bimodal, equiaxed or Widmanstätten structures. Similar results for Widmanstätten microstructure tested in HCF or VHCF at three-point bending fatigue test were obtained by Bao et al [6].

The influence of loading frequency on fatigue properties depending on the microstructure is negligible as reported in the work of Papakyriacou et al [7]. Stress ratio and its effect on HCF and VHCF are discussed in work Liu et al [8] and Huang et al [9]. Both authors show that asymmetry cycle parameter R = -1 or -0.5 is represented by S-N curve with horizontal asymptote shape and S-N curves with asymmetry cycle parameter R = -0.1, 0.1 and 0.5 have step-wise or duplex character. No matter what the R is, the character of S-N curves for titanium alloy has a sharp decrease at a higher number of loading cycles due to changing of failure mode and different crack initiations. When the stress ratio is higher, the alloy shows decreasing of a surface without facets crack initiation sites, on the contrary to increasing numbers of the surface with facets crack initiation sites. Factors as $\alpha_p$ (primary $\alpha$) content and its grain size, the $\alpha$ lamellae width also have an impact on fatigue life of Ti6Al4V alloy [10].

Many recent studies as mentioned above are deal with three elementary microstructures (bimodal, equiaxed and lamellar) and its influence on HCF or VHCF properties of Ti6Al4V alloy. Most of the fatigue tests were done with the parameter of cycle asymmetry R = -1. Our goal was to investigate the fatigue life of specimens made of Ti6Al4V alloy after heating over $\beta$-transus temperature with dwell time 3 hours and followed by water quenching. The specimens with the result $\alpha'$ martensitic microstructure were subjected to three-point bending fatigue test. Obtained results were compared to results achieved from three-point bending fatigue test on as received alloy without quenching to see changes in fatigue life if any.

## 2 Experimental material and methods

Experimental material – titanium alloy Ti6Al4V was supplied from BIBUS Metals AG, Brno CZ. Chemical composition (wt. %) and selected mechanical properties according to the material list are in table 1, the alloy was in the annealed condition.

**Table 1.** The chemical composition (wt. %) and selected mechanical properties of Ti6Al4V alloy

| Fe | C | N | H | O | Al | V | Ti |
|---|---|---|---|---|---|---|---|
| 0.059 | 0.012 | 0.0044 | 0.001 | 0.097 | 6.04 | 4.0 | Bal. |
| UTS [MPa] | | Yield strength 0.2% [MPa] | | Elongation [%] | | Hardness HV | |
| Longitudinal | 1011 | 951 | | 15 | | 266 | 346* |
| Transverse | 1023 | 971 | | 14 | | | |

* Own hardness measurements HV10/10

Heat-treatment of specimens consisted of heating on 1050°C with dwell time 3 hours followed with water quenching (cooling rate was 500°C.s$^{-1}$). After heat-treatment, the set of 15 un-notched specimens with simple blocky dimensions 10 x 9 x 50 mm were polished and subjected to three-point bending fatigue test.

The fatigue test was carried out on resonance test machine ZWICK/ROELL Amsler 150HFP 5100, figure 1, according to standard STN 42 0363. Parameters of the fatigue test were set as follows: the mean stress was $\sigma_{m.}$ = -555.5 MPa, stress amplitude varies $\sigma_a$ = 277.7 – 444.4 MPa, frequency f = 69 Hz and parameter of cycle asymmetry R = 0.5 – 0.8 (R < 1). The number of cycles was set 2.0 x $10^7$ – specimens whose withstand this number of cycles were considered as run-out.

Specimens for metallography evaluation and SEM specimens after fatigue test were prepared via regular metallographic procedures consisted of wet sanding and wet polishing on Struers TEGRASYSTEM [11, 12] etched by 10% HF or 1,5ml HF + 2ml $HNO_3$ + 10ml $H_2O$ (this reagent provides better results). Microstructures were observed on NEOPHOT 32 optical microscope and SEM observation and fractography was done on scanning electron microscopy TESCAN VEGA LMU II.

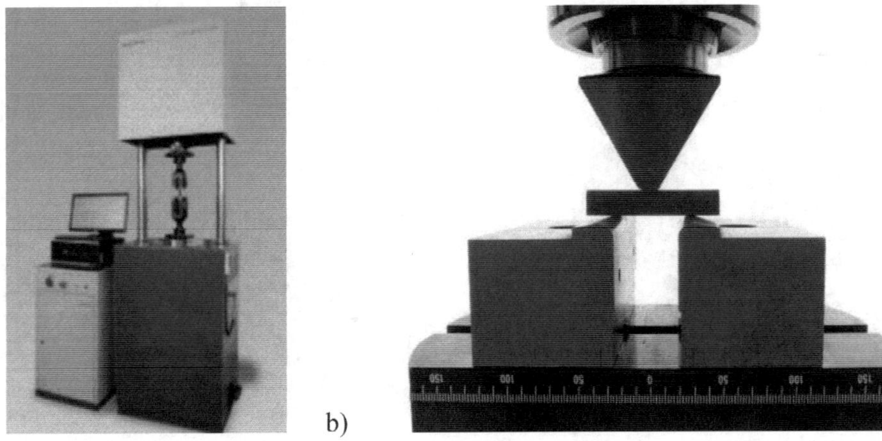

a) b)

**Figure 1**. Fatigue testing equipment; a) ZWICK / ROELL Amsler 150HFP 5100 pulsator, and b) specimen setting during the fatigue test.

## 3 Results and discussion

Metallography observation of as-received state and after applied heat-treatment is in figure 2a and 2c. The microstructure of as-received alloy corresponds to regular polycrystalline wrought α+β Ti6Al4V lamellar structure with α-phase lamellae in prior β-phase grains. Width of α-phase lamellae varies from 1.64 μm to 3.67 μm. Applied heat-treatment over β-transus temperature, which is for α+β Ti6Al4V alloy 1000°C, and followed by water quenching leads to the formation of martensitic structure. It is nonequilibrium supersaturated α-type structure created by rapid cooling rates (in our case $500°C.s^{-1}$) [13]. There are two different types of martensite in titanium alloys. One of them, α′ (figure 2b and 2d) crystalize in a hexagonal lattice and is presented only at quenching, while the second type α″ has an orthorhombic lattice and may be occurred during quenching as well as during external mechanical stresses – was not observed during analysis [14].

A Vickers hardness HV measurements were done as well. The loading of 98.1 N and dwell time of 10s was used to verify hardness given in material list from BIBUS Metals AG supplier. The hardness was calculated from 10 measurements and the average value is 346 HV10/10, which is about 30% higher than hardness given in the material list. Based on hardness measurements the UTS was calculated according to equation (1) [15 - 17]. Calculated UTS = 1332 MPa, which is again higher about 30% compared to the material list. This calculation is important for setting appropriate parameters for the fatigue test.

$$\text{UTS} = k \times HV \quad [MPa] \tag{1}$$

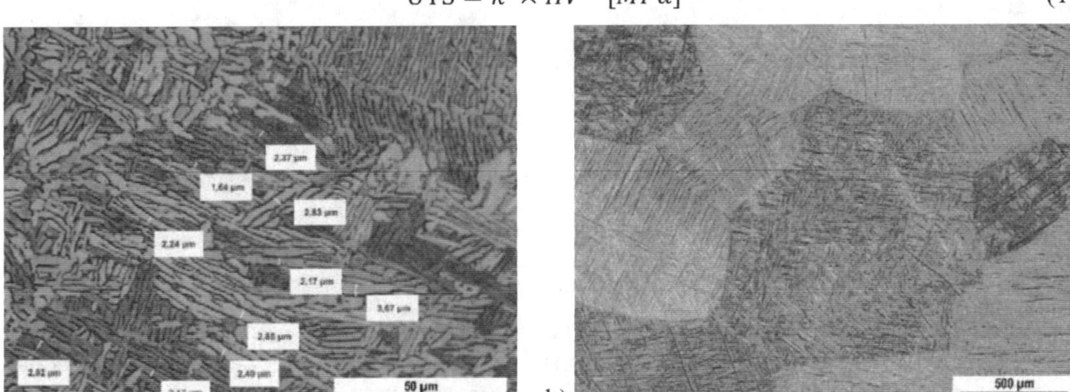


a) b)

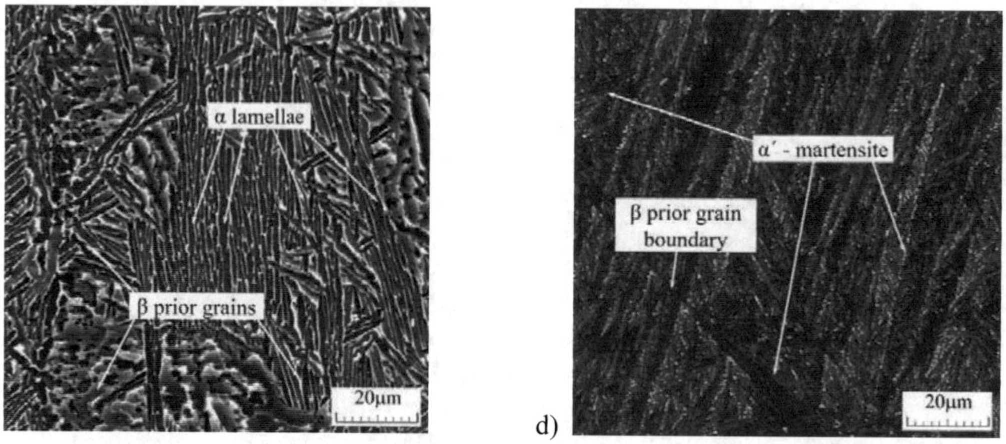


c) d)

**Figure 2.** The microstructure of TiAl6V4 alloy; a) as received α+β structure, α - lamellae measurements, b) microstructure after water quenching, α′-martensite and prior β-phase grains, polarised light, c) detail of (a) – SEM micrograph, d) detail of (b) – SEM micrograph after water quenching (cooling rate 500°C-s$^{-1}$), etch. a – b 10% HF, c – d 1,5ml HF + 2ml $HNO_3$ + 10ml $H_2O$.

Results of three-point bending fatigue test, the S-N curve, is shown in figure 3. The maximum bending stress of specimens at run-out was $\sigma_o$ = 847 MPa and bending stress amplitude at run-out was $\sigma_a$ = 292 MPa. Results show decreasing of fatigue life (bending stress amplitude at run-out) on specimens with α′-martensite structure about 47% in comparison with lamellar microstructure [17], which were $\sigma_a$ = 431 MPa. Even at 1.6 x $10^5$ number of cycles has lamellar microstructure about 24% stress amplitude higher compared to α′-martensite structure ($\sigma_a$ = 552 MPa at 1.6 x $10^5$ for lamellar and $\sigma_a$ = 444 MPa at 1.6 x $10^5$ for α′-martensite). Results are in good correlation with the work of Shital et al, Nalla et al, Fan et al and Bao et al [3 - 6] and confirm the fact that material with lamellar or Widmanstätten structures has higher fatigue characteristics at a higher number of cycles, even compared to α′-martensitic ones.

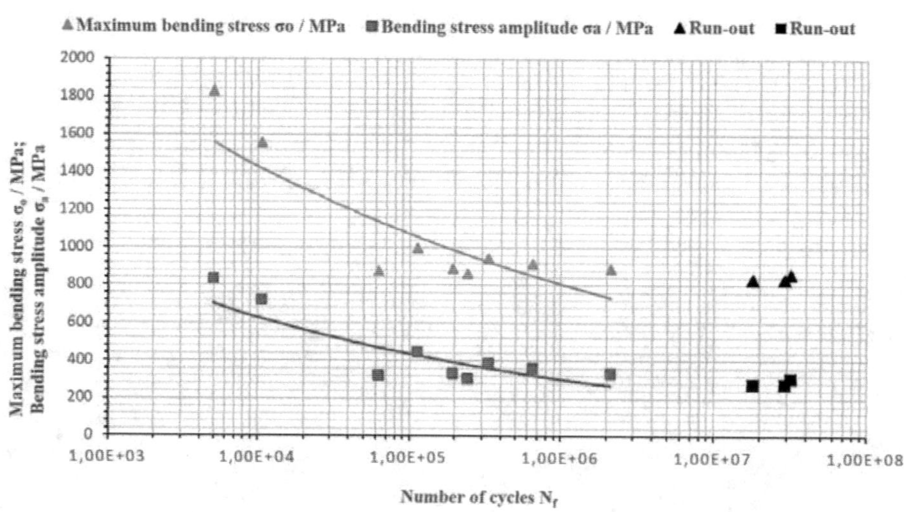


**Figure 3.** The S-N curve of water quenched Ti6Al4V alloy with R < 1.

The SEM fractography observation shows that fatigue crack has initiated via surface with facets mechanism, figure 4 a, b. The fatigue crack propagation mechanism was a mixture of intercrystalline cleavage at α′ - β prior grains phase boundary and transcrystalline propagation where the secondary fatigue cracks were presented, figure 4c, d. The fatigue part of the failure was very narrow and after initiation of fatigue crack the static failure comes very quickly. This fact is due to α′-martensite microstructure nature where the UTS is high and ductility of the alloy is decreasing. That is the reason why at static failure part of breaking surface is also intercrystalline and transcrystalline damage occurred, figure 4e, f.

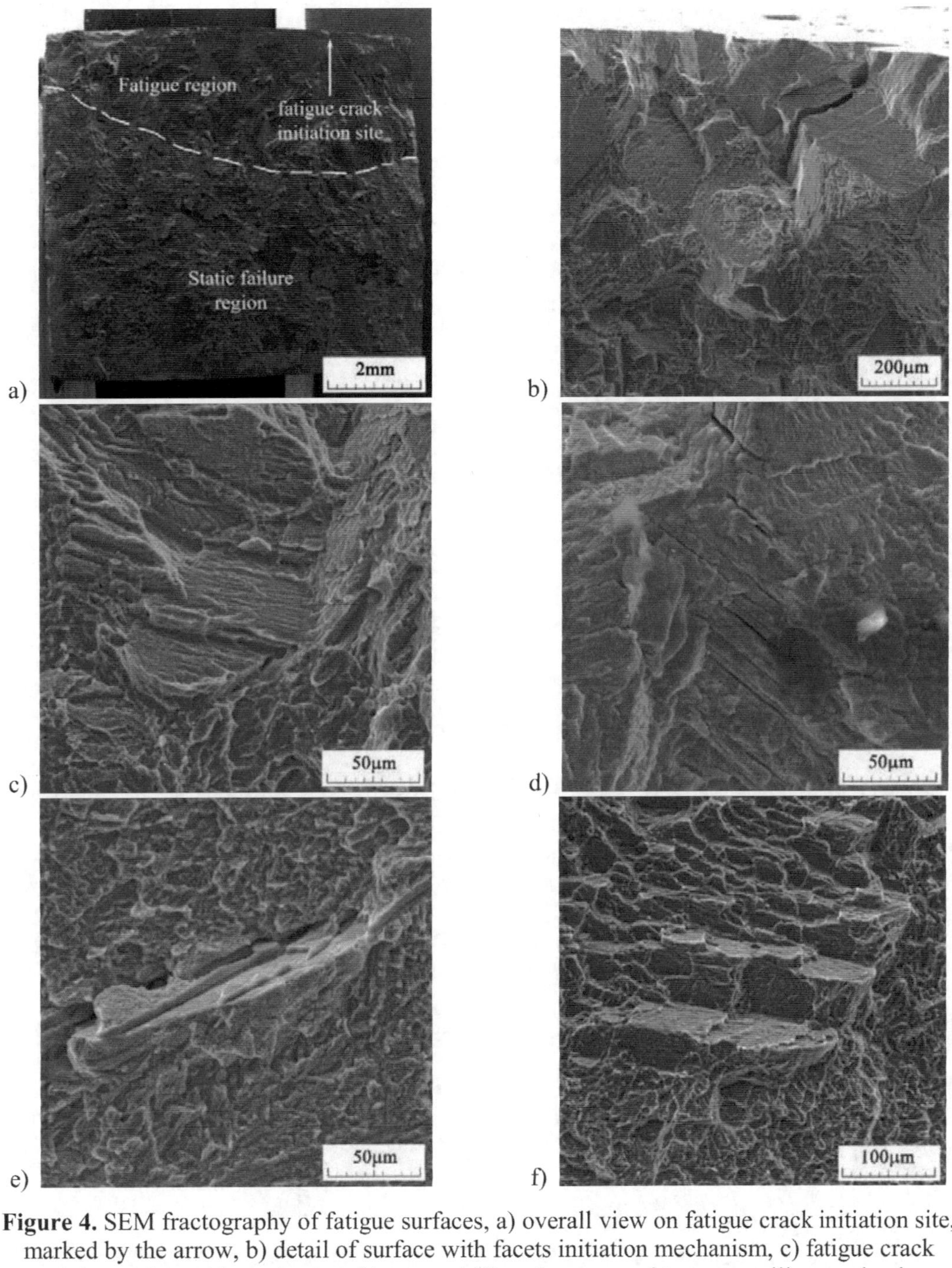


**Figure 4.** SEM fractography of fatigue surfaces, a) overall view on fatigue crack initiation site, marked by the arrow, b) detail of surface with facets initiation mechanism, c) fatigue crack propagation region with a mixture of intercrystalline cleavage and transcrystalline mechanism, d) detail of secondary fatigue cracks, e) and f) static part of fatigue failure.

## 4 Conclusions

Titanium alloy Ti6Al4V in heat-treated condition was used as an experimental material for three-point bending fatigue test at room temperature with the parameter of cycle asymmetry R < 1. Heat-treatment consist of heating over β-transus temperature 1050°C and dwell 3 hrs. followed by water quenching (500°C.s$^{-1}$). Microstructure after heat-treatment is formed by α′-martensite and prior β grains. This microstructure changing caused decreasing of fatigue life about 47% compared to non-treated lamellar structure ($\sigma_a$ = 431 MPa to $\sigma_a$ = 292 MPa) at set run-out 2.0x10$^7$ cycles and about 24% at 1.6x10$^5$ number of cycles ($\sigma_a$ = 552 MPa to $\sigma_a$ = 444 MPa). According to fatigue test results, we may conclude that applied heat-treatment significantly decreases fatigue life and presence of α′-martensite in structure in ineligible even when applied heat-treatment increases UTS = 1820 MPa what is increase about 36% comparing with the as-received state (UTS = 1332 MPa)

**Acknowledgement**

This work was supported under the project of Operational Program Research and Innovation: Research and development activities of the University of Žilina in the Industry of 21st century in the field of materials and nanotechnologies, No. 313011T426. The project is co-funding by the European Regional Development Fund. The authors also acknowledge the KEGA projects 016ŽU-4/2020 and 012ŽU-4/2019 for the financial support of this work.

## References

[1] Leyens C, Peters M 2003 Titanium and Titanium Alloys. Fundamentals and Applications ed. C Leyens and M Peters (Weinheim: WILEY-VCH Verlag GmbH & Co. KGaA) p 513
[2] Moiseyev V N 2006 Titanium alloys: Russian Aircraft and Aerospace Applications vol 5 ed. J N Fridlyander and D G Eskin (Boca Raton: CRC Press Taylor & Francis) p 207
[3] Shital J, Amit P, Sandip P, Ashish S, Bhagwan F and Rajkumar S 2017 *Materials Science and Engineering* **201** (2017) 012035 p 1-6
[4] Nalla R K, Boyce B L, Campbell J P, Peters J O and Ritchie R O 2002 *Metall and Mat Trans A* **33A** (2002) p 899-918
[5] Fan Y, Tian W, Guo Y, Sun Z and Xu J 2016 *Advances in Materials Science and Engineering* **2016** (2016) 7278267 p 1-9
[6] Bao X, Cheng L, Ding J, Chen X, Lu K and Cui W 2020 *Materials* **13** (1) 68 (2020) ma13010068 p 1-16
[7] Papakyriacou M, Mayer H, Pypen C, Plenk H and Stanzl-Tschegg S *2001 Materials Science and Engineering A* **308** (2001) p 143-152
[8] Liu X, Sun Ch and Hong Y 2015 *Materials Science and Engineering A* **622** (2015) p 228-235
[9] Huang Z Y, Liu H Q, Wang H M, Warner D, Khan M K and Wang Q Y *2016 International Journal of Fatigue* **93** (2016) p 232-237
[10] Wu G Q, Shi C L, Sha A X, Sha H R and Jiang H R 2013 *Materials and Design* **46** (2013) p 668-674
[11] Uhríčik M, Palček P, Chalupová M, Oravcová M and Frkáň M 2017 *Arch. Metall. Mater.* **62** (3) (2017) p 1615-24
[12] Zatkalíková V, Markovičová L, Uhríčik M and Hanusová P *Material Science and Engineering* **726** (2020) 012017 p 1-8
[13] Ahmed T and Rack H J 1999 *Materials Science and Engineering A* **243** (1999) p 206-211
[14] Gammon L M, Briggs R D, Packard J M, Batson K W, Boyer R and Domby Ch W 2004 ASM Handbook, Metallography and Microstructures vol 9, ed. G F Vander Voort (Ohio: ASM International) p 899-917
[15] Tabor D 2000 *The Hardness of Metals* 1st. ed. Oxford Classic Texts in the Physical Sciences (Oxford University Press, UK) p 192
[16] Zhang P, Li S X and Zhang Z F 2011 *Materials Science & Engineering A* **529** (2011) p 62-73
[17] Belan J, Kuchariková L, Tillová E and Chalupová M 2019 *Advances in Materials Science and Engineering* **2019** (2019) 2842416 p 1-11

# Analysis of structure and properties of multilayer polymer foil

**Marcin Bilewicz**

Silesian University of Technology, Faculty of Mechanical Engineering, Gliwice, Poland

marcin.bilewicz@polsl.pl

**Abstract**. Light and cheap composites are the main goal of manufacturing in many branches. Polymers, containing low weight and high relative strength, are suitable materials as matrix of composites reinforced with different metallic or ceramic fillers in the shape of fibres, plates, blocks or other types. This work takes into consideration optimisation and manufacturing of multilayer foil with different composition. To obtain the film has been used advanced blow molding process equipped with three extruders feeding material to the machine. Feeding of material is precisely controlled through gravimetric dosimeters. Final product of this technology is the polymer film which is obtained as combination of three films adherent together in the advanced head of blow molding machine.

Keywords: polymer composites, advanced polymer processing, multilayer foil, blow molding, thermoplastics.

## 1 Introduction

Recent years film manufacturing has evolved in different directions - monolayer films, multilayer films and also ultra-thin films with nanoparticles. Coextrusion has developed thanks to the development of computer possibilities, including simulations, economic needs, machine and process development, advantages in polymer chemistry. Nowadays multilayered structures, including films are used for many applications e.g. packaging, lamination and materials with special properties like barrier and UV resistance. Such n-layer structure of the film gives much more possibilities to design film with specific and advanced properties also for airplanes plating [1 -5].

## 2 Experimental procedure

In the work has been used blow molding machine with advanced head to control the process of manufacturing film. The innovative head can prepare film by combining together materials from different extruders in aim to obtain 3-layer polymer film (fig. 1). Manufacturing of the film combine the advantages of individual layers to achieve multilayer composite. Particular stages of the work include control of the process to obtain different thickness of the final 3-layered film. Full index of experimental plan of samples is presented in table 1. Material used in the experiment was LDPE.

**Figure 1.** Blow molding machine with advanced head to obtain 3=layerd polymer films

**Table 1.** Experimental plan

| Sample designation | Sample description |
|---|---|
| 1 Lay 60 | 1-layer film with thickness 60 μm |
| 3 Lay 60 | 3-layer film with thickness 60 μm |

IOP Conf. Series: Materials Science and Engineering 1178 (2021) 012007 doi:10.1088/1757-899X/1178/1/012007

## 3 Results

All films were tested on universal testing machine Zwick 020 at room temperature. For each test were used at least 5 samples to get average results. Curves obtained in the test are presented on figures below (fig 2-4). All values and average values of tensile strength test are presented in the table 2 and on the figure 5.

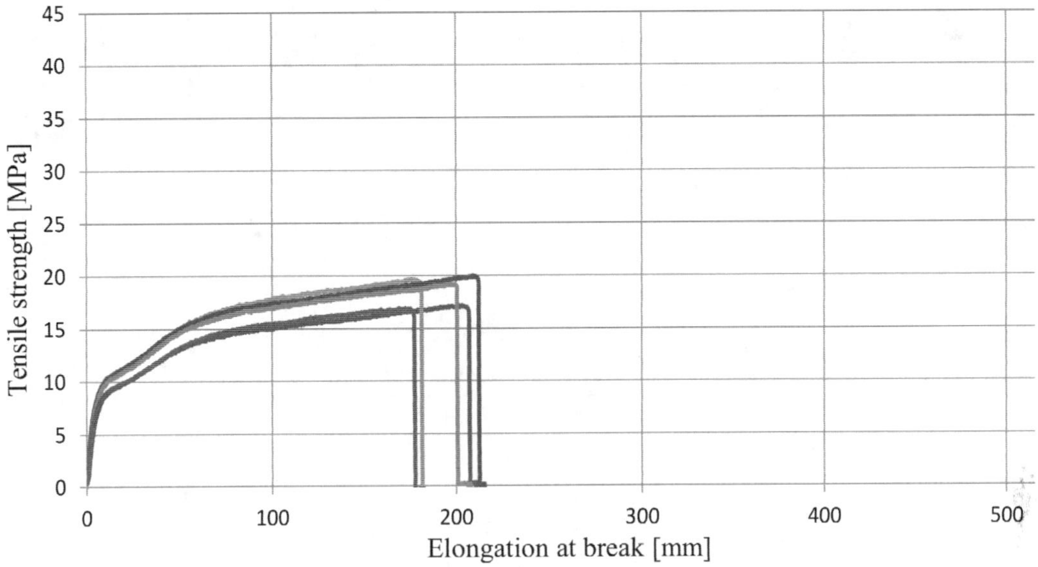


**Figure 2.** Tensile curves for 1-layer film with 60 μm thickness

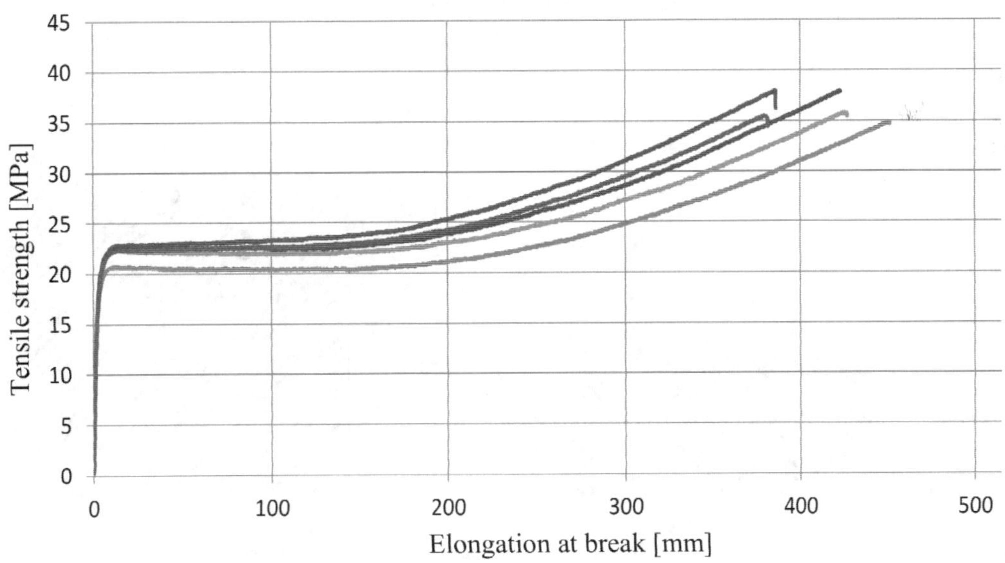


**Figure 3.** Tensile curves for 3-layer film with 60 μm thickness.

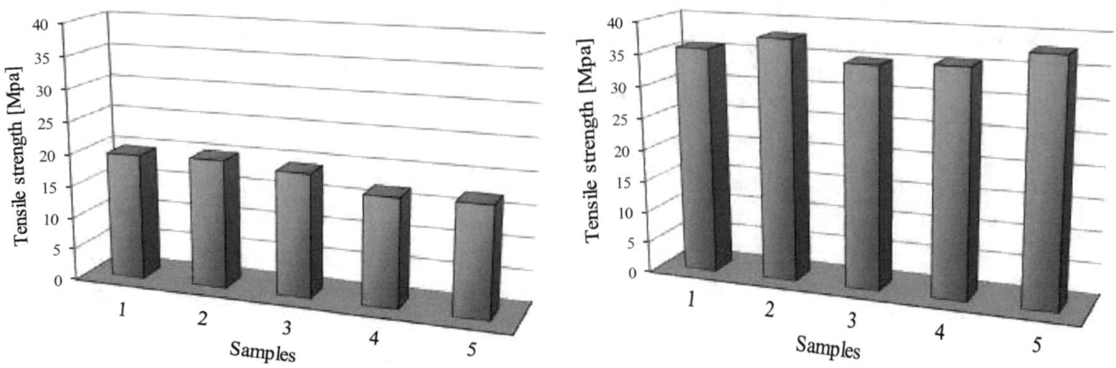


**Figure 4.** Results of tensile strength for 1-layer (left) and 3-layer film (right) with 60 μm thickness.

**Table 2.** Tensile strength of particular samples

| Sample | [MPa] | |
|---|---|---|
| | 1 Lay 60 | 3 Lay 60 |
| 1 | 19,68573952 | 35,75850805 |
| 2 | 19,99214172 | 37,96298981 |
| 3 | 19,20668602 | 34,89029566 |
| 4 | 16,92331505 | 35,47321637 |
| 5 | 17,17913055 | 37,97039986 |
| Average | 18,59740257 | 36,41108195 |

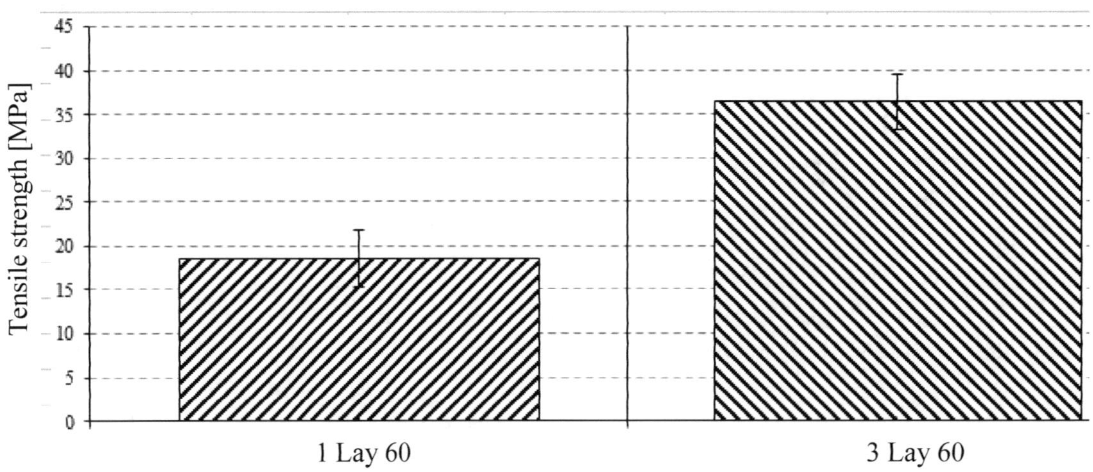


**Figure 5.** Average values of tensile strength for 1-layer and 3-layer films with 60 μm thickness

Both 1-layer and 3-layer samples are made of LDPE, however comparing of the tensile test curves of these materials shows, that samples from 3-layer LDPE have double elongation. The difference is in the structure which is created from 3 different layers obtained from 3 separated extruders. Typical structure of polymeric material contains core and skin which reinforce mostly the structure. Joining together 3 layers creates the stronger structure

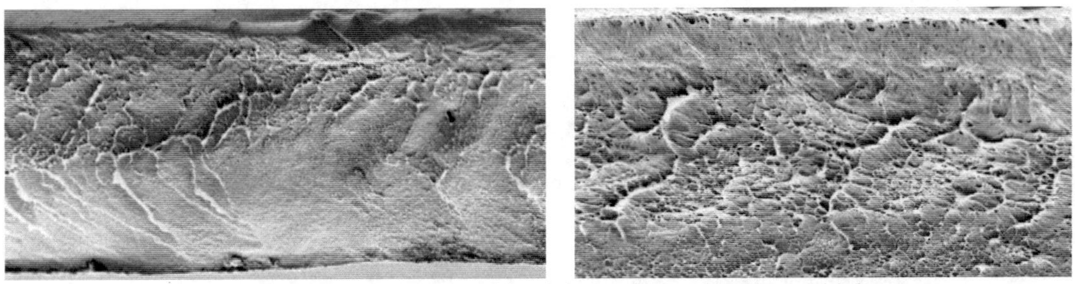

**Figure 6.** Cross section of the 1-layer film with 60 μm thickness

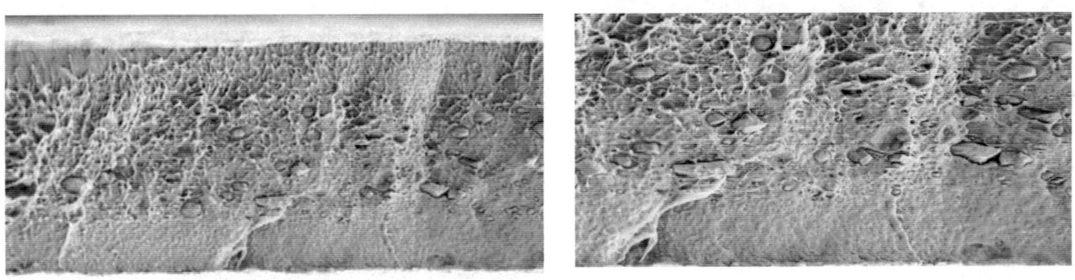

**Figure 7.** Cross section of the 3-layer film with 60 μm thickness

Observation of the cross section of 1-laer and 3-layer structure shows higher absorption of energy for the 3-layer structure. There is also visible plastic character of the fracture. Good adherence between the layers brings reinforcement of the structure.

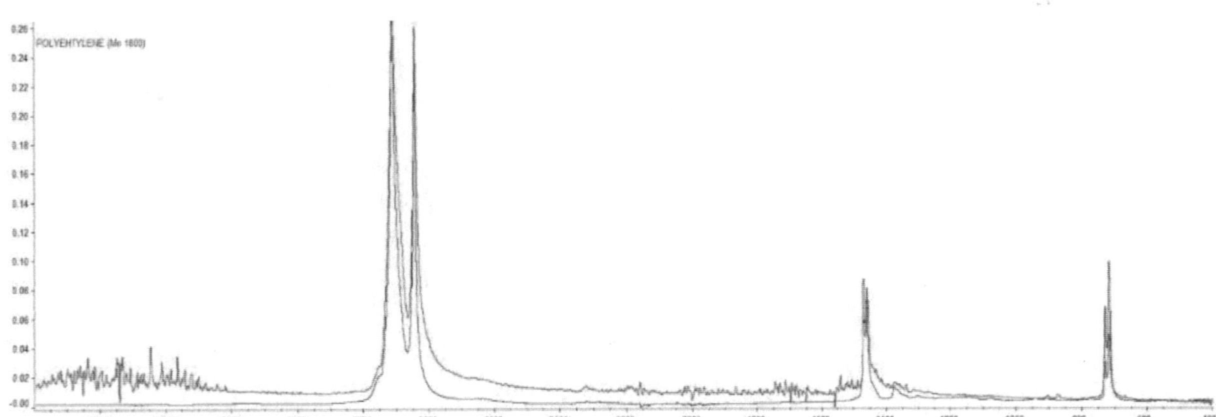


**Figure 8.** Spectrum of 3-layer film using FTIR..

Spectrum done using FTIR spectrometry shows confirmation of material used in the experiment and recognized as PE - visible overlapping of the peaks accordingly to database spectrum.

## 4 Summary

Samples of 1-layer film present more fragile fracture character than 3-layer film. On SEM pictures can be visible flat surface. Cross section of 3-layer film present more plastic fracture character where more energy was absorbed my material, which results in increased strength of material. Comparing structure with tensile curves can be visible that also elongation of material is much longer which correlates with picture of structure confirming higher absorption of energy by material. Comparison of tensile strength of 3-layer sample to 1-layer results in 196% of increment, that is double higher stiffness of the samples. Further investigation is necessary to bring closer phenomena of reinforcing of the structure.

## References

[1] Andrew W, Manufacturing and Novel Applications of Multilayer Polymer Films, 2016
[2] Fink J K, Handbook of Engineering and Specialty Thermoplastics, 2010
[3] Jarka P, Tański T, Matysiak W, Krzemiński Ł, Hajduk B, Bilewicz M, 2017 *Applied Surface Science 424*, 206-212
[4] A. Katunin A, Krukiewicz K, Turczyn R, Sul P, Łasica A, Bilewicz M., 2017 *Composite Structures 159* 773-783

# Effect of heat treatment on mechanical properties of duplex steel SAF 2507 manufactured by DED

**M Brázda, P Salvetr, S Rzepa, D Melzer, J Vavřík**
COMTES FHT – Dobřany, Czech Republic, EU

michal.brazda@comtesfht.cz

**Abstract**: This work is focused on the influence of heat treatment on the mechanical properties of duplex steel SAF 2507 produced by direct laser deposition (DLD). Duplex steels combine the advantages of ferritic and austenitic steels and reach excellent combination of mechanical and corrosion properties. The best properties are achieved by approximately equal phase ration of ferrite and austenite. Therefore, the proportion of austenite and ferrite before and after heat treatment was investigated within the study. Direct laser deposition technology was used for sample production. The DLD works on the principle of laser melting of the feedstock material (powder), which is carried by a protective gas into the nozzle. The material is successively deposited on the build platform. This technology can be used as an additive manufacturing process for surface repairs and component modifications. In this experimental study, the samples were deposited and subsequently subjected to the heat treatment. Solution annealing was chosen as the heat treatment method. The microstructure was characterized by LM, SEM and EBSD analyses. The mechanical tests were performed by a tensile testing on miniaturized specimens extracted from the experimental material in three orientations. The heat treatment had a fundamental effect on the proportion of austenite and ferrite, but also on the resulting mechanical properties. From the mechanical properties the maximum values of YS/UTS ~ 680/830 MPa were reached and after heat treatment ~ 540/760 MPa were reached.

## 1 Introduction

Duplex steels contain a balanced proportion of ferrite and austenite. For the best mechanical properties and corrosion resistance, the optimal proportion is 1:1. These steels are most often used in the marine, petrochemical, oil, gas and shipbuilding industries. They have one of the highest chromium contents of the entire group of cast stainless steels [1], [2]. The chromium levels range from 21 to 27%. Corrosion resistance is increased by alloying with molybdenum (up to 6%) and nitrogen. With such compositions, the two-phase structure is achieved by nickel additions at 5.5 to 8.5%. Nickel and nitrogen are the austenite-forming elements that compensate for the high content of ferrite-forming elements (chromium, molybdenum). If a certain proportion of ferrite and austenite contents is intended to be achieved, the levels of ferrite-forming and austenite-forming elements must be balanced as well. A small change in the composition can cause a significant change in the resulting microstructure and thus in the properties of the steel. However, the proportion of both constituents depends not only on the chemical composition but also on applied heat treatment. Before use, duplex steels have to be heat treated. Duplex steels are usually subjected to solution annealing at about 1100 °C to convert unwanted carbides and intermediate phases into a solid solution [3]-[4][5] [6]. Localized heating sources used in additive manufacturing lead to residual stresses, structural instability and crack formation. Residual stresses in the built structure are important to consider in order to ensure that they remain within the safe limits for the structure [7].

Preheating is one of promising methods to reduce residual stress [8]. DED technology can even handle extraordinary material combinations, such as Ni with Ti [9], Ti6Al4V with TiC, Invar with TiC, many others as per the application requirement [10].

## 2 Material and methods

The Insstek MX-600 machine was used for the deposition process. Its energy source is a 2 kW fibre-guided Ytterbium laser. The whole process took place under a protective atmosphere of argon 4.8. A powder designated as SAF 2507 with a particle size of 45-150 μm from Sandvik Osprey LTD was the feedstock material. The chemical composition of the powder is shown in Table 1 that is stated chemical composition from the material data sheet supplied by the manufacturer. The additive manufacturing parameters are listed in Table 2 [5]. A heat treatment was applied for solution annealing of the material. The heat treatment regime was a sequence of heating up to 1100 °C, then the material keeping at the temperature for 60 minutes followed by cooling in water. Figure 1 shows a cut design of the samples.

**Table 1** Chemical composition of SAF 2507 [wt. %]

| Fe | C | Cr | Ni | Mo | Mn | Si | N | O | Cu | Al | P | S |
|---|---|---|---|---|---|---|---|---|---|---|---|---|
| Bal. | 0.02 | 25.2 | 7.0 | 4.02 | 0.9 | 0.5 | 0.3 | 0.05 | 0.02 | 0.01 | 0.008 | 0.006 |

**Table 2.** Deposition parameters

| Module | Layer height | Beam diameter | Feed rate | Hatch distance | Deposition strategy |
|---|---|---|---|---|---|
| SDM 800 | 250 μm | 800 μm | 849 mm/min | 0.5 mm | ZigZag |

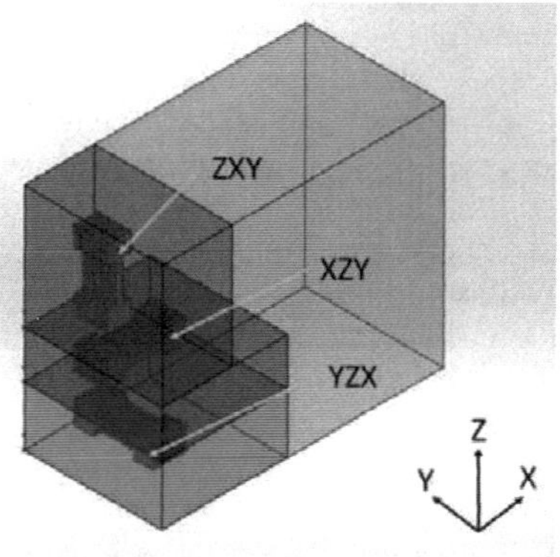


**Figure 1** Additively manufactured duplex steel SAF 2507 - cut design

The samples were then cut away from the platform and prepared for metallographic analysis. Metallographic sections were prepared by grinding and polishing. The microstructure was revealed by etching in Beraha II with $K_2S_2O_5$. It was observed under a light microscope (Nikon ECLIPSE MA200) and in a scanning electron microscope (JEOL IT 500 HR). The EBSD camera (EDAX Hikari Super) provided information on the proportion of ferrite and austenite on area 800x900 μm with step by 1 μm.

The tensile properties of the experimental material were investigated based on the results of miniaturized tensile test (MTT). The reliability of the method was documented in other studies [11]. The specimens were manufactured by wire electric discharge machine (WEDM) and subsequently polished. The tests were conducted under quasi-static conditions at room temperature in accordance to the standard ISO 6892-1 using a universal testing machine TiraTest with a linear drive and load capacity of 10 kN. At least four specimens per condition were tested. The deformation was tracked by contactless extensometer Mercury RT based on Digital Image Correlation technique. Before starting the testing procedure, the specific dimensions of each specimen (width and thickness) were measured by a micrometer and noted. Then, the specimens were cleaned using alcohol-based agent and a stochastic pattern was created on the surface for DIC measurement. After testing, the specific dimensions were measured again by stereomicroscope and the tensile characteristics calculated (yield strength - YS, ultimate tensile strength – UTS, uniform elongation – Ag, total plastic deformation – A and Reduction of Area – RA).

## 3 Results

### *3.1 Microstructure*

The microstructure was observed on LM and SEM after etching in Beraha II reagent. Ferrite and austenite phases in different ratios were observed in the microstructure before and after heat treatment, see in Figure 2. In the as built state, the elongated grains of ferrite grow through several layers and there are fine-grained areas along the boundaries of the meltpool. The austenitic phases in three variants forms during deposition and following cooling. Grain boundary austenite (GBA) was formed as first along ferrite grain boundaries at the highest temperatures between 1350 °C and 850 °C, widmanstätten austenite (WA) creates thin needles, whereas the intergranular austenite (IGA) precipitates as fine particles within ferrite grains at the lowest temperature. All types of austenite are visible in Fig1b as-deposited state. Ferrite grains in the etched samples grew across several layers of the build. The meltpool boundaries are lined with heat affected zones (HAZ). [3], [6], [12], [13]. GBA predominates inside meltpools and IGA predominates in HAZ. Meltpool is marked in Figure 2 (1a) in orange and in Figure 3 by black arrows. HAZ is marked in red in Figure 2 (1a) and in purple arrows in Figure 3. The austenitic grains (mostly IGA) were roughened after heat treatment and the IGA was evenly distributed. Furthermore, the fine-grained area of ferritic grains disappeared.

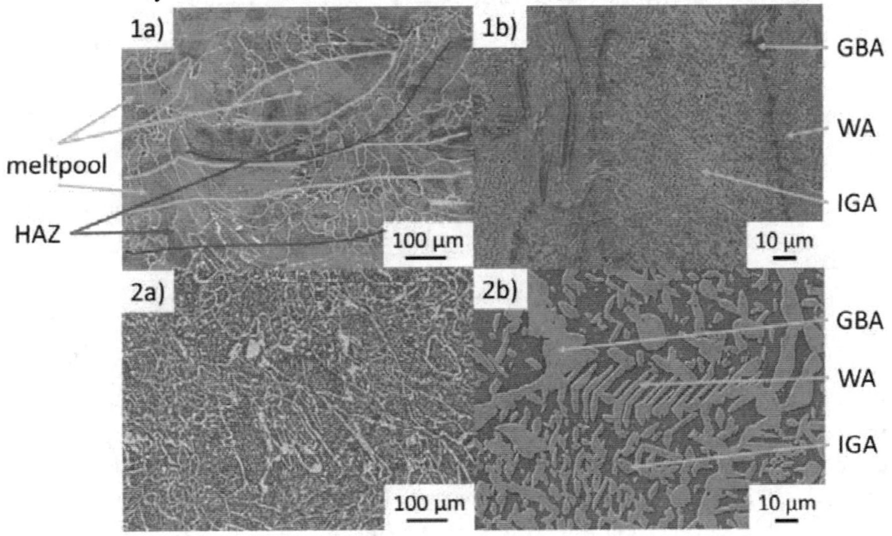


**Figure 2** Microstructure of SAF 2507 - 1) as built, 2) heat treated, a) LM - 100x, b) SEM - 1000x

### *3.2 EBSD analysis*

In the as built state (Figure 3a)), there was austenite along the boundaries of ferritic grains (GBA) and meltpools (IGA). GBA was uniformly distributed, whereas IGA was mainly found along the meltpool boundaries and in the heat affected zone (HAZ) [5]. Furthermore, the ratio of austenite and ferrite before and after heat treatment was measured by EBSD. After the heat treatment, the austenitic grains grew evenly throughout the volume of the material (Figure 3 b)). The proportion of ferrite and austenite was determined by EBSD and the results are shown in Table 3.

IOP Conf. Series: Materials Science and Engineering 1178 (2021) 012008 doi:10.1088/1757-899X/1178/1/012008

**Table 3** Austenite and ferrite content [%]

| | **Austenite** | **Ferrite** |
|---|---|---|
| **As built** | 21.1 | 78.9 |
| **Heat treated** | 52.4 | 47.6 |

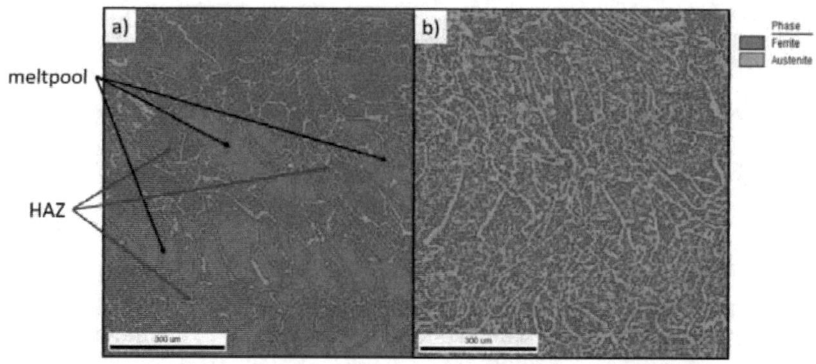


**Figure 3** EBSD analysis results, a) as built, b) heat treated

### *3.3 Mechanical properties*

The specimens were tested in as built and heat-treated states in three different orientations designated as XZY, YZX and ZXY according to the document ASTM WK49229 [14]. The obtained strength values, elongation and reduction of cross-sectional area are summarized and compared in Figures 4 and 5. Table 4 shows typical mechanical properties of as built 2507-AM powder L-PBF (Laser beam - Powder Bed Fusion) in heat treated condition (solution annealing 1040-1110 °C followed by air or water cooling) evaluated at room temperature. [15]

**Table 4** Mechanical properties of SAF 2507 (manufactured by L-PBF) [15]

| **Condition** | **Direction** | **Proof strength** | **Tensile strength** | **E-modulus** | **Elongation** | **Area reduction** |
|---|---|---|---|---|---|---|
| | | $R_{p0.2}$ **[MPa]** | $R_m$ **[MPa]** | | **A [%]** | **Z [%]** |
| Heat treated | Horizontal | 627 | 956 | 207 | 39 | 74 |
| | Vertical | 626 | 923 | 202 | 43 | 74 |

The applied solution annealing significantly affected the tensile characteristics of AM-ed material. The specimens subjected to the heat treatment yielded lower strength values (~540/760 MPa of YS/UTS) and almost twice higher elongation (~30%) in comparison to the as built material. The as built specimens tested in ZXY orientation were distinguished by the highest yield and uniform tensile strength (~680/830 MPa) over all tested specimen batches. In contrast, the heat-treated ZXY-specimens exhibited the lowest ultimate tensile strength (~750 MPa). The elongation values (uniform elongation at UTS point and total plastic elongation at the end of the test) were rather uniform within each batch, nevertheless, the results for as built material were characterized by increased uncertainty due to a relatively high data scatter observed. In the case of the annealed specimens, the orientation effect appeared to be negligible [6]. In comparison with the measured results and values from Table 4, it is assumed that lower values were achieved. The differences in the "as build" state may be due to the smaller powder particles for L-PBF and after the heat treatment the differences depend on the heat treatment parameters. The mechanical properties in the "as build" state will depend on the ratio of austenite and ferrite, when a smaller amount of austenite achieves higher strength. [16]

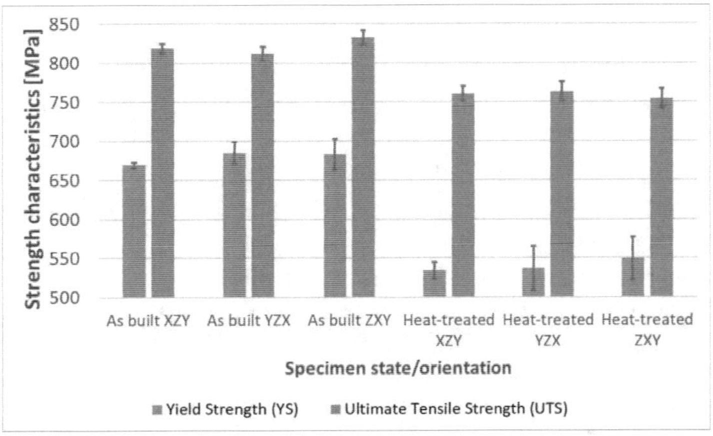


**Figure 4** MTT results of AM-ed duplex steel SAF 2507 in as built and solution annealed states, tested in three orientations – yield and ultimate tensile strength values.

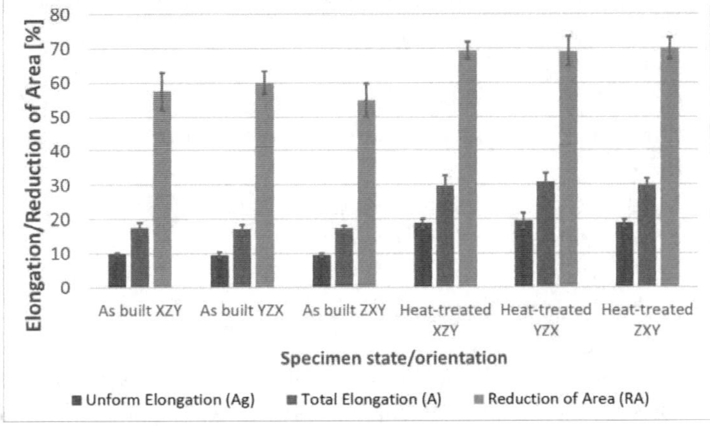


**Figure 5** MTT results of AM-ed duplex steel SAF 2507 in as built and solution annealed states, tested in three orientations – uniform and total elongation, reduction of cross-sectional area.

## 4 Conclusion

This paper deals with the influence of heat treatment on the resulting microstructure and mechanical properties of duplex steel SAF 2507. After the heat treatment, the amount of austenite and the uniform ratio of austenite to ferrite in the steel increased. The mechanical properties changed after the heat treatment and the reduction of YS/UTS from ~ 680/830 MPa to ~ 540/760 MPa was recorded. On the contrary, the ductility of the material increased almost twice. Due to an increase in the austenite content of the duplex steel after heat treatment was observed, it is likely to affect the mechanical properties because austenite is characterized by lower strength and higher elongation than ferrite.

## Acknowledgements

This paper was developed thanks to the project of Pre-Application Research of Functionally Graduated Materials by Additive Technologies, No.: CZ.02.1.01/0.0/0.0/17_048/0007350, financed by the Ministry of Education of the Czech Republic.

## References

[1] Z. Zhang, H. Zhao, H. Zhang, J. Hu and J. Jin, "Microstructure evolution and pitting corrosion behavior of UNS S32750 super duplex stainless steel welds after short-time heat treatment," *Corrosion Science,* vol. 121, pp. 22-31, 2017.

[2] X. Zhang, K. Wang, Q. Zhou, J. Ding, S. Ganguly, G. Marzio, D. Yang, X. Xiangfang, P. Dirisu and S. W. Williams, "Microstructure and mechanical properties of TOP-TIG-wire and arc additive manufactured super duplex stainless steel (ER2594)," *Materials Science and Engineering: A,* vol. 762, no. 138097, 2019.

[3] S. Saravanan, K. Raghukandan and N. Sivagurumanikandan, "Pulsed Nd: YAG laser welding and subsequent post-weld heat treatment on super duplex stainless steel," *Journal of Manufacturing Processes,* vol. 25, pp. 284-289, 2017.

[4] A. Eghlimi, M. Shamanian and K. Raeissi, "Effect of current type on microstructure and corrosion resistance of super duplex stainless steel claddings produced by the gas tungsten arc welding process," *Surface and Coatings Technology,* vol. 244, pp. 45-51, 2014.

[5] M. Brázda, P. Salvetr, J. Dlouhý and J. Vavřík , "Influence of laser power in direct laser deposition on the proportion of ferrite and austenite in duplex steel SAF2507," *METAL 2020 - 29th International Conference on Metallurgy and Materials, Conference Proceedings,* pp. 539-544, 2020.

[6] M. Patrik, P. Somayeh and B. O. Isgor, "Effects of heat treatment and applied stresses on the corrosion performance of additively manufactured super duplex stainless steel clads," *Materialia,* vol. 14, no. 100878, 2020.

[7] A. Kiran, J. Hodek, J. Vavřík and M. Urbánek, "Numerical Simulation Development and Computational Optimization for Directed Energy Deposition Additive Manufacturing Process," *Materials,* vol. 13, 2020.

[8] J. Vavřík and J. Hodek, "Residual stress investigation in additively manufactured samples," *IOP Conference Series: Materials Science and Engineering,* vol. 723, 2020.

[9] P. Salvetr, J. Vavřík, A. Školáková and P. Novák, "Formation of intermetallics layers in Ni-Ti system prepared by direct energy deposition," *Manufacturing Technology,* vol. 19, no. 4, pp. 674-679, 2019.

[10] A. Yakovlev, E. Trunova, D. Grevey, M. Pilloz and I. Smurov, "Laser-assisted direct manufacturing of functionally graded 3D object," *Surface and Coatings Technology,* vol. 190, no. 1, pp. 15-24, 2005.

[11] Dzugan J and Sibr M 2017 Mechanical properties determination of AM components. *IOP Conference Series: Materials Science and Engineering,* 179, 012019

[12] Y. Zhang, S. Cheng, S. Wu and F. Cheng, "The evolution of microstructure and intergranular corrosion resistance of duplex stainless steel joint in multi-pass welding," *Journal of Materials Processing Technology,* vol. 277, no. 116471, 2020.

[13] J. Singh and A. S. Shahi, "Metallurgical and corrosion characterization of electron beam welded duplex stainless steel joints," *Journal of Manufacturing Processes,* vol. 50, pp. 581-595, 2020.

[14] ASTM WK49229, "Standard Guide for Orientation and Location Dependence Mechanical Properties for Metal Additive Manufacturing. Work in Progress," 2017.

[15] Sandvik LTD, "OSPREY® 2507-AM SUPER-DUPLEX," [Online]. Available: https://www.metalpowder.sandvik/siteassets/metal-powder/datasheets/osprey-2507-am-super-duplex.pdf. [Accessed 7 10 2020].

[16] J. Kunz, A. Boontanom, S. Herzog, P. Suwanpinij, A. Kaletsch and C. Broeckmann, "Influence of hot isostatic pressing post-treatment on the microstructure and mechanical behavior of standard and super duplex stainless steel produced by laser powder bed fusion," *Materials Science and Engineering: A,* vol. Volume 794, 9 September 2020.

# Effects of the process parameters on the formation of the Zn-enriched surface layer of Mg by thermochemical treatment in a medium containing $ZnCl_2$

**T Bucki[1] and M Konieczny[1]**

[1] Kielce University of Technology, Kielce, Poland, EU
E-mail: tbucki@tu.kielce.pl

**Abstract**. The fabrication of the Zn-enriched surface layer of Mg involved annealing of the Mg specimens in a contact with the paste containing $ZnCl_2$. Annealing of the specimen at 440 °C for 120 min resulted in the formation of a surface layer with a complex structure. In the outer zone, dendrites of a solid solution of Zn in Mg were observed at a eutectoid matrix (an MgZn intermetallic phase and a solution of Zn in Mg). The inner, transition zone, which was located at the interface adjacent to Mg substrate, was composed of a solid solution of Zn in Mg. The use of various period of annealing time (30, 60, 90 and 120 min) showed that increasing the annealing time leads to an increase in the thickness of the layer. A similar structure of the layer was observed for each variant of the annealing time. The analysis for the specimen fabricated at 410 °C for 120 min showed that the use of lower temperature resulted in a formation of a thin layer, eutectoid in structure. When higher annealing temperature (470 °C) was used, the thick surface layer with an uneven thickness and containing macroscopic unevenness was produced on the specimen surface.

## 1 Introduction

The application of magnesium as a structural metal is an effective method to reduce the weight of structures in such industries as automotive, aviation, and electronics. However, the use of magnesium alloys is limited, mainly due to their low resistance to corrosion and abrasion [1,2]. The literature data shows that these properties can be efficiently improved by the surface treatment of magnesium products. Positive results are achieved by enriching the surface with chemical elements that form intermetallic phases with magnesium, i.e. Al, Zn, Cu, Ni, Si and a combination of a few elements, such as Al/Si, Al/Zn, Al/Cu, Al/Mn, Al/Ir, Ni/Cu/Al and Zr/Al/Ni/Cu [3–15]. There are many methods of a heat treatment of magnesium and its alloys. Thermo-chemical treatment is a method which advantages are simple technology and low production cost. It involves annealing of the alloy in contact with a solid or liquid medium. The most common solid media are powders (e.g. aluminum powder [12]), that lead to the formation of a surface layer as a result of diffusion processes. The thermo-chemical treatment with liquid medium can be performed using molten salts, such as $AlCl_3$ [9]. As a consequence, a surface layer is formed due to the reaction of a liquid substance with an alloy surface and the diffusion between the components. In the literature, there are also studies with the author's contribution [13,14], which show that surface treatment of magnesium or its alloys can also be carried out by annealing of specimens coated with a paste containing $ZnCl_2$. The results shows that the process led to the formation of a layer containing Zn-rich phases on the surface of a pure Mg and AZ91 alloy. The authors noted also that produced layers were characterized by high microhardness.

IOP Conf. Series: Materials Science and Engineering 1178 (2021) 012009 doi:10.1088/1757-899X/1178/1/012009

The aim of presented work was to analyze how the parameters of the process, i.e. time and temperature of annealing, affect the formation of the Zn-enriched surface layer of Mg, fabricated by thermochemical treatment in a medium containing $ZnCl_2$. The study involved annealing of the specimens covered with a paste containing $ZnCl_2$ in a chamber furnace. Four variants of the annealing time (30, 60, 90 and 120 min) and three different values of the temperature (410, 440 and 470 °C) were applied in the experiment. The microstructure of the surface layers was analyzed with an optical microscope and a scanning electron microscope.

## 2 Experimental procedure

The specimens with dimensions of 20x15x10 mm were machined from an Mg ingot (99.9 % Mg). Before the application of the paste, the surfaces of the specimens were ground using abrasive papers and then cleaned with ethanol. The upper, largest surface of the specimens (20x15 mm) was coated with a paste containing $ZnCl_2$, and then the specimens were annealed in a chamber furnace. No protective atmosphere was used during the process. The specimens were cooled in the air after they were removed from the furnace. The paste used was a mixture containing $ZnCl_2$, KCl, pine rosin and ethanol. The composition of the mixture was selected on the basis of liquidus and solidus temperatures which are associated with the possibility of a reaction between $ZnCl_2$ and the Mg surface and formation of the surface layer. The results of previous studies [13,14] were also used to develop the proportion of ingredients. However, slight modifications to the paste composition were made to improve the uniformity of the layer thickness. The method of producing a surface layer on magnesium and its alloys using a paste containing $ZnCl_2$ is the subject matter of the patent of Republic of Poland (PL230625-B1).

In order to analyze the effect of the parameters on formation of the layer during the process, four variants of the annealing time and three different values of the temperature were used. The specimens were annealed at the temperature of 440 °C for: 30, 60, 90 and 120 min as well as for 120 min at various temperatures, equal to: of 410, 440 and 470 °C.

The specimens for microscopic observations were prepared by using a standard metallographic procedure. The final polishing was carried out with a colloidal silica. The microstructural analysis was conducted using a Nikon ECLIPSE MA200 optical microscope (OM) and a JEOL JSM-7100F scanning electron microscope, which was equipped in a LINK ISIS 300 energy dispersive X-ray spectrometer (SEM/EDS).

## 3 Results and discussion

Figure 1 presents the microstructure of the surface layers fabricated as a result of annealing of the specimen in contact with a paste containing $ZnCl_2$ in a furnace heated to 440 °C. Specimens were placed in a furnace for 30 min, 60 min, 90 min and 120 min. In the specimen produced when the shortest annealing time was applied (figure 1a), the thickness of the surface layer was about 50 μm. The layer was composed of two zones: outer – lamellar in structure and inner – transition zone. The longer annealing time period of 60 min (see figure 1b) resulted in the formation of a surface layer with a thickness of about 130 μm. In its structure, outer and inner zones can be also distinguished. Gray particles with dendritic shapes were also noticeable in the outer zone on the lamellar matrix background. Further extension of the annealing time period to 90 min (figure 1c) and 120 min (figure 1d) resulted in the increase in thickness of the surface layer to approximately 200 μm and 230 μm, respectively. A complex structure of the layer, similar to the previous variants, was also observed in these cases.

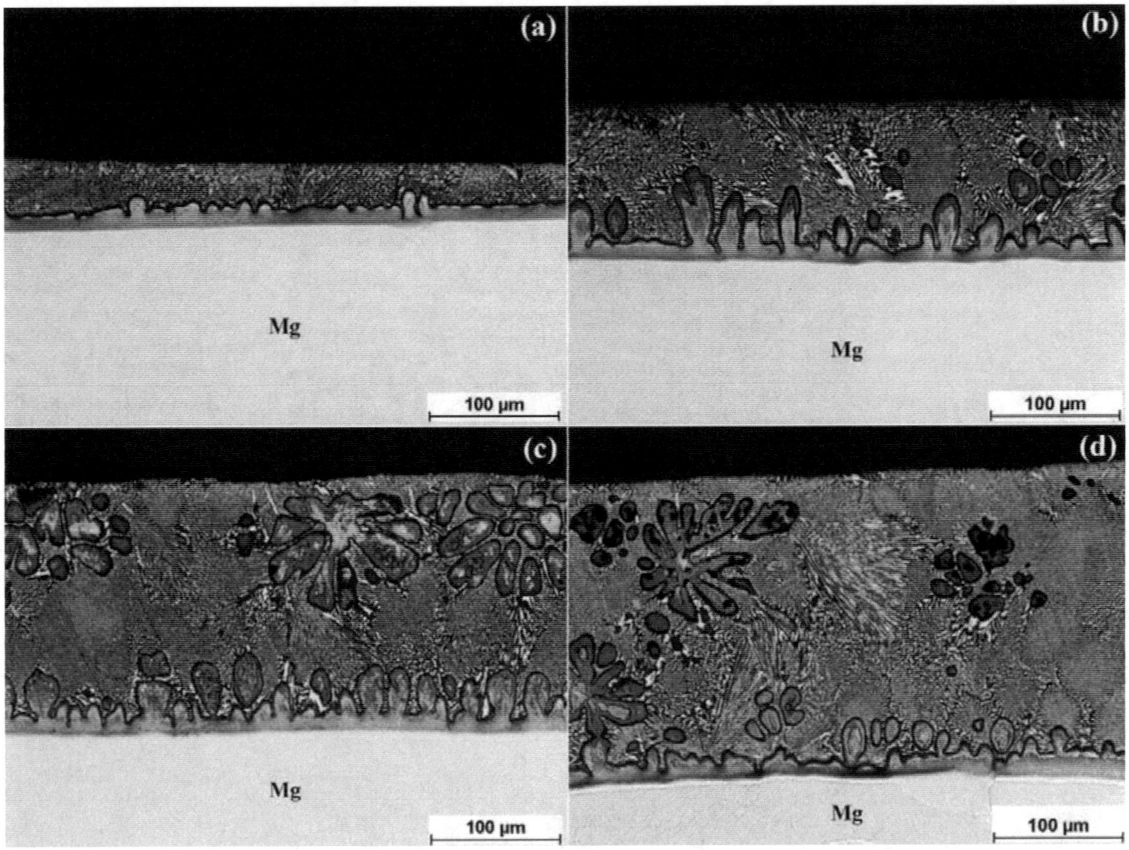


**Figure 1.** OM micrograph of the surface layer fabricated by annealing of the Mg specimens coated with the paste at 440 °C for: (a) 30 min, (b) 60 min, (c) 90 min, (d) 120 min.

Figure 2 shows the SEM micrograph of the surface layer fabricated by annealing of the Mg specimen coated with a paste containing $ZnCl_2$ at 440 °C for 120 min. A linear analysis was performed along the line marked in the figure. The distribution of Zn and Mg show that the surface layer of the specimen was enriched in a Zn. In the transition zone marked as 1, a gradually increasing Zn content was observed towards the outside of the specimen. A detailed analysis of the phase composition of the surface layer, which was carried out using the EDS analysis and confirmed by XRD (X-Ray Diffraction) method, was presented in the previous work of authors [14]. According to the results, the transition zone of the surface layer was composed of a solid solution of Zn in Mg. The microstructure of the outer part of the surface layer observed in the SEM at a high magnification is shown in Figure 3. From the results, it is found that dark particles with dendritic shape (marked as 2 in figure 3) are composed of a solid solution of Zn in Mg. The results for the two-phase matrix of the layer (point 3) indicated a eutectoid consisting of an MgZn intermetallic phase and a solid solution of Zn in Mg. Locally distributed bright particles, marked as 4, were also observed in the matrix of the layer. The analysis of the chemical composition showed that they are composed of an $Mg_2Zn_3$ intermetallic phase.

The analysis of the results and the Mg-Zn system [16] shows that annealing of the specimens in contact with the paste containing $ZnCl_2$ causes the enrichment of Mg surface in Zn. At the elevated temperature applied during the process, a thin zone composed of a solid solution of Zn in Mg is formed. When the Zn content exceeds the solubility limit, the Zn richer phases are created. Then, during cooling of the specimen, the eutectoid (an MgZn phase and a solid solution of Zn in Mg) as well as the dendrites of a solid solution of Zn in Mg crystallize in the surface layer of specimen.

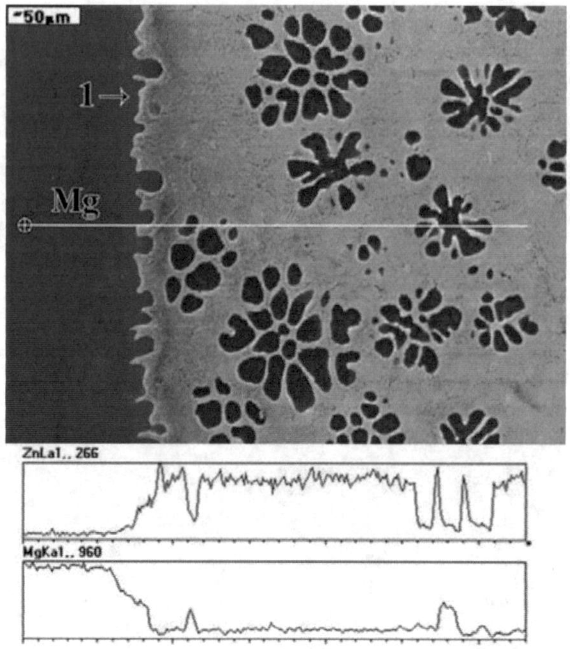


**Figure 2.** SEM micrograph and corresponding EDS line scan results for the surface layer fabricated by annealing of the Mg specimen coated with the paste at 440 °C for 120 min.

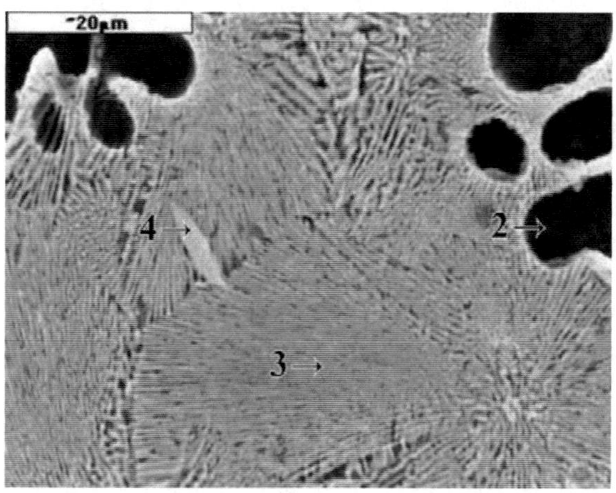


**Figure 3.** Details of the microstructure of the surface layer fabricated by annealing of the Mg specimen coated with the paste at 440 °C for 120 min.

Figure 4 presents the OM microstructure observed at high magnification, for the case when the surface layer was produced at a lower temperature, equal to 410 °C, for 120 min. The results show that annealing of the specimen at 410 °C resulted in the formation of the surface layer with a eutectoid structure. The layer was characterized by relatively thin thickness (about 25 µm).
Figure 5 shows the microstructure of the layer fabricated when the higher annealing temperature (470 °C) was applied. In this case, the Zn-enriched surface layer with a microstructure corresponding to that produced in 440 °C was formed. Increasing the annealing temperature of the specimens, however, resulted in the formation of the surface layer which was thick, but uneven in thickness (300-700 µm). The macroscopic unevenness were also clearly visible on the specimen surface.

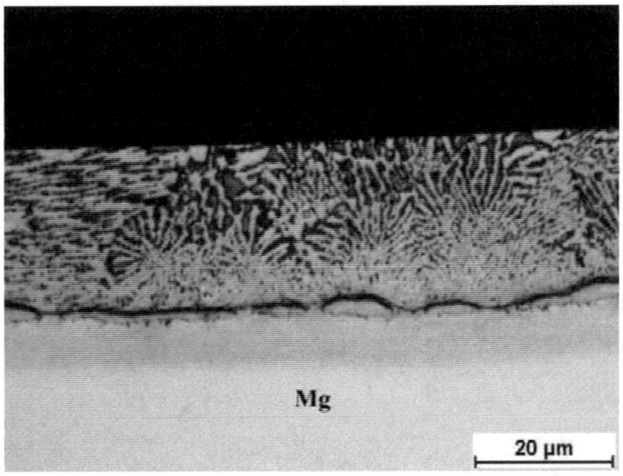


**Figure 4.** OM micrograph of the surface layer fabricated by annealing of the Mg specimens coated with the paste at 410 °C for 120 min.

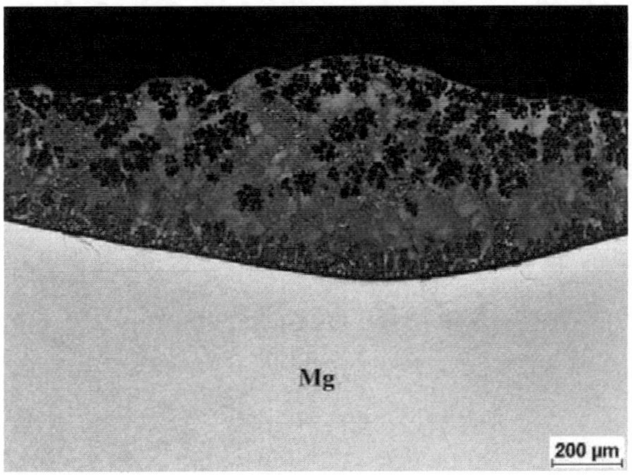


**Figure 5.** OM micrograph of the surface layer fabricated by annealing of the Mg specimens coated with the paste at 470 °C for 120 min.

The analysis of all the variants of annealing temperature show that the temperature of 440 °C was the most appropriate for the production of the Zn-enriched surface layer. This condition led to the formation of the continuous and thick layer, which was uniform in a thickness. It is also evident that the thickness of the layer depends on the annealing time and obviously it is related to the properties of the specimen.

## 3 Conclusions

The Zn-enriched surface layers of Mg were fabricated by annealing of the Mg specimens coated with the paste containing $ZnCl_2$.

The experiment showed that when the specimen was annealed at 440 °C for 120 min, a surface layer with a thickness of about 230 μm was formed on the Mg substrate. The microscopic observations indicated that the microstructure of the produced surface layer was complex. Their outer zone contained a eutectoid matrix (an MgZn intermetallic phase and a solid solution of Zn in Mg) with some dendrites of a solid solution of Zn in Mg and fine particles of an $Mg_2Zn_3$ intermetallic phase. The inner zone of the surface layer was composed of a solid solution of Zn in Mg.

IOP Conf. Series: Materials Science and Engineering 1178 (2021) 012009 doi:10.1088/1757-899X/1178/1/012009

The results also showed that the annealing time affects the thickness of the produced surface layer. The layer formed by annealing of the specimen at 440 ºC for 30 min had the lowest thickness (50 µm), but the thickness increased with increasing the annealing time of the specimen (60, 90 and 120 min, respectively). All the layers produced under these process conditions had a similar microstructure.

The application of a lower process temperature (410 ºC) resulted in the formation of a layer about 25 µm in thickness that was eutectoidal in structure. The layer observed for the specimen annealed at 470 ºC was relatively thick and its microstructure corresponded to that produced in 440 ºC. However, in this case, the thickness of the layer was not uniform (300-700 µm) and the macroscopic unevenness was observed on the surface of specimen.

## References


[1] Mola R and Bucki T 2018 The microstructure and properties of the bimetallic AZ91/AlSi17 joint produced by compound casting *Arch. Foundry Eng.* **18** pp 71–6

[2] Mola R, Bucki T and Dziadoń A 2017 Effects of the pouring temperature on the formation of the bonding zone between AZ91 and AlSi17 in the compound casting process *IOP Conf. Ser. Mater. Sci. Eng.* **179** pp 1–6

[3] Kutschera U and Galun R 2000 Magnesium Alloys and their Applications ed K U Kainer (New Jersey: John Wiley & Sons, Ltd) pp 330–5

[4] Majumdar J D, Maiwald T, Galun R, Mordike B L and Manna I 2002 Laser surface alloying of an Mg alloy with Al + Mn to improve corrosion resistance *Laser Eng.* **12** pp 147–69

[5] Gao Y, Wang C, Pang H, Liu H and Yao M 2007 Broad-beam laser cladding of Al–Cu alloy coating on AZ91HP magnesium alloy *Appl. Surf. Sci.* **253** pp 4917–22

[6] Chen C, Wang M, Wang D, Jin R and Liu Y 2007 Laser cladding of Al+Ir powders on ZM magnesium base alloy *Rare Met.* **26** pp 420–5

[7] Yue T M and Su Y P 2007 Laser multi-layer cladding of Zr65Al7.5Ni10Cu17.5 amorphous alloy on magnesium substrates *J. Mater. Sci.* **42** pp 6153–60

[8] Yue T M and Li T 2008 Laser cladding of Ni/Cu/Al functionally graded coating on magnesium substrate *Surf. Coat. Technol.* **202** pp 3043–9

[9] He M, Liu L, Wu Y, Tang Z and Hu W 2009 Improvement of the properties of AZ91D magnesium alloy by treatment with a molten AlCl3–NaCl salt to form an Mg–Al intermetallic surface layer *J. Coat. Technol. Res.* **6** pp 407–11

[10] Hirmke J, Zhang M-X and StJohn D H 2011 Surface alloying of AZ91E alloy by Al–Zn packed powder diffusion coating *Surf. Coat. Technol.* **206** pp 425–33

[11] Mola R 2014 Fabrication and microstructure of diffusion alloyed layers on pure magnesium substrate *Arch. Metall. Mater.* **59** pp 1409–12

[12] Mola R and Jagielska-Wiaderek K 2014 Formation of Al-enriched surface layers through reaction at the Mg-substrate/Al-powder interface *Surf. Interface Anal.* **46** pp 577–80

[13] Bucki T and Mola R 2016 Metal 2016: 25th Anniversary International Conference on Metallurgy and Materials (Ostrava: Tanger Ltd) pp 1357–61

[14] Mola R, Bucki T and Gwozdzik M 2018 Surface alloying of magnesium and AZ91 by thermochemical treatment in a medium containing zinc chloride and potassium chloride *Materialwiss. Werkstofftech.* **49** pp 1006–14

[15] Mola R and Cieslik M 2019 Microscopic analysis of layers containing mg2si and mg17al12 phases fabricated on az91 through thermochemical treatment *Arch. Foundry Eng.* **19** pp 119–24

[16] Agarwal R, Fries S G, Lukas H L, Petzow G, Sommen F, Chart T G and Effenberg G 1992 Assessment of the Mg-Zn system *Z. Metallkd.* **83** pp 216–23

# Comparison of laser and electron beam weldability of high-strength steels

**J Capek[1], J Kec[2], S Nemecek[3], K Trojan[1], N Ganev[1] and I Cerny[2]**

[1] Faculty of Nuclear Sciences and Physical Engineering, Czech Technical University in Prague, Trojanova 13, 120 00 Prague 2, Czech Republic
[2] SVÚM a.s., Tovární 2053, 250 88 Čelákovice, Czech Republic
[3] RAPTECH s.r.o., U Vodárny 473, 330 08 Zruč-Senec, Czech Republic

E-mail: Jiri.Capek@fjfi.cvut.cz

**Abstract**. Pressure vessels made of high-strength steels are usually welded by arc methods. The paper deals with the development of laser and electron thick-walled beam welding technology for the mentioned applications. The mechanical and structural properties of laser and electron beam welds of plates with a thickness of 5–20 mm are compared both with and without additional filling material. The study is completed with a detailed analysis of residual stresses and their relation not only to the used welding technology but also to the results in the weld area. Laser and electron beam welds achieve high mechanical values and sufficient fatigue life. Therefore, the technology has been verified and applied for welding the test pressure vessel.

## 1 Introduction

Welding is a widely used method for joining materials in many branches of industry. Therefore, high demands are laid on mechanical properties and durability of welds used to connect two or more components. Up to now, laser, electron and conventional arc welding are some of the most widely used welding methods. Due to lower heat input, laser and electron beam welding give better mechanical properties and the favourable surface state of residual stresses [1] compared with conventional MAG (metal active gas) technique. Nevertheless, due to the heterogeneous application of energy and localized fusion, which occur during the welding process, high undesirable residual stresses (RS) can be present in the region near weld and in the weld itself. These RS occur because of the superposition of thermal and transformation processes [2] and could reach high values and subsequently could cause fatigue life reduction or promote crack growth [1, 3].

In terms of service life, the microstructural changes of plastically deformed crystalline materials are necessary to study. Fusion welding causes to decrease toughness, ductility and corrosion resistance of ferritic steel due to the grain coarsening, carbides precipitation and martensite formation occurred in their microstructure [4]. These so-called microstructural notches are the potential critical areas for the potential crack initialization, e.g. in the fusion zone.

The purpose of this investigation is to analyse strengthening behaviour and reveal relationship between microstructure of weld metal, mechanical properties and surface residual stresses.

## 2 Experimental techniques

The investigated specimens were prepared by welding of plates 300×150 mm$^2$ in size and 5–20 mm in thickness made of P355NL1 and P460NL1 (marked as "N" and "H") fine-grained ferritic-perlitic steel used for pressure vessels, heat exchangers and reactors. The laser and electron beam welded specimens (marked as "L" and "E") were prepared using a high-power diode laser by the company RAPTECH s.r.o. and at Faculty of Materials Science and Technology in Trnava, respectively. The welding parameters were as follows: power 3 kW, welding speed 5.5 mm·s$^{-1}$, continuous mode, laser wavelength 900–1080 nm for laser beam welding and welding speed 30 mm·s$^{-1}$, voltage 55 kV, current 220 mA for electron beam welding. 20 mm plates were laser beam welded according Fig. 1a. The butt part was double-side welded, where top sides "U" were welded as the first and bottom sides "D" as the second one. Further, the V-shape groove was filled using G3Si1 filling wire. All plates welded by the electron beam and plates with thickness of 5 mm welded by laser beam were single-side welded from the top sides "U", see Fig. 1b. The others plates were double-side welded. Using these marks, "N8LD" sample denotes laser welded 8mm plate made of P355NL1 analysed on the bottom side.

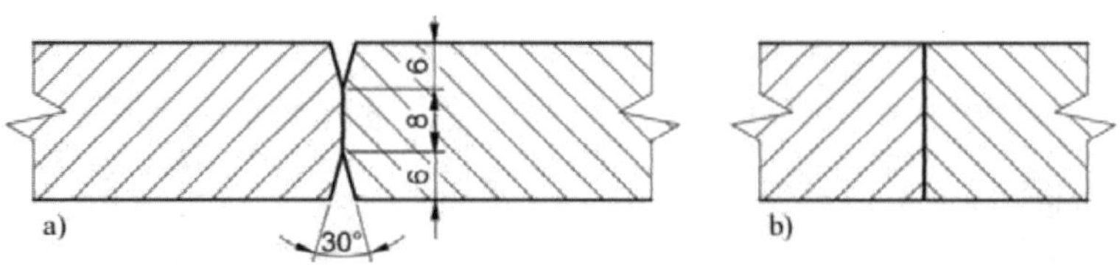


**Figure 1.** a) The double V-groove shape of laser beam welded 20 mm plates geometry b) the others

Specimens for metallographic analysis were hot embedded into conductive bakelite resin with the carbon filler. After the embedding, the specimens were prepared by machine grinding and polishing. Microstructure was revealed by etching in 2 % Nital (98 ml ethanol + 2 ml $HNO_3$). Microstructure was investigated by optical microscope *Zeiss Axio Observer Z1M.*

The tensile test was carried out at room temperature according to ČSN EN ISO 6892-1 on the *EUS 40* testing machine. Charpy V-notch impact test was performed in accordance with ČSN EN ISO 148-1 standard on *PSWO 30* impact testing machine at 0 °C.

The *PROTO iXRD COMBO* diffractometer in ω-goniometer set-up with chromium anode was used for determination of lattice deformations of α-phase (ferrite, bainite, and martensite). Diffraction angles $2\theta^{hkl}$ were determined using *Gaussian function* and *Absolute peak* method from the peaks of the diffraction lines $K\alpha_1$ of the *{211}* planes. To describe the state of residual stresses (RS), the *Winholtz & Cohen* method [5] and X-ray elastic constants $\frac{1}{2}s_2$ = 5.75 TPa$^{-1}$, $s_1 = -1.25$ TPa$^{-1}$ were used. The specimens were analysed in the direction perpendicular "T" and parallel "L" to the welds (welding direction) on both sides of the specimens. Welded plates were analysed in the middle part in 18 points (x = 0–30 mm from the axis of the weld). The irradiated area was approx. 11×1 mm$^2$ with the average effective penetration depth of the X-ray radiation approx. 4 μm [6].

## 3 Results and discussion

### *3.1 Microstructure*

Comparison of laser and electron beam weld macrostructure is shown in Figs. 2 a) and b), respectively. The laser beam weld is almost three times wider than the electron beam weld, which is a consequence of the different power input and welding speed of the technologies.

Scanning electron microscope (SEM) micrographs of the base material of P355NL1 and P460NL1 steels are shown in Figs. 3. The microstructure of both steels consists of polygonal ferrite (PF) and pearlite (P). PF is fine-grained, but the case of P460NL1 volume fraction of P is occurred in the case of P355NL1. P is in typical lamellar morphology.

Microstructure in the heat-affected zone (HAZ) and weld metal (WM) is mostly affected by temperature input and cooling rate. The highest cooling rate is in WM, which led to the formation of non-diffusion ferrite morphologies. Allotriomorphic ferrite (ALF) is occurred along the primary austenitic grains and the rest of primary austenite transformed to the acicular ferrite (AF) and bainite ferrite (BF), see Fig. 4 c). The boundary between WM and HAZ, well-known as a fusion zone (FZ), is formed by coarse bainite. As can be seen from Fig. 4 d), the mixture of BF, PF and P was found in the HAZ. In areas where the base material was heated to a temperature higher than A1, lamellar perlite transformed to globular, see Fig. 4 b).

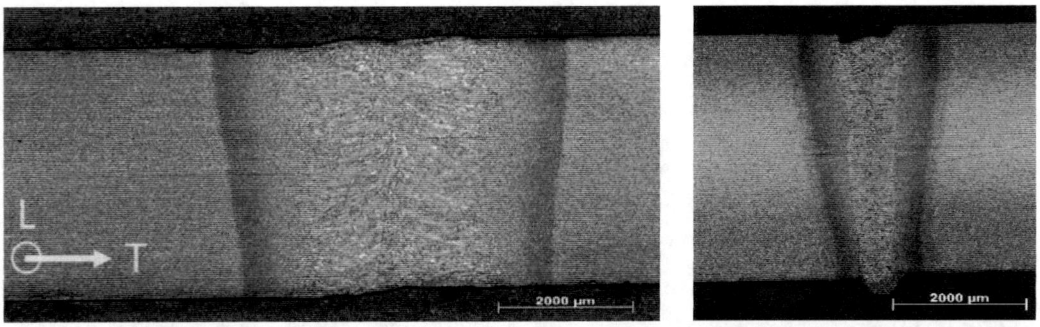


**Figure 2.** Macrostructure of a) laser beam weld and b) electron beam weld

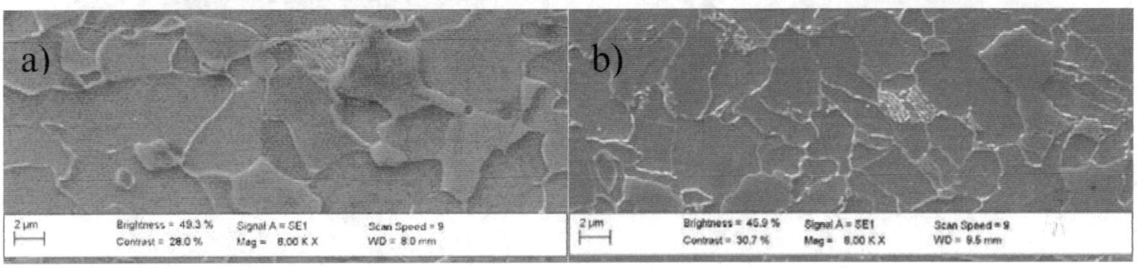


**Figure 3.** Microstructure of a) P355NL1 and b) P460NL1 steel

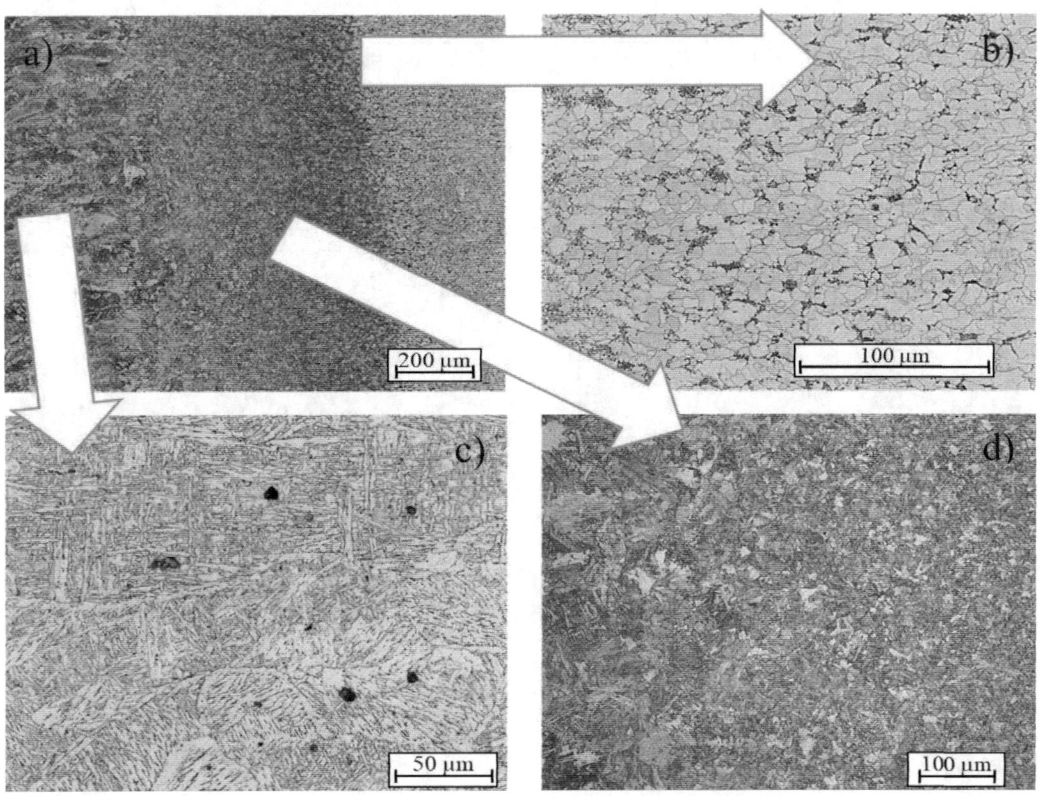


**Figure 4.** Microstructures in the WM and HAZ

The results of the tensile strength of welded steels are shown in Figs. 5. During the tensile tests, a minimum scatter was observed and almost all specimens reached the minimum strength of the base material, which is 490 MPa and 570 MPa for P355NL1 and P460NL1 steels, respectively. The final fracture was almost always observed far from the WM and the HAZ. Unsatisfactory results were found in specimen H5E, where a significant lack of fusion was found on the fracture surface, which led to a reduction of tensile strength and localization of the final fracture to the weld area.

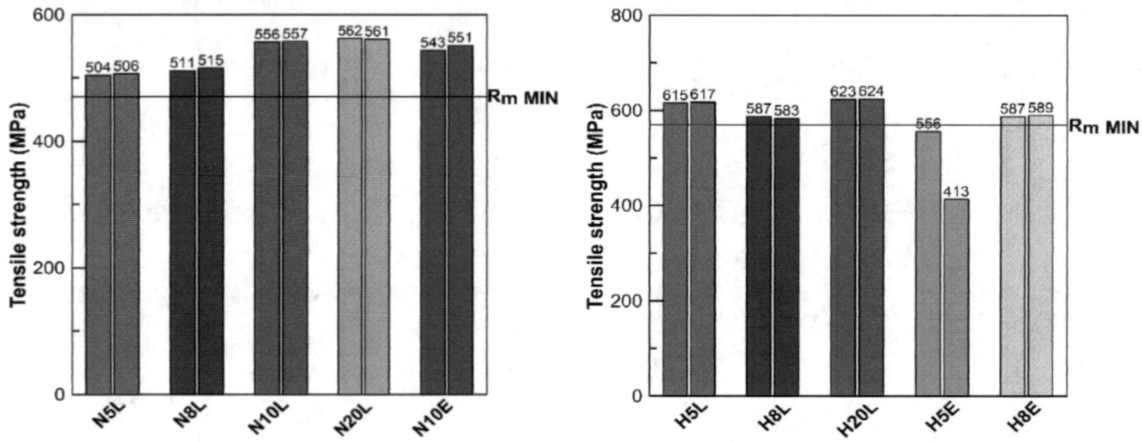


**Figure 5.** Tensile strength of laser beam welded, electron beam welded P355NL1 and P460NL1 steel

The results of the Charpy V-notch impact test ($KV_2$) are shown in Figs. 6. A significant scatter was observed between individual specimens, which is caused by a different amount of porosity in the welds and mainly by the deviation of the main crack from WM to HAZ and base material. Deviation of crack from WM causes that the values of absorbed energy are artificially increased, this issue is also widely discussed in all technologies that create narrow weld beads such as laser and electron beam welding. Many articles [e.g. 7, 8] discuss this problem in detail and it is generally called fracture path deviation (FPD). FPD arises not only in $KV_2$ but also in static three-point bend tests and fracture toughness tests. Different fracture micromorphology was also observed depending on FPD - dimple morphology occurred on specimens with higher $KV_2$ energy and cleavage facets on specimens with lower $KV_2$ energy, see Fig. 7.

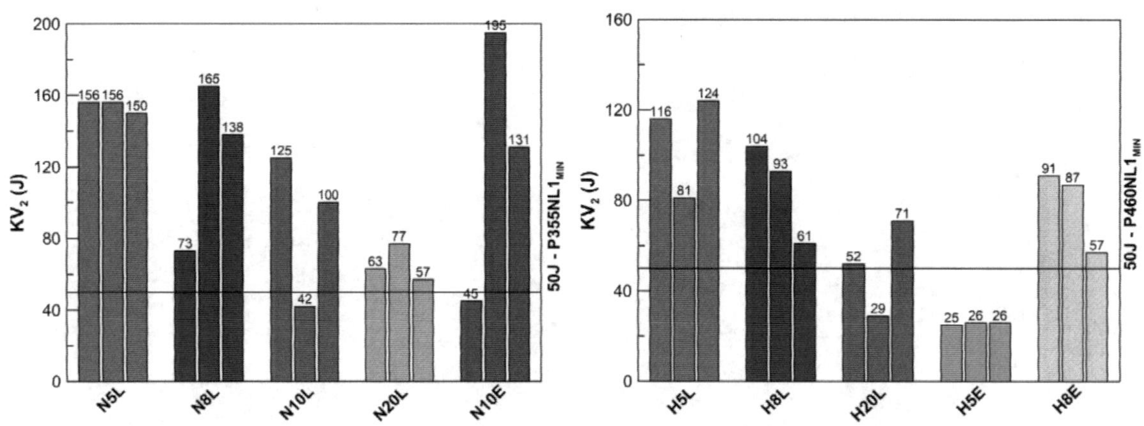


**Figure 6.** $KV_2$ energy of laser and electron beam welded P355NL1 and P460NL1 steel

**Figure 7.** Effect of FPD on fracture propagation of N8L sample

### *3.2 X-ray diffraction*

A selected residual stresses (RS) dependences for both unconventional technologies are presented in Figs. 8, where N, H, 8, L, E, U, and D indicate P355NL1 steel, P460NL1 steel, the thickness of the plate in millimetres, laser beam welded specimen, electron beam welded specimen, the side which was welded first, and the other side, respectively. Since the main stress component of the longitudinal weld of the pressure vessel is in the direction perpendicular to the weld, i.e. the T direction.

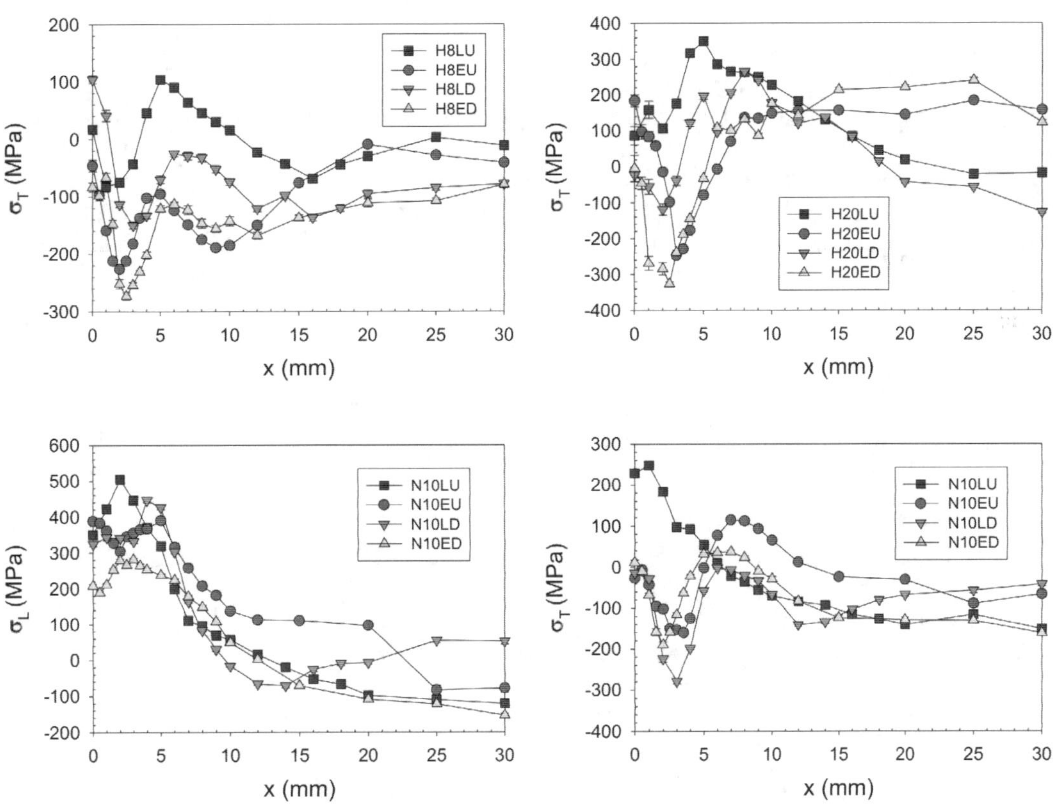


**Figure 8.** Surface residual stresses in relation to the analysed area of the welded specimen, where x denotes the distance from the axis of the weld

Comparison of surface RS of specimens after different welding techniques shows, that the type of welding has a significant influence on the state of RS, compared to [2]. Presented trends follow a combination of anisotropy shrinkage during cooling and volume-limited phase transformation from austenite to ferrite and/or martensite and/or bainite. In direction T, higher tensile RS were found for the laser beam weld on the top side, while lower compressive stresses are on the bottom side. These stresses are probably generated due to the filling the groove during laser beam welding and the associated deformation during cooling, or generally due to accumulation of heat of double-side weld. In this case, the effect of shrinkage did not prevail along the weld, which resulted in a high tensile RS. Another effect could be lateral warping of the plate when welding the bottom side. These RS near the weld could be higher than the yield strength, see Fig. 8 N10LU specimen in the L direction. This effect could be caused by the higher hardness of weld, which indicates the occurrence of hard phases with a higher yield strength.

For electron beam welding, the favourable compressive RS were analysed in the HAZ in the distance of 1–3 mm from the weld axis in the T direction. Stress affected zone is approx. 15–20 mm in the distance from the axis of the weld. Because of the amount of heat, the laser and electron beam welds exhibit comparably narrower HAZ in the comparison with e.g. MAG welding.

## 4 Conclusion

The most critical areas for the potential initialization of surface fatigue crack is FZ because of higher hardness, microstructural and surface notches. The microstructural notch of FZ is formed by coarse bainite. From surface RS point of view, the area with high tensile RS is the most critical. The positive finding was that the final fracture was localized in the base material during the tensile tests. The combination of welds properties (microstructural, RS, hardness, imperfections etc.) exceeds the properties of base materials.

Despite the scatter of $KV_2$ energy between individual specimens, there are no significant differences between laser and electron beam welded specimens. Moreover, higher $KV_2$ energy was observed for thinner specimens. This trend could be caused by a different amount of porosity in the WM and high tensile RS. The main reason for tensile RS generation in the WM and HAZ is a higher amount of thermal energy, i.e. using the filling wire and double-side welding.

Therefore, the laser and electron are a suitable tool for welding of pressure vessels steels, nevertheless, the biggest advantage of laser compared with electron beam welding is that there is no need to have a vacuum chamber. Contrarily, the electron beam welds have narrower WZ which could improve the mechanical properties of the weld.

**Acknowledgements**

This work was supported by the project TH02010664 of the Technology Agency of the Czech Republic and by the Grant Agency of the Czech Technical University in Prague, grant No. SGS 19/190/OHK4/3T/14.

## References

[1] Čapek J, Ganev N, Trojan K, Němeček S and Kolařík K 2019 *Adv. X-Ray Anal: Proc. Denver X-Ray Conf.* vol 63 (USA: ICDD)
[2] Nitschke-Pagel T and Digler K 2014 *Mater. Sci. Forum* **783–786** 2777
[3] Radaj D 2012 *Heat effects of welding: temperature field, residual stress, distortion* (Berlin: Springer Science & Business Media)
[4] Khorrami MS, Mostafaei MA, Pouraliakbar H and Kokabi AH 2014 *Mater. Sci. Eng. A* **608** 35
[5] Welzel U, Ligot J, Lamparter P, Vermeulen AC and Mittemeijer EJ 2005 *J. Appl. Crystallogr.* **38** 1
[6] Čapek J and Pala Z 2015 *Mater. Struct.* **22** 78
[7] Ohata M, Morimoto G, Fukuda Y, Minami F, Inose K and Handa T 2015 *Weld. World* **59** 667
[8] Üstündağ Ö, Gook S, Gumenyuk A and Rethmeier M 2020 *J. Mater. Process. Tech.* **275** 116358

# Analysis of deformation mechanisms in a textured AZ31 magnesium alloy

**J Dittrich[1], J Čapek[1,2], M Knapek[1] and P Minárik[1]**

[1]Charles University, Department of Physics of Materials, Prague, Czech Republic, EU
[2]Paul Scherrer Institute, Laboratory for Neutron Scattering and Imaging, Villigen, Switzerland

E-mail: dittrich.jan.cz@gmail.com

**Abstract**. A combination of advanced in-situ and ex-situ methods providing complementary information was employed in order to reveal the active deformation mechanisms during deformation of a heavily textured commercial magnesium alloy AZ31. Three sets of samples for both compression and tension testing were prepared from the rolled sheet with respect to its identified strong basal texture – normal direction (ND), rolling direction (RD) and 45° between RD and ND. During both compressive and tensile deformation of these samples, the signal of acoustic emission (AE) was concurrently measured. Afterwards, electron backscattered diffraction (EBSD) was used to study the microstructure after the selected stages of deformation. The EBSD analysis revealed that in the samples favourably oriented for extension twinning it played an important role during the plastic deformation, whereas its role in the unfavourably oriented samples was minor. This orientation and deformation mode dependency of extension twinning activation resulted in considerable differences in the deformation behaviour, which also influenced the AE response. A consistent link between the AE signal and the microstructural changes investigated by EBSD was established. Furthermore, to study the dynamics of plastic deformation and twin growth, high-speed camera footage of the sample surface was captured during compressive deformation. The obtained results were in complete accordance with the predicted behaviour, thus providing direct proof of the theory and extending the understanding of the deformation mechanisms.

## 1 Introduction

The deformation behaviour of magnesium (and its alloys) is rather complex due to its hexagonal close-packed structure with the *c/a* ratio, 1.623, close to the ideal value of 1.633. In the case of strongly textured materials, such as rolled sheets, the anisotropy of the hcp unit cell translates into anisotropy of the mechanical properties of the sheet, which negatively affects their formability [1-3].

There are two basic modes of deformation, playing critical roles in the deformation of magnesium at room temperature – dislocation slip and mechanical twinning. The slip system with the lowest critical resolved shear stress (i.e. the first one to be activated) is generally the $(0001)\langle 11\bar{2}0\rangle$ basal slip. The next slip system to be activated is typically the $\{10\bar{1}0\}\langle 11\bar{2}0\rangle$ prismatic slip [4]. The combination of these deformation mechanisms, however, provides only four independent slip systems, none of which provides elongation in the $\langle c\rangle$ direction, and thus fails to meet the Von Mises criterion [5]. To fulfil the criterion, either the $\{11\bar{2}2\}\langle 11\bar{2}3\rangle$ 2nd order pyramidal slip or deformation twinning needs to be activated as well. The critical resolved shear stress for the 2nd order pyramidal slip at the room temperature (which is the case of this study) is high, and therefore in the grains oriented favourably to do so, the $\{10\bar{1}2\}\langle10\bar{1}1\rangle$ extension twinning (i.e. providing elongation along the *c*-axis) is preferentially activated [4].

The acoustic emission (AE) phenomenon is defined as a transient elastic wave generated by a rapid release of energy within the deformed material [6]. The AE signal originates from active defects, such as formation and propagation of cracks, nucleation of twins and an avalanche-like movement of dislocations. The non-continuous (burst) AE signal can typically be ascribed to the nucleation of twins (and the initiation and propagation of cracks in the final stages of deformation). While nucleation of twins provides an excellent source of AE, their growth is too slow to be detected [6-8]. The continuous signal originates from the movement of dislocations. In order to be detectable, an extensive number of dislocations need to move rapidly over a sufficient distance.

For the purposes of this work, a so-called hit-based approach to the evaluation of AE events will be employed. In this approach, the AE event is defined by a set of criteria (e.g. threshold level, number of counts, duration). Once the events are separated from the raw signal, their record is systematically evaluated and described by multiple parameters (e.g. amplitude, count rate, energy rate). These parameters are subsequently examined to reveal the particular mechanism of the AE event origin. This approach has successfully been applied by other authors, e.g. [9,10].

The choice of magnesium-based sheet alloys is currently somewhat limited – the most common commercially available sheets are those of the AZ31 alloy. There has been extensive research conducted focusing on the anisotropy in the deformation behaviour of the rolled AZ31 alloy, such as [4] and more recently [11], dealing with the roles of non-basal slip mechanisms and mechanical twinning. The AE technique has also previously successfully been employed to investigate the mechanisms mediating the deformation of AZ31 alloy [12, 13]. The novelty of this study lies in the utilisation of the combination of the acoustic emission technique, providing sensitive yet indirect observations of the dynamical processes during the deformation, and high-speed camera imaging of the loaded sample surface, enabling direct observations of such processes. These observations are complemented by the EBSD-based microstructural analysis of the deformed samples.

## 2 Material and experimental methods

All experiments were conducted using a rolled sheet of the commercial AZ31 magnesium alloy (3% wt. Al, 0.8% wt. Zn, 0.2% wt. Mn). The sheet exhibited a strong basal texture in the normal direction (figure 1). Sets of samples with three different orientations with respect to the detected initial structure for both compressive and tensile tests – the selected orientations being normal direction (ND), rolling direction (RD) and 45° between RD and ND (denoted 45) – were prepared. All samples were of a rectangular cross-section, where the dimensions of samples for compressive tests were 10 mm x 6.5 mm x 6.5 mm, and the active area dimensions of the tensile samples were 6 mm x 3 mm x 2 mm.

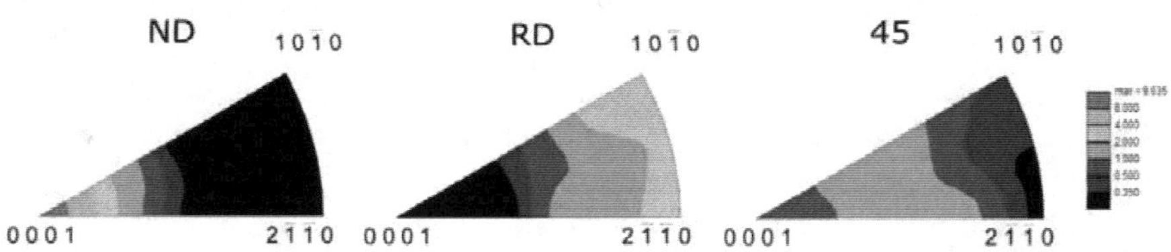


**Figure 1.** Inverse pole figures (IPF) of the ND, RD and 45 oriented samples.

All deformation tests were performed at room temperature with the initial strain rate of $10^{-3}$ $s^{-1}$ using the Instron 5882 universal deformation machine. Concurrently with the deformations tests, the AE signal was measured. To record the broad AE spectra, the PAC PICO AE sensor was employed using the threshold level of 22 dB and 24 dB for the compressive and tensile tests. The signal was preamplified by the PAC 2/4/6 preamplifiers with a 40 dB gain in the frequency band 100 – 1200 kHz and recorded with the sampling rate of 5 MHz using the PAC PCI-2 device. The microstructure of samples deformed to selected strain (i.e. 1% plastic strain in compression and 1%, 5% and 10% plastic strain in tension) was investigated using electron backscattered diffraction (EBSD) with the EDAX EBSD camera installed in the ZEISS Auriga Compact scanning electron microscope. The high-speed camera footage of the sample surface during compressive deformation using the MTI SEMTester 1000 deformation table was captured via the PHOTRON FastCam SA-Z 2100K camera. Samples for both high-speed camera footage and EBSD were mechanically ground down to 0.25 μm diamond paste. In the case of EBSD samples, this was followed by ion-polishing using the Leica EM RES102 machine.

## 3 Results and discussion

The deformation curves from both compressive and tensile tests of all orientations of samples are shown in figure 2 (marked by dashed lines). For each curve, its yield point is marked (denoted $\sigma_{0.2}$). From the texture of the material (figure 1), it follows that the RD samples are favourably oriented for the $\{10\bar{1}0\}\langle 11\bar{2}0\rangle$ prismatic slip and for the $\{10\bar{1}2\}\langle 10\bar{1}1\rangle$ extension twinning in compression. The 45 samples are preferentially oriented for the $(0001)\langle 11\bar{2}0\rangle$ basal slip in both modes of deformation. The preferred slip system for the ND samples is the $\{11\bar{2}2\}\langle 11\bar{2}3\rangle$ 2nd order pyramidal slip and namely the $\{10\bar{1}2\}\langle 10\bar{1}1\rangle$ extension twinning in tension [14].

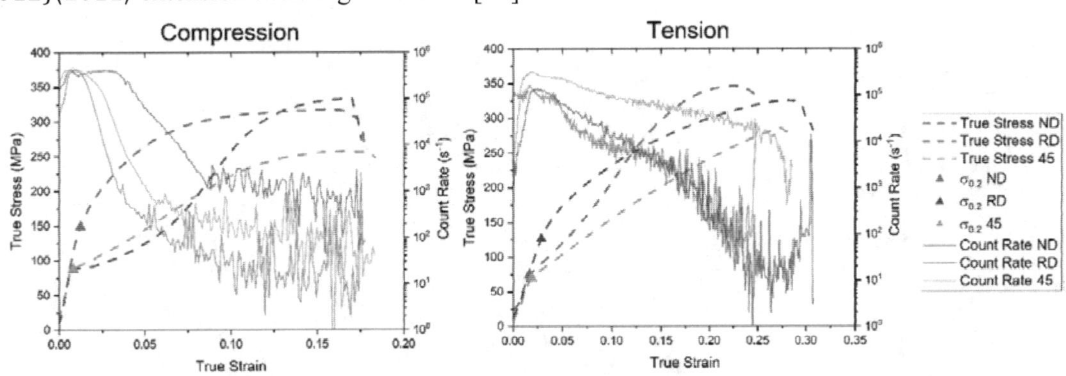


**Figure 2.** AE count rates and deformation curves for the ND, RD and 45 oriented samples deformed in compression and tension.

In accordance with these expectations, the deformation curves, namely of the RD, partly also of the 45 oriented samples in compression and ND (45) samples in tension exhibit sigmoidal S-shape evolution typical for deformation governed by the $\{10\bar{1}2\}\langle 10\bar{1}1\rangle$ extension twinning [15]. The samples oriented unfavourably for twinning, i.e. ND in compression and RD in tension, exhibit a convex deformation curve indicating predominant slip character of deformation. The values of yield strength for both modes of deformation are significantly higher for the samples oriented unfavourably for twinning due to the necessity to activate the 2nd order pyramidal slip to achieve homogenous plastic deformation.

The AE count rate (i.e. the frequency of oscillations exceeding the threshold level) as a function of true strain is shown for all modes of deformation and orientations of samples in figure 2 and marked by (partially transparent) full lines. Higher AE count rate values indicate more pronounced and rapid activity of deformation mechanisms. The maxima of the AE count rate for all samples in both deformation modes were achieved near their respective yield points. The slightly more pronounced response from the samples oriented favourably for extension twinning can be explained by higher AE energy for twin nucleation than for dislocation slip [8].

However, while in compression the AE response diminishes rather abruptly after reaching its peak, the decline in tension is much more gradual. This phenomenon can be explained by a different manner of twinned volume increase – during compression, most of the $\{10\bar{1}2\}\langle10\bar{1}1\rangle$ extension twins nucleate around the yield point, and with further deformation, the already nucleated twins grow laterally (undetected by AE). On the other hand, during continued tensile deformation, the twinned volume increases via nucleation of new twins, which produces a detectable AE signal and thus prolongs its decline. This explanation of the so-called compression-tension asymmetry has been experimentally confirmed by combining the AE and neutron diffraction techniques [16].

The EBSD observations of the deformed microstructure, shown in figure 3, agree with the above-stated conclusions. The RD sample deformed in compression (figure 3b) exhibits a significant amount of $\{10\bar{1}2\}\langle10\bar{1}1\rangle$ extension twins, many of which penetrate grain boundaries and stretch across multiple grains due to their adjacent stress fields [17]. Muránsky et al. [18] argue that this phenomenon (so-called autocatalytic nucleation of twins) causes the surge in the AE activity near the macroscopic yield point (cf. figure 2). The twinning-wise unfavourable ND and partially favourable 45 samples (figures 3a and 3c) then exhibit only a small amount of heavily twinned grains. Their deformation is thus being governed by dislocation slip mechanisms.

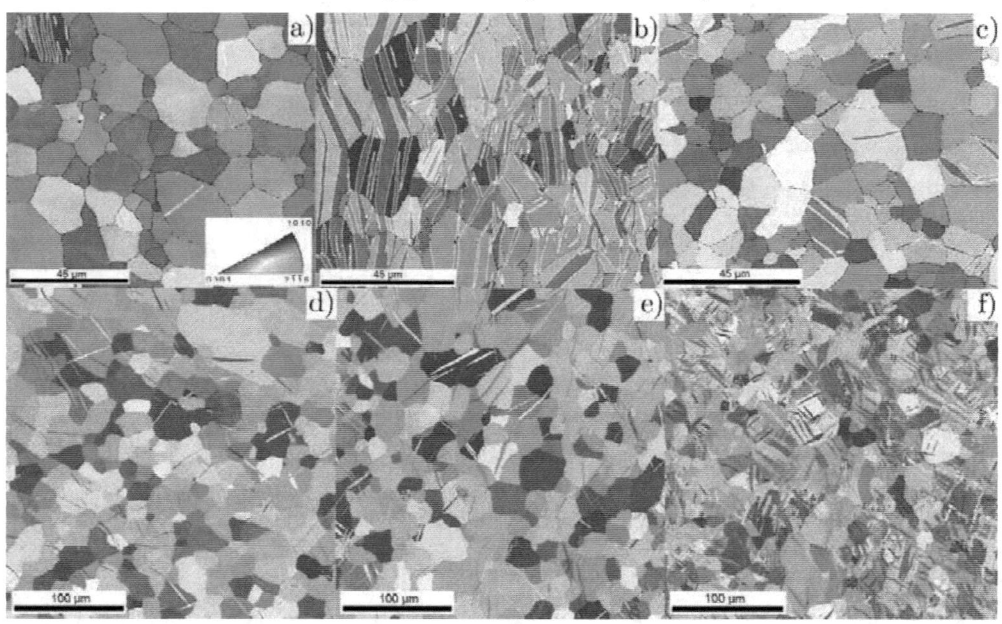


**Figure 3.** IPF orientation maps of the microstructure of a) ND, b) RD and c) 45 oriented samples deformed in compression up to 1% plastic strain and the ND sample deformed in tension up to d) 1%, e) 5% and f) 10% of plastic strain. Deformation was applied along the out-of-plane axis in fig a) and in the horizontal axis in figs b) to f).

The microstructural evolution of the ND sample deformed in tension depicted in figure 3 confirms that with advancing plastic deformation, the twinned volume increases preferentially via nucleation of new twins rather than growth of the already nucleated ones (cf. figures 3d and 3e). The ND tensile sample deformed up to 10% of plastic strain (figure 3f) then exhibits heavily deformed microstructure where all deformation mechanisms, including the $\{10\bar{1}1\}$-$\{10\bar{1}2\}$ double twinning were activated (as was confirmed by the reorientation analysis). At these later stages of deformation, the dislocation mean free path is reduced due to increasing dislocation density and newly formed twin boundaries, which causes the decrease of the AE activity (cf. figure 3).

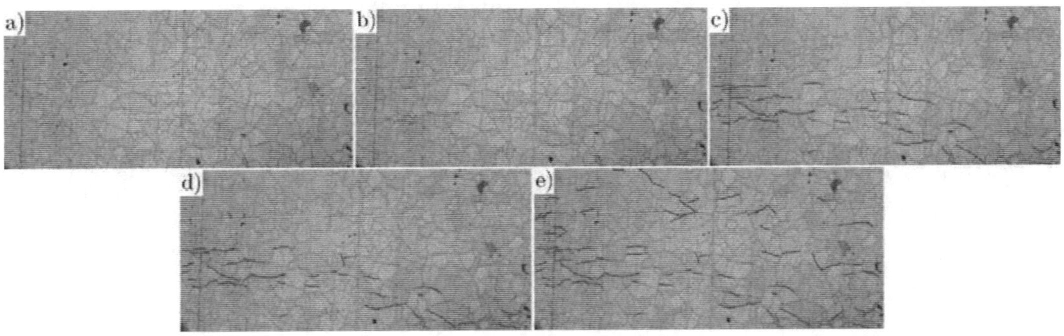


**Figure 4.** High-speed camera images of a) microstructure of the RD sample compressed before the yield point, b) twinning band nucleated near the yield point, c) the twins forming this band marked by red colour, d) another twinning band nucleated ca. 1.5 MPa after the yield point and e) twins forming the second band marked by blue colour. Deformation was applied along the vertical axis.

The high-speed camera footage was captured during the compressive deformation of the samples. The footage confirmed that the deformation of the ND samples was governed by dislocation slip systems since no deformation twinning was observed. The 45 samples tended to retain more homogenous microstructure, while in the case of the ND samples, isolated bands of heavily deformed microstructure alternated with bands of significantly less deformed microstructure. In the RD sample, the $\{10\bar{1}2\}\langle10\bar{1}1\rangle$ extension twins nucleated mainly in the vicinity of the macroscopic yield point. The nucleation of twins was non-continuous and occurred in bands oriented perpendicularly to the loading direction, as is illustrated in figure 4. Barnett et al. [19] observed similar twinning bands in the extruded AZ31 alloy related to the Lüders phenomenon. The bands observed in this study are not likely to be caused by the Lüders behaviour since no plateau in the stress-strain curve following yielding was detected. However, in the said study [19], numerous twin population propagating from the sample edges perpendicularly to the loading direction with the increase of load was detected even in a coarse-grained sample, which exhibited no Lüders behaviour. The authors reason this behaviour by the preferential nucleation of extension twins on the sample edges due to higher stress values caused by friction and possible irregularities. Unlike the Lüders-like band propagation, however, in order for further twins to nucleate and thus accommodate the strain, an increase in stress is necessary, which means that no yield plateau is formed. It seems likely that analogous argumentation can be applied to explain the behaviour observed in this study.

The detected heterogeneous character of deformation may skew the results obtained by highly localised techniques, such as EBSD, especially if only one sample surface area is inspected. Therefore, such observations should either be performed using a significant statistical dataset or accompanied by methods taking information originating from more extensive sample areas into account.

## 4 Conclusions

The main results of this investigation, employing the combination of deformation tests with simultaneous AE recording, EBSD measurements and high-speed camera imaging to study the deformation behaviour of a rolled magnesium alloy AZ31, can be summarised as follows:

- The deformation behaviour followed the favourability for the activation of the $\{10\bar{1}2\}\langle10\bar{1}1\rangle$ extension twinning, arising from the different sample orientations with regards to the detected basal texture of the rolled sheet.
- A compression-tension asymmetry of the deformation behaviour was detected in the AE response and consistently linked to the EBSD microstructural observations.
- A heterogeneous character of deformation, displayed by the formation of deformation bands oriented perpendicularly to the loading direction, was observed during the compressive deformation by the high-speed camera imaging.

**Acknowledgements**
The authors are grateful for the financial support of the Czech Science Foundation under the contract 19-08937S. J. D. is grateful for the financial support of the Grant Agency of Charles University under contract 1172120. P.M. acknowledges partial financial support by ERDF under project No. CZ.02.1.01/0.0/0.0/15_003/0000485.

## References

[1] Kelly E W and Hosford W F 1968 Deformation characteristics of textured magnesium *Tran. Metall. AIME* **242** (4) p 654
[2] Avedesian M M and Baker H 1999 Magnesium and Magnesium Alloys (ASM Speciality Handbook) (ASM International)
[3] Yukutake A, Kaneko J and Makoto S 2013 Anisotropy and non-uniformity in plastic behavior of AZ31 magnesium alloy plates *Materials Transactions* 44 (4) pp 452-457
[4] Agnew S R and Duygulu O 2005 Plastic anisotropy and the role of non-basal slip in magnesium alloy AZ31B *Int. J. Plast.* **21** (6) pp 1161-1193
[5] Mises R von 1928 Mechanik der plastichen formänderung von kristallen *Zamm – Z. Angew. Math. Me.* **8** (3) pp 161-185
[6] Heiple C R, Carpenter S H and Carr M J 1981 Acoustic emission from dislocation motion in precipitation-strengthened alloys *Met. Sci.* **15** (11-12) pp 587-597
[7] Heiple C R and Carpenter S H 1987 Acoustic Emission Produced by Deformation of Metals and Alloys - A Review: Part I *J. Accoust. Emiss.* **6** (2) pp 177-204
[8] Heiple C R and Carpenter S H 1987 Acoustic Emission Produced by Deformation of Metals and Alloys - A Review: Part II *J. Accoust. Emiss.* **6** (4) pp 215-237
[9] Chmelík F, Klose F B, Dierke H, Sachl J, Neuhauser H and Lukáč P 2007 Investigating the Portevin-Le Chatelier effect in strain rate and stress rate controlled tests by the acoustic emission and laser extensometry techniques *Matter. Sci. Eng.* A **462** (1-2) pp 53–60
[10] Dobroň P, Bohlen J, Chmelík F, Lukáč P, Letzig D and Kainer K U 2007 Acoustic emission during stress relaxation of pure magnesium and AZ magnesium alloys *Matter. Sci. Eng.* A **462** (1-2) pp 307-310
[11] Hou D, Liu T, Chen H, Dongfeng S, Chunhua R and Pan F 2016 Analysis of the microstructure and deformation mechanisms by compression along normal direction in a rolled AZ31 magnesium alloy *Materials Science & Engineering: A* **660** pp 102-107
[12] Bohlen J, Chmelík F, Dobroň P, Kaiser F, Letzig D, Lukáč P and Kainer K U 2004 Orientation effects on acoustic emission during tensile deformation of hot rolled magnesium alloy AZ31 *Journal of Alloys and Compounds* **378** pp 207-213
[13] Meza-García E, Dobroň P, Bohlen J, Letzig D, Chmelík F, Lukáč P and Kainer K U 2007 Deformation mechanisms in an AZ31 magnesium alloy as investigated by the acoustic emission technique *Materials Science & Engineering: A* **462** pp 297-301
[14] Čapek J, Knapek M, Minárik P, Dittrich J and Máthis K 2018 Characterisation of deformation mechanisms in Mg alloys by advanced acoustic emission methods *Metals* **8** (8) pp 644–655
[15] Prasad K E, Li B, Dixit N, Shaffer M, Mathaudhu S N and Ramesh K T 2014 The dynamic flow and failure behavior of magnesium and magnesium alloys *JOM* **66** (2) pp 291–304
[16] Čapek J, Máthis K, Clausen B, Strásková J, Berna P and Lukáš P 2014 Study of the loading mode dependence of the twinning in random textured cast magnesium by acoustic emission and neutron diffraction methods *Materials Science & Engineering: A* **602** pp 25-32
[17] Šiška F, Stratil L, Čížek J, Ghaderi A and Barnett M 2017 Numerical analysis of twin thickening process in magnesium alloys *Acta. Matter.* **124** pp 9–16
[18] Muránsky O, Barnett M R, Carr D G, Vogel S C, Oliver E C 2010 *Acta Materialia* **58** pp 1503-1517
[19] Barnett M R, Nave M D, Ghaderi A Yield point elongation due to twinning in a magnesium alloy 2012 *Acta Materialia* **60** pp 1433-1443

# Characterisation of structure, mechanical and corrosion properties of pure magnesium prepared by powder metallurgy route

**Drahomír Dvorský[1,2], Jiří Kubásek[1], Dalibor Vojtěch[1]**

[1]Department of Metals and Corrosion Engineering, Faculty of Chemical Technology, University of Chemistry and Technology Prague, Technická 5 166 28 Praha 6 – Dejvice, Czech Republic
[2]Institute of Physics, Czech Academy of Science, Prague, Na Slovance 1999/2 182 21 Praha 8, Czech Republic

dvorskyd@vscht.cz

**Abstract**. Powder metallurgy is a progressive method for materials production. Final products are usually characterised by improved strength and corrosion properties compared to materials prepared by conventional methods like casting and subsequent thermomechanical processing. The presented work studies the effect of extrusion temperature on the microstructure and properties of products prepared from pure Mg powder. It has been observed that higher extrusion temperature led to the formation of larger grains and massive oxides at interparticle boundaries. Tensile yield strength, compressive, and tensile ultimate strengths were increased at the cost of lower plasticity. The observed mechanical properties were mainly affected by texture, oxides, and grain size. The high corrosion rate was attributed to the presence of a relatively high amount of oxides and impurities which facilitate localized corrosion.

## 1. Introduction

Magnesium is a light metal suitable for application in the automotive and aviation industry. Its mechanical and corrosion properties are also advantageous for application in medicine as a material for biodegradable implants [1]. Such materials should have appropriate mechanical and corrosion properties which can be improved by alloying. However, in the case of biodegradable implants, alloying elements should also be biocompatible. The most advanced used materials for medicine are alloys with rare earth (RE) elements [2]. Even though the cytotoxicity tests and in vivo tests revealed no harmful effect of those elements, the long-term impact of those elements is yet unknown. Therefore, there is an effort to reduce the amount of alloying elements or to use pure Mg. Nevertheless, pure Mg is prone to fast corrosion if dangerous impurities like Fe, Ni, or Cu are present in the amount exceeding allowed limits (Fe: 35–50 ppm, Ni: 20–50 ppm, Cu: 100–300 ppm) [3].

The mechanical and corrosion properties might be improved by the processing of the material. Powder metallurgy seems to be a suitable way to produce such materials. It was investigated that materials prepared by powder metallurgy exert superior mechanical and corrosion properties compared with casted ingot [4-6]. Especially, if it is combined with thermomechanical processing such as extrusion. Material is severely deformed and dynamic recrystallization may occur due to the elevated temperature. The final desired structure is characterised by fine uniaxial grains [7].

Nevertheless, extrusion of materials with hcp structure may resolve in the specific texture in the material. Magnesium is usually characterised with typical basal texture for rolling or fibre texture

after extrusion, which causes anisotropy of mechanical properties [8, 9]. The orientation of basal planes of magnesium hexagons is parallel to the extrusion direction after extrusion. This orientation is favorable for the twinning in compression and unfavorable for the twinning in tension. Twinning requires less energy to occur in textured materials than the slip mechanism in the unfavorably oriented basal planes, and therefore, there are great differences in the measured compressive and tensile yield strengths of materials with strong texture [9, 10].
This paper deals with the effect of extrusion temperature on the final mechanical and corrosion properties of powder metallurgy products.

## 2. Materials and methods

### *2.1. Material preparation*

Pure Mg atomized powder (90 ppm of Fe, 20 ppm of Cu, 10 ppm of Ni – identified by ICP-MS Elan DRC-e) with round-shaped particles was obtained from the commercial supplier with the particle size ranging between 0.1 and 250 µm. The powder size distribution is in the figure 1. The powder was pressed on the universal LabTEST 250SP1-VM machine with a pressure of 255 MPa for 5 min. The green compact sample was then extruded at the same machine at 200 °C, 300 °C, 400 °C, and 500 °C. The final rods had 6 mm in diameter. The extrusion ratio was 10 and the extrusion rate 5 mm·min$^{-1}$.

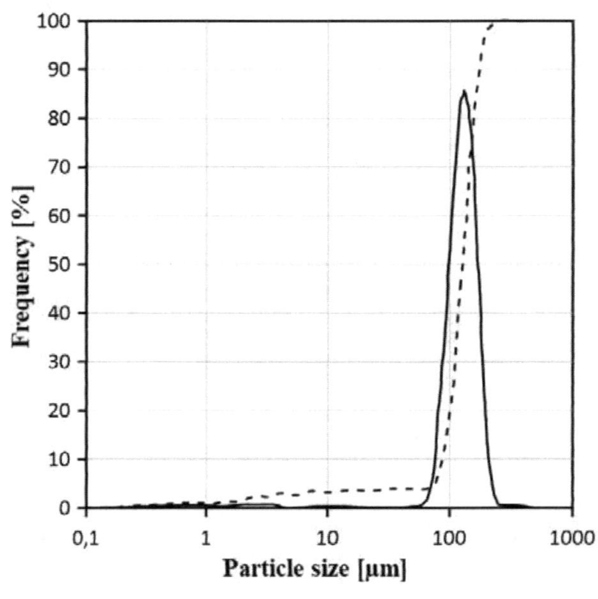


Figure 1: Powder size distribution.

### *2.2. Microstructure characterisation*

Characterisation of the samples was performed using a TescanVEGA3 LMU scanning electron microscope (SEM) equipped with energy dispersive spectroscopy (EDS). Samples were ground on the SiC papers P80-P4000, afterward, they were polished on diamond pastes D2 and D0.7. The final polishing was performed on the Eposil F suspension. Samples were etched in the solution of 30 ml of acetic acid, 10 ml of nitric acid, 40 ml of water, and 120 ml of ethanol. EBSD analysis was performed on the same microscope with NordlysMax electron back-scattered diffraction (EBSD) detector, and AZtecHKL software). Parameters selected for EBSD analysis: working distance of 15 mm, accelerating voltage of 20 kV, dwell-exposure time of 0.26 s/pt, step size of 0.25 µm.

### *2.3. Mechanical properties*

Compressive and tensile properties were measured using LabTest 5.250SP1-VM at room temperature at a strain rate of 0.001 s$^{-1}$. Cylindrical samples with a diameter of 5 mm and 7.5 mm high were used for compressive tests. Compressive yield strength (CYS) and ultimate compressive strength (UCS) were determined from compressive curves. Tensile tests were performed on dog bone specimens with a gauge length of 25 mm and diameter equal to 3.5 mm. Tensile yield strength (TYS), ultimate tensile strength (UTS), and elongation to fracture (A) were determined from compressive curves. Three measurements were performed for each material.

### *2.4. Corrosion test*

Corrosion behaviour was studied in the simulated body fluid SBF according to Müller [11]. 100 ml of SBF was used for 1 cm$^2$ of the sample surface. Samples were exposed to SBF in non-hermetically closed plastic containers for 14 days at 37 °C. The corrosion rate was evaluated based on the weight changes after the removal of corrosion products in the solution of 200 g/l $CrO_3$, 10 g/l $AgNO_3$, 20 g/l $Ba(NO_3)_2$. The difference in weight was divided by density, exposed surface, and number of days of exposure and multiplied by 365 in order to obtain corrosion rate in mm·a$^{-1}$. Three measurements were performed for each material.

## 3. Results and discussion

### *3.1. The microstructures*

Microstructures of extruded products are displayed in figure 2. All materials were characterised by rows parallel to the extrusion direction containing oxides. Generally, the porosity of the material after extrusion is close to zero according to the image analysis, however, some oxides might be removed during polishing and etching which created some pores during metallography preparation. Presented rows represent the original surface of Mg particles, therefore, the powder particles were deformed and stretched in the extrusion direction. Those borders were not eliminated probably due to the low extrusion ratio. Cavojsky et al. [12] prepared pure Mg with an extrusion ratio of 16 and they were able to disturb original powder particles. Nevertheless, oxides in the structure were ordered in rows parallel to the extrusion direction. One can see that the grain size increases with the extrusion temperature as well as the amount of oxides in the structure. Similar results were observed by Perez et al. [13] who prepared pure Mg with an extrusion ratio of 18 at various extrusion temperatures. Oxides were originally on the surface of each particle in a small amount, which is visible on the sample extruded at 200 °C. However, during heating before extrusion, the residual air between particles of green compact increased the amount of oxides on the surface of the original particles.

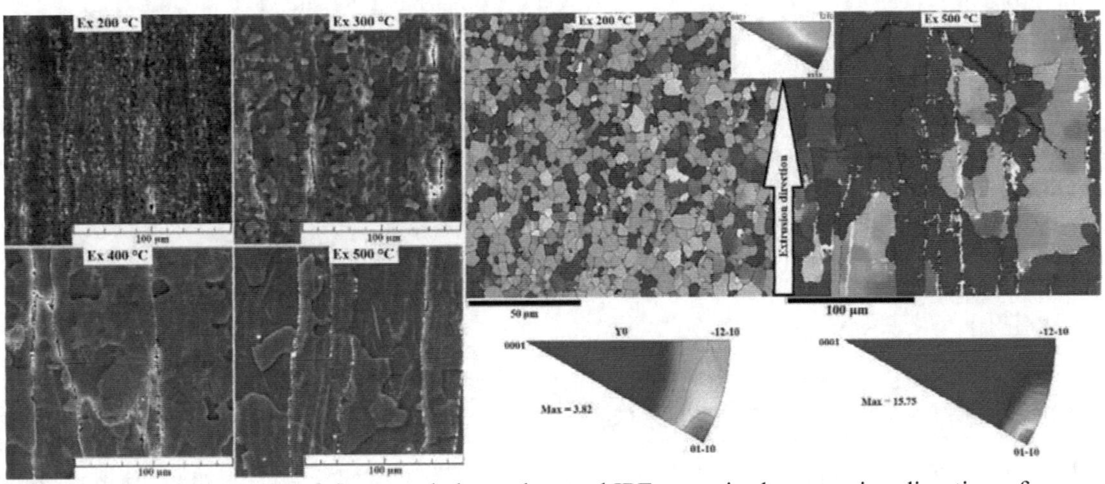


**Figure 2**: Microstructures of the extruded powders and IPF maps in the extrusion direction of extruded products.

EBSD analysis revealed information about the grain size and grain orientation in the material. The average grain sizes were 2.7 µm, 6.4 µm, 11.7 µm, and 23.4 µm for materials extruded at 200 °C,

300 °C, 400 °C, and 500 °C, respectively. Those are similar or slightly smaller grain sizes compared to those prepared at the equal extrusion temperature by powder metallurgy [12, 13]. In comparison with materials prepared by extrusion of an ingot, there is a lower grain size of powder metallurgy products [14, 15]. Figure 2 represents the results of EBSD analysis at the two extreme conditions (extrusion at 200 °C and 500 °C). White areas in figure 2 represent unindexed points due to the defects or oxides while red linear grain is probably sign of twinning. One can see a large difference in grain size and primarily in grain orientation. There is a five times stronger basal texture in the case of extrusion at 500 °C compared to the extrusion at 200 °C. The same results were measured by Perez et al. [13] who performed extrusion temperatures of 250 °C, 325 °C, 400 °C, and 450 °C. The strong texture was observed also by Cavojsky et al. [12] for powder extruded at 390 °C and even by Kubasek et al. [14] for extruded ingot at 200 °C.

### *3.2. Mechanical properties*

Compressive (solid line) and tensile (dash line) properties are summarized in figure 3 and table 1. At first sight, there are visible some trends, that the compressive yield strength (CYS) and plasticity (A) decreases with higher extrusion temperature. Contrary, the ultimate compressive strength (UCS) and tensile yield strength (TYS) increased with the extrusion temperature.

Mechanical properties of pure Mg are affected by several factors such as grain size, texture strength, and volume and character of oxides. There was an increased amount of oxides with higher extrusion temperature, which may result in the enhancement of strengths. On the contrary, its presence deteriorates plasticity. Smaller grains shall exert the improvement of yield strengths according to the Hall-Patch relation. Therefore, higher extrusion temperature should lead to lower CYS and TYS. This is true only for the CYS as there is a third factor – texture. The presence of texture may justify the anisotropy of mechanical properties, as there are great differences between CYS and TYS values of prepared products. Moreover, there are contradictory trends in CYS and TYS related to the extrusion temperature.

The strongest texture was measured for the sample extruded at 500 °C as well as the greatest difference between CYS and TYS, which was 130 MPa. This is caused by preferential grain orientation which is favourable for the twinning mechanism in compression. Moreover, a larger grain size intensifies the twinning mechanism. This is visible on the compressive curve (black solid line in figure 3), which has the typical "S" shape with a well visible area of strengthening. Twinning requires less energy in textured materials than a slip mechanism in a unfavourably oriented basal planes, and therefore, the measured CYS is relatively low. On the other hand, tensile yield strength is increased as the twinning mechanism is suppressed in tension due to the presented texture, and the deformation is primarily provided by the slip mechanism. Large grains have a long slipping distance, which allows strengthening in tension as well, which is also visible on the tensile curve (dash black line in figure 3).

In the case of extrusion at the lowest temperature, there was CYS higher than TYS. Such behaviour is related to the defects or inhomogeneities in a material that affects more tensile properties than compressive. Relatively small grains together with weaker specific texture are not suitable for twinning with low activation energy and consequently, the value of CYS is increased. Low amount of oxides, lower specific texture strength manifest in lower TYS and better ductility for sample extruded at 200 °C.

The obtained mechanical properties are better than properties of as-casted materials [16] and even slightly better than extruded ingot [14]. This is associated with the grain size and also by MgO reinforcement, which is not present in the casted ingot. On the contrary, other authors achieved higher strengths [12, 13], however also lower plasticity, which can be contributed to the stronger texture of the material. A higher strength might also be connected with a higher extrusion ratio. Milling of the powder prior to extrusion resulted in higher strengths, lower anisotropy of mechanical properties and slightly lower ductility [17].

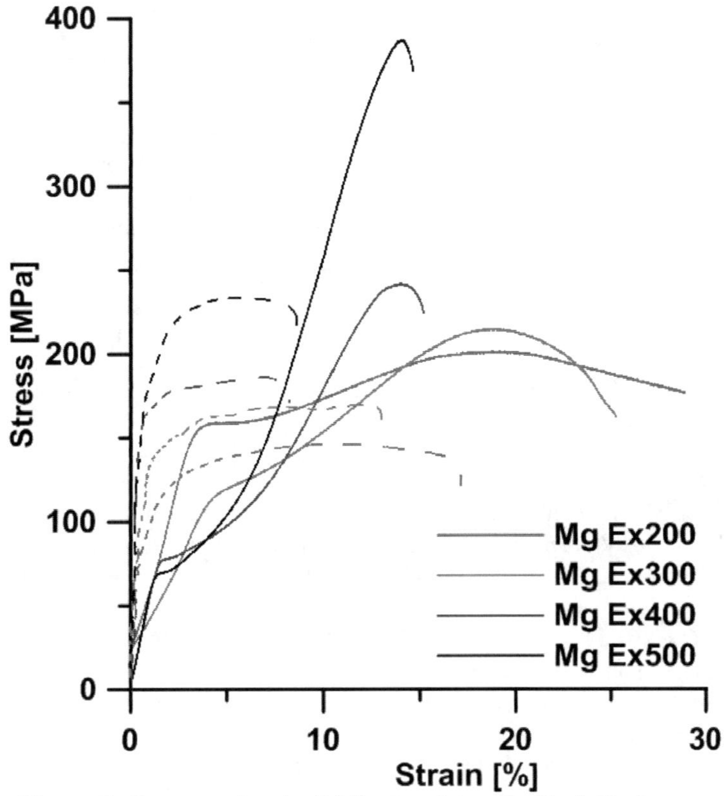

Figure 3: Compressive (solid line) and tensile (dash line) curves.

**Table 1:** Mechanical properties of prepared samples.

| | Ex 200 °C | Ex 300 °C | Ex 400 °C | Ex 500 °C |
|---|---|---|---|---|
| CYS [MPa] | 151 ± 2 | 112 ± 2 | 76 ± 1 | 70 ± 1 |
| UCS [MPa] | 200 ± 3 | 210 ± 8 | 235 ± 9 | 382 ± 10 |
| TYS [MPa] | 89 ± 4 | 132 ± 3 | 163 ± 4 | 174 ± 5 |
| UTS [MPa] | 147 ± 8 | 168 ± 7 | 187 ± 9 | 233 ± 9 |
| A [%] | 17 ± 2 | 13 ± 1 | 8 ± 1 | 7 ± 1 |

### *3.3. Corrosion properties*

Corrosion properties were measured in SBF for 14 days at 37 °C and the values calculated based on the weight changes are summarized in table 2. Measured corrosion rates are generally high even for pure magnesium. A high corrosion rate is associated with the presence of oxides on the boundary between the original particles [18]. Such places are prone to corrosion and the corrosion front spreads easily there. A similar finding was observed in the material prepared by spark plasma sintering [19]. In this case, the amount of oxides increased with higher extrusion temperature like the corrosion rate. The corrosion rate was also enhanced by the increased amount of Fe in the structure.

Corrosion rate might be reduced after disruption of continuous oxide borders between particles for example by a higher extrusion ratio or ECAP [20]. Another possibility is to avoid the creation of oxides at all by hot vacuum pressing (HVP) which eliminates the air between particles. Nevertheless, there would be still some oxide on the surface of the powder. It could be eliminated by pre-treatment of the powder for example by HF [21].

**Table 2: Corrosion rate in SBF.**

| | Ex 200 | Ex 300 | Ex 400 | Ex 500 |
|---|---|---|---|---|
| $v_{cor}$ [mm·a$^{-1}$] | 7.4 ± 1.0 | 8.2 ± 0.5 | 8.7 ± 0.6 | 9.9 ± 0.7 |

## 4. Conclusion

Extrusion of pure magnesium in the form of compacted tablets from atomized powder resulted in a non-uniform microstructure. Individual powder particles surrounded by oxides have been observed. Higher extrusion temperature increased the amount of oxides in the microstructure as well as the grain size. The texture was strongest in the material extruded at 500 °C, which caused great anisotropy of mechanical properties, where the difference between CYS and TYS was almost 130 MPa. Contrary, no anisotropy, and the low texture strength were measured for the extruded product at 200 °C. Microstructure with oxides surrounding each particle together with high Fe content resulted in a very high corrosion rate for all materials, which could be partly reduced by a higher extrusion ratio or by pre-treatment of the powder.

**Acknowledgment**
This work was supported by the grant of Specific university research – grant No. A1_FCHT_2020_003.

## References

[1] Wang H, Shi Z M and Yang K 2009 Magnesium and magnesium alloys as degradable metallic biomaterials. In: *4th International Light Metals Technology Biennial Conference (LMT2009),* ed M S Dargusch and S M Keay (Gold Coast, Queensland: Trans Tech Publications Inc.) pp 207-10
[2] Levorova J, Duskova J, Drahos M, Vrbova R, Vojtech D, Kubasek J, Bartos M, Dugova L, Ulmann D and Foltan R 2018 *J. Biomater. Appl.* **32** 886-95
[3] Witte F, Hort N, Vogt C, Cohen S, Kainer K U, Willumeit R and Feyerabend F 2008 *Curr. Opin. Solid State Mater. Sci.* **12** 63-72
[4] Kubásek J, Dvorský D, Čavojský M, Vojtěch D, Beronská N and Fousová M 2017 *J. Mater. Sci. Technol.* **33** 652-60
[5] Cabeza S, Garcés G, Pérez P and Adeva P 2015 *J. Mech. Behav. Biomed.* **46** 115-26
[6] Yoshihito Kawamura K H, Akihisa Inoue, Tsuyoshi Masumoto 2001 *Special Issue on Platform Science and Technology for Advanced Magnesium Alloys* **42** 1172-6
[7] Zhang X, Yuan G, Niu J, Fu P and Ding W 2012 *J Mech Behav Biomed Mater* **9** 153-62
[8] Elsayed A, Umeda J and Kondoh K 2011 *Mater. Des.* **32** 4590-7
[9] Kleiner S and Uggowitzer P J 2004 *Mater. Sci. Eng., A* **379** 258-63
[10] Dvorský D, Kubásek J, Voňavková I and Vojtěch D 2019 *Mater. Sci. Technol.* **35** 520-9
[11] Müller L and Müller F A 2006 *Acta Biomater.* **2** 181-9
[12] Čavojský M, Trembosova V, Beronská N, Nagy S and Nosko M 2019 *Metallic Materials* **57** 371-6
[13] Pérez P, Garcés G and Adeva P 2007 *Journal of Materials Science* **42** 3969-76
[14] Kubásek J, Vojtěch D and Dvorskỳ D 2017 *Mater.Technol.* **51** 289-96
[15] Lei W, Zhu D, Wang H and Liang W 2019 *Journal of Wuhan University of Technology-Mater. Sci. Ed.* **34** 1193-6
[16] Gu X-N and Zheng Y-F 2010 *Front. Mater. Sci.* **4** 111-5
[17] Dvorský D, Kubásek J and Vojtěch D 2020 *Manufacturing Technology Journal* **20** 708-13
[18] Narita K, Hiromoto S, Kobayashi E and Sato T 2021 *Metals* **11** 227
[19] Dvorsky D, Kubasek J and Vojtech D 2018 *Mater. Lett.* **227** 78-81
[20] Minárik P, Král R, Pešička J, Daniš S and Janeček M 2016 *Mater. Charact.* **112** 1-10
[21] Dvorský D, Kubásek J and Vojtěch D 2019 *Manuf. Technol.* **19** 740-4

IOP Conf. Series: Materials Science and Engineering 1178 (2021) 012013 doi:10.1088/1757-899X/1178/1/012013

# Creep behaviour of ultrafine-grained CP titanium processed by multi-pass rolling

**J. Dvorak [1,2], A.G. Kadomtsev [3], V.I. Betekhtin [3], V. Sklenicka [1,2], P. Kral [1,2] and M. Kvapilova [1,2]**

1 CEITEC – IPM, Institute of Physics of Materials, Czech Academy of Sciences, Zizkova 22, 616 62 Brno, Czech Republic

2 Institute of Physics of Materials, Czech Academy of Sciences, Zizkova 22, CZ-616 62 Brno, Czech Republic

3 Ioffe Physical-Technical Institute, Russian Academy of Sciences, Politechnicheskaya 26, 194 021 St.Petersburg, Russia

E-mail: dvorak@ipm.cz

**Abstract**. Two commercially pure (CP) Ti heats containing the same total impurity content have been used under investigation. Both materials were subjected to the processing procedure method of severe plastic deformation (SPD). This procedure consists of two steps of hot rolling (at 673 K) and the final rolling step at ambient temperature. The difference in individual chemical contents of impurities resulted in two states, both with ultrafine-grained (UFG) microstructures. For comparison reason, coarse-grained (CG) state was prepared by annealing of UFG state at temperature 823 K/1h. Microstructure investigations were performed using scanning electron microscope equipped with EBSD unit (SEM/EBSD) and transmission electron microscopy (TEM). Constant load tensile creep tests were conducted at temperature 673 K and applied uniaxial stress of 200MPa. Creep tests were run up to the final fracture of the creep specimens. The main objective of this work was to evaluate how the synergistic effect of SPD and the resulting microstructure affects the creep behaviour and properties of the studied materials.

## 1 Introduction

It is well known that heavy deformation, e.g. by rolling, extrusion, forging or drawing, can result in significant refinement of the microstructure of metals and alloys [1]. In recent years, severe plastic deformation (SPD) processing has been developed as a new method of manufacturing bulk specimens having ultrafine/nano-crystalline grained structures. Using the idea of this grain refinement effect, recently various bulk nano-structured materials processed by several methods of SPD were prepared [1-4]. It is known that the metals produced by these processes have very small average grain sizes of less than 1μm, with grain boundaries of mostly high-angle misorientation. The high-angle grain boundaries (HAGBs) in particular allow additional deformation mechanisms to operate including grain-boundary sliding during low-temperature straining. So far, the most detailed investigations have focused on fcc metals such as Al, Cu and their alloys [3-6]. However, modern technological progress still requires light materials with special mechanical properties e.g. for application at elevated and high temperatures. Nowadays, titanium is an element which is frequently used in design of modern high temperature components. Increase of the fraction of HAGBs contributes to enhance some mechanical properties of UFG Ti. Microstructure refinement and formation of UFG structures in pure titanium result in considerable improvement in mechanical properties, such as the tensile strength limit [7], the compressive strength limit [8], and fatigue strength [9]. A new interesting example of grain boundaries (GB) engineering is investigation of mechanical properties of UFG Ti produced by combined SPD processing in the shape of long-length rods. In this work, we suggest this way of producing a stable submicrocrystalline structure: SPD by longitudinal rolling in combination with helical rolling under different conditions [10,11]. The main objective of this work was to evaluate how the combined way of SPD influences the creep properties and to investigate the effect of microstructure on creep behaviour.

## 2 Experimental Materials and Methods

Two commercially pure Ti heats (marked as State 1 and State 2) with the same total impurity content of 0.3 wt % (0.004C, 0.003N, 0.12Fe, 0.143O, 0.0008H, 0.001Al, 0.002Si) have been used in this study. The main distinction between the two states, as established by energy-dispersion X-ray spectroscopy, was that the content of carbon in State 1 was twice that in State 2.

Both titanium heats were received in the form of the rods of 30 mm in diameter. The rods were processed under the same conditions consisting helical rolling at 673 K; lengthwise rolling at 673 K; and again helical rolling to final 8 mm diameter at ambient temperature [10,11]. The difference in chemical content of impurities resulted in two states both with UFG microstructures. To relieve internal stresses and improve the stability of the structure, titanium samples were annealed at 623 K for 3 h. For comparison, these two states were compared with CP Ti of conventional grain size. Coarse-grained state was obtained by annealing of UFG state at temperature 823 K for 1 hour. Creep specimens were cut from SPD processed and coarse-grained (CG) billets. The orientation of the tensile axis corresponded to the rolling direction. Flat tensile specimens had the gauge lengths of 25 mm and the cross-section of 3 mm$^2$. Constant tensile load creep tests were conducted at temperature 673 K under different levels of the applied uniaxial stress and under protection atmosphere of argon. Creep tests were run up to the final fracture of the creep specimens. For each level of the applied stress UFG specimens and CG specimen were tested. Microstructure investigations were performed using microscope Tescan Lyra 3 XMU FEG/SEM-FIB equipped with EBSD unit and transmission electron microscopy (TEM).

## 3 Results and discussion

### *3.1 Creep behaviour*

Representative creep curves are shown in Fig. 1. All of these plots were obtained at an absolute temperature of 673 K (~ 0.35 $T_m$; $T_m$-melting temperature of the material) and at applied tensile stress of 200 MPa. Fig. 1a show standard ε vs. t curves for coarse-grained state and for the UFG materials. To identify individual stages of creep these standard strain ε versus time t creep curves can be easily replotted in the form of the instantaneous strain rate dε/dt versus time t (as shown in Fig. 1b) and/or in the form of the instantaneous strain rate dε/dt versus strain ε (Fig. 1c). As demonstrated by figures, there are differences in creep behaviour between the two states of materials. First, State 1 exhibits markedly better creep resistance in comparison with State 2. The creep life of State 1 specimen is seen to be longer than that of State 2 by a factor of more than 3. Second, comparing of UFG and CG states, UFG states exhibit better creep life than CG one.

IOP Conf. Series: Materials Science and Engineering 1178 (2021) 012013 doi:10.1088/1757-899X/1178/1/012013

Further, the minimum creep rate for the CG material is about one or more order of magnitude higher than that of the UFG counterparts. Finally, fracture plasticity (elongation) is not improved by applying SPD methods. Lower values of elongation measured for CG State 1 are probably caused by scatter of results. The creep curves are clearly different in appearance between the two states as illustrated in Fig.1. Supposing that the creep rate dε/dt at given stress is a certain measure of the "softness "of the microstructure, than the dε/dt–ε plots (Fig.1c) reveal the strain evolution of this "softness". In UFG condition, hardening was observed at the beginning of creep exposure followed by a rapid softening. The secondary stage of creep is there very short. For State 1, gradual course of tertiary creep is finished by rapid accelerating of the last tertiary part of creep exposure shortly before fracture, while State 2 has steady softness progress. By contrast, in CG materials the primary creep stage is markedly longer followed by a distinct tertiary stage with gradually increasing creep rate.

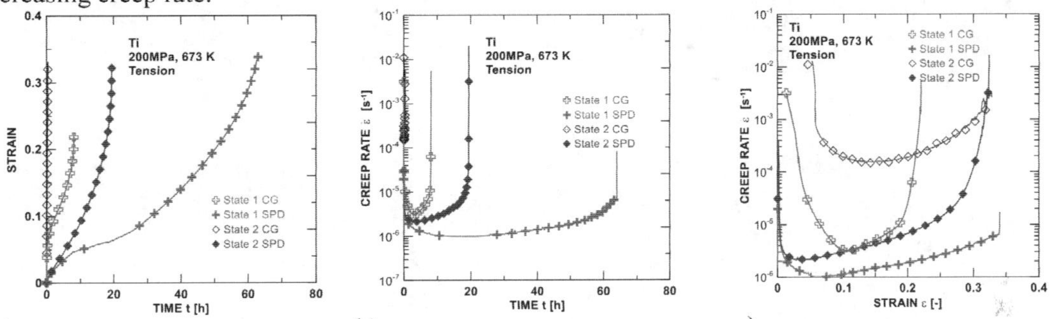


a) b) c)

**Figure 1.** Creep curves for UFG and for CG Ti samples: (a) standard creep curves, (b) creep rate vs. time, (c) creep rate vs. strain.

### 3.2 *Microstructure before creep testing*

Two types of microstructure were achieved by application of three steps of SPD rolling operation. Due to the effect of SPD deformation the grain size is reduced to nanometer range for both states of material. The EBSD results showed inhomogeneous microstructure with a small submicron elongated grains along the rolling direction for the State 1 (Fig. 2a). The mean grain size was measured about 0.2 μm. State 2 is represented by a similar microstructure (Fig. 2b). The mean grain size was measured as ~ 0.25 μm. Further, EBSD measurements were taken to determine the grain boundary misorientations. The mean value of the relative fraction of high-angle (θ> 15°) grain boundaries (HAGB) was estimated to be ~ 64%. TEM figures revealed high dislocation density in the interior of grains created by large plastic deformation imposed into material predominantly by the last step of rolling operation at room temperature (Fig.2d).

The microstructure of CP titanium in the coarse grain (CG) state is shown in Fig. 2c. This state was obtained from UFG states by annealing at temperature 823 K for one hour. It contains a relatively homogenous microstructure with a presence of large bands of equiaxed grains. This annealing lead to coarsening of grains and the mean grain size measured by EBSD analysis was about 1.2 μm with presence about 50% of HAGB. The annealing process led to complete recovery of dislocations formed during last rolling process at room temperature.

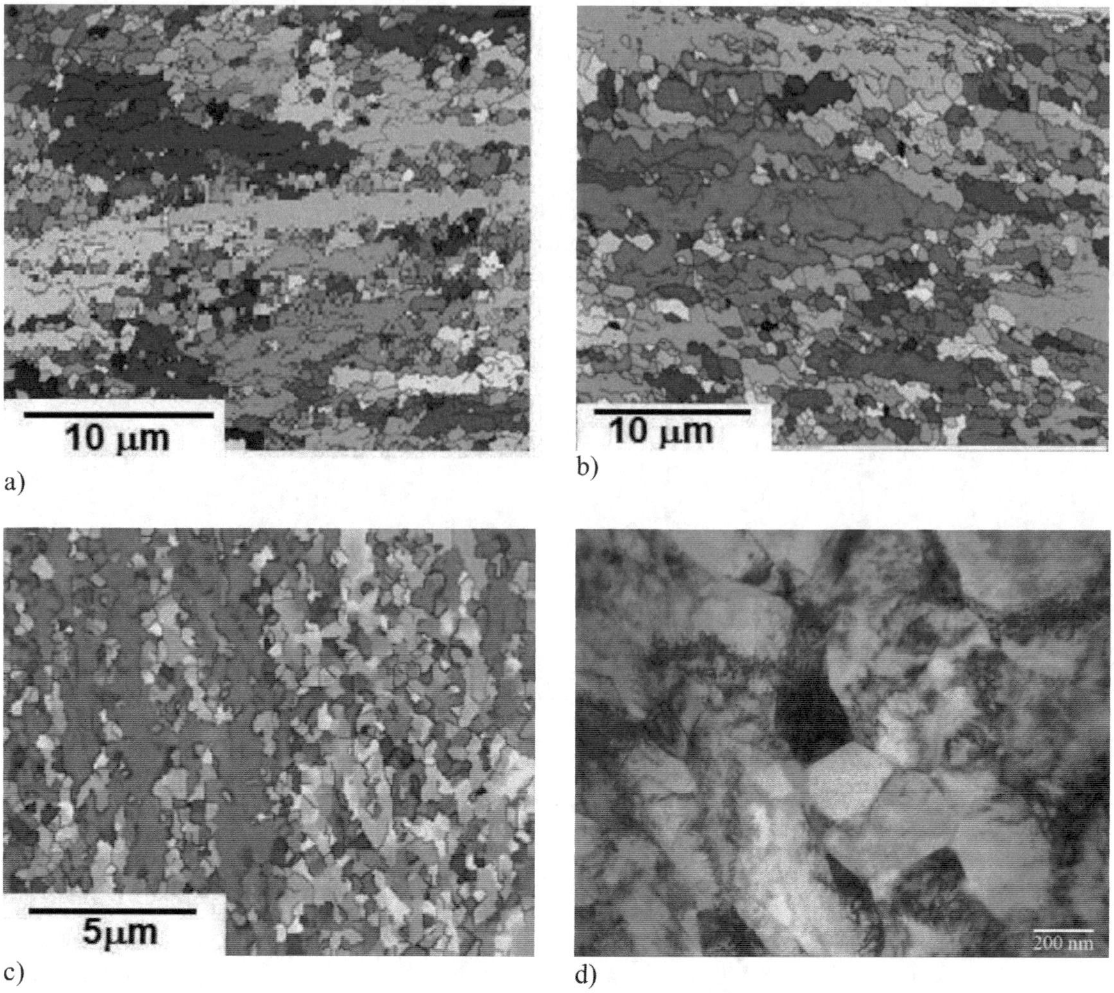


**Figure 2.** Microstructure of SPD-processed titanium: a) State 1, b) State 2, c) CG state created by annealing of State 2 at temperature 823 K/1h and d) TEM – dislocation substructure in State 1.

### *3.3 Microstructure after creep testing*

After creep exposures at 673 K and 200 MPa the microstructures of State 1 and 2 contain clusters of grains with nearly the same orientation. In microstructure are still visible bands of elongated grains. Creep loading also leads to coarsening of grain size up to 1.75 μm and 1.45μm for State 1 and State 2 (Fig. 3a), respectively.

The microstructure of CG state after subsequent creep exposure show a change in the mean grain size to 2 μm with occurrence mixed small and large predominantly equiaxed grains (Fig. 3b). In compare with as-annealed state, bands of elongated grains was not observe. Finally, no expressive change has been observed on fracture surface using SEM fractography. In all states typical mix mode of transgranular and intergranular fracture was observed (Fig. 4).

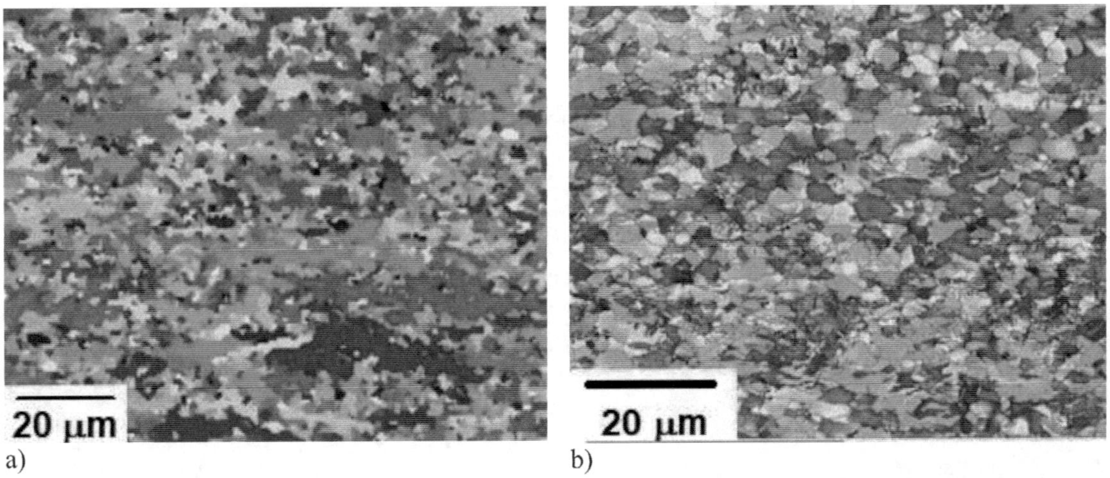


**Figure 3.** Microstructure of SPD-processed titanium after creep exposure (673 K/200 MPa): a) State 1, b) CG State 2.

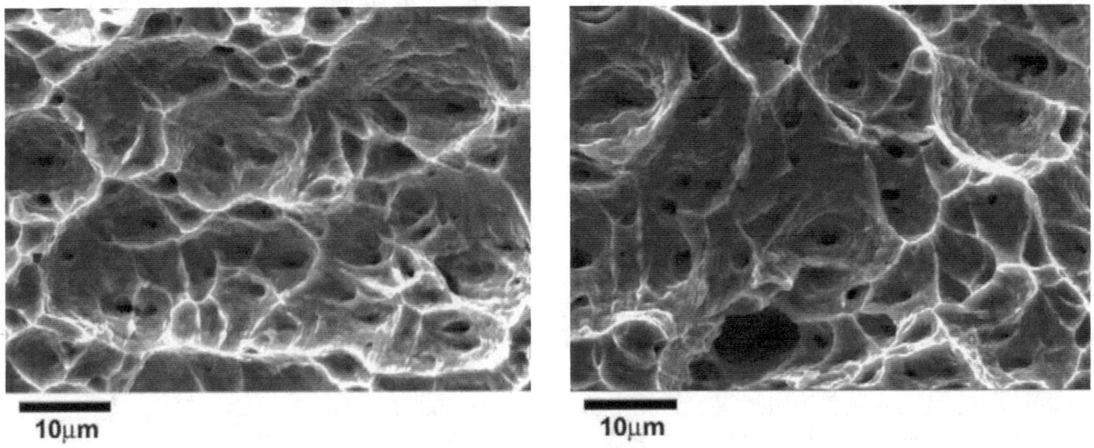


**Figure 4.** Creep fracture surface of UFG titanium (673 K, 200 MPa): (a) State 1, (b) State 2.

## 3 Discussion

Since both titanium States 1 and 2 were fabricated under the same SPD conditions creating almost the same structure, the proved difference between their the high temperature static and long-term characteristics is thought to be caused by their defect substructure, nanoporosity, and the presence of nonmetallic disperse carbide type inclusions. Higher resistance observed for State 1 can be attributed to increased content of carbon which create titanium carbide disperse particles [12,13] and thus can be responsible for more effective particle hardening. In connection with this effect, during static tests and tension under high temperature creep conditions, the expected negative effect of nanoporosity is likely to be effectively compensated just by the hardening induced by disperse particle (carbides). The presence of such inclusions could explain the high static strength and long-term strength during creep of titanium in State 1.

The results also show that creep life at 673 K is increased in microstructure with an occurrence of large elongation grains. Thus, the attained microstructure is created by large degree of straining during the last cold steps of SPD process. When the frequency of elongated grains in microstructure is increased, the total area of grain boundaries in a constant volume is decreased and this can enhance the creep resistance due to reducing of grain-boundary mediated deformation processes activities.

Static annealing of CG state at 823 K /1h does not lead to expressive differences in structure of UFG states. Generally, higher creep resistance of UFG states can be also explained by higher dislocation density. During heating and soaking on creep temperature (673 K), imposed dislocations probably are not fully recovered and thus dislocations could strengthen materials by deformation hardening. By contrast, high annealing temperature (823 K) of CG state can be associated with completely recovery of dislocations.

Operating deformation mechanism is probably not qualitatively different for all three states of materials under investigation as evidenced by similar values of fracture elongation. The high values of elongation suggest that the fracture proceeds due to the loss of plastic stability matrix of the material (local necking of the specimen section) and the resulting final fracture is mixed intergranular and transgranular mode.

## *4* Conclusions

Commercial Ti were processed by method of SPD by means of two rolling hot steps and one step at room temperature. The microstructure changes after application of SPD, annealing and creep testing were investigated. It was observed that short-term annealing of UFG state led to the coarsening of grain size and fully recovery of dislocation situated in the interior of grains. It led to degradation of creep resistance. The higher content of carbon in State 1 and potential formation of highly dispersed carbide particles was found to enhance the thermal stability of SPD titanium.

## Acknowledgements

The authors acknowledge financial support from the Czech Science Foundation (grant No. 20-14450J), the joint project running within Russian and Czech Academies of Sciences. This research was carried out under the project CEITEC 2020 (LQ1601) with financial support from the Ministry of Education, Youth and Sports of the Czech Republic under the National Sustainability Programme II.

## References

[1] Valiev, R.Z., et al., *Progress Materialia. Sci.* 2000 **45** 103.
[2] Kral, P., Dvorak. J., Sklenicka, V., Langdon T.G., 2019 *Mater. Trans.* **60** 1506.
[3] Kral P., et al., 2020 *Mater. Sci. Eng. A* **785** 139383.
[4] Furukawa, M., Horita, Z., Langdon, T.G., 2003 *Metals and Materials International* **9** 141.
[5] Lugo N., et al, 2008 *Mater. Sci. Eng. A* **477**, 366.
[6] Sklenicka, V., Dvorak, J., Svoboda, M., 2004 *Mater. Sci. Eng. A* **387– 389** 696
[7] Kim, W.J., at al., 2006 *Scripta Materialia*, **54**, 1745.
[8] Zeng, Z., et al., 2009 *Materials and Design* **30** 3105.
[9] Semenova, S.I., et al., 2009 *Mater. Sci. Eng. A* **503** 92.
[10] Betekhtin V.I, et al., *Technical Physics* 2011 **56** 1599.
[11] Betekhtin V.I, et al., *Physics of the Solid State* 2017, **59** 960.
[12] Betekhtin V.I, et al., *Technical Physics* 2015 **60** 66.
[13] Pandey U., et al., 2017 *Materials Today: Proceedings* **4** 5452.

IOP Conf. Series: Materials Science and Engineering 1178 (2021) 012014 doi:10.1088/1757-899X/1178/1/012014

# Boronizing in Arms Industry

**J. Hajek [1] V. Coufal [2]**

[1] COMTES FHT a. s., Prumyslova 995, 334 41 Dobrany, Czech Republic

[2] Regional Technological Institute, University of West Bohemia, Univerzitni 8, 306 14 Pilsen, Czech Republic

jiri.hajek@comtesfht.cz

**Abstract.** Comparison is made between several thermochemical treatment methods, with focus on boronizing. Boronizing has the potential to become a substitute for chromium plating or nitriding (gas nitriding) in the arms industry. This experimental study explored the chemical composition, hardness, corrosion resistance, tribological properties and toughness of several surface layers and coatings. The experimental substrates were EN 32CrMoV12-10 and EN 31CrMoV9 steels. Boronizing is intended for the gas tube and piston of a specific firearm. In the boride layer, the hardness of 1900 HV0.1 was reached. In the chromium and nitrided layers, hardness was 1100 HV0.1. In these experiments, original rifle tubes and disc samples were used. The thicknesses were as follows: the chromium layer: up to 70 μm, boride layer: 100 μm and nitride layer: 290 μm. Wear properties were compared using the Pin-on-Disc wear test. Fracture toughness according to Palmquist was the highest in the nitride case and the lowest in the chromium layer. Across all the layers, the range was from 2.5 to 9.5 $MN/m^{3/2}$. Some differences were found between the boride layer with post-process heat treatment and layers without heat treatment. As a result of the research, it was found that Boronizing of the gas tube proved a suitable substitute for hard chrome plating or nitriding.

## 1. Introduction

Technological way of boronizing could be different, such as molten salt boriding, gas boriding, plasma boriding, paste boriding, and pack boriding [1]. In the arms industry, chromium-molybdenum steel is commonly used for many parts where high toughness is required, for bolts, screws, crankshafts etc. Sen and colleagues, [2], Boronized EN 42CrMo4 steel in the salt bath. [3]Borided same steel using plasma-assisted and gas boriding methods, with single-phase ($Fe_2B$) on the surface of the steel. There are other investigations have been carried out on the properties of borided steels. [4,5]

## 2. Materials and methodology

The materials chosen for this study were EN 32CrMoV12-10 "gun-barrel steel" and medium-alloy hardening chromium-molybdenum-vanadium EN 31CrMoV9 grade. Specimens were prepared from these two steels.

The EN 32CrMoV12-10 steel is used for making components for CZ BREN 2 assault rifle, including the gas tube and piston. The nitriding grade EN 31CrMoV9 was chosen because of its similar composition, mechanical properties, better availability and, above all, lower price. If it delivers equivalent or better performance, it may be recommended as a substitute material for relevant components. Coatings were created on specimens which were then examined and measured using the following tests. Hardness testing of the surface layer was performed using the Vickers indenter. Fracture toughness testing of the surface layer by forcing a diamond right pyramid indenter with an apex angle of 136° into the surface and measuring the resistance to crack initiation and propagation through the coating. Pin-on-Disc wear test. Test was carried out using 10N load and $Al_2O_3$ ball as a counterpart. This wear test simulates the action of the piston in the gas tube. After the gun is fired, the pressure of combustion gasses from the burnt composition causes linear reciprocating movement of the piston which involves friction between the piston and the gas tube. Finally, a corrosion test was designed and carried out to simulate the weather conditions in regions with high humidity. Corrosion test was Carried out in a ClimaCORR CC 1000-TL FR (40/80) cyclic corrosion chamber. The test parameters: Demi water/40°C/96hours. Corrosion test was Carried out in a ClimaCORR CC 1000-TL FR (40/80) cyclic corrosion chamber. The test parameters: demi water/40°C/96hours. For dimensional changes measurement, passameter were used.

The chemical composition of the materials in weight per cent is given in Table 1. These steels are normally supplied in the hardened condition. 32CrMoV12-10 were Austenitized at 920°C/40min/ quenched in oil. Tempering temeperature was 590°C/90min/air. 31CrMoV9 were Austenitized at 860°C/40min/ quenched in oil. Tempering temeperature was 590°C/90min/air. The materials were therefore hardened to the following hardness levels: 459HV30 (32CrMoV12-10) and 371HV30 (31CrMoV9).

**Table 1.** Steel compositions in wt. %

| | Steels composition % | | | | | | | | |
|---|---|---|---|---|---|---|---|---|---|
| | **C** | **Si** | **Mn** | **Cr** | **Ni** | **Mo** | **V** | **P** | **S** |
| **32CrMoV12-10** | 0.31 | 0.25 | 0.46 | 3.13 | 0.19 | 0.81 | 0.26 | 0.02 | 0.02 |
| **31CrMoV9** | 0.32 | 0.25 | 0.59 | 2.68 | 0.05 | 0.18 | 0.12 | 0.017 | 0.017 |

Deposition parameters and coating characteristics are given in the following sections.

*Hard Chrome Plating*

The chromium coating was deposited by hard chrome plating in a bath with Heef25 catalyst. [6, 7, 8]The chromium coating contained distinct pores across its entire thickness.

On 32CrMoV12-10 specimens, the chromium coating thickness was 60 μm. The chromium coating thickness on the 31CrMoV9 specimens was 70 μm. The increase in the specimen and part dimensions fully matched the coating thickness (0.06 mm and 0.07 mm, respectively). The chromium coating had a hardness of 1067±28 HV0.1 on 32CrMoV12-10 steel. On the 31CrMoV9 grade, it was 1088±61 HV0.1.

*Boronizing*

*Boronizing without post-process heat treatment*

The specimens to be boronized were placed in a container and buried in Durborid boronizing compound. To prevent the ingress of air during the diffusion-based process, the container lid was sealed using a special sealant. At 900°C, boronizing in argon took 12 hours. The container then cooled in the furnace to 700°C. After that, it was removed from the furnace and cooled in still air.

The 32CrMoV12-10 specimens developed a $Fe_2B$ boride layer with a wedge-like morphology and a mean thickness of 75 μm (Figure 1a). On the 31CrMoV9 steel grade, FeB boride layer had an approximate thickness of 7 μm and transitioned to a $Fe_2B$ layer with a wedge-like structure and a mean thickness of 90 μm (Figure 2). [9]

Measurement of dimensional changes revealed that the increment is equal to approximately 33% of the layer thickness on 32CrMoV12-10 and 20% of the layer thickness on 31CrMoV9 grade. The impact of the higher chromium level in 32CrMoV12-10 on the thickness and morphology of the boride layer is apparent. A higher level of chromium leads to a thinner coating. It also compromises the anchorage of the boride layer because it loses its wedge-like morphology.

a) b)

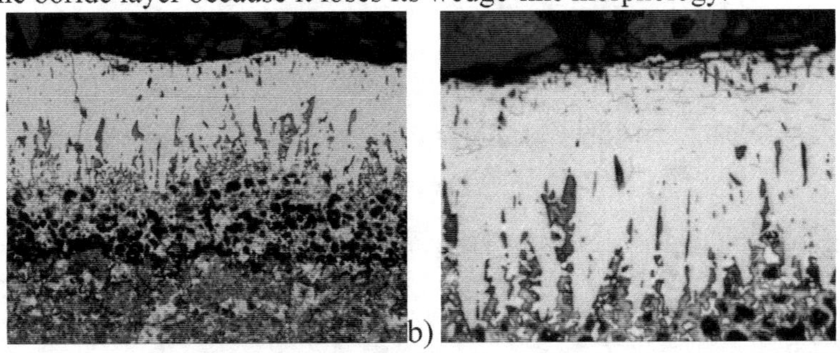

**Figure 1.** Boride layer on 32CrMoV12-10 steel without post-process heat treatment (left: magnification 500×, right: magnification 1000×)

a) b)

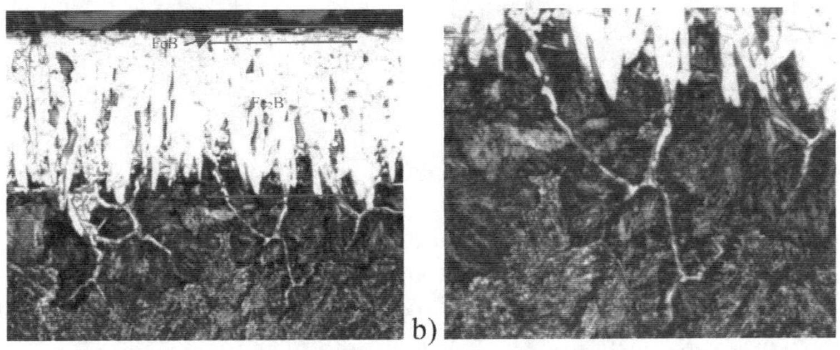

**Figure 2.** Boride diffusion layer on 31CrMoV9 steel without post-process heat treatment (left: magnification 500×, right: magnification 1000×)

Molybdenum has similar effects on the formation of the layer (0.8% in 32CrMoV12-10, and 0.2% in 31CrMoV9). The hardness of these layers exceeded 1500 HV0.1 by a large margin (32CrMoV12-10=1913 HV0.1 and 31CrMoV9=1710 HV0.1).

*Boronzing and post-process heat treatment*

The boronizing parameters were identical to those in the previous case. However, there was an additional post-process heat treatment (HT) whose role was to achieve the normal strength in the core of the part.

HT parameters:

Quenching temperature 920°C, soaking time 40 minutes, quenching in oil bath, no bath circulation.

Tempering temperature 620°C, holding time 90 minutes, cooling in still air.

*Nitriding*

Nitriding was carried out in ammonia/nitrogen atmosphere at 495°C for 20 hours. It was followed by gradual cooling in the furnace atmosphere to 100°C. Then the specimens were removed and left to cool in ambient air.

The layers had a thickness of 0.26 mm (31CrMoV9) and 0.29 mm (32CrMoV12-10). The corresponding dimensional changes were 0.008 mm and 0.01 mm (32CrMoV12-10).

In both nitride layers, the nitrides in the matrix were embedded in tempered martensite. No continuous nitride network was formed. The white layer thickness has not exceeded 10 μm in any location. The microstructure at the core remained virtually intact upon nitriding. It comprised tempered martensite with the visible orientation of fresh martensite structures, particularly in the 31CrMoV9 grade. The layer on the 32CrMoV12-10 steel exhibits higher hardness. As with boronizing, this is due to its higher chromium level which contributes to hardness.

## 3. Results and discusion

### *3.1. Wear testing*

In this test, the volume loss from the specimen due to a ball being forced into the surface is measured. The specimen is affixed to a revolving plate.

The circular wear track was measured according to ASTM 99 standard. The ball was made from $Al_2O_3$.

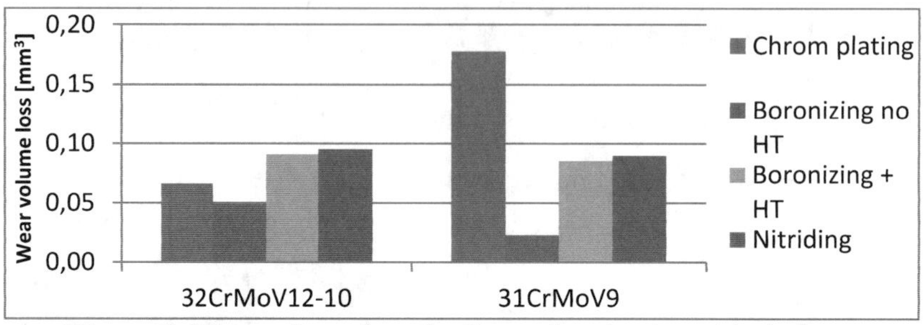


**Figure 3.** Wear volume loss for the surface treatment techniques

The boride layer with no post-process heat treatment showed the best wear resistance. The lowest resistance was found in the nitride layer (Figure 3). The wear resistance is related to the hardness of the layer and its resistance to delamination. Poor adhesion may lead to a sudden increase in wear, as seen in the data for the test on 31CrMoV9 chromium coating.

### *3.2 Fracture toughness*

Fracture toughness testing was carried out using a diamond right pyramid with a 136-degree apex which was forced into the surface layer at a load of 50 kg and produced cracks whose lengths were measured. The chromium coating and the boride layer without post-process heat treatment suffered delamination. These hard layers had poor adhesion to the substrate. The causes are given by the very deposition process. The cracks induced by indentation were shallow (Palmqvist cracks). Fracture toughness was calculated using Palmqvist equation according to ISO 28079. The measured and calculated values were plotted in the graph shown in Fig. 4.

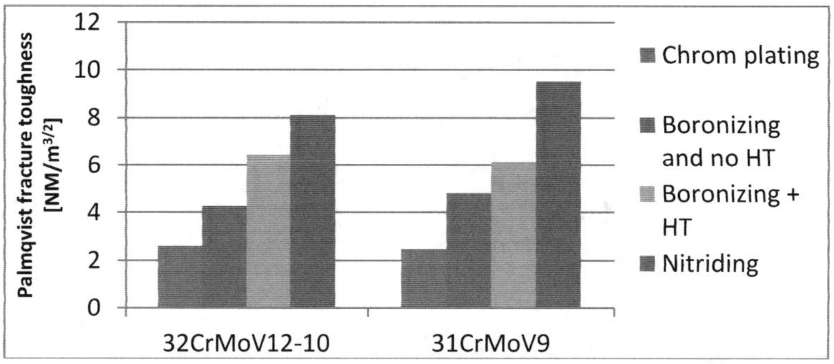


**Figure 4.** Palmqvist fracture toughness

As the results show, the chemical composition of the steel is not the primary factor here. The chromium coating has the lowest fracture toughness. The values are in agreement with the wear test data. The toughness of the nitride layer is almost four times that of the chromium coating. Heat treatment after boronizing led to a slightly higher toughness of the layer.

### *3.3 Corrosion test*

The corrosion test of parts of 32CrMoV12-10 and 31CrMoV9 steels was carried out in ClimaCORR CC 1000-TL FR (40/80) cyclic corrosion test chamber. The specimens were examined and appearance rating (RA) based on the corrosion deterioration area A was assigned and listed in Table 2.

**Table 2.** Surface area effect

| Treatment | Steel | Surface Area Effect [%] | Degree [$R_A$] |
|---|---|---|---|
| No treatment | 32CrMoV12-10 | $2{,}5 < A \leq 5$ | 4 |
| | 31CrMoV9 | $25 < A \leq 50$ | 1 |
| Chromium plating | 32CrMoV12-10 | Bez vad | 10 |
| | 31CrMoV9 | $2{,}5 < A \leq 5$ | 4 |
| Boronizing and no HT | 32CrMoV12-10 | $A > 50$ | 0 |
| | 31CrMoV9 | $A > 50$ | 0 |
| Boronizing and HT | 32CrMoV12-10 | $10 < A \leq 25$ | 2 |
| | 31CrMoV9 | $25 < A \leq 50$ | 1 |
| Nitriding | 32CrMoV12-10 | $0 < A \leq 0{,}1$ | 9 |
| | 31CrMoV9 | $0 < A \leq 0{,}1$ | 9 |

The highest corrosion resistance on 32CrMoV12-10 is with the chromium coating and the nitride layer. On 31CrMoV9 steel, the highest resistance is found in the nitride layer followed by the chromium coating. On 32CrMoV12-10 and 31CrMoV9 steels, boride layers perform poorly, as the almost entire surface was attacked by corrosion.

### *3.4 Discussion*

Thermochemical treatment and surface treatments explored in this experiment led to the enlargement of the parts. Understandably, this must be taken into account at the product design stage. Where chromium plating is to be substituted with boronizing, the initial dimensions prior to thermochemical treatment must be corrected by 0.04 mm. In terms of manufacturability, this substitution is seamless. The alternative without post-process heat treatment is more convenient, as heat treatment may cause distortion.

In the chromium-plated specimens, there is a step-change in hardness between the surface and the core. The hardness of the surface layer is 1027–1091 HV0.1. The reason for the step change is that electrochemical deposition causes chromium to attach to the surface without any substantial diffusion layer being formed. Clearly, this poses a risk of delamination under load. The hardness of boride layers on these steels is approx. 1900 HV0.1. With these layers, there is a steep hardness gradient as well, where hardness decreases towards the core. After boronizing, the anchorage of the layer to the substrate greatly depends on the chemical composition of the latter. Metallographic examination revealed that in these steels, the anchorage is strong thanks to a transition zone of several dozen micrometres, as opposed to high-alloy tool steels where anchorage is virtually non-existent. Since the present steels are essentially nitriding steels, the nitriding process can produce hardnesses over 1000 HV0.1. The characteristic properties are attained thanks to hard nitrides embedded in a hardened matrix, which delivers high toughness, the main benefit of this thermochemical treatment method. From the perspective of wear resistance, the boride layer is the best choice. It is due to the combination of surface hardness and adhesion. However, once delamination occurs in a chromium or boride layer, the wear rate soars. Provided that proper adhesion of the $Fe_2B$ intermetallic is achieved, the boride layer is certain to deliver the best wear resistance. In fracture toughness tests,

IOP Conf. Series: Materials Science and Engineering 1178 (2021) 012014 doi:10.1088/1757-899X/1178/1/012014

an indentation of the surface layer produced cracks and caused delamination. The impact of chemical composition was minimal between the steels chosen for this experiment. The post-process quenching and tempering did not impair the adhesive-cohesive behaviour of the boride layer. Corrosion testing highlighted the clear benefits of chromium plating and nitriding as these surfaces were virtually intact after the test – in contrast to the boride layer.

## 4 Conclusion

Boronizing of the gas tube proved a suitable substitute for hard chrome plating or nitriding. The acceptable substrate material is 31CrMoV9+QT steel. Sound layers formed on both steels in the experiment but chemical composition makes 31CrMoV9 a better candidate than 32CrMoV12-10. Lower chromium and molybdenum levels help adhesion. The chemistry of the steel affects not only the morphology and composition of the surface layer but the composition and microstructure of the transition zone as well.

Process planning will require changes, as the dimensions related to the gas tube diameter need to be corrected by almost 0.1 mm on production drawings. From the wear resistance viewpoint, the outcome of the comparison between the layers is dictated by the occurrence of delamination of the hard chromium coating. In the wear tests, the boride layer clearly dominated. It is because of its high hardness and good adhesion.

On the other hand, its weakness lies in its poor corrosion resistance. It can be eliminated by using the borochromizing process. This is an essentially boronizing process where a chromium coating is created on the final surface. The materials for pack boronizing is ferrochromium.

**References**


[1] Keddam, M., Chentouf, S.M.: A diffusion model for describingthe bilayer growth (FeB/$Fe_2B$) during the iron powder-packboriding. *Appl. Surf. Sci.*252, 393–399 (2005)

[2] Sen, S., Sen, U., Bindal, C.: Tribological properties of oxidisedboride coatings grown on AISI 4140 steel. *Mater. Lett.*60, 3481–3486 (2006)

[3] Kuper, A., Qiao, X., Stock, H.R., Mayr, P.: A novel approach togas boronizing. *Surf. Coat. Technol.*130, 87–94 (2000)

[4] Akshay A., Hosmani S. , Pack-Boronizing of AISI 4140 Steel: Boronizing Mechanism and the Role of Container Design, 2014, *Materials and Manufacturing Processes* vol. 29, pp. 1062 - 1072

[5] Šuchmann P., Hradil D., Gregor T., Moni V., Improving the service life of excavation teeth, 2019, 27th National Conference on Heat Treatment

[6] Lausmann, G. A. Electrolytically depostied hardchrome, *Surface and Coatings technology*, Elsevier 1996, pp. 814 – 820

[7] Valeš M., Zkoušení povrchových úprav a ochran, 2011, *MM Průmyslové spektrum* [online]., 4, 60 [Accessed 11/11/2018]. Online: https://www.mmspektrum.com/clanek/zkouseni-povrchovych-uprav-a-ochran.html

[8] Lausmann, G. A., Unruh, J. N. M., Die galvanische Verchromung, 1. Auflage, Bad Saulgau, Eugen G. Leuze Verlag, 1998, *Schriftenreihe Galvanotechnik und Oberflachenbehandlung*

[9] Moni V., Hradil D., Miletič T., Blata J., Řehoř M., Design of excavators tooth with refinement technical parameters, 2019 *International Multidisciplinary Scientific GeoConference Surveying Geology and Mining Ecology management, SGEM*, 2019

IOP Conf. Series: Materials Science and Engineering 1178 (2021) 012015 doi:10.1088/1757-899X/1178/1/012015

# Thermal conductivity of SLM deposited H13 steel

**K Halmešová[1], J Džugan[1], M Ackermann[2], M Koukolíková[1] and Z Trojanová[1]**

[1]COMTES FHT a.s., Průmyslová 995, 334 41 Dobřany, Czech Republic

[2]Laboratory of Rapid Prototyping, Centre for Nanomaterials, Advanced Technologies and Innovation, Technical University of Liberec, Studentská 1402/2, 460 01 Liberec 1, Czech Republic

Corresponding author's e-mail address : kristyna.halmesova@comtesfht.cz

**Abstract**. Hot-worked H13 tool steel was deposited using selective laser melting (SLM) process in different deposition conditions. The samples were deposited in 2 different directions: parallel and perpendicular to the base plate of SLM device. Thermal conductivity of the deposited samples was calculated on the base of laser flash measurements up to 700°C. Furthermore, the thermal expansion, specific heat, and thermal diffusivity of the deposited samples and their variations with different deposition conditions are also presented. To explain the measured differences in thermal properties, some metallographic studies together with porosity investigation are also provided. It was shown that all measured thermal properties of the samples are affected by temperature and that the microstructural defects play an important role for the resulting thermo-physical properties.

## 1 Introduction

Selective laser melting (SLM) process is one of the complex, most popular, and fastest developing additive manufacturing techniques, where the controlled laser beam scans the selected part of the powder bed and melts the solid material in the form of powder and forms the resulting part adding layer by layer. The production method enables the production of parts with high geometrical requirements that are technically impossible with traditional processing and almost all metals can be processed by SLM [1]. Similar to welding, the residual internal stresses resulting from high temperature gradients, large amounts of thermal expansion, and shrinkage, or nonuniform plastic deformation during the heating and cooling cycles [2] are the major drawbacks of this technique and lead to cracks or distortion of the parts [3,4]. Applying postheat treatment, heating of the substrate plate or rescanning by the laser beam results in the reduction of residual stresses [5].

Another weakness of SLM is the reduced density due to an occurrence of pores, lack of fusion in processed materials. Eventually, these defects act as stress concentration sites that initiate cracks and lead to failure of printed parts [6].

Thermal properties of SLM deposited materials are rarely discussed. H13 tool steel is widely used in the injection moulding industry, especially of aluminum and manganese, because it offers high strength at elevated temperatures, high hardness, and it also has the advantages of high resistance to thermal shock and thermal fatigue, high abrasion resistance and heat resistance. In SLM applications, H13 can also achieve in-situ hardening during SLM manufacture, which can potentially reduce the need for postbuild hardening heat treatment [7]. However, to effectively apply H13 tool steel in SLM manufacture of injection moulds, it is necessary to carefully manage SLM processing parameters which is discussed in detail in [8].

The thermal properties are interesting especially for hot work tools. Thermal diffusivity is an important transient thermal property, required for such purposes such as design applications, determination of safe operating temperature, process control, and quality assurance [9]. The higher thermal conductivity allows more efficient cooling of the tool and hence shortening of the production time.

In this paper, the specific heat capacity, thermal expansion, density, and thermal diffusivity of SLM prepared H13 specimens were measured and these results were used for the calculation of thermal conductivity. The effect of different deposition parameters on the thermal properties was also studied.

## 2 Experimental

All samples were manufactured at Technical University of Liberec using SLM280HL machine (SLM Solution AG, Lübeck, Germany). H13 tool steel powder was also provided by the SLM Solutions AG company and the chemical composition (in weight %) is 0.42C, 0.44Mn, 0.85Si, 5.22Cr, 1.50Mo, 1.04V. The laser power for appropriate powder melting was 175 W and the scanning speed was 610 mm/s during the building of the samples. Inner part of the specimen was processed using the stripe hatch strategy. The laser tracks were rotated 90° for each layer to minimize inner porosity. Nitrogen gas was used as a protective atmosphere during the process. The building plate was preheated to 200 °C to minimize the residual stresses in the built samples.

Afterwards, the samples were subjected to heat treatment to remove the internal stresses; annealing at 600 °C for two hours in a protective gas was followed by slow cooling.

All samples were directly manufactured by SLM technique in two different building directions. Only supports were manually removed and no other machining or sample preparations were required. Figure **1** shows the specimen orientation and its designation.

Thermal diffusivity results ($a$), together with related values of specific heat capacity ($c_p$) and density ($\rho$) values, can be used in many cases to derive the thermal conductivity ($\lambda$), according to the relationship [9]:

$$\lambda = a\, c_p\, \rho. \tag{1}$$

The Laser Flash technique (Linseis LFA 1000) was applied for nondestructive and noncontact determining of the thermal diffusivity and specific heat from room temperature up to 700 °C under vacuum. The integrated motorized sample charger allows measurement of 6 samples at the same time (coin shape with 12.7 mm in diameter and 2 mm in thickness) and the measurement was repeated three times. Then the median, mean, and standard deviation were calculated. The thermal diffusivity value is determined from the specimen thickness and the time required for the rear side to reach 50% of its maximum temperature increase. Pure molybdenum specimen was used as a reference for the specific heat determination based on the comparison method of reference and measured specimen.

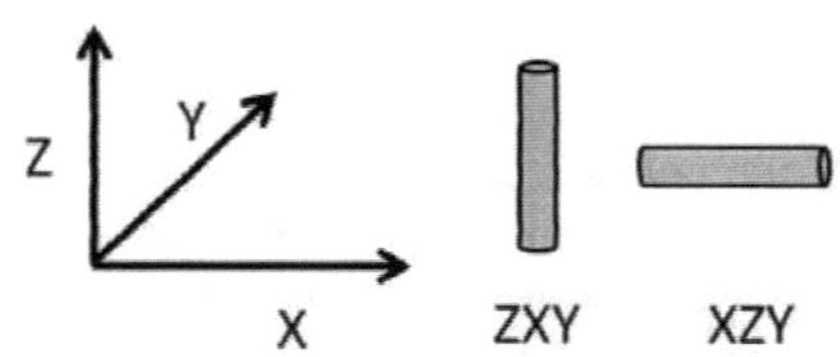


**Figure 1:** Identification of the dilatometry samples according to the building orientation.

The coefficient of thermal expansion was measured using a dilatometer L75PT (Linseis) up to 1000 °C in an argon protective atmosphere at a cooling/heating rate of 10 °C/min. Correction measurement with $Al_2O_3$ samples was done to remove the thermal expansion of the instrument. Knowledge of the linear thermal expansion coefficient of the sample dimension (α) and assuming the isotropic behavior, then the sample density relationship over the temperature can be written as:

$$\rho = \rho_{20} / [1 + \alpha(T-20)]^3, \quad (2)$$

where $\rho_{20}$ is the density at the room temperature, which was estimated by the hydrostatic weighting. The microstructure and porosity of the specimens were examined after etching with Villela-Bain chemical agent and using a light microscope NIKON ECLIPSE MA200. Detailed material microstructure was also observed using scanning electron microscope (SEM) JEOL IT 500 HR in the secondary electron regime.

## 3 Results and discussion

Figure 2a shows the measured dilatometry curves. Phase transformation above 863 °C during heating represents the austenitization of the steel and the transformation during cooling below 310 °C characterized by volume expansion refers to martensitic transformation. Post 700 °C Curie transformation is clearly seen, above this temperature, the relative elongation no longer exhibits a linear character, and therefore the average linear thermal expansion coefficient, $\alpha_{20}$, between 20 °C and a given temperature, was calculated from the linear part of the dilatometry curve during heating (Figure 2b). Despite the fact that the effect of different building directions seems to be negligible, the thermal expansion is a little lower for sample ZXY at higher temperatures. Using the quasi-harmonic model for atomic vibration, the volumetric thermal expansion coefficient can be described as [10]:

$$\alpha_T^V = \frac{C_{V,T}^V \gamma}{BV}, \quad (3)$$

where $\alpha_T^V$ is the volumetric thermal expansion coefficient, $C_{V,T}^V$ the heat capacity at constant volume, $\gamma$ the Grüneisen parameter, $B$ the bulk modulus, and volume $V$. The lower values of the average linear thermal expansion coefficient for XZY samples may be probably an effect of lower density of the material because of the higher presence of pores (see Figure 4), which originates from unoptimized process parameters. To support this statement, the microstructure of the samples was also provided.

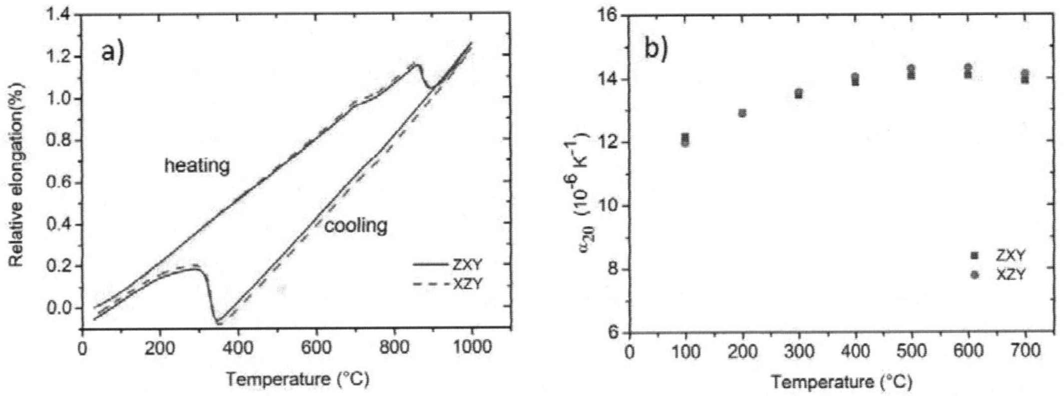


**Figure 2:** Relative elongation (a) and the average thermal expansion coefficient (b) and their behaviors with the temperature of SLM prepared samples in different directions ZXY and XZY.

Figure 3 shows the microstructure of the samples XZY and ZXY in the plane parallel to the building direction after heat treatment. No important differences in microstructures were observed. Microstructure consists of the tempered martensite with the ferritic matrix and fine carbides. Figure 4 displays the defects in the microstructures (mostly lacks of fusion in our case). It is evident that the specimen ZXY contains a higher fraction of defects. The maximum porosity measured in the worst position in the metallographic sample was 0.02% for XZY and 0.1% for ZXY.

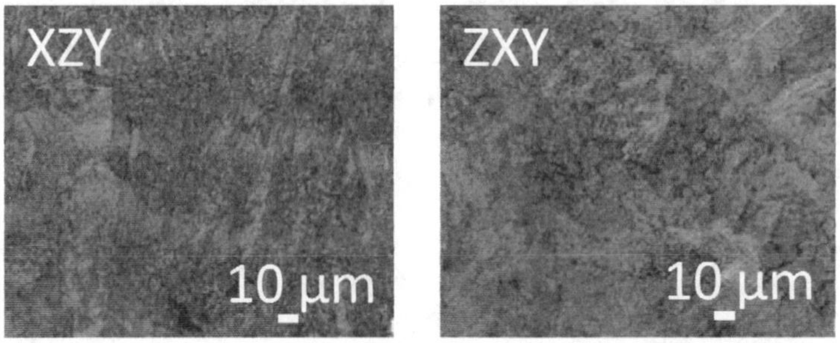


**Figure 3**: Microstructure of the specimens XZY and ZXY taken from plane parallel to the building direction, after heat treatment.

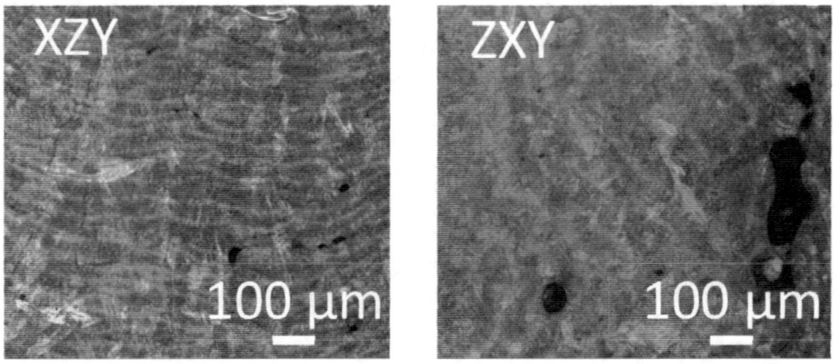


**Figure 4**: Microstructure defects of the specimens XZY and ZXY taken from plane parallel to the building direction, after heat treatment.

The thermal diffusivity, *a,* together with the values of the specific heat, $c_p$, measured from room temperature up to 700 °C are displayed in Figure 5. As seen, the diffusivity values decreased with an increasing temperature for all samples while the specific heat tendency was increasing with rising temperature. These results were used for calculation of the thermal conductivity that is shown in Figure 6.

The thermal conductivity of the material is attributed to the heat transport by electrons and by phonons and can be considered as independent entities. The electron contribution is dominant in pure metals. Even in a concentrated alloy like stainless steel, the electrons carry most of the heat except in the region of the maximum in the phonon part, roughly at $T=0.1\Theta_D$, where the phonons give a significant contribution [11]. For most materials, the Debye temperature is between 200-400 K; $\Theta_D$ (Fe) =470 K. The measured thermal conductivity in this work can be described by the electronic contribution to thermal conductivity. Since free electrons are responsible for both electrical and thermal conduction in metals, the two conductivities are related by the Wiedemann-Franz law:

$$\lambda = \sigma L T, \tag{4}$$

where σ is the electrical conductivity and L is a constant (Lorenz number). Therefore, the same factors that affect the electrical conductivity also affect the thermal conductivity. A lower thermal conductivity for ZXY building direction could be caused by higher microstructural defects in the sample as was already mentioned. Ramírez et al. [12] studied the effect of porosity on the thermal properties of aluminum alloy and observed that the thermal conductivity and specific heat capacity decreased with the increase in porosity.

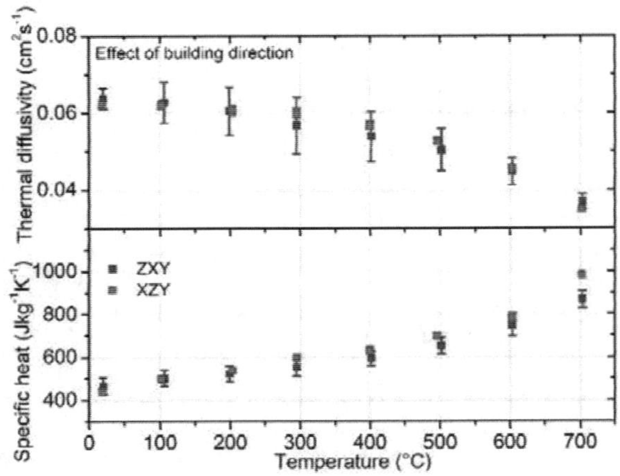


**Figure 5**: Thermal diffusivity and specific heat as a function of the temperature of SLM prepared samples in different directions ZXY and XZY. The standard deviations are also provided.

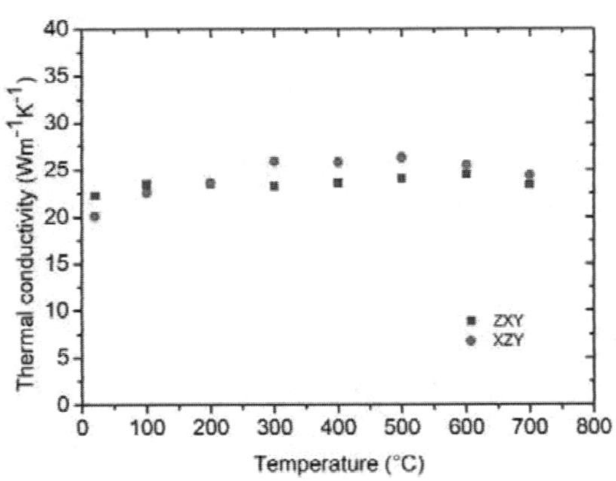


**Figure 6**: Temperature dependence of the thermal conductivity of SLM prepared samples in different directions ZXY and XZY.

More detailed resutls of these inevstigations can be found in [13]

## 4 Conclusions:

The samples in two different building directions ZXY and XZY were successfully prepared with selective laser melting technology from H13 tool steel powder. Heat treatment afterwards reduced the internal stress and homogenized the microstructure. Microstructure studies proved that there were a present a lot of microstructural defects resulting from not-well optimized process parameters. The sample ZXY prepared with higher number of layers in z direction suffered from higher concentration of defects compared to the XZY sample. These material imperfections led to lower values of thermal conductivity, which connects the effect of selective laser melting building direction together with the thermal properties of the printed parts.

**Acknowledgement**
The paper was supported from the project entitled 3D COVER – Metallic materials in the process chain of additive manufacturing, Project No.: 185, financed by the ERDF.

## References

[1] Kruth J P Mercelis P Vaerenbergh J Froyen L Rombouts 2005 M Binding mechanisms in selective laser sintering and selective laser melting *Rapid Prototyping Journal* **11** pp26-36
[2] Mercelis P Kruth J P 2006 Residual stresses in selective laser sintering and selective laser melting *Rapid Prototyping Journal* 12 pp254-265
[3] Yadroitsev I Yadroitsava I 2015 Evaluation of residual stress in stainless steel 316L and Ti6Al4V samples produced by selective laser melting *Virtual and Physical Prototyping* pp67-76
[4] Kruth J P Deckers J Yasa E Wauthlé R 2012 Assessing and comparing influencing factorsof residual stresses in selective laser melting using a novel analysis method, Proceedings of the Institution of Mechanical Engineers, Part B: *Journal of Engineering Manufacture* 226 pp980-991
[5] Shiomi M Osakada K Nakamura K Yamashita T Abe F 2004 Residual stress within metallic model made by selective laser melting process, *CIRP Annals & Manufacturing Technology* **53** pp195-198
[6] Brandl E Heckenberger U Holzinger V Buchbinder D 2012 Additive manufactured AlSi10Mg samples using Selective Laser Melting (SLM): microstructure, high cycle fatigue, and fracture behavior *Mater. Des.* pp159-169
[7] Lee J-H Jang J-H Joo B-D Yim H-S Moon Y-H 2009 Application of direct laser metal tooling for AISI H13 tool steel. *Transactions of Nonferrous Metals Society of China* 19 pp284–287
[8] Mazur M Leary M McMillan M Elambasseril J Brandt M 2016 SLM additive manufacture of H13 toolsteel with conformal cooling and structural lattices Rapid Prototyping Journal Vol. **22** Iss 3 pp504-518
[9] ASTM E1461-13 2013 Standard Test Method for Thermal Diffusivity by the Flash Method ASTM International, West Conshohocken, PA, www.astm.org
[10] Zhang X Sun S Xu T 2019 Temperature dependent Grüneisen parameter *Sci. China Technol.* Sci. **62** pp1565–1576
[11] Grimvall G 1999 Thermophysical properties of materials Elsevier, North Holland Amsterdam ISBN 978-0-444-82794-4
[12] Ramírez A M Beltrán F J E Yáñez-Limón J M Vorobiev Y V González-Hernández J and Hallen J M 1999 Effects of porosity on the thermal properties of a 380-aluminum alloy *J. Mater. Res.* vol. 14 no. 10 pp3901–3906
[13] Džugan J, Halmešová K, Ackermann M, Koukolíková M and Trojanová Z 2020 Thermo-physical properties investigation in relation to deposition orientation for SLM deposited H13 steel, *Thermochim. Acta*, 683(2) 178479

# Investigation of premature failure of hip joint replacement

**P Hanusová, P Palček, M Roszak, V Zatkalíková**

University of Žilina, Faculty of Mechanical Engineering, Univerzitná 8215/1, 010 26 Žilina, Slovakia

E-mail: patricia.hanusova@fstroj.uniza.sk

**Abstract**. Hip joint replacement is one of the most frequently used surgical procedures worldwide. Currently, more than 340 of the total 4431 implants per year are reoperated in Slovakia. Despite the excellent properties of the titanium alloy, endoprosthesis often fails and hip replacement is necessary. Common causes are inflammation and aseptic loosening. Other causes of failure include injury, implantation failure, manufacturing inaccuracies, and aseptic loosening. The case study of a failed endoprosthesis is carried out in cooperation with the Orthopaedic Clinic of the University Hospital in Martin, Slovakia. The patient necessarily needed THA and this paper deals with the causes of failure of his total hip joint replacement.

## 1 Introduction

Titanium alloys are considered the most attractive metallic materials for biomedical applications because of their biocompatibility, high corrosion resistance, and appropriate mechanical properties [1]. Biomaterials are materials intended for direct contact with the internal tissues of the human body. Both the recipient and the implanted material must be carefully studied, as foreign substances are placed in the chemically active environment of the body. Biomaterials are classified into four categories according to the interaction with the recipient. An inert material that elicits a favorable reaction, such as growth. A viable biomaterial that includes living cells at implantation that the host perceives as normal and that is actively absorbed. The implant material is the original material that is cultured in vitro from cells obtained from a patient and subsequently implanted. And a biomaterial that serves as a replacement for part of the human tissue in the support system [2].

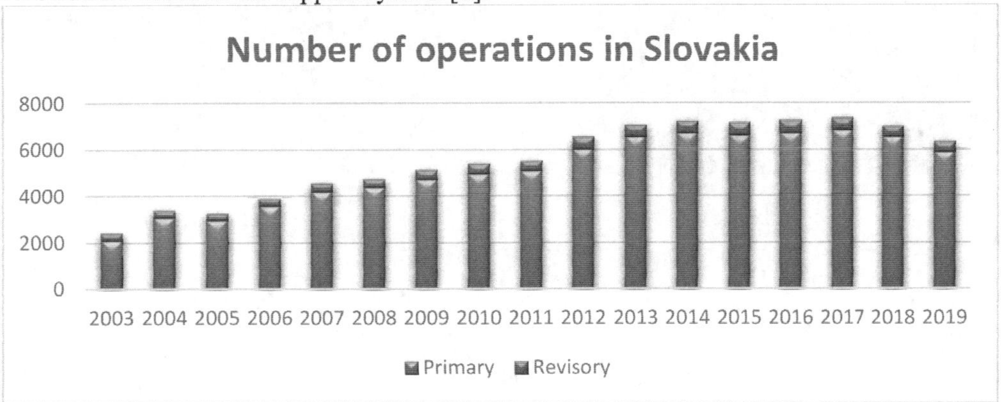


**Figure 1.** Number of operations in Slovakia since 2003

At present, Ti-based alloys consist of biocompatible alloy bases that replace both hard tissues and dental ligaments. Ti-based alloys are also used to repair soft tissue, such as blood vessels [3]. During biomedical applications, various ways of compressive, tensile, and shear stresses may occur in the material. Sometimes there is a combination of all three. In a typical tensile mode, the same force is applied perpendicular to the surface of both ends of the sample in opposite directions, while during compression the force is applied perpendicular to both ends in the opposite direction. In shear loads, the forces acting tangentially on opposite surfaces. Due to differences in properties, biomaterials are divided into four groups: metals, ceramics, and polymers, and composites [4]. The generation of polyethylene wear debris is a major cause of aseptic loosening, and the most common reason for THA revisions. The interaction of particles and cells in periprosthetic tissue and surrounding bone contributes to the deregulation of bone homeostasis resulting in osteolysis at the bone–implant interface [5].
In 2018, 458 patients underwent revision surgery (Figure 1), of which 60 at Martin Hospital. In collaboration with physicians from the Orthopedic Clinic of the University Hospital in Martin, where the patient underwent the necessary reoperation, we analyzed the possible causes of failure of the titanium endoprosthesis (Figure 2), to prevent other possible complications [6]. A 61-year-old patient underwent a left hip replacement in 2008 at the Orthopaedic Clinic of the University Hospital in Martin. After two years, he underwent surgery to replace his right hip joint. The man's BMI at that time was 27.8, so his height was 180 cm and his weight was 90 kg. In 2018, the patient began to feel pain in his left hip, could not walk, and had a fever of 38.1 ° C. The indicator of the inflammatory response of CRP reached the value of 170.5 mg l-1, the physiological value of which is up to 10 mg l-1.

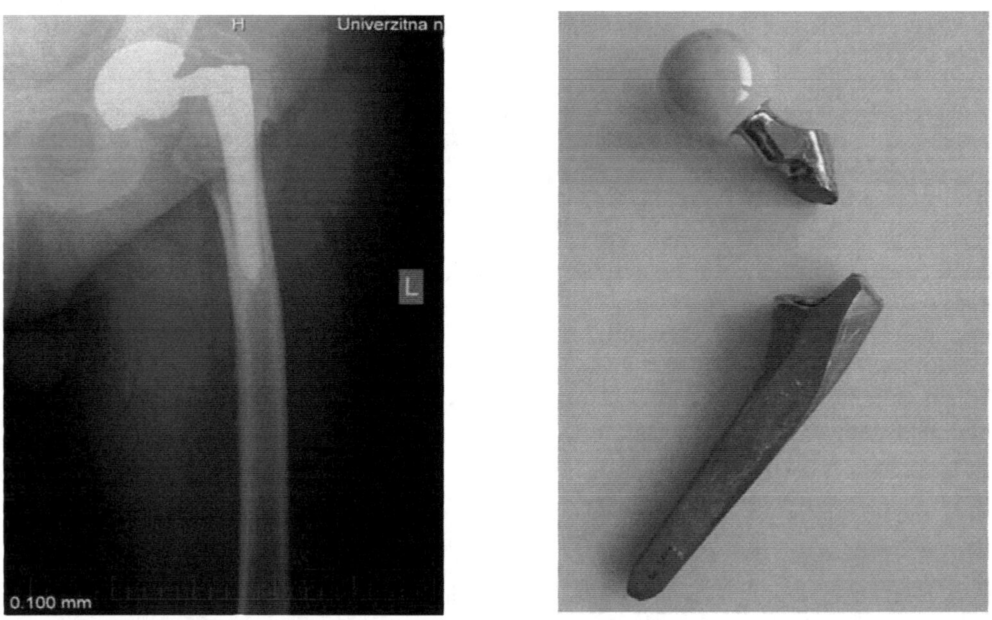


a) X-Ray view b) after extraction

**Figure 2.** Prematurely failed endoprosthesis

## 2 Materials and methods

The chemical composition of the failed endoprosthesis was examined using a SPEKTROMAXx device (Table 1). The spectral analysis confirmed that the implant was made of Ti6Al4V alloy. After the metallographic preparation, the microstructure was evaluated on a NEOPHOT 32 light microscope and

TESCAN VEGA LMU2 scanning electron microscope, which includes an EDX analyzer. Faults in the defectoscopy analysis were investigated using VGStudio MAX 2.2 software from Volume Graphics.

**Table 1**. Chemical components of endoprosthesis

| Spectral analysis of endoprosthesis [wt %] | | | | | | | |
|---|---|---|---|---|---|---|---|
| **Al** | 6,47 | **V** | 3.94 | **Mn** | 0.002 | **Ti** | 89.20 |
| **Sn** | 0,0073 | **C** | 0.0358 | **Cr** | 0.0128 | **Nb** | 0.0236 |
| **Mo** | 0,0156 | **Si** | 0.0194 | **Ni** | 0.0137 | **Fe** | 0.219 |

The Beznoska Trio implant was made of Ti6Al4V alloy according to ISO 5832-3. It is made in three variants - cemented, uncemented, and uncemented with the possibility of using different types of modular necks. The uncemented variant is sprayed in the upper third by plasma spraying with a bioactive layer of titanium oxide, which ensures biocompatibility. There is also a version with hydroxyapatite, which is the main component of bone [7].
The length of the stem depends on the individual needs of the patient and ranges from 115 mm to 150 mm. The width of the implant exists in 10 possible alternatives from 12 mm to 16.5 mm, depending on the width of the patient's femur. The examined implant was size 3, and thus its length was 125 mm and width 14 mm.

## 3 Results and discussion

### *3.1 Microstructure*

Evaluation of the microstructure of the material was performed on samples from both parts of the prosthesis. The samples were metallographically prepared (Figure ) by the method used for titanium alloys on a TegraPol-15 Struers automatic instrument. Samples were polished and etched with 10% HF for structure analysis.

a) optical microscope b) SEM

**Figure 4**. Microstructure of stem

The study of the microstructure confirmed the differences in the parts of the implant. The neck part of the endoprosthesis (Figure 5) was made by casting. The microstructure of the sample was polyhedral with fine grains. There was a predominantly equiaxed α phase with a small amount of irregularly shaped β phase at the grain boundaries of the α phase. From the present knowledge, the dependence of the fatigue life on the lamellar α phase is known. The larger the lamella dimensions are, the lower the fatigue life of the alloy is [8]. The lamellae α had the same shape and size. On the contrary, the stem part of the prosthesis was made by forging. On the microstructure, this was reflected in grain elongation and visible line spacing even at lower magnifications.

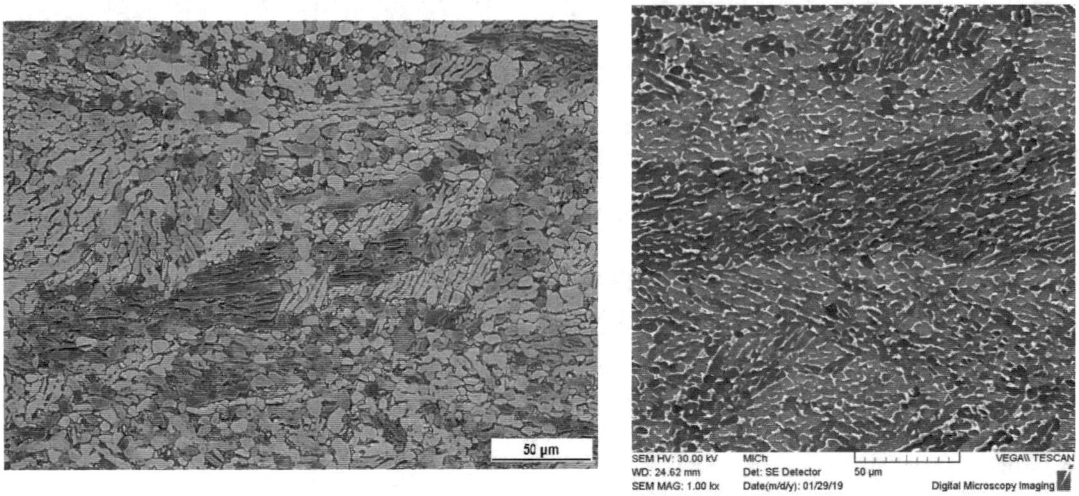


a) optical microscope b) SEM

**Figure 5.** Microstructure of neck

### *3.2 Defectoscopy*

In the images, cracks are visible in the material occurring in an angular range of approximately 140 °, on the side facing the hip socket (Figure ). An example of one of the cracks is marked by a red cursor in three planes and in space. The crack was formed at the place with the highest concentration of stress. Production inaccuracies caused imperfect fitting of endoprosthesis components.

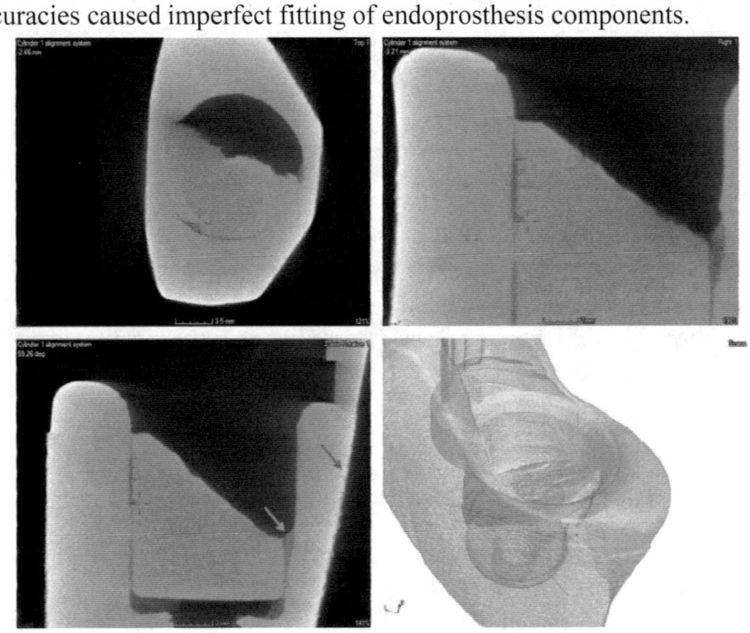

**Figure 6.** Description of the selected crack

There are strict requirements for accuracy in the production of a cone. However, a cavity is evident at the interface between the femoral neck and the stem, confirming that the two segments do not fit together correctly. The micro-movements in the conical part of the neck led to movements under load, which disrupted the passivation layer of the titanium alloy and initiated the crack.
Another material marked with orange arrows is also visible (Figure ). It is a physiological material from the patient's body that has reached the interface between the neck and the stem. Either by insufficient drying during implantation or along a crack in the micro-movements of the neck. The green arrow indicates beam hardening, which is an artifact created during scanning when the absorption of soft radiation on a surface gives the impression that the surface has a higher density. The defect is shown in the lower part of the joint, marked by an orange arrow (Figure 7). The defect extends into the neck, as well as into the hip part of the prosthesis.
The above analysis shows that there are no defects in the monitored area that the software could identify. The cracks described are either surface-exposed or too "thin" to be identified. Due to the lack of information on the initial state, it is not possible to confirm from the point of view of CT analysis whether they were already there before implantation. However, according to the direction of their orientation and their distribution, it can be assumed that they were formed during use by cyclic stressing of the material. Other defects represent damage to both parts of the endoprosthesis, as well as defects at the interface of the prosthesis components.

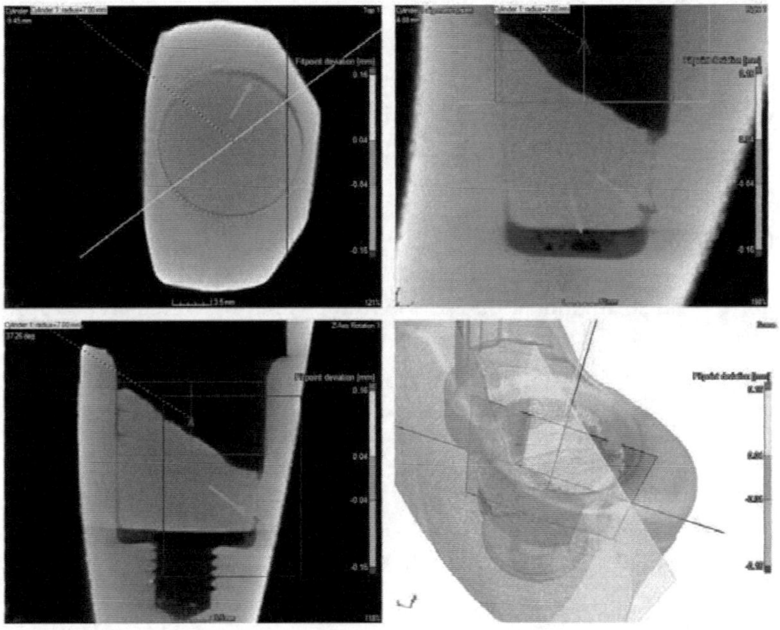

**Figure 7.** The interface of the two parts of the endoprosthesis

## 4 Conclusion

The results of the dissertation project suggest that despite the excellent mechanical properties, hip endoprostheses made of titanium alloys are damaged. An analysis of the current issue of premature failure replacements leads to the conclusion that titanium alloys have the most suitable application as replacements in the human body.

- Spectral and EDX analysis confirmed that the experimented material was a Ti6Al4V alloy.
- Analysis of the microstructure of the femoral stem and modular neck confirmed differences in their microstructure. The resulting visible differences were caused by different processing of the

material. The linear row structure represents the forged stem part of the endoprosthesis. The neck part is represented by a cast microstructure with dendritic elements.

- The main causes of prosthesis failure include manufacturing inaccuracies and micro-movements at the neck-stem interface. At these points, the stress is concentrated, and a fatigue crack is formed, which leads to material failure. The largest micro-movements were observed at the lateral edge of the conical connection of the stem and neck, which is consistent with the location of the cracks. In the case of normal walking, the life of the damaged implant is long, but at higher loads (running, climbing stairs) the life of the implant is significantly shorter.
- The location of crack is on tensile side of the femoral component.

**Acknowledgement**

This research was supported by the grant agency KEGA Project no. 013ŽU-4/2019 and also Visegrad scholarship grant under the V4EAP PhD. program.
This work was also supported within the project Operational Program for Science and Research: Research and Development of Activities of the University of Žilina in Industry in the 21st Century in the Field of Materials and Nanotechnologies, no. 313011T426. The project is co-financed by the European Regional Development Fund.

**References**

[1] Zatkalíková V 2016 Analysis of fractured screw shaped Ti6Al4V dental implant (Materials Today: Proceedings vol 3) chapter 4 pp 1216-1219
[2] Pilkey W D 1994 Mechanical properties and testing of engineering materials (Formulas for Stress, Strain, and Structural Matrices)
[3] Findik F 2017 Titanium based biomaterials (Eng. Biosci vol 7) chapter 3 pp 1-3
[4] Basu B 2017 Biomaterials for musculoskeletal regeneration (Berlin: Springer) pp 262
[5] Baxter R M 2011 Do Tissues From THA Revision of Highly Crosslinked Liners Contain Wear Debris and Associated Inflammation? (Clin Orthop Relat Res) chapter 469, pp 2308–2317
[6] Hanusová P 2020 A study of hip joint replacement failure (Materials Today: Proceedings)
[7] Beznoska, s.r.o. (http://www.beznoska.cz/)
[8] Belan J 2019 Three-Point Bending Fatigue Test of TiAl6V4 Titanium Alloy at Room Temperature (Advances in Materials Science and Engineering)

# Analysis of equipment for biaxial material testing

**P Hassasroudsari[1], V Kubec[1, 2] and A Prantl[1]**

[1] COMTES FHT a.s., Prumyslova 995, 334 41 Dobrany, Czech Republic
[2] Corresponding Author's E-mail: vaclav.kubec@comtesfht.cz

**Abstract.** This paper aims to analyze the operation of the equipment for biaxial testing. The equipment's stiffness is one of the main factors influencing the magnitude of reaction forces in the device's guidance, consequently affecting the measurement accuracy. Based on intensive research on the topic of biaxial testing devices, the present paper deals with the analysis and optimization of the equipment's current design.

## 1. Introduction
Various mechanical examinations on different materials and specimens help us understand their mechanical properties or, in other words, their behavior under stress states and strain. It is well known that, in order to accurately characterize the failure properties of materials, to formulate related constitutive and evolutionary damage laws, it is necessary to perform an experimental analysis [1]. One of the most important actions is the biaxial test of butterfly specimens. Planar biaxial testing applies force in two directions, allowing an analysis of the anisotropic behaviors of materials. From this point of view, the issue of ductile failure and biaxial tests are applied to wide areas in mechanical engineering, including, the automotive, aerospace, as well as electrical industry. Therefore, in order to select the suitable material for an application, it is very important to perform a measurement correctly with minimum error or inaccuracy which makes the testing process more difficult than common material testing.
For biaxial loading, there are various measurement methods, distinguished mainly by the type of testing equipment, the shape of a sample, and measurement conditions. They can be generally divided into tensile tests with a special shape of the testing specimen or tests with a simpler specimen geometry but with a complex equipment design which makes the biaxial loading possible. Due to this fact testing equipment is one of the main keys to accomplishing successful biaxial testing of specimens. This paper focuses on specific testing equipment that allows biaxial testing on a butterfly shape sample.

## 2. The biaxial equipment
The biaxial testing equipment owned by COMTES FHT (Figure 1), was developed in the framework of the project identified as FR-TI2/279, between COMTES FHT, Skoda JS Company, and CVUT University. This Biaxial equipment was designed to apply planar biaxial loading to the specimen during the uniaxial loading of the equipment itself. The equipment (Figure 1) is designed for standard testing machine and it consist of a bottom and an upper welded frame. The welded frames are connected by standard bushings and columns. The bushing were lubricated according to the manufacturer's instructions. The grips are placed on a bottom and upper welded frame and allow the sample to rotate. The rotation is by 5 ° in the range of 0 ° to 90 °, this rotation makes biaxial loading of the specimen possible.

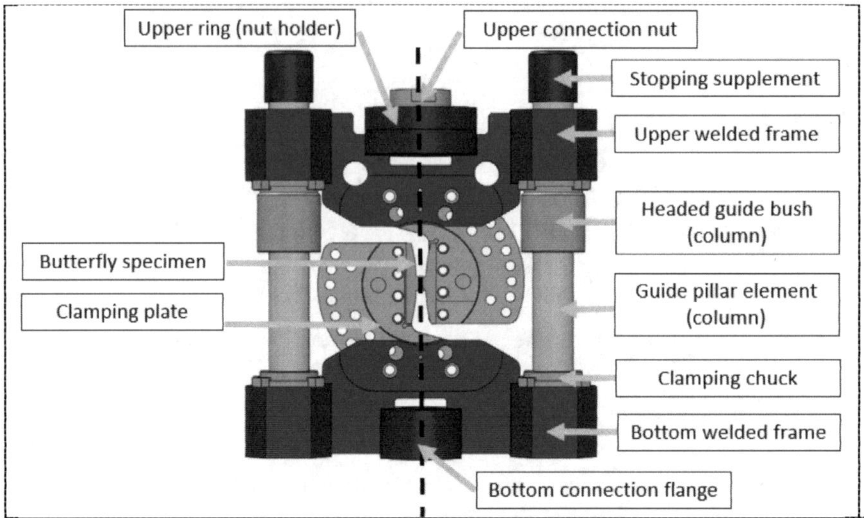


**Figure 1.** Biaxial equipment's basic design.

## 3. Equipment analysis

After several tests on this equipment and comparing the result with numerical simulation, it was found that the equipment has a negative effect on the testing process. The results were not reliable and the testing was not accurate. In this section, the main goal was to find the reason behind the device's inaccuracy and determine the amount of the device's deformation and the reaction force which applies during its movement. By analyzing the results, it will be possible to decide whether it is necessary to modify the equipment and which specific parts should be modified in case of necessity. To design the experiment the following subsections must be defined:

### *3.1. Equivalent specimen*

The main shape of the butterfly specimen is problematic from the production point of view. Different curvature in different directions and thickness changes in the center of the sample where the main stress concentration point is realized makes the production difficult. To overcome this problem, equivalent samples were designed with a similar shape to the main specimen, but in a shape that each one has different stiffness in the middle part (Figure 2). Five equivalent samples were designed, with similar main dimensions as thickness, width, and length. The tool steel 1.2842 was chosen as the material of all specimens.

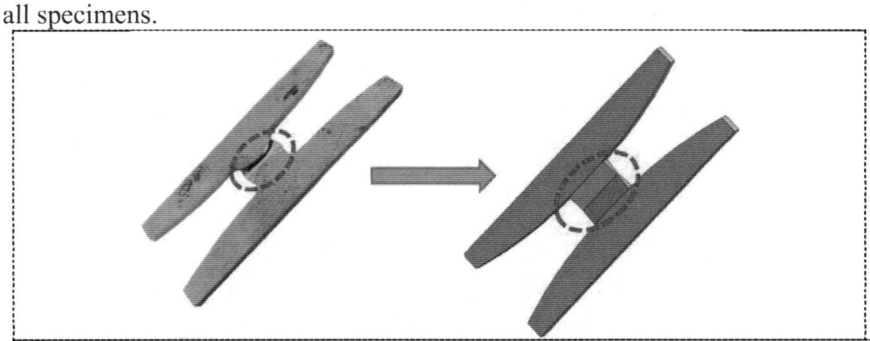

**Figure 2.** Design of equivalent specimen [1]

### *3.2. Loading condition*

In the second step for designing the experiment, the loading condition of the specimen during its testing must be defined. The loading condition was chosen, in which the sample causes the greatest reaction force in a radial direction to the guide pillars (columns) of the equipment. FEM analysis has been carried out based on this consideration. A simple FEM analysis was performed by loading a

sample that was loaded at different angles (corresponding to the device). Reaction forces in boundary conditions were monitored in the vertical (along the pillars axis) and transverse (radial to the pillar axis) directions.
Results of the simulation in Figure 3 verified the assumption that the greatest load value in the radial direction was for angles 45° and 60° and the angle 45° was used for following analysis. The higher angles (75°and 90°) are irrelevant based on sample shape and loading condition.

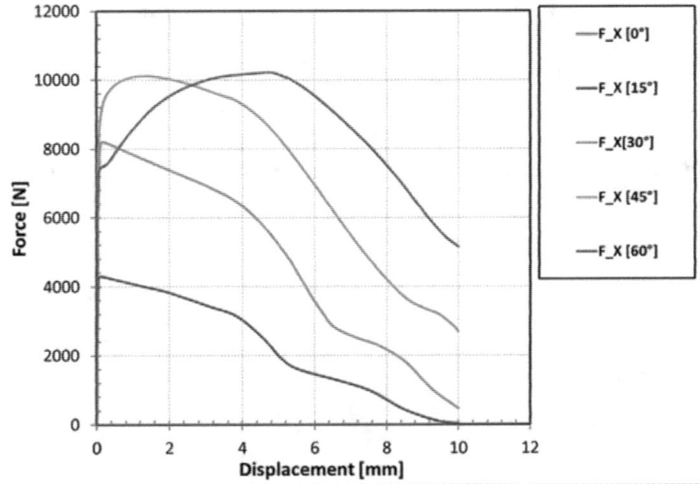


**Figure 3.** Relation of reaction force on displacement from FEM simulation [1]

*3.3. Measuring instruments*
The experiment was performed on an MTS 810 (Material Test System) machine. Basic operating data from the machine which are; the force required to move the working piston and movement of the working piston were recorded. These basic data were supplemented by other methods of measuring the deformation of the equipment. Using the 3D scanner the total deformation of the device was measured and the dial distance indicator measured the lateral deformations in the left side of the device.

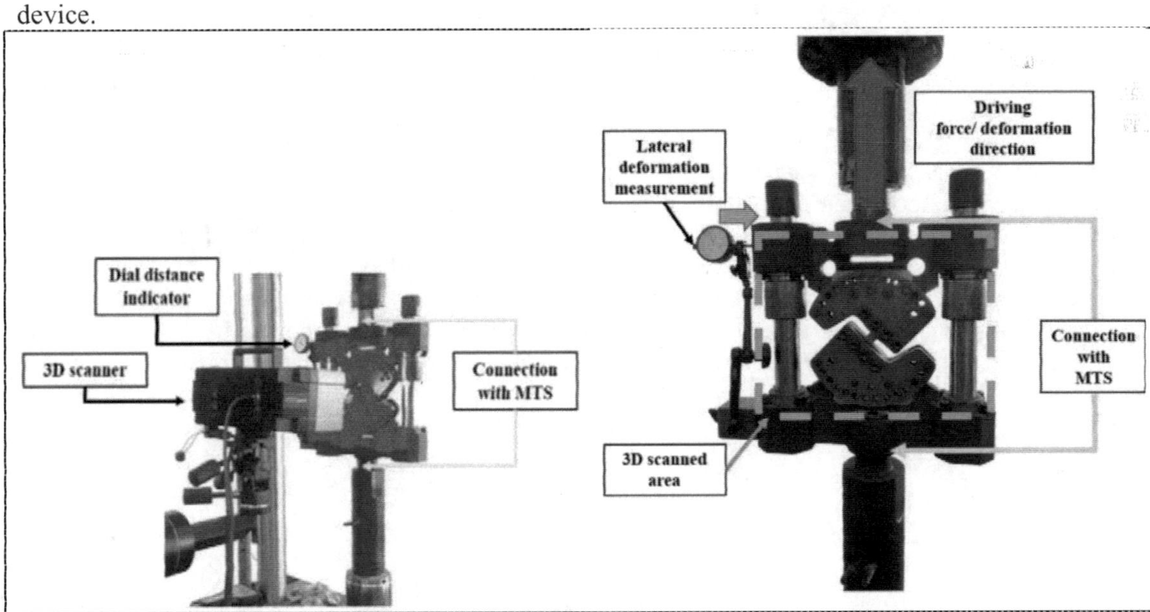


**Figure 4.** Mounted equipment on the MTS machine and measuring instruments [1]

## 4. Experiment results

The device's movement was driven by input force from the MTS machine. The amount of the input force was calculated so that the experiment will occur only in the elastic area. This strategy was chosen because of the linear behavior of materials in the elastic area. Due to the experimental mistakes, it was not possible to apply this strategy for all the specimens. Figure 4 (right) shows the method of the result analysis. The experiment was carried out in the two following main phases.

### *4.1. Hysteresis analysis (cyclic movement without specimen)*

These measurements were applied to analyze the behavior of the equipment in case of repeated measurements or loading. The results of the measurements are defined as the hysteresis curve.

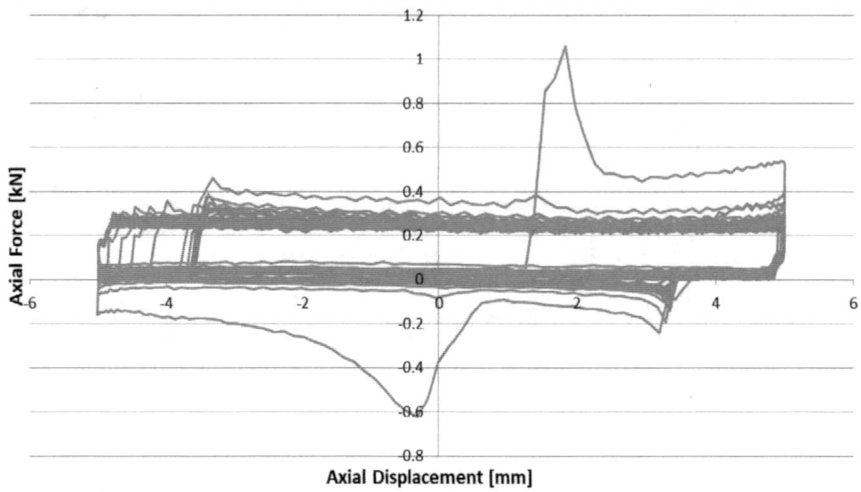


**Figure 5.** Cyclic movement results of the equipment [1]

From the graph in Figure 5, can be seen that at the beginning of the experiment, during quasi-static stretching, the equipment forces a resistive force of up to 1000N, then during compressing, its amount reduces to 600N, and finally, it can be seen that a constant reaction force stays around 300N. The main source of this force is the frictional resistance in the sliding because it is the only moving kinematic pair in the equipment.

### *4.2. With equivalent specimen's analysis*

During the device's movement, the lateral deformation of the fixture was measured under maximum force and stroke for each sample and it was measured by the dial gauge. The force-displacement dependency was measured from the MTS machine. The testing result from Sample V2 is chosen as a representative result among other samples. V2 was the only specimen that cracked during the experiment due to the large force that was applied to it. Sample V2, as shown in Figure 6, reached a maximum force of 3kN. According to this graph, it can be surely claimed that in the area marked with a red rectangle, the sample was cracked and therefore it stopped applying any force inside the equipment.

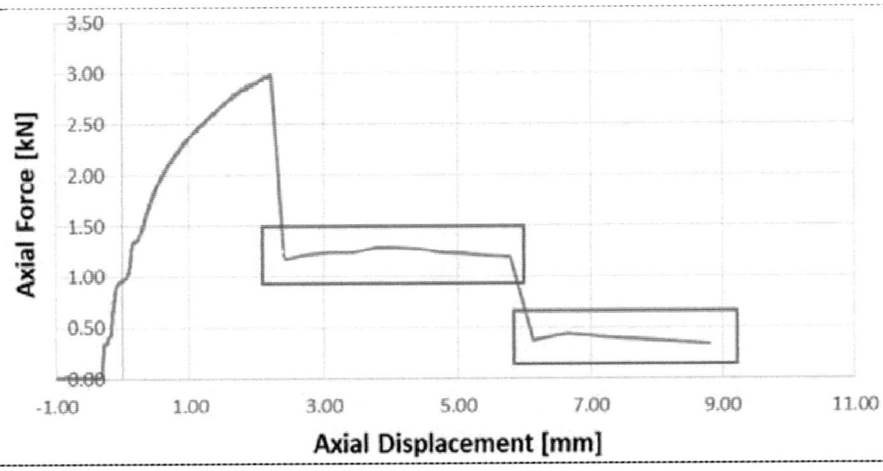


**Figure 6.** Complete experimental results of sample V2 [1]

For analyzing results from a 3D scanner, from the five samples, the data from the measurement of sample V1 was selected as a reference. The sample shape was designed for its maximum stiffness, so it is possible to measure the equipment deformation under loading (the sample has a constant thickness of 3 mm). The first series of scans took place before the loading began when the specimen was clamped to the fixture. The second series was scanned during the experiment when the maximum load was applied to the sample. Comparison of these two states was performed in GOM Inspect software, where Scans of both series were imported and then compared with each other. The overall displacement of the equipment could be evaluated from the comparison (Figure 7). The values obtained show that the equipment's deformations were relatively small and the device is sufficiently rigid. The only significant displacement was recorded for the parts in the direction of axial displacement, which corresponds to the function of the equipment.

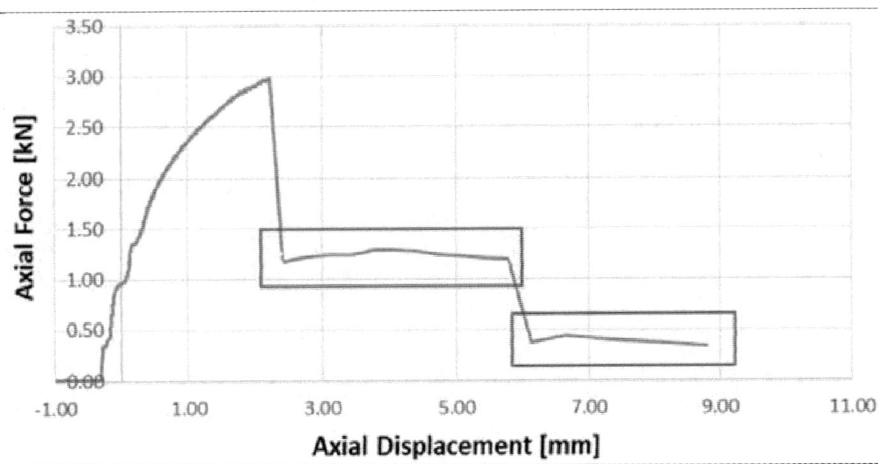


**Figure 7.** Deformation results based on the 3D scan – comparison unloaded and loaded state of equipment [1]

*4.3. Experiment discussion*

The results from hysteresis analysis (Figure 5) confirmed that the equipment applies 300N reaction force while cyclic movement without the specimen. Also results from form sample V2 show that the axial force recorded in the red marked region (Figure 6) is the clear resistive force of the equipment and this coincides with the behavior of the device during cyclic movement (Figure 5). Both values are around 300N. The source of this problem is the high value of friction in guide elements of the

IOP Conf. Series: Materials Science and Engineering 1178 (2021) 012017 doi:10.1088/1757-899X/1178/1/012017

equipment which impedes the movement of the headed guide bush. This value is in some cases 10% of the maximum force value and therefore the influence of this frictional resistance cannot be neglected. During the experiment, it was also found out that the equipment's movement in the loading direction is not stable. This cause that negatively affects the equipment is due to the instability of the upper nut (Figure 1) and a large clearance between this nut and the ring which holds the nut on the equipment. Based on this discussion it is necessary to modify the equipment's parts which caused negative effects during the experiment.

## 5. Solution Variants

According to considerations obtained in the previous section, it was attempted to suggest variations that improve the construction of this product. The main aim of these variants is to eliminate or reduce frictional forces in the guide elements of the equipment. One variant separately focuses on the upper nut of the equipment and its method of connection to the test devices.

Two following variants are proposed to minimize frictional forces in guide elements of the equipment. 1st variant focuses on changing the headed guide bush, and 2nd variant aims at changing the design of the entire fixture by removing the entire guide from the chain. As the second option reduces the rigidity of the equipment while the first option is easier to operate, the 1st variant is suggested as the sufficient one.

Also to remove the instability of the equipment, the ring modification is suggested to connect the equipment with the testing machine instead of using a connection nut. This modification requires producing a connection ring with the same dimensions and shape as the current ring but with a threaded hole in the middle of the ring. Drilling the threaded hole according to the linkage part that COMTES FHT usually uses for connecting to the testing machines, allows the linkage to be directly connected to the equipment. This direct connection with the upper frame of the equipment without using any additional nut connection could significantly stabilize the equipment's movement during testing and should not apply any additional reaction forces.

## 6. Conclusion

Biaxial testing equipment for butterfly-shaped specimen's fixture performs biaxial tests on various materials, mainly steel types. The special geometry of this sample is very useful for describing the characterization of the stress states of a material.

Significant friction value in the guide elements of the current equipment, prevents it to be fully functional and restrain measurement to be sufficiently accurate. This work aimed to find problem points based on the experiment and analysis of the current state, then to determine its value and verify the existence of unsuitable reaction forces that may arise during the test. Subsequently, variants were proposed for solving problematic nodes.

Based on an experiment conducted at COMTES FHT, it has been determined that the guidance of this device, which allows the clamping plates to move concerning each other, causes an undesirable reaction force. Another adverse effect was the instability of the upper nut of the device. This instability mainly affected the continuity and eventual jumps, which were recorded during measurement. By using a 3D scanner, it was verified that the rigidity of the entire existing fixture is sufficient and there is no need (contrary to the original assumptions) to change the main concept of design in this fixture.

## 7. References


[1] Hassasroudsari P 2020 *The equipment for biaxial examination* (bachelor thesis University of West Bohemia Pilsen)

[2] Růžička J 2015 *Metodika kalibrace nesvázaných modelů a stochastický přístup v problematice tvárného porušování* (disertation thesis Czech Technical University in Prague Prague)

[3] Hůlka J 2014 *Výpočtová predikce tvárného porušování* (disertation thesis Brno University of Technology Brno)

[4] Španiel M, Prantl A, Džugan J, Růžička J, Moravec M, and Kuželka J 2014 *Calibration of fracture locus in scope of uncoupled elastic–plastic-ductile fracture material models* Advances in Engineering Software **72** Amsterdam 95-108

# Relationship between 316L steel powder characteristics and residual stress in additive manufactured builds

**J Hodek[1] and J Vavřík[2] and M Brázda[3]**
COMTES FHT - Dobřany, Czech Republic

jhodek@comtesfht.cz

**Abstract**. This investigation involved cube samples created by an Additive Manufacturing (AM) process. Four commercially available AISI 316L powders with a particle size range of 50-150 μm were used to produce four cube-shapes samples using Direct Energy Deposition (DED) in the industrial metal 3D printer InssTek MX-600. The contour method (CM), a destructive stress relaxation method, was used to determine the residual stress distribution on the cross-section. The results illustrate the impact of the powders on residual stress distribution after DED additive process.

## 1 Introduction

Today, metal additive manufacturing (AM) is a widely-used production method for parts to be used in a number of industries, such as aviation, the automotive industry, space research, medicine and others. Using AM, one can produce special shapes and parts which were impossible or too expensive to manufacture until now. Most metal AM processes rely on purposeful melting of metal powder in a location where a new part is to be built. The part is produced incrementally. The nature of the process is very different from other processes, such as casting. The state of matter and the temperature of the input material vary with time and space. Residual stress [1] generated by this process depends on a number of factors. The question arises as to what impact the use of powders from different manufacturers may have on the resulting residual stress after AM processing with identical parameters. In professional literature, this topic is covered in [2] a [3] in regard to Selective Laser Melting. There is no information available in the literature on the impact of different powders on residual stress in DED. Residual stress may be studied using a number of methods. Residual stress in the interior of a part can be found using the contour method [4] and [5]. It is a destructive experimental-computational method which relies on superposition.

## 2 Experimental procedures

In this study, austenitic 316L stainless steel powders from four producers (Sandvik Osprey LTD, Höganäs, H. C. Starck, and Prax Air) were used as the feedstock. For all the powders, the size range was 53-150 μm, except the one from Prax Air, where it was 45-106 μm. The chemical composition of the powders is shown in Tab 1.

**Tab 1.** Chemical composition of powders and base plate (wt. %)

| | Fe | Cr | Ni | Mo | Mn | Si | C |
|---|---|---|---|---|---|---|---|
| Sandvik | Bal. | 17.2 | 10.4 | 2.3 | 1.3 | 0.8 | |
| Höganäs | Bal. | 16.9 | 12.8 | 2.6 | 1.6 | 0.7 | 0.014 |
| H. C. Starck | Bal. | 17.1 | 11.8 | 2.3 | 1.5 | 0.7 | |
| Prax Air | Bal. | 17.0 | 12.1 | 2.5 | 1.4 | 0.6 | |
| Base plate | Bal. | 16.24 | 10.49 | 2.14 | 1.12 | 0.44 | |

The experimental specimens from powders (Figure 1) were built in the DED system MX600 from InssTek. The specimen was a cube 35×35×35 mm in size built in the centre of a base plate with a size of 95×95×10 mm. Residual stresses in the base plates were relieved by annealing at 400°C for four hours. A schematic depiction of the specimen is shown in Figure 2, including the location of the cut.

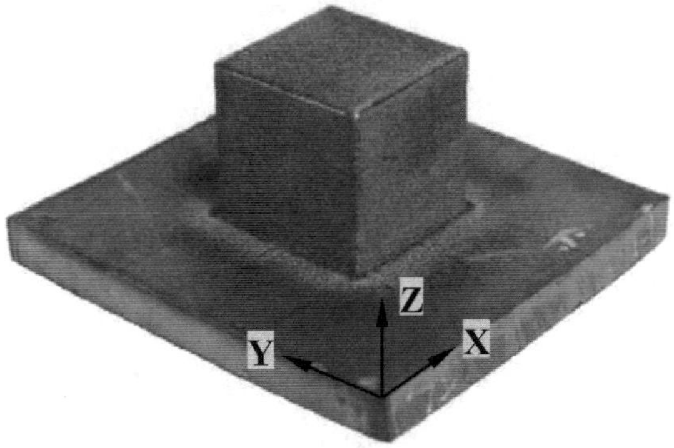


**Figure 1**. The build from Sandvik powder

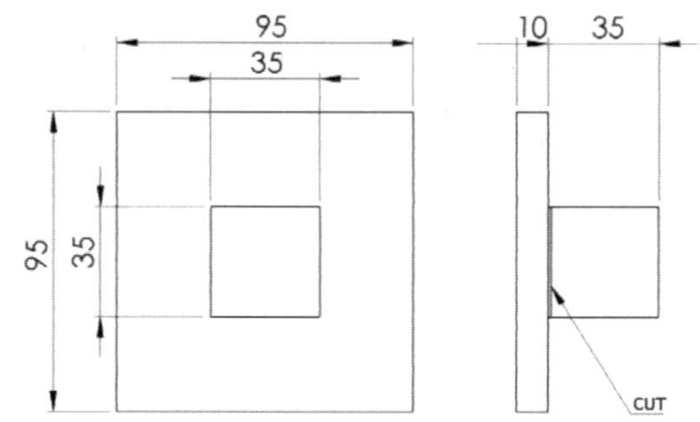


**Figure 2** Dimensions of the specimen and the base plate with the cut location indicated

The height and width of the weld track were 0.25 mm and 0.5 mm, respectively. The build comprised a total of 140 layers. The mean laser power was 400 W. The protective atmosphere consisted of 4.8-purity argon. The building process took 8 hours. The deposition strategy comprised the contour path and "zig–zag" interior and involved 90° rotation between layers. The scanning strategy is indicated schematically in Figure 3 .

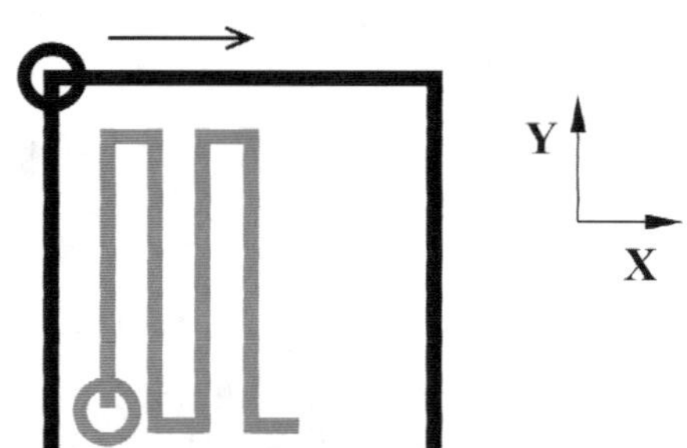


**Figure 3** Specimen deposition strategy (contour – shown in black with the initial position indicated; zig-zag interior – in blue with the initial position indicated)

The finished built-up specimens were cut off by wire EDM (wire diameter: 0.25 mm) 1 mm above the base plate (see Figure 4) and the coordinates of both surfaces were measured with Optical Precision Measuring Machine (OPMM) ATHOS Capsule made by GOM.

**Figure 4** A specimen prepared for OPMM measurement

The coordinates were processed with a user script in Octave 4.4.1. environment. Smoothing, averaging and inversion were applied. The results were input as boundary conditions into the FE model. Elastic analysis of both counterparts yields the magnitude of normal stress on the plane of cut before actual cutting. The model was developed in MSC MARC software as a linear elastic analysis problem, using Young's modulus E=193 GPa and Poisson's ratio ν=0.299. The mesh was built with eight-node isoparametric type 7 elements with a size of 1 mm. Calculated normal stress values (along Z-axis) on the cut plane for Höganäs powder are shown in Figure 5 and Figure 6.

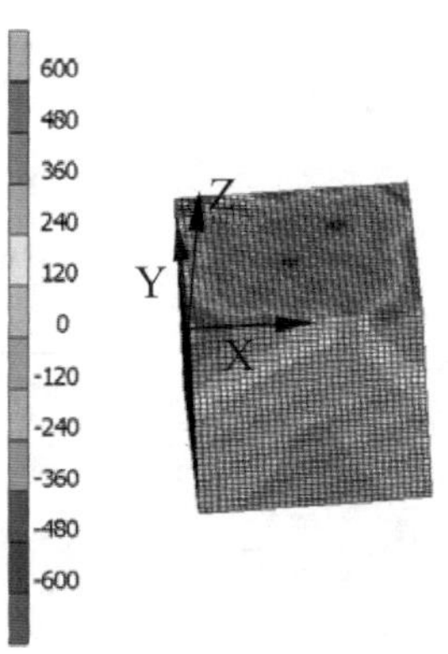


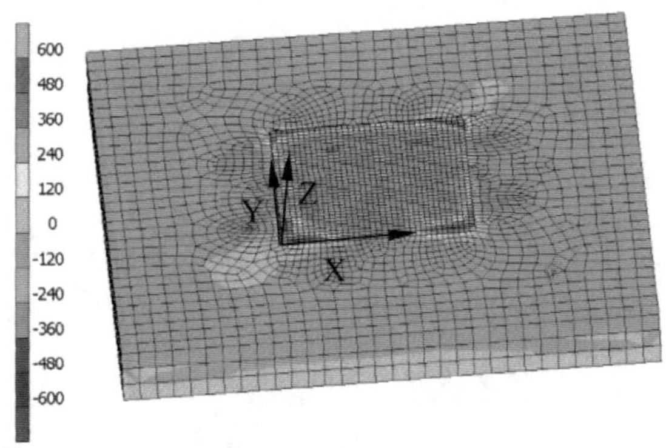


**Figure 5** Stress in the build along Z-axis [MPa] **Figure 6** Stress in the base plate along Z-axis [MPa]

The resulting normal stress on the cut plane is calculated as an average of the stresses in the counterparts reflected in their different shapes and sizes [6].

## 3 Results

The calculated mean stresses were determined for X and Y directions on the cut plane for all the builds. The position and orientation of X and Y lines are indicated in Figure 7.

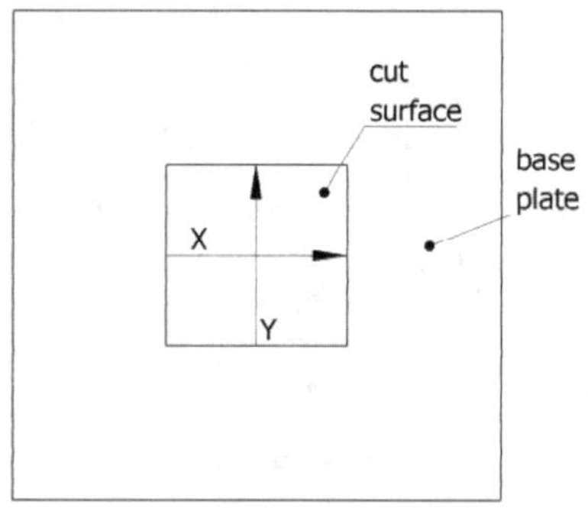


**Figure 7** Position of X and Y lines on the section plane

Stress levels along line X Figure 8 and line Y are plotted in Figure 9. In the centres of all the builds, the stress is compressive. Closer to the surface, the stress is tensile. The stress profile along line X has a symmetric U shape in all the specimens. By contrast, stress levels of 400-600 MPa at the end of line Y are higher than at its start. This is probably due to wire EDM cutting since the specimen was sectioned along line Y. The waviness of the curves is curious but its cause was not found. It is not clear whether it was the effect of a particular powder or another cause. The chemical compositions and particle size distribution of the powders were similar, as revealed by SEM analysis.

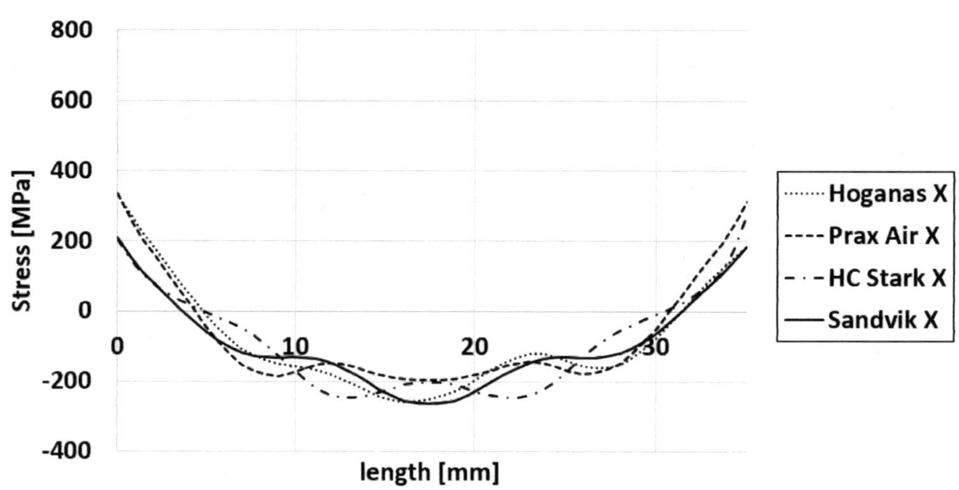


**Figure 8** Residual stress in the direction of Z axis measured along line X

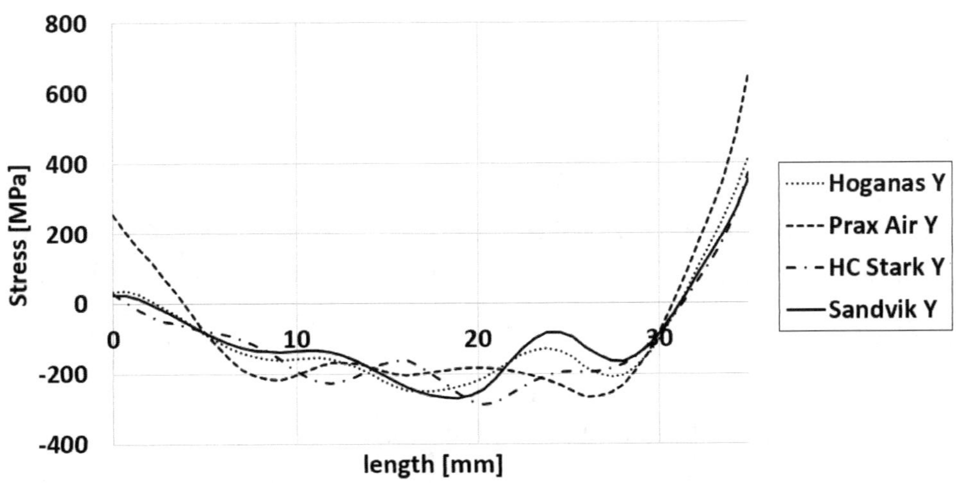


**Figure 9** Residual stress in the direction of Z axis measured along line Y

## 4 Conclusions

This work explores the relationship between the powder and residual stress in DED builds. Four builds were made from powders from different manufacturers. Residual stress in the specimens was determined along two lines on their cross-section using the contour method. The stresses in the specimens were of the same nature. In the specimen interior, there were compressive stresses which changed to tensile stresses closer to the surface. The values suggest a certain contribution from wire EDM cutting. The waviness of the stress profiles is worth attention but its causes were not identified. The authors will continue to explore this theme in the future.

**Acknowledgements**

The paper was supported from ERDF Pre-Application Research of Functionally Graduated Materials by Additive Technologies, No. CZ.02.1.01/0.0/0.0/17_048/0007350.

## References

[1] C. Li, Z. Y. Liu, X. Y. Fang, and Y. B. Guo, "Residual Stress in Metal Additive Manufacturing," in *Procedia CIRP*, 2018, vol. 71, doi: 10.1016/j.procir.2018.05.039.

[2] D. Boisselier and S. Sankaré, "Influence of Powder Characteristics in Laser Direct Metal Deposition of SS316L for Metallic Parts Manufacturing," *Phys. Procedia*, vol. 39, pp. 455–463, 2012, doi: https://doi.org/10.1016/j.phpro.2012.10.061.

[3] C. Pleass and S. Jothi, "Influence of powder characteristics and additive manufacturing process parameters on the microstructure and mechanical behaviour of Inconel 625 fabricated by Selective Laser Melting," *Addit. Manuf.*, 2018, doi: 10.1016/j.addma.2018.09.023.

[4] J. Hodek, A. Prantl, J. Džugan, and P. Strunz, "Determination of Directional Residual Stresses by the Contour Method," *Metals (Basel).*, vol. 9, no. 10, p. 1104, Oct. 2019, doi: 10.3390/met9101104.

[5] J. Vavřík and J. Hodek, "Residual stress investigation in additively manufactured samples," in *IOP Conference Series: Materials Science and Engineering*, 2020, doi: 10.1088/1757-899X/723/1/012032.

[6] A. Achouri, P. Bouchard, S. Kabra, and F. Hosseinzadeh, "Asymmetric Cuts In The Contour Method For Residual Stress Measurement," 2017.

IOP Conf. Series: Materials Science and Engineering 1178 (2021) 012019 doi:10.1088/1757-899X/1178/1/012019

# Analysis of the relationship between changes in mechanical properties and precipitation in SUPER 304H steel during its exposure

**J Horváth[1,2], M Horváthová[1], L Pilsová[1], V Smola[1], E Chvostová[3], J Janovec[1]**

[1] ČVUT FS, Ústav materiálového inženýrství, Karlovo nám. 13, 121 35 Praha 2

[2] UJP PRAHA a.s., Nad Kamínkou 1345, 156 10 Praha – Zbraslav

[3] COMTES FHT a.s. Průmyslová 995, 334 41 Dobřany

E-mail: jakub.horvath@fs.cvut.cz, horvathj@ujp.cz

**Abstract**. Aim of the article is to provide properties comparison between initial state and two thermally exposed states of creep resistant steel SUPER 304H by usage of conventional and miniaturised tensile test specimens. Tests were performed at room temperature. Thermally exposed material was aged by 15 and 30 thousand hours at 675°C. This exposition which simulated accelerated operation exposition caused precipitation of intermetallic phases which affect mechanical properties like ultimate strength and elongation. The performed tests showed a decrease in elongation and ultimate strength between the first exposed state and initial state followed by partial increase of both tested values for the second exposed state. Sigma phase precipitation was confirmed for the exposed samples. The presence of the sigma phase also led to a change in the fracture morphology from full ductile (initial state) to combined ductile with the presence of brittle particle cleavage (exposed states). A comparison of the tests performed on classical and miniaturized test samples proved the possibilities of using miniaturized samples both for the basic material and for the exposed ones.

## 1 Introduction

One of the main efforts of the world's coal energy sector is to improve its efficiency. A possible way to reach this goal, which also means a reduction in the ecological burden of nature, is to increase the working parameters of steam. However, together with this increase, the application of appropriate materials is necessary. SUPER 304H steel is one of the suitable steels for heat exchange surfaces of supercritical blocks [1].

SUPER 304H steel, as shown for example in articles [2-4], shows significant microstructural changes during its temperature exposure, which affect its mechanical properties. It affects the ultimate strength and yield strength limits and the decrease in absorbed energy values. Current methods for testing mechanical properties using a minimum amount of material are increasingly being developed, where these changes could lead to skewed results. This paper evaluates mechanical and microstructural parameters of thermally exposed steel SUPER 304H by performed tensile tests at conventional and miniaturized tensile samples in combination with microscopy evaluations.

IOP Conf. Series: Materials Science and Engineering 1178 (2021) 012019 doi:10.1088/1757-899X/1178/1/012019

## 2 Methodology

The methodology chapter give to possibility of tests and examination repetition. Methodology is divided into two separated subchapters based on laboratory testing methods.

The experimental material used for evaluation in this paper is austenitic creep-resistant complex alloy steel SUPER 304H. The used states of this steel were the initial state and then the temperature-exposed states at a temperature of 675 °C for 15 and 30 thousand hours in laboratory furnaces.

### *2.1 Tensile testing and hardness measurement*

Two types of test specimens were used for testing, namely classic short tensile specimens of cylindrical cross-section with a diameter of 5 mm clamped in threaded heads and miniaturized tensile specimens of rectangular cross-section. Drawing of miniaturized specimen is evident in the Figure. 1. The miniaturized specimens were clamped pneumatically over flat jaws. In both cases, a constant crosshead feed rate of 0.5 mm / min was used for testing. The measured length was 4 mm for miniaturized tensile samples and 25 mm for conventional tensile samples. At least three samples were measured for each measured group of samples.

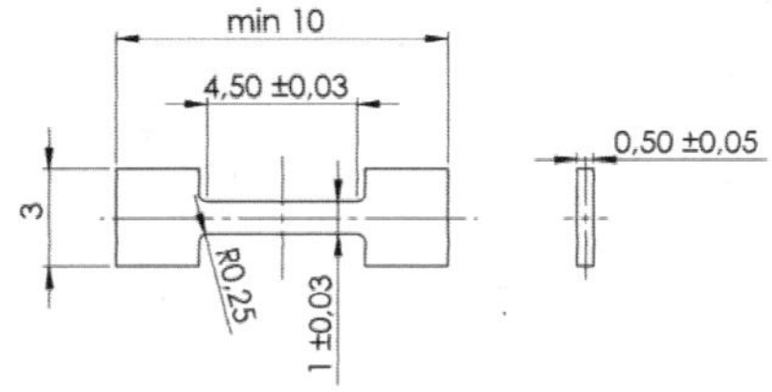


**Figure 1.** Dimension of used miniaturized tensile test sample.

Hardness maps were measured on samples prepared for microstructure observation. The load HV1 was used for conventional tensile samples and HV0.1 for miniaturized specimens due to their thickness. In all cases, three rows were measured with the smallest possible step from the fracture surface to the clamping head of the tensile specimen.

### *2.2 Microstructure analysis*

Light and electron microscopy was used for microstructural evaluation. Samples for microstructural evaluation were hot-pressed and subsequently grinded and polished. Two methods of etching were used for observation, namely selective color etching electrolytically in 10% NaOH solution (3V DC, 6s) and etching of a surface mixture of acids called glyceregia [5]

Optical microscopy was focused on the evaluation of the microstructure in the area of the final refraction of tensile samples with a focus on precipitates such as sigma phase. Furthermore, the development of fracture with respect to particles and grain of the material was evaluated.

Scanning electron microscopy was focused on the observation of fracture surfaces on tensile samples in combination with the detection of backscattered electrons together with the performance of chemical composition analyzes using energy dispersive spectroscopy (EDS). The images used for this paper are a combination of the observed fracture surface and the chemical composition map, where chrome is shown in green, iron in red and niobium in darker blue.

## 3 Measurement and observations results

### *3.1 Tensile testing*

Tensile tests were performed in order to compare the behavior of the material in tests of both conventional and miniaturized tensile samples. A comparison of representative tensile diagrams for each group of tested samples is shown in Figure 2, where MTT denotes miniaturized tensile samples and TT conventional tensile samples. Marking 0 corresponds to the initial material, 1 to the exposed state for 15,000 hours and marking 2 to the exposed state for 30,000 hours.

IOP Conf. Series: Materials Science and Engineering 1178 (2021) 012019 doi:10.1088/1757-899X/1178/1/012019

From the comparison of the measurements, it is evident that with the increasing thermal exposure of the material, its ductility decreases, which is in accordance with the ongoing precipitation processes inside the material. For miniaturized samples, the ductility decreases from 52.4% for the initial state over 41.3% to 38.7% for the exposed state after 30,000 hours. A similar trend can be observed for conventional tensile samples where the ductility decreases from 54.5% for the initial state to 32.6% for the material exposed for 30,000 hours. Furthermore, it is possible to observe a decrease for the ultimate strength from the initial state for the exposed state after 15,000 hours which is followed by a gradual increase for the exposed material after 30,000 hours. For miniaturized samples, this change is from an initial value of 615 MPa through 599 MPa to a final value of 607 MPa. In the case of conventional tensile samples, the trend is the same, from 761 MPa through 707 MPa to the value of 736 MPa. The overall trend of changes documented by means of conventional and miniaturized specimens is in both cases the same with an offset of approximately 130 MPa.

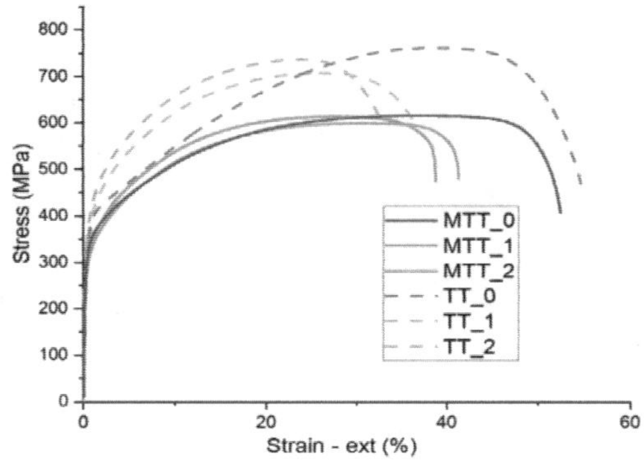


**Figure 2.** Comparison of SUPER 304H TT and MTT tensile diagrams for base and exposed state.

The performed measurements of hardness maps, both for classical samples and for miniaturized samples, are summarized in Figures 3a to 3d. In all cases, it is demonstrable that the deformation of the specimen body and thus its strengthening occurs evenly over the test part. Furthermore, it can be stated that the reinforcement of the material is corresponding both for the initial state between the classical sample (from 375 to 186 HV1) and miniaturized samples (from 396 to 169 HV0.1). The exposed material after 15,000 hours also shows the same hardening manifested on the hardness which has a measured range for conventional samples from 417 to 202 HV1 and for miniaturized samples from 416 to 197 HV0.1.

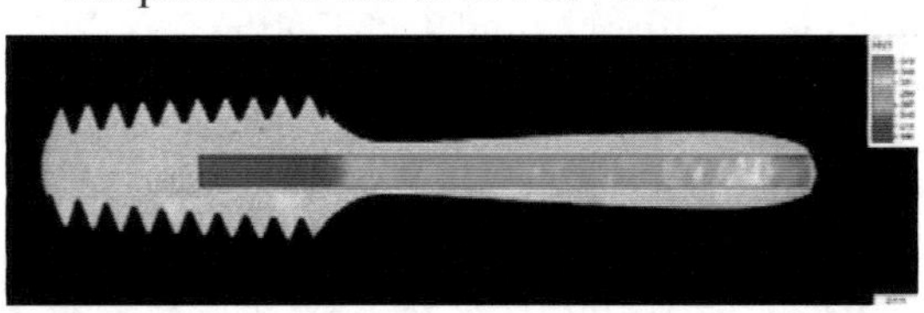

**Figure 3a.** Hardness measurement map of base material, conventional tensile test sample.

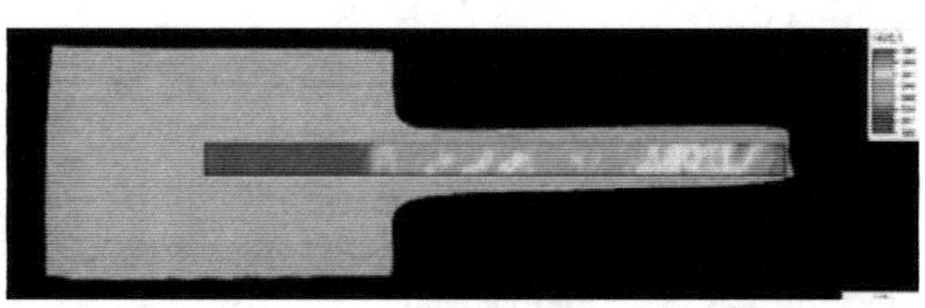

**Figure 3b.** Microhardness measurement map of base material, miniaturised tensile test sample.

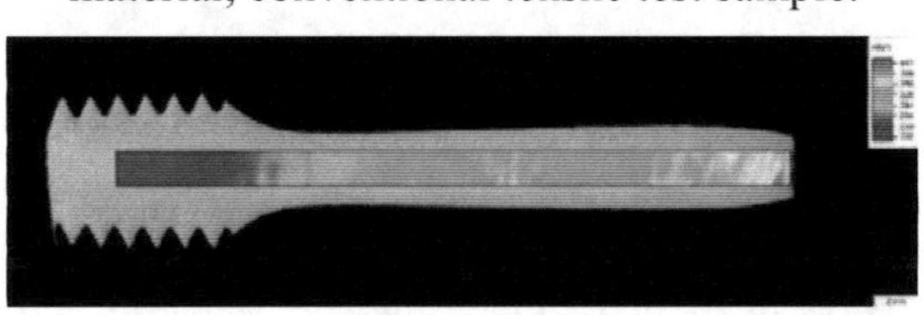

**Figure 3c.** Hardness measurement map of exposed material, conventional tensile test sample.

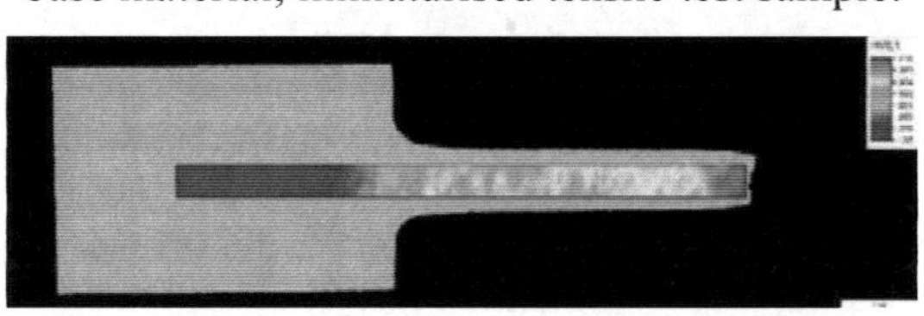

**Figure 3d.** Microhardness measurement map of exposed material, miniaturised tensile test sample.

### *3.2 Microstructure analysis*

To verify the presence of the sigma phase on the fracture surfaces, the chemical composition was measured by point analysis using EDS. The marked sites of analysis were selected based on the assumption that the sigma phase is a brittle intermetallic phase [6]. The measurement sites are shown in Figure 4 and the results of the weight percentages are given in Table 1. The measured chemical composition corresponds to the sigma phase composition in this steel type and the results given in [2,3,4,6].

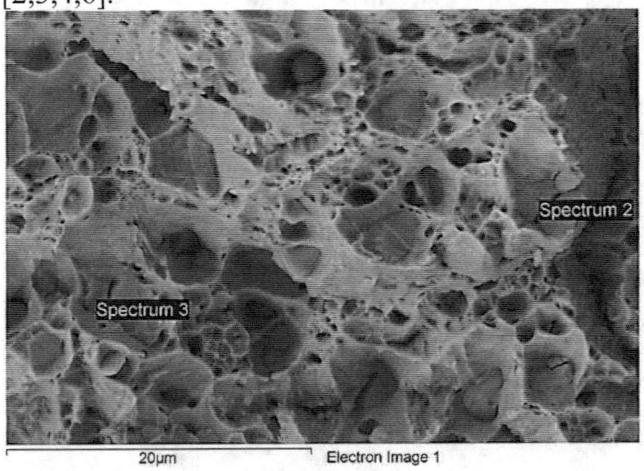


**Figure 4.** EDS mesasurement position for sigma phase confirmation.

**Table 1.** Measured chemical composition (wt. %).

| | Spectrum 2 | Spectrum 3 |
|---|---|---|
| Si | 0.57 | - |
| P | 0.57 | - |
| Cr | 37.61 | 39.38 |
| Fe | 55.92 | 53.48 |
| Ni | 3.69 | 1.89 |
| Cu | - | 5.25 |
| Mo | 1.64 | - |

The following two figures (Figures 5a and 5b) document the fracture surfaces prepared as a metallographic sample using subsequent surface etching and color etching to highlight the sigma phase. In both cases, grain deformation and the presence of the sigma phase in the microstructure are evident. For both types of samples (exposed for 15,000 hours), a gradual development of failure is observable, combined with the predominant transcrystalline propagation of the fracture using weakened sigma phase sites.

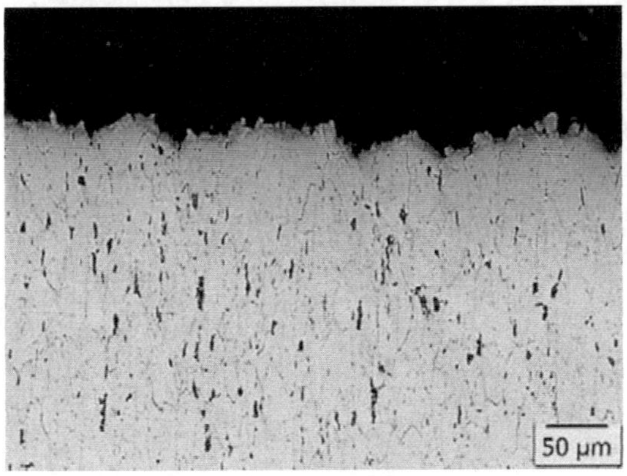


**Figure 5a.** Microstructure of fracture surface, conventinal tensile sample, expose state.

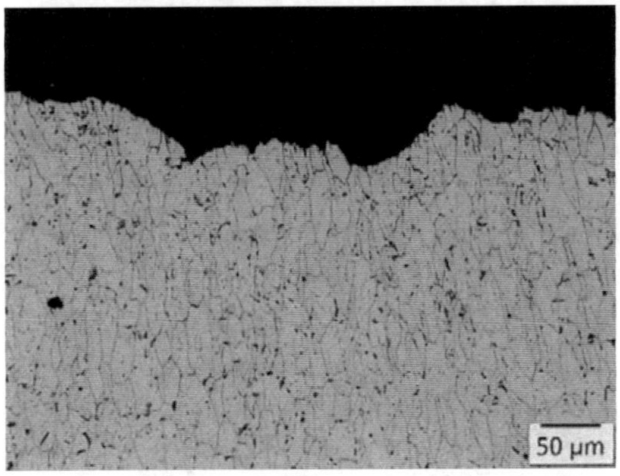


**Figure 5b.** Microstructure of fracture surface, miniaturized tensile sample, expose state.

The following figures (Figures 6a to 6d) document the fracture areas together with the chemical composition maps. Chromium-rich areas are shown in green. Iron-rich areas are shown in red. Niobium-rich areas are shown in dark blue. The areas shown only in grayscale of the original scanning electron microscopy image are places where, due to the fragmentation of the fracture surface, there was a shadow of secondary electron detection for EDS analysis.

The fracture morphology for the initial state without exposure is shown in Figure 6a for the conventional sample and in Figure 6b the miniaturized sample. In both cases, the fracture surface can be assessed as fully ductile with a pit morphology. The images also show the presence of niobium-rich regions, which represent niobium carbonitrides.

The fracture surfaces of the conventional sample (Figure 6c) and the miniaturized sample (Figure 6d) for samples after temperature exposure for 15,000 hours also show the same morphology. In both cases, it is possible to observe a combination of ductile fracture of the material and brittle cracking of particles on the fracture surface, which have been identified as sigma phases. The sigma phase is highlighted in these two images due to the higher chromium content compared to the steel matrix. Furthermore, as with samples without temperature exposure, it is also possible to observe niobium-rich areas highlighted in dark blue.

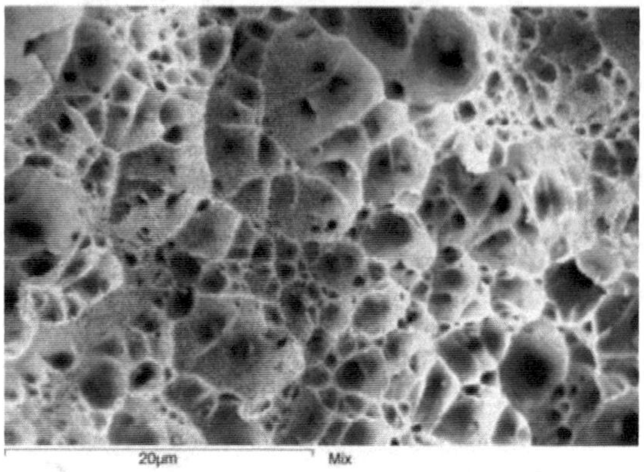


**Figure 6a.** Fracture surface of conventional tensile test sample of base material combinated with EDS chemical maps.

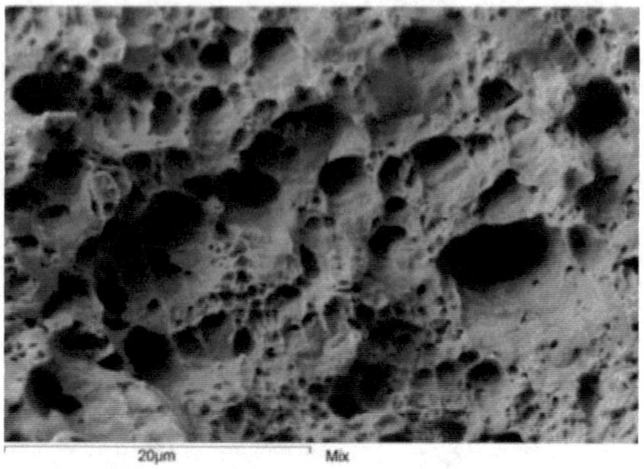


**Figure 6b.** Fracture surface of miniaturized tensile test sample of base material combinated with EDS chemical maps.

**Figure 6c.** Fracture surface of conventional tensile test sample of exposed material combinated with EDS chemical maps.

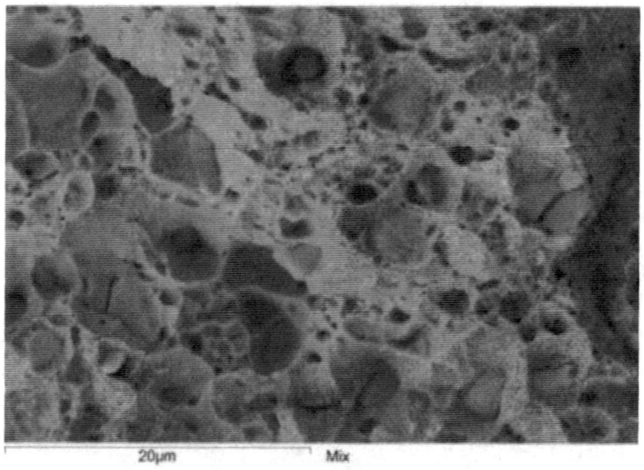


**Figure 6d.** Fracture surface of miniaturized tensile test sample of exposed material combinated with EDS chemical maps.

## 4 Discussion

Laboratory exposition at a constant temperature of 675 °C for 15 and 30 thousand hours was used to accelerate the achievement of microstructural changes that take place in SUPER 304H steel during its operational exposure. Microstructural changes are described, for example, in [2-4]. In general, it can be stated that chromium carbides precipitate and subsequently the brittle sigma intermetallic phase during thermal exposition. Confirmation of these microstructural changes was performed by chemical composition EDS analysis of exposed material (Figure 4 and Table 1).

The results from the evaluation of the initial and exposed material showed a very good conformity from the point of view of the morphology of the fracture surface. For the origin material, the ductile pit morphology of fractures on tensile samples was observed for both conventional and miniaturized samples. The exposed samples also showed the same fracture morphology by a combination of ductile disruption and cleavage facets through the sigma phase. From the point of view of measuring hardness maps, the conformity between conventional and miniaturized samples for the basic and exposed state was also proved. From the point of view of the materials parameters inside the samples, a methodical conformity is proved, which confirms the possibility of using miniaturized samples for tensile tests of the basic and exposed state of SUPER 304H steel.

The trend of changes in mechanical properties is also similar for both conventional and miniaturized samples. The only documented difference was in the direct measured values obtained from the tensile tests. In the measured values summarized in Chapter 3.1, the miniaturized specimens show lower overall values, both in terms of yield strength and elongation. The reason for this difference does not stem from differences in the microstructure, at most the results could be influenced by the number of grains with respect to the fracture surface. The next possible reason can be in the difference of initial (engineering) strain rate for TT and MTT samples.

## 4 Conclusions

The main goal of the performed analyzes was to compare the difference in testing results using conventional and miniaturized tensile samples. It was a matter of creating data on whether it is possible to apply test methods to superheater tubes made of SUPER 304H not only in the initial state but also in the degraded or operated material, which will involve microstructural changes.

The performed tests showed a decrease in elongation and ultimate strength during first half of exposition followed by partial increase of both tested values for the second half of exposition. The presence of the sigma phase led to a change in the fracture morphology. In general documented changes are same in both case conventional and miniaturized samples. These assumptions give the possibilities of using miniaturized samples both for the initial material and for the exposed ones. Future experimental work will deal with comparison of tensile samples with same strain rate because of the initial strain rate for MTT samples with gauge length 4 mm is different in comparison with TT samples with gauge length 25 mm.

**Acknowledgement**

The authors acknowledge support from the Ministry of industry and trade project No. FV40166. and from the Grant Agency of the Czech Technical University in Prague within the grant no. SGS 19/162/OHK2/3T/12

**References**

[1] Abe F, Kern T U and Viswanathan R 2008 Creep-resistant steels. (England: Woodhead Publishing Limited and CRC Press LLC) ISBN 978-1-84569-178-3.

[2] Horváth J, Král P, Janovec J and Sklenička V 2015 The effect of sigma–phase formation on long-term durability of SUPER 304H steel *24th International conference on Metallurgy and materials-conference proceedings* (Ostrava: Tanger) pp 505–10

[3] Horváth J, Janovec J and Král P 2016 The Effect of sigma -Phase Formation on Long-Term Durability of Welding Joints in SUPER 304H Steels *Acta Physica Polonica A* (Warszawa: Polish Academy of Sciences) pp 960–2

[4] Horváth J, Janovec J and Junek M 2017 The Changes in Mechanical Properties of Austenitic Creep Resistant Steels SUPER 304H and HR3C Caused by Medium-Term Isothermal Ageing *Solid State Phenomena* 258 (Zurich: Trans Tech Publications Inc.)

[5] ASTM E3-11(2017), Standard Guide for Preparation of Metallographic Specimens (West Conshohocken: ASTM International)

[6] Hsieh C C and Weite W 2012 ISRN Metallurgy Overview of Intermetallic Sigma (σ) Phase Precipitation in Stainless Steels. p.16

IOP Conf. Series: Materials Science and Engineering 1178 (2021) 012020 doi:10.1088/1757-899X/1178/1/012020

# Properties of protective NiCr coatings, deposited by Cold Spray technology

**Š Houdková[1], P Šulcová[1], A. Keslová[1], K Lencová[1]and J Duliškovič[1]**

[1] Výzkumný a zkušební ústav Plzeň, s.r.o., Tylova 46, Plzeň, 301 00, Czech Rep.

houdkova@vzuplzen.cz

**Abstract**. The NiCr coatings, deposited by modern Cold Spray technology, and by well-established Twin Wire Arc Spraying and High Velocity Oxygen Fuel technologies, was studied in terms of microstructure, mechanical, tribological and corrosion properties. Compared to other thermal spraying technologies, CS coatings achieve higher cohesive strength and associated tribological properties. The dense structure is responsible for the coatings´ high corrosion protection ability. The influence of CS technological parameters was found negligible, except the usage of nozzle for internal diameter spraying, which results to significant increase of coatings porosity.

## 1 Introduction

The Ni-based alloys are used in the most demanding environments to beat the corrosion, high temperature oxidation, wear and their mutual combinations namely in power industry [1,4-7]. Due to their high costs, they are often applied as coatings, providing the superior surface protection to the coated components made from less noble materials. Various technologies, such laser [2] or Plasma Transferred Arc (PTA) [3] overwelding or thermal spraying [4-6] are utilized for deposition of Ni-based protective coatings. The final properties of coatings then varied in dependence on specifics of used deposition technology and also in dependence on the chemical composition of specific alloy.

From the group of Ni-based alloy, the Ni-20%Cr alloy is one of the possible candidates for protective coating. It has been designed to withstand oxidation and corrosion to temperatures up to 900 °C. The HVOF sprayed Ni20Cr coating indeed provide protection for steel pipes at high temperatures [4], as well as excels with high protection again solid particle erosion [5].

The Cold Gas Dynamic Spraying (CGDS) or simply Cold Spraying (CS) belongs to the family of thermal spraying technologies. From them, it differs by eliminating the high temperature for melting of deposited particles during deposition. Instead of temperature, it uses supersonic velocities of particles stream. After impacting the substrate, the kinetic energy of the impact transform into the heat, and cause not only the plastic deformation of the particles, but is also responsible for creation of local microwelds on the boundaries. The principle of CS is described in detail in [8].

Although the portfolio of applied materials is growing rapidly, thanks to the principle of application, it is still used primarily for materials with a good ability to deform plastically, such as copper, aluminium, and its alloys.

The Ni-based alloys can be also successfully deposited by the Cold Spray technology [7, 9,10]. The low spraying temperature can help to avoid the occurrence of oxides inclusions, commonly present in thermally sprayed coatings, deposited by Twin Wire Arc Spraying or Atmospheric Plasma Spraying. Thanks to the high impact energy, the amount of porosity can be also significantly decreased.

Although some information regarding the properties of cold sprayed NiCr coatings are available, the cold sprayed coatings application to the high-valuable components of power or aircraft industry should be preceded by further research. The more detailed study is needed to reveal the influence of various factors, that can take part during cold spraying.

Presented study focusses on the influence of basic technological parameters on the microstructure and selected mechanical and tribological properties. While the process parameters were kept constant, the size of the NiCr particles, deposition angle and distance were varied to simulate the non-ideal deposition condition. Moreover, the use of nozzle for deposition of coating in the inner diameter of pipes (ID nozzle) were tested and the resulted coatings microstructure is presented.

## 2 Experimental

### *2.1 Coatings deposition*

The cold sprayed coatings were prepared in Impact Innovations company in Germany, from Sandwik Ospray NiCr powder (-32 +10 and -25+5 μm mesh). The surface of carbon steel substrates (200 × 100 × 5 mm) was grit blasted by $Al_2O_3$ using Impact Innovation Cold Spray system. The previously optimized spraying parameters and nitrogen as a carrier gas were used to deposit the coating with the final thickness around 1 mm. After the deposition, the coated substrate was cut to obtain testing samples with required dimensions. The denomination of samples with respect to the variable deposition parameters are summarized in Table 1.

**Table 1.** Denomination of Cold Sprayed samples

| | NiCr powder mesh | Angle of deposition | Deposition distance | Type of the nozzle |
|---|---|---|---|---|
| **CS1** | -32+10 μm | 90° | 30 mm | Impact Gun 5/11 |
| **CS2** | -25+5 μm | 90° | 30 mm | Impact Gun 5/11 |
| **CS45** | -32+10 μm | 45° | 30 mm | Impact Gun 5/11 |
| **CSDS** | -32+10 μm | 90° | 15 mm | Impact Gun 5/11 |
| **CSDL** | -32+10 μm | 90° | 45 mm | Impact Gun 5/11 |
| **CSID** | -32+10 μm | 90° | 5 mm | Impact ID Gun |

For comparison, the properties of NiCr coatings deposited by TWAS and HVOF technologies were subjected to the similar experimental procedure. The HVOF and TWAS sprayed samples were produced in Výzkumný a zkušební ústav Plzeň s.r.o. (The Research and Testing Institute in Plzen, Ltd.), using Oerlicon Metco Smart Arc and Praxair HP/HVOF TAFA JP5000 spraying guns and Metco 8450 wire and Metco AMDRY 4535 powder. The 600 μm and 350 μm final coatings thicknesses were reached for TWAS and HVOF coatings, resp.

### *2.1 Coatings characterisation*

The microstructures were evaluated on the coating's cross sections, prepared by the standard metallographic procedure. Both the optical (OM) and scanning electron (SEM) microscopy was used for microstructure evaluation. The microhardness was measured on the coatings polished cross-sections using the HV0.3 method. For each coating, at least 7 indents were done, and the average value is reported. The adhesive – cohesive strength of the coatings was evaluated by tensile test in accordance to ASTM C-633. For each coating, 4 tests were provided, and the average values are reported.

The wear resistance against three-body abrasion and solid particle erosion were measured using the Dry Sand/Rubber Wheel test in accordance to ASTM G-65 and centrifugal erosion test [11]. The impact angle of erosive media ($Al_2O_3$; F70) varied between 15°-90°. The coatings volume loss was determined from mass loss using the values of density, previously determined by Archimedes method. For each coating at least 3 independent measurements were done, and the average value is reported. The worn surfaces were subsequently observed by SEM.

The protective properties of the coatings wear tested by salt fog corrosion test in accordance to ČSN EN ISO 9227 and also in high temperature aggressive environment of $Na_2SO_4Fe_2(SO_4)_3$. The details of high temperature test are described in [12].

## 3 Results and discussion

The optical micrographs of cold sprayed coatings cross section are shown in the Figure 1 and 2.

The structure of cold sprayed coatings deposited at optimized parameters, is dense, without any pores, regardless the size of the original powder (Fig. 1a and 1b). Small amount of porosity appeared if the coating was deposited at 45° declination of the nozzle (Fig. 1c), as well as if the deposition distance was too short (Fig.1d). The longer deposition distance does not contribute to the change of porosity (Fig. 1e). On the other hand, use of nozzle for internal diameters has led to a significant increase of porosity (Fig. 1f). Nevertheless, except the CSID, the differences are negligible compare to TWAS (Fig. 2a) and HVOF (Fig. 2b).

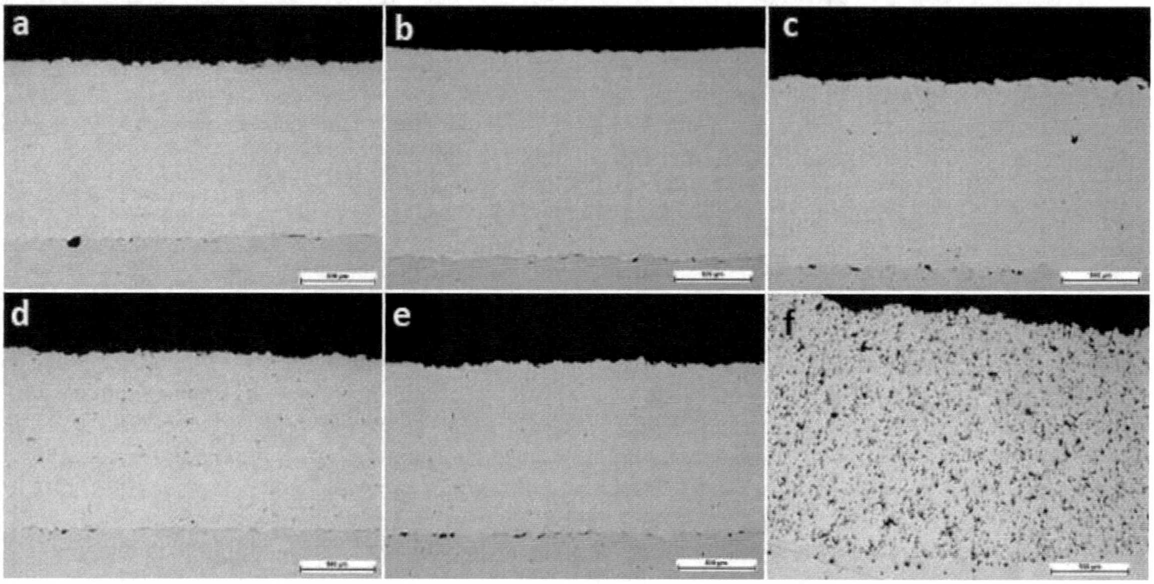


**Figure 1.** Cross section microstructures of CS1 (a); CS2 (b); CS45 (c); CSDS (d); CSDL (e) and CSID (f) Cold Sprayed NiCr coatings

The adhesive-cohesive strength of the cold sprayed samples exceeds in all cases the strength of used glue, and reached almost 60 MPa. On the contrary, the cohesive strength of TWAS sprayed samples was measured to be 40 ± 13 MPa, as well as the HVOF sprayed coatings 40 ± 5 MPa. In agreement with expectations, the measured microhardness values corresponded to the coatings microstructures. The higher the obvious porosity, the lower the microhardness value and higher value scatter. The CS coating microhardness varied between 330 and 430 HV0.3, being highest for CS2, and lowest for CSID (339 HV0.3) and CS45 (393 HV0.3). The TWAS NiCr coating microhardness reached only 231 HV0.3 and HVOF 345 HV0.3 microhardness values.

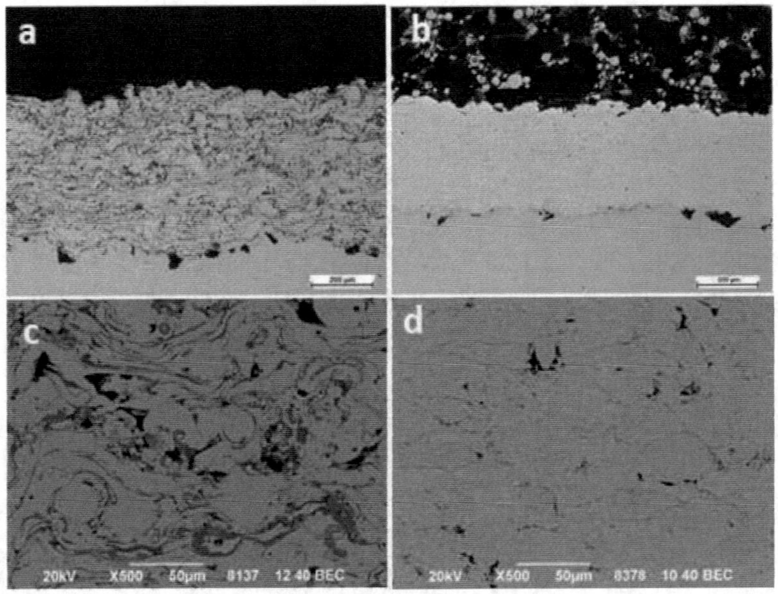


**Figure 2.** Cross section microstructures of TWAS sprayed (a, c) and HVOF sprayed (b,d) coatings

The results of wear tests are shown in the graphs in the Figure 3. The Dry Sand Rubber Wheel tests result (Fig. 3a) of CS samples did not varied at all – the technological parameters had no noticeable influence. All evaluated CS coatings shown higher wear rate, compare to TWAS coating and the HVOF coating, which performed the best. Such unexpected observations are not in agreement with the results of tensile tests and microhardness measurements and the reason for such behavior has to be further studied. On the other hand, the solid particle erosion test (Fig. 3b) pointed out to significantly higher resistance of CS coatings to erosion wear, compare to TWAS and HVOF sprayed coatings. Moreover, the test revealed a positive influence of finer original powder (CS2) and shorten deposition distance (CSDS). The results didn't include the highly porous coating, deposited by nozzle for internal diameters, which become a subject of following in-detail research study. The volume loss of TWAS and HVOF coatings was multiplied higher, as well as the scatter of the values.

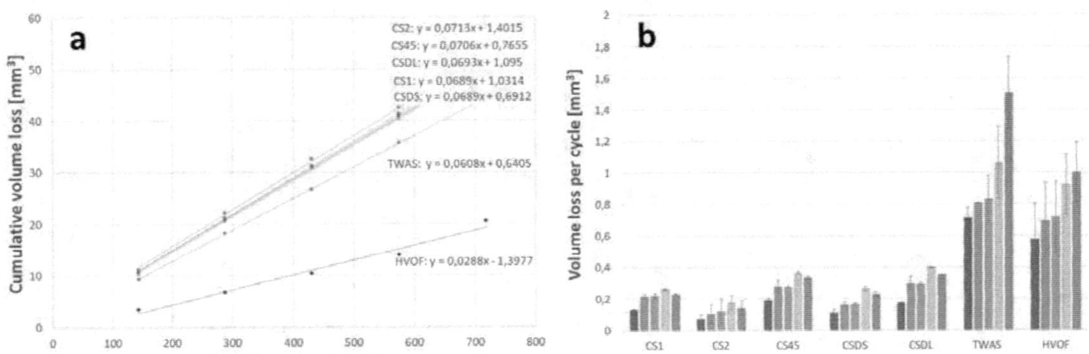


**Figure 3.** The three-body abrasive wear (a) and solid particle erosion wear (b) resistance

The SEM of the worn surfaces (Fig. 4) shown the mechanism of erosion wear: in the case of CS, the surface is plastically deformed, number of erosive particles are embedded in the coating surface, namely in the case of perpendicular particles impact. In the case of TWAS, as well as HVOF, the delamination of whole splats took place after the repeated perpendicular impact of solid particles. If the particles impact the surface at low angle, the major abrasion-like mechanism, demonstrated by grooving of the surface, can be observed. Nevertheless, namely in the case of TWAS, the whole splat delamination also took place.

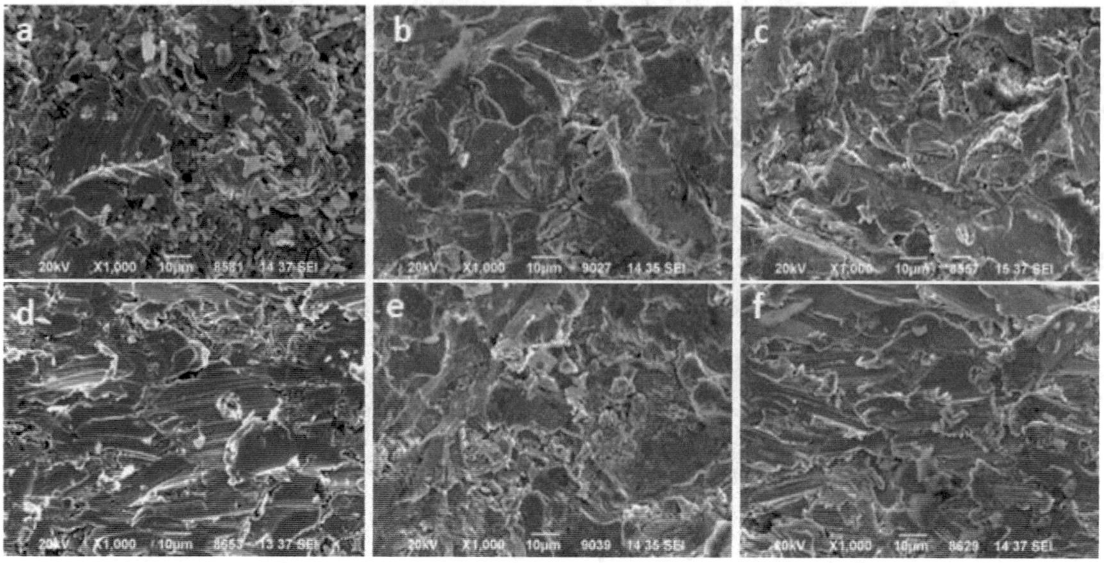


**Figure 4.** The SEM of eroded surfaces: CS2-90°(a); TWAS-90°(b); HVOF-90°(c); CS2-15°(d); TWAS-15°(e); HVOF-15°(f)

The salt fog test of coating corrosion resistance proved the protective properties of CS coating. Even though the NiCr alloy itself is corrosion resistant, the TWAS and HVOF sprayed coatings does not fully protect the underlaying substrate from the influence of surrounding media (Fig. 5). The remaining porosity can enable the penetration of surrounding atmosphere to the interface between substrate and coating and cause the substrate corrosion. Nevertheless, the application of coatings decrease the corrosion attack compare to uncoated surface significantly.

**Figure 5.** The coatings surfaces after the salt fog tests CS2 (a); TWAS (b) and HVOF (c)

The high-temperature corrosion test in aggressive salts shown similar results regardless the used deposition technology. The mass gain, representing the amount of coatings surface oxidation, was comparable for all tested coatings. The ability to protect the substrate was not evaluated, as the substrate material for the test is corrosion resistant.

## 4 Conclusion

The influence of technological parameters on the properties of cold sprayed NiCr coatings was evaluated. Based on the realized tests, only a small variation of evaluated properties was found regarding the deposition angle and distance. The microstructure and microhardness of coating, deposited by nozzle for internal diameters spraying, was significantly lower. The influence of increased porosity on the functional properties, namely on ability to protect the substrate from corrosion, is the subject of current research. Compare to the NiCr coating, deposited by TWAS and HVOF deposition technologies, the CS coatings provides higher resistance against the solid particle erosion and offer better corrosion protection of the underlaying substrate.

**Acknowledgements**
The paper has originated in the framework of the project of Technology Agency of the Czech Republic no. TN01000007 - National Center for Energy.

## References

[1] Kaushal G 2011 High-Temperature Erosion-Corrosion Performance of High-Velocity Oxy-Fuel Sprayed Ni-20 Cr Coating in Actual Boiler Environment *Metall and Mat Trans A* **42** pp 1836–1846
[2] Song B 2016 Laser Cladding of Ni50Cr: A Parametric and Dilution Study *Physics Procedia* **80** pp 706-715
[3] Das C R 2007 Effects of dilution on microstructure and wear behaviour of NiCr hardface deposits, *J. Mat.Sci. Technol.* **23** pp 771-77
[4] Abu-warda N 2020 Ni20Cr coating on T24 steel pipes by HVOF thermal spray for high temperature protection *Surf. Coat. Technol.* **381** p 125-133
[5] Sidhu H S 2007 Solid particle erosion of HVOF sprayed NiCr and Stellite-6 coatings *Surf. Coatings Technol.* **202** pp 232–238
[6] Kumar S 2019 Comparative study of high temperature oxidation behavior and mechanical properties of wire arc sprayed NiCr and NiAl coatings *Eng. Fail. Anal.* 106 p 104173
[7] Kumar M 2015 Development of nano-crystalline cold sprayed Ni–20Cr coatings for high temperature oxidation resistance *Surf. Coatings Technol.* **266** pp 122–133
[8] Yin S 2018 Cold spray additive manufacturing and repair: Fundamentals and applications, *Additive manufacturing* **21** pp 628-650
[9] Srinivasan D 2016 Characterization of Cold-Sprayed IN625 and NiCr Coatings. *J Therm Spray Tech* **25** pp 725–744
[10] Bala N 2010 High Temperature Corrosion Behavior of Cold Spray Ni-20Cr Coating on Boiler Steel in Molten Salt Environment at 900 °C. *J Therm Spray Tech* **19** pp 110–118
[11] Deng T 2008 A comparison of the gas-blast and centrifugal-accelerator erosion testers: The influence of particle dynamics *Wear* **265** pp 945–955
[12] Česánek Z 2019 High-temperature corrosion behavior of selected thermally sprayed coatings in corrosive aggressive environment *Mat. Res. Expr.* **6** 016426

# Impact of semi-solid processing and cryogenic treatment on wear resistance of X210Cr12 tool steel

**D Hradil[1], M Duchek[1], H Jirková[2] and K. Rubešová[2]**

[1] COMTES FHT a.s., Průmyslová 995, 334 41 Dobřany, Czech Republic
[2] University of West Bohemia, Regional Technological Institute, Univerzitní 8, 301 00 Pilsen, Czech Republic

E-mail: dhradil@comtesfht.cz, mduchek@comtesfht.cz, hstankov@rti.zcu.cz, krubesov@rti.zcu.cz

**Abstract**. Semi-solid processing is type of process where metastable microstructure with carbide network may be obtained. Metastable austenite is embedded in carbide-austenite network even in ordinary tool steels. The paper describe semi-solid processing of the ledeburitic steel X210Cr12, with 2% carbon and 12% chromium. The semi-solid processing temperature was 1250 °C with dwell time 60 minutes. Two states for comparison were evaluated, with or without subsequent deformation. Another process used in this study is cryogenic treatment. Cryogenic treatment defined as an add-on process to conventional heat treatment, during which the material is brought to a temperature close to the liquid nitrogen temperature. The cryogenic treatment were used in this study for improving and comparing of impact on wear resistance combined with semi-solid processing. Wear resistance were evaluated based on pin-on-disc testing at ambient temperature.

## 1 Introduction

It is well known that even commonly used steels may gain unusual microstructures by some types of unconventional processing, thus expanding their field of application. One of the new investigated processes is semi-solid processing. By this process technology metastable microstructure with carbides can be obtained. The microstructure is significant due to their multi-phase structures that form in the part based on different concentration of chemical elements in liquid and solid phases [1-3]. Microstructures produced by semi-solid processing typically consist of polyhedral austenite grains embedded in a ledeburite network. In this state, the microstructure is very brittle. Consequent refinement, disperse or decomposition its fragment uniformly across the part is needed. The final microstructure will be fine-grained and homogeneous, containing a fine dispersion of carbides with appropriate mechanical properties [4, 5].

Tool steels produced by a conventional metallurgical route contain mostly sharp primary carbide of the $M_7C_3$ type, these type of carbides is very difficult to eliminate by conventional processes. Subsequent forming follows semi-solid processing. It can remove sharp-edged chromium carbides by converting them into a very fine dispersion in a martensitic matrix [2-5]. The experimental steel for these experiments is ledeburitic cold work tool steel X210Cr12 [6]. The steel has relatively wide solidification range, appropriate for semi-solid processing [5].

Deep cryogenic treatment (DCT) is defined as an add-on process to conventional heat treatment, with process temperature up to -196 °C, with holding times in range of hours to days. It is generally used for increasing the hardness, reduction amount of retained austenite, stabilizing martensite for dimensional and geometric stability of the precision parts, to improve wear resistance for most of steel grades and lead to longer lifetime of tools. DCT is in most cases applied between quenching and tempering process [7-9].

DCT leads to several microstructural changes, below -130 °C, retained austenite transforms to martensite same way as in cold treatment (-80 °C). Cooling down to deepest temperature (up to -196 °C) causes decomposition of primary martensite by time-dependent transformation. It resulted in the nucleation of plenty coherent nanometric carbides [10, 11]. These processes are starting during DCT, however the process is done during tempering period. The considerable improvement in wear resistance is supposedly caused by newly-formed carbides [8, 11, 12].

## 2 Experimental program

The material which was studied in these trials was X210Cr12 tool steel. Chemical composition of experimental steel is shown in

**Table 1**. Experimental specimens with bar shape were 75 mm in diameter and 150 mm in length. The bar with diameter 65 mm from steel X210Cr12 were inserted into containers made of SJ355 low-carbon steel. The containers steel has higher melting point than our experimental steel. Containers were composed of a tube which was sealed with lids on both ends and welded closed. Initial microstructure were in annealed state, the microstructure was compound by large sharp chromium carbides with very fine cementite in feritic matrix (**Figure 1**). The hardness was 220 HV10.

**Table 1.** Chemical composition of experimental steel (wt. %)

| C | Cr | Mn | Si | Ni | P | S |
|---|---|---|---|---|---|---|
| 2.0 | 12.0 | 0.3 | 0.35 | max. 0.5 | max. 0.03 | max. 0.035 |

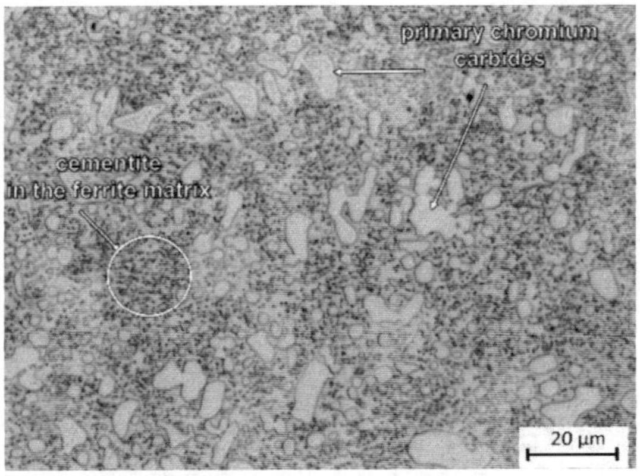


**Figure 1.** Microstructure of initial state

In this paper, two main production ways were used. All samples were heated to the semi-solid temperature that was choose to 1250 °C. The temperature were determined based on previous research [2-5]. Soaking time was 60 minutes. Subsequent treatment were divided to 2 categories. 1$^{st}$ category was cooling to room temperature in water (samples are labelled as NoDef). 2$^{nd}$ category was included subsequent cooling for period of 35 seconds in water and then heating to forging temperature 1080 °C (recommended forging temperature according data sheet). After 30 minutes of soaking time there were 10× deformation to final thickness (40 mm) performed between flat dies in a hydraulic press. The final process was rapid cooling to room temperature in water (these category is labelled as Def). Scheme of forging is shown in **Figure 2**. Specimens subjected by semi-solid processing with subsequent cooling or forging were treated identically by different manners of post-processing procedures. In the experimental program were used these processes: tempering at 300 °C for 2 hours (labelled as Def_temp or NoDef_temp) and DCT at -160 °C for 24 hours in variant with or without tempering at 300 °C for 2 hours (labelled as Def_cryo, NoDef_cryo, Def_c+t and NoDef_c+t).

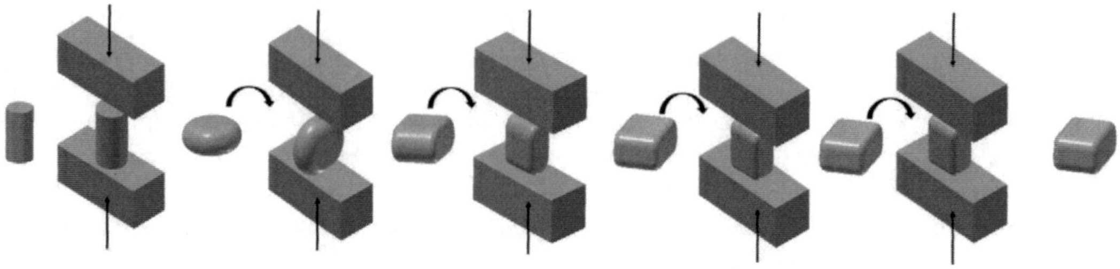

**Figure 2.** Scheme of the forging process – vizualization of 5× deformation (10× deformation is performed analogously)

## 3 Discussion and Results

Labelling of these specimens as well as its hardness is shown in **Table** *2*. The results shown, that 10× deformation leads to significant growth of hardness values, subsequent DCT leads to increase as well. There is expected increased amount of small carbides leading to higher values of hardness. The effect of additional DCT is more visible for specimens after forging, where is due to deformation refined grain. Polyhedral austenite grains with uniformly disperse ledeburite network is creating during transition by semi-solid state, lower magnification is shown in detail (austenite grains embedded by ledeburite network). Microstructure after forging consist broken ledeburite network and a fine dispersion of chromium carbides and cementite particles combined with martensite-austenite matrix with carbides. Both microstructures examined using optical microscope (Def and NoDef state) we can see on **Figure 3**.

**Table 2.** Experimental processes, labelling, deformation and input hardness

| *Label* | *Deformation* | *Description* | *Hardness HV30* |
|---|---|---|---|
| Def | 10× | No other post-processing | 780 |
| Def_temp | 10× | Tempering 300 °C/2 hrs | 707 |
| Def_cryo | 10× | DCT -160 °C/24 hrs | 906 |
| Def_c+t | 10× | DCT -160 °C/24 hrs + tempering 300 °C/2 hrs | 781 |
| NoDef | Without | No other post-processing | 388 |
| NoDef_temp | Without | Tempering 300 °C/2 hrs | 386 |
| NoDef_cryo | Without | DCT -160 °C/24 hrs | 402 |
| NoDef_c+t | Without | DCT -160 °C/24 hrs + tempering 300 °C/2 hrs | 402 |

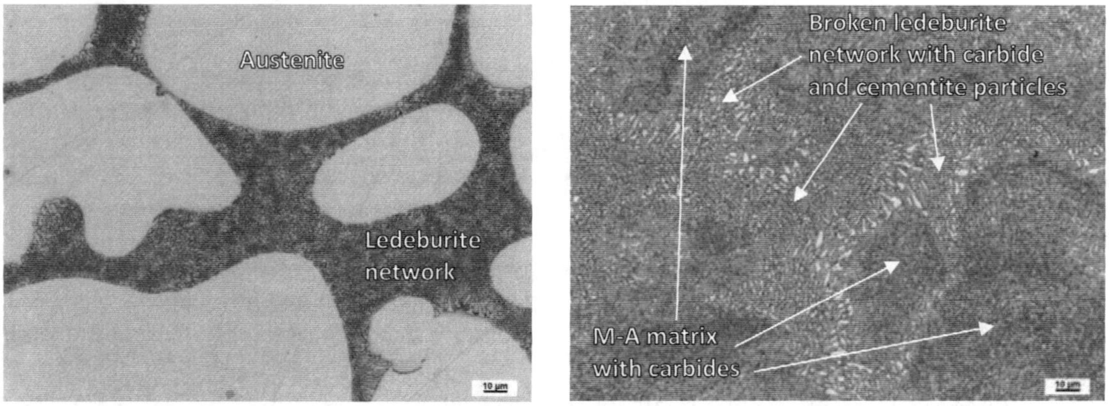


**Figure 3.** Microstructure after semi-solid processing – NoDef (left) and after forging – Def (right)

### 3.1 *Wear resistance*

Deep cryogenic treatment is typically used for expected improvement in wear resistance. For testing of wear resistance were used method pin-on-disc. The principle of pin-on-disc test is forcing a ball into the surface of a rotating flat specimen. The ball indenter was pressed by a defined force (exerted by a weight) against the specimen. Testing was conducted at ambient temperature. Each testing for each manner of production were done 2× for each sequence. The test surfaces were grinded to hardness approx. with Ra (Arithmetical mean height) 0.5 μm. The testing device was high temperature tribometer by company CSM Instruments (Switzerland).

The test parameters for all specimens were as follows:

| | |
|---|---|
| - 6 mm diameter ball indentor ($Al_2O_3$) | - 1.5 mm radius of wear track |
| - 15 N load | - ambient temperature |
| - 240 rpm | - 15000 cycles / 141.375 m (number of cycles / path travelled by the ball indenter) |

The wear rate was calculated from the measured and selected test data using (1). The wear track volume was evaluated using a contact profilometer. Its value was found as the mean from 8 readings around the track. The results show difference between state after deformation (Def) and state without deformation (NoDef). The main aspect is microstructure of tested specimens, which correspond to the hardness of investigated sequences. Sequences with deformation has finer microstructure and approx. double hardness values. It is main reason why the wear rate is lower for these sequence (**Figure** *4*). Percentage difference is approx. 32 % between specimens labelled as Def and NoDef. When we compare states without any other post-processing, the difference is approx. 45 %. The results of wear rate after additional processes (tempering, DCT or DCT with tempering) lead to different behaviour depending on the initial microstructure. In case of specimens after forging process, all additional processes lead to increase of wear rate. The specimen after DCT has lower wear rate comparing with sequences with tempering. Specimens without deformation with conducted semi-solid treatment has higher or equal values of wear rate comparing with other additional processes. Tempering leads to positive decrease of wear due to presence of newly created fine carbides around the polyhedral austenite grains. The amount of specimens / repetition is not enough for obtaining significant statistical file. The results showed the presumably tendencies of wear rates.

$$W = \frac{Weartrack\ volume[\mu m^3]}{Load[N] \cdot Path\ travelled\ by\ ball\ indenter[m]} \quad (1)$$

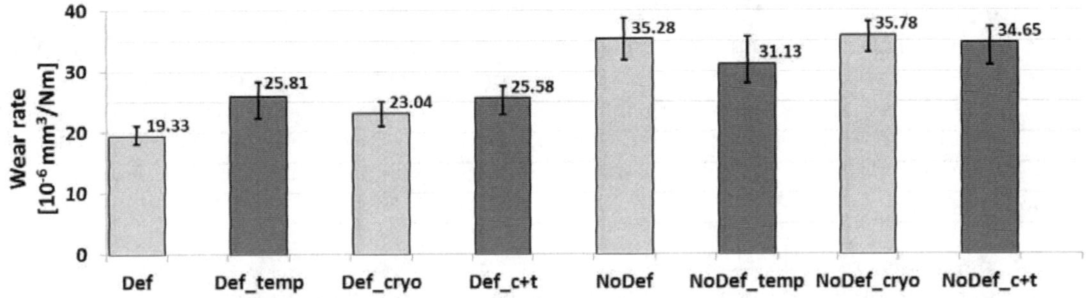


**Figure 4.** Results of wear rate, sequences Def and NoDef with additional processes

Wear tracks were analysis by optical microscope. Comparison of state Def vs. NoDef is shown in Chyba! Nenalezen zdroj odkazů.. The detail of worn surfaces shows the width of wear track. For deformed specimen (Def) the width is 507 μm in maximum, the state after semi-solid processing (NoDef) has width up to 610 μm. Whole wear tracks are shown on left side in **Figure 5.**

The average wear track depths for forged specimens (Def) was approx. 8 μm, for states after semi-solid processing (NoDef) it was approx. 11 μm.

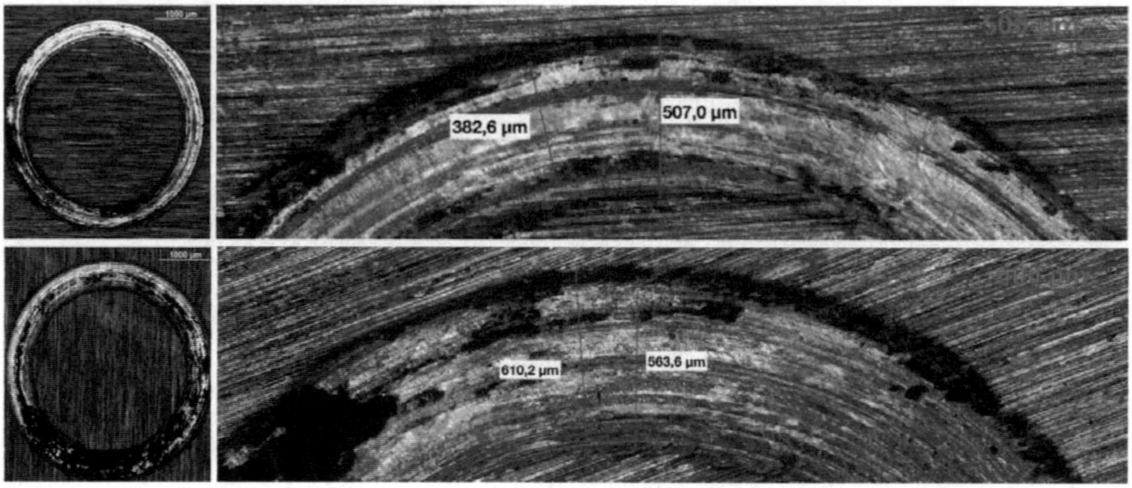


**Figure 5.** Comparison of wear tracks width - Def (upper) and NoDef (bottom)

## 4 Conclusion

Experimental semi-solid processing with or without subsequent forging was carried out on X210Cr12 cold work tool steel. Forging was conducted with 10× deformation. The both types of specimens were treated identically by different manners of post-processing processes. In the paper is presented state after tempering, cryogenic treatment and after cryogenic treatment with subsequent tempering. Specimens including covering container with 75 mm in diameter and 150 mm in length were processed at semi-solid temperature 1250 °C, subsequent forging temperature was 1080 °C.

The hardness measured after performed procedures show significant difference between specimens with or without forging (Def or NoDef). The hardness is approx. double higher after forging. The hardness in state after no post-processing is 780 HV30 (Def), after semi-solid the hardness is 388 HV30 (NoDef). Cryogenic treatment applied as additional process leads to increase of hardness values. Forging state with DCT had hardness up to 906 HV30 (Def_cryo), semi-solid state after DCT had 402 HV30 (NoDef_cryo). Wear resistance of treated specimens were executed by pin-on-disk method at ambient temperature. Highest resistance was measured on specimen after forging without additional post-processing, comparing to process without forging and any additional process the increase in resistance (wear rate) is approx. 45 %. When we compare both performed procedures and their wear rates (including additional process) the percentage difference is approx. 32 %. Metallography of wear tracks shows wider track for specimen in state just after semi-solid processing (NoDef), the dimension difference is about 100 μm (after forging (Def) it is approx. 500 μm, after semi-solid processing (NoDef) it is approx. 600 μm). The average wear track depths for forged specimens (Def) was approx. 8 μm, for states after semi-solid processing (NoDef) it was approx. 11 μm.

**Acknowledgements**
This research was funded by the Technology Agency of the Czech Republic under the project TJ02000182 „Improvement of wear resistance of tool steels by a combination of semi-solid treatment with subsequent hot forming and deep cryogenic treatment".

**References**

[1] Puttgen W, Hallstedt B, Bleck W, Uggowitzer PJ 2007 On the microstructure formation in chromium steels rapidly cooled from the semi-solid state Acta Mater. **55**, p 1033–1042

[2] Jirková H, Rubešová K, Opatová K, Peković M and Mašek B 2017 Effect of heating temperature during semi-solid processing on structure of X210Cr12 steel. Acta Metallurgica Slovaca. **23**(3), p 229-235

[3] Jirková H, Aišman D, Rubešová K, Opatová K and Mašek B 2017 Semi-solid processing of high-chromium tool steel to obtain microstructures without carbide network IOP Conference Series Materials Science and Engineering

[4] Jirková H, Rubešová K, Konopík P and Opatová K 2018 Effect of the Parameters of Semi-Solid Processing on the Elimination of Sharp-Edged Primary Chromium Carbides from Tool Steel. Metals. **8**(9), p 1-15

[5] Rubešová K, Peković M, Jirková H and Bystrianský M 2020 Effects of Deformation on the Behaviour of Chromium Carbides in Tool Steel Studied by Use of Semi-Solid Forming. IOP Conf. Series: Materials Science and Engineering. **723**

[6] ČSN 41 9436. Ocel 19436 chromová Praha: Český normalizační institut, 1994

[7] Podgornik B, Paulin I, Zajec B, Jacobson S and Leskovšek V 2016 Deep cryogenic treatment of tool steels. Journal of Materials Processing Technology. **229**, p 398-406

[8] Hradil D, Duchek M, Hrbáčková T and Ciski A 2018 Gas nitriding with deep cryogenic treatment of high-speed steel. Acta Metallurgica Slovaca. **24**(2), p187–193.

[9] Šuchmann P and Nižňanská J 2015 Deep cryogenic treatment of H11 hot working tool steel, Journal of Achievements in Materials and Manufacturing Engineering. **73**(1), p 21-26

[10] Stratton FP 2007 Optimizing nano-carbide precipitation in tool steels, Materials Science and Engineering **A449-451**, p 809-812

[11] Das D, Dutta KA, Toppo V and Ray KK 2007 Effect of deep cryogenic treatment on the carbide precipitation and tribological behavior of D2 steel, Materials and Manufactirung Processes. **22**(4), p 474-480

[12] Das D, Dutta KA, Toppo V and Ray KK 2009 Inconsistent wear behavior of cryotreated tool steels: the role of mode and mechanism, Materials Science and Technology. **25**(10), p 1249-1257

IOP Conf. Series: Materials Science and Engineering 1178 (2021) 012022 doi:10.1088/1757-899X/1178/1/012022

# Material Properties of Steel with NiCr-Cold Spray Coating

**M Chocholousek, Z Spirit, J Brom**

Centrum Vyzkumu Rez, CVR, Hlavni 130, Husinec-Rez, Czech Republic

E-mail: michal.chocholousek@cvrez.cz

**Abstract**. The work is aimed on the NiCr-Cold spray coating applied on GOST 08CH18N10T steel and GOST 22K steel. Cold spray coating shows a high cohesion even as a few millimeter thick layer, therefore it has potential for the repairs not only of the energy industry components. Applied on the steel substrate, the cold spray can be considered as a bimetallic material for the purpose of the mechanical properties evaluation. Performed comparison based on tensile mechanical properties at room temperature and at 350 °C of separate steels and cold spray versus bimetallic combination is discussed. The main goal is to find the optimal approach of bimetallic material evaluation. Additional metallographic and microstructural analysis is attached to show the differences between the used combinations. Influence of possible heat treatment on the mechanical and adhesive properties is also included.

## 1 Introduction

Cold spray is a promising method of coating formation and the subject of interest of the project solved in Centrum Vyzkumu Rez (CVR). Cold spray is one of many names describing a process for strengthening metals by exposing the substrate to a high-velocity stream of small solid particles accelerated by a supersonic gas nozzle [1]. If the potential of a given method for repairing components by replacing the original material by spraying is evaluated, it is necessary to determine the mechanical properties not only of the coating itself, but also of the substrate-coating combination. The cold spray coating exhibits sufficient strength and adhesion to be tested both alone and in combination with the substrate.

The aim of the project is to verify the applicability of standardized procedures to samples made of tubes and sprays with sufficient thickness and to adapt the methodology for testing thinner layers together with the substrate in cases where it is not possible to produce standard samples. It is also necessary to verify whether the measured mechanical properties of the separate spray correspond to the properties measured and evaluated simultaneously with the substrate. Due to the potential use of the methodology for the energy industry in the Czech Republic, the specimens are also adapted to the European standard EN ISO 6892 [2] and steels used in the energy industry are chosen as substrates. The thickness of the coating can possibly influence the material properties, where the higher thickness can lead to lower ultimate stress due to higher probability of defect occurrence.

Tensile mechanical properties of cold spray layer can be evaluated by various methods [3]: (1) Micro-flat test (MFT) specimens can be manufactured due to the sufficient thickness layer. MFT specimens can be tested according to ASTM E8 standard [4]. (2) Schmidt & coll. developed specific Tubular coating test (TCT), where a pair of tubes connected with coating is tested. (3) 3- or 4-point bend test can be performed to evaluate the bend mechanical properties of the coating.

## 2 Material and Methodology

Structural carbon steel GOST 22K and austenitic steel GOST 08CH18N10T were used as substrates for cold spray (CS) deposition. All the cold spray coatings were sprayed at uniform conditions – substrate surface preparation, NiCr powder [5] of chosen grain size (5-25 μm), nitrogen gas, angle of deposition and the spraying parameters (intellectual property of Impact Innovations GmbH company).

The tensile specimens were prepared as proportional specimens according to EN ISO 6862 from substrate materials and for substrate-cold spray combination (figure 1), where the cold spray was sprayed as approximately 1.2 mm thick layer on 4 mm diameter substrate. Flat specimens of 3.5 mm thickness were used for material properties evaluation of the cold spray without substrate. Flat specimens were evaluated as proportional too. The shape of specimens was chosen due to the suitability for cold spraying and due to the simple evaluation (i.e. the thickness measurement before and after the coating) compared to the shape of tube segments.

The specimens were heat treated (HT1) at 640 °C (limit temperature for the austenitic pipe lines due to the risk of microstructural changes) for 1 hour (i.e. holding time 10 min/1 mm of specimen thickness), the heating speed 80 °C/hour and the cooling speed 30 °C/hour. Considering low influence of HT on mechanical properties of steel-CS specimens after tensile tests, flat specimens of cold spray without substrate were heat treated with 12 hours holding time (HT12). The maximal temperature for HT was set due to considering theoretical application in industry.

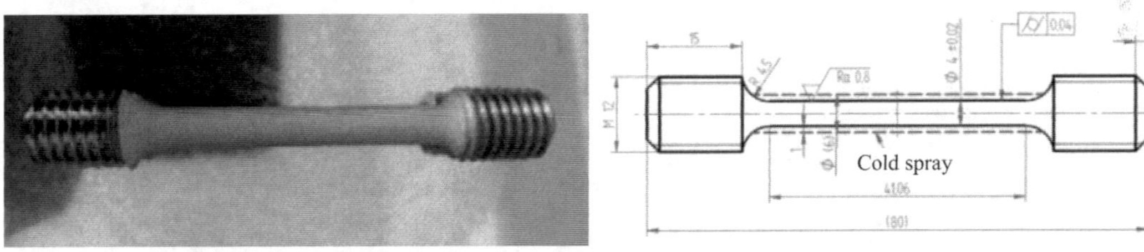


**Figure 1.** Proportianal specimens of substrate-cold spray for tensile testing.

The tensile tests were performed on the universal testing machine Zwick/Roell Z250 with the laser extensometer at temperatures 24 °C and 350 °C. Material properties of substrates were evaluated according to standards. The tensile test of specimens substrate-cold spray were stopped after coating's rupture.

The specimen substrate-cold spray can be ideally considered as two parallel specimens from different materials, those are under uniaxial tension stress. Both materials have equal deformation (i.e. elongation in the tensile direction) due to specimen's compactness. Total force is given by the sum of partial forces in both materials (i.e. $F = F_1 + F_2$, where index 1 is for substrate and index 2 is for cold-spray coating). By relating the force to the initial cross-section $S_0$, where $S_0 = S_{01} + S_{02}$ (sum of the cross-sections of both materials), the total stress of the tested pair can be expressed as

$$\sigma = \sigma_1 \frac{S_{01}}{S_0} + \sigma_2 \frac{S_{02}}{S_0}. \tag{1}$$

According to equation (1), the total stress of the sample is given by the weighted average of individual stresses at a given deformation and the weight of the stress contribution is given by the ratio of the initial cross-section of the given material to the total initial cross-section. After measuring the total fracture stress and knowledge of the stress-strain characteristics of the substrate, it is possible to determine the stress of the substrate for the given deformation and evaluate the theoretical stresses of the cold spray layer $\sigma_2$.

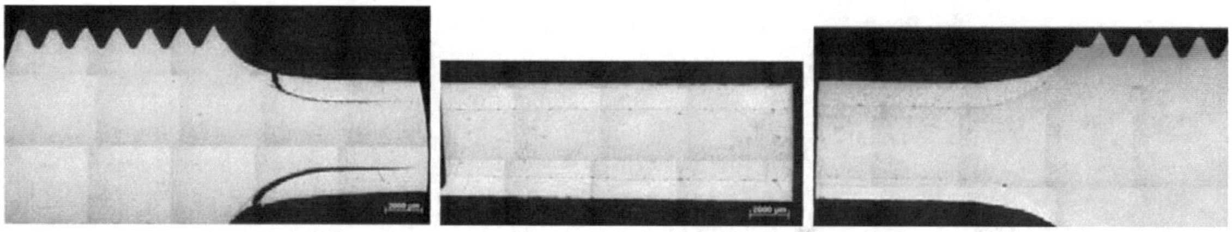

**Figure 2.** Cold spray coating's crack location after tensile test (steel 22K-NiCr cold spray, light microscope longitudinal-section view).

## 3 Results and discussion

In the case of threaded rod-shaped specimen (figure 2), the cold spray layer usually breaks at the point where the radius of transition from the test part of the body to the threaded head is. The cold spray layer is thinned there and due to the changing cross-section of both the substrate and the cold spray layer, the resulting stresses (table 1) may be lower than expected. Informative values for calculated stresses of the cold spray layer are given in table 2.

**Table 1.** Ultimate tensile strength (UTS) of tested materials (substrates with and without cold spray).

| Material | T (°C) | UTS (MPa) | UTS standard deviation (MPa) | $\sigma_{0.1}$ (MPa) | $\sigma_{0.1}$ standard deviation (MPa) |
|---|---|---|---|---|---|
| **08CH18N10T** | 24 | 552 | 4 | 249 | 3 |
| **08CH18N10T** | 350 | 405 | 3 | 211 | 7 |
| **08CH18N10T (HT1)** | 24 | 587 | 1 | 249 | 6 |
| **08CH18N10T (HT1)** | 350 | 426 | 0 | 171 | 3 |
| **22K** | 24 | 537 | 1 | 389 | 13 |
| **22K** | 350 | 513 | 1 | 262 | 1 |
| **08CH18N10T + CS** | 24 | 577 | 21 | - | - |
| **08CH18N10T + CS** | 350 | 443 | 21 | - | - |
| **22K + CS** | 24 | 585 | 12 | - | - |
| **22K + CS** | 350 | 511 | 27 | - | - |
| **08CH18N10T + CS (HT1)** | 24 | 633 | 28 | - | - |
| **08CH18N10T + CS (HT1)** | 350 | 532 | 25 | - | - |
| **22K + CS (HT1)** | 24 | 682 | 8 | - | - |
| **22K + CS (HT1)** | 350 | 584 | 14 | - | - |

Despite the fact that the chosen specimen shape is not the optimal for the evaluation, the cold spray layer shows compactness and higher tensile stress values than the substrates. The values calculated using equation (1) need precise stress value of substrate steel at cold spray rupture as an input for evaluation. The stress value is given by the particular deformation.

Plastic elongation of cold spray layer without heat treatment is low (approx. 0.1 %), thus the substrate values of stress at 0.1 % elongation is given too (table 1).

Comparing the substrate and cold spray stresses, the mechanical properties of the coating are significantly higher. The specimen with cold spray is a compact material capable to withstand higher stresses than the substrate itself.

The elongation of NiCr cold spray without heat treatment is negligible (approx. 0.1 %) due to brittleness of the coating. The used heat treatment increases the elongation at rupture to 0.6 % at room temperature for specimens from steel 22K with NiCr cold spray, however the tests at 350 °C shows decrease to 0.2% elongation. The equal heat treatment (640 °C, 1 hour holding time) used on specimens from steel 08CH18N10T with cold spray coating had negligible influence on elongation results.

**Table 2.** Ultimate tensile strength (UTS) of cold spray (from specimens with and without substrate).

| Material | T (°C) | UTS (MPa) | UTS standard deviation (MPa) |
|---|---|---|---|
| **NiCr (CS)** | 24 | 494 | 55 |
| **NiCr (CS)** | 350 | 425 | 30 |
| **NiCr (CS, HT12)** | 24 | 779 | 7 |
| **NiCr (CS, HT12)** | 350 | 684 | 8 |
| **NiCr (08CH18N10T + CS)** | 24 | 783 | 33 |
| **NiCr (08CH18N10T + CS)** | 350 | 615 | 33 |
| **NiCr (08CH18N10T + CS, HT1)** | 24 | 869 | 38 |
| **NiCr (08CH18N10T + CS, HT1)** | 350 | 733 | 44 |
| **NiCr (22K + CS)** | 24 | 706 | 21 |
| **NiCr (22K + CS)** | 350 | 662 | 41 |
| **NiCr (22K + CS, HT1)** | 24 | 862 | 13 |
| **NiCr (22K + CS, HT1)** | 350 | 771 | 22 |

## 4 Conclusion

Tensile tests on 08CH18N10T and 22K steel specimens with and without NiCr cold spray coating were performed to evaluate mechanical properties of the cold spray layer. Tensile test of flat specimens were performed on NiCr cold spray without steel substrate. All tests were performed at room temperature and 350 °C. Influence of heat treatment at 640 °C on cold spray layer was confirmed, however the influence decreases with increase of tensile test temperature.

NiCr cold spray is suitable material to be tested (according to tensile ASTM/ISO standards) without substrate in case of several millimeter thicknesses. Though NiCr cold spray without heat treatment shows higher mechanical properties when it is tested with substrate steel. Therefore the specimens substrate-cold spray are appropriate for tensile testing, especially in case of coating thickness insufficient for standard specimens manufacturing.

The shape of substrate-cold spray specimen is recommended to be flat or tube segment. The standard rod-shape specimen with threads is not recommended due to potential rupture of cold spray in radius.

NiCr cold spray increases mechanical properties of used steels. It confirms the compactness of the cold spray coating.

**Acknowledgements**

*The presented work has been realized within Institutional Support by Ministry of Industry and Trade. The submitted work was created thanks to a project supported by the TA CR No. TN01000007 within the Program of the National Center of Competence - 1st PC.*

**References**

[1] Papyrin A, Kosarev V, Klinkov S, Alkhimov A and Fomin V 2007 Cold spray technology *Elsevier*
[2] ISO 6892-1:2019(en) Metallic materials — Tensile testing — Part 1: Method of test at room temperature
[3] Villafuerte J (Ed.) 2015 Modern Cold Spray (Windsor, Ontario, Canada: Springer International

Publishing)
[4] ASTM E8 / E8M-21, Standard Test Methods for Tension Testing of Metallic Materials, ASTM International, West Conshohocken, PA, 2021, www.astm.org
[5] Srinivasan D., Chandrasekhar V., Amuthan R. et al. Characterization of Cold-Sprayed IN625 and NiCr Coatings. J Therm Spray Tech 25, 725–744 (2016). https://doi.org/10.1007/s11666-016-0382-z

# Development of a methodology for measuring fracture toughness using sub size test specimens

**E. Chvostová[1], J, Džugan[1], M. Rund[1],**

[1]COMTES FHT a.s., Dobřany, Czech Republic, EU

E-mail: eva.chvostova@comtesfht.cz

**Abstract**. To obtain real data of mechanical properties - fracture toughness - to describe the behaviour of cracks in the material, standard 1T-CT test specimens are usually used that, according to the standard, have the dimensions of 50x62.5 and a thickness of 25 mm. In the case of lack of material obtained for example by 3D printing, SPD methods, or by semi-destructive sampling from a real component to evaluate its residual life, these tests cannot be performed using a standard sample, but can be performed using a designed miniaturized test specimen marked as 0.16T-CT with dimensions 10x10 and thickness 4 mm. This paper contains the development of a methodology for testing fracture toughness using a sub-sized test specimens and a comparison of the obtained results with the results of tests obtained using a standard specimen. The techniques based on miniaturized test specimens for determining fracture toughness have been shown here as a tool to provide a deeper insight into the fracture behaviour of a material and to understand better the behaviour of the material in cases where a limited number of experiments are available.

## 1 Introduction

Fracture toughness properties provide very important information about the behaviour of material, in which the crack with sharp tip is present. This information is in many cases crucial for deciding whether to use or decommission the component.

However, often the lack of experimental material is an obstacle, but the assessment can be performed using a miniaturized test specimen. An example where a very small volume of experimental material is available are the assessment of the residual life of components in operation, where experimental material can be obtained only by semi-destructive approach, development of new materials, development of nano-materials by extreme deformation methods (SPD-Severe Plastic Deformation), development of special mechanical-thermal treatments using a physical simulator, and recently also in the evaluation of parts produced by additive manufacturing techniques. The article presents the use of sub-dimension test specimens for the fracture toughness evaluation. The experimental program was performed on Cr-Ni-Mo-V ferritic-martensitic refractory low-alloy steels marked as GOST 15CH2NMFA.

## 2 Test samples

To evaluate fracture toughness, the tests were performed using standard test specimens and sub-dimensioned test specimens.

### 2.1 Fracture toughness tests on standard test specimens

The fracture toughness tests of standard bodies were performed in accordance with ASTM E 1820-17 – Standard Test Method for Measurement of Fracture Toughness. Six 1T-CT test specimens were tested for the Master curve concept.

The testing procedure consists of the preparation of the test specimens by a fatigue crack propagation of required length using a Rumul Testronic magnetoresonance testing machine. The final oscillated crack length is given by the ratio a / W = 0.5 and the value of the stress intensity factor $K_{final}$ = 16 MPa√m. Subsequently, the samples were provided with side notches. This preparation was followed by the testing procedure itself, which took place on a servo-hydraulic testing machine MTS 810. The crack mouth opening displacement (CMOD) of the sample was measured using a clip-on gauge COD extensometer.

### 2.2 Fracture toughness tests on sub-size test specimens

Recently, much attention has been focused on the geometry of mini-CT (0.16T-CT) test specimens, which can also be made from a broken half of a standard Charpy sample [1-4].

In 2017, a round robin test was performed aimed at verifying the reliability of experimental mini-CT data. The test results confirmed that the mini-CT test specimens offer a very attractive opportunity to derive the reference temperature $T_0$ comparable to the larger fracture toughness samples and the Mini-Charpy (KLST) samples [5]. Mini Charpy and 0.16T CT test specimens were selected for testing. The tests were performed according to the same test standards as the standard test specimens. All test specimens were subjected to the fatigue crack propagation procedure and side notches machining. The tests were performed using a servo-hydraulic testing machine Si-PLAN. A comparison of the test specimens is shown in FIGURE 1a). The drawing of the 0.16T CT test specimen is shown in FIGURE 1b).

a)

b)

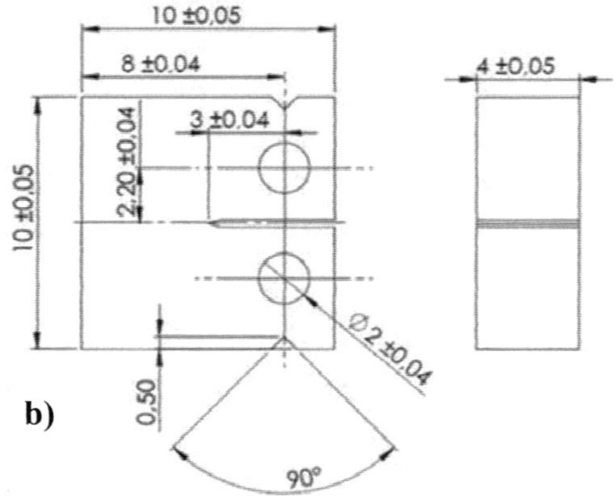


**Figure 1 a)** Comparison of used test specimens **b)** Drawing of test specimen 0.16T - CT

## 3 Fixture for performing fracture toughness tests using 0.16T - CT test specimen

To perform the test using a 0.16T-CT specimen with dimensions of 10x10 mm and a thickness of 4 mm, a device which ensures the exact position of the test specimen in the standard test machine and at the same time allows the measurement of the deformation - opening of the test specimen have to be used.

This problem is solved by a specially designed fixture, which consists of a clamping device for the test specimen gripping during the test and tips attached to the arms of the extensometer to measure

IOP Conf. Series: Materials Science and Engineering 1178 (2021) 012023 doi:10.1088/1757-899X/1178/1/012023

the CMOD. This device has been designed in such a way that it is possible to precisely position the miniaturized test specimen in the load axis of the test machine, and at the same time it allows measuring the crack opening of the test specimen during the test with a COD extensometer. The device must also allow a reliable loading of the test specimen during the test without the effect of friction, which could affect the evaluation of the results of fracture toughness - energy balance of crack growth.

Due to its structural fixture, the designed device enables the realization of not only the own test of static and dynamic fracture toughness, but also its preparation, i.e. crack propagation of the real fatigue crack in the test specimen in accordance with the boundary conditions specified by the standard. Furthermore, the design also eliminates side effects – friction. During the test, it allows sensing the opening of the sample using a COD sensor, and thus obtaining evidence of the behaviour of the inclined real crack in the monitored test sample.

The device allows to perform a fracture toughness test using a miniaturized test specimen and thus evaluate the parameters of the J-R curve or KIC of the investigated material. This approach is suitable for describing of the real cracks behaviour, even in the absence of material for the production of standard test specimens. It enables a significant refinement of the prediction of crack behaviour in real components and modern 3D printed materials, which do not allow the use of standard test specimens. The essence of the technical solution is a jig for testing the test mini-CT test specimen by eccentric tensile (Compact Tension test), which consists of the upper and lower grips, pins, mini-CT test specimen and extensometer with special tips. Compared to a standard CT test specimen, this is another technical solution that allows to scan the CMOD from the outside, using a standard extensometer equipped with specially designed tips inserted into the outside machined grooves of the mini-CT test specimen. Thanks to the larger initial measured length ($L_0$), this solution ensures greater measurement accuracy for fracture toughness tests (determination of KIC and J-R curves), or higher frequency for fatigue crack propagation velocity tests. The extensometer tip is designed as a self-centering detachable joint with an optimized tip angle abutting the notch of the mini-CT test specimen. The grips are designed to allow maximum opening of the mini-CT test specimen while protecting the extensometer against exceeding its maximum measured length and the resulting destruction. The test specimen is clamped in the grips with clearance by means of pins. The grips are screwed into the rods that are part of the testing machine. The grips are stiff enough not to distort the measurement results. After setting up the test specimen and resetting the force, the test process can be started. After reaching the prescribed test parameters according to the standard, the test is completed, the sample is removed and the device can be used repeatedly. A schematic representation of a device for testing fracture toughness using miniaturized CT test specimens 0.16T CT is shown in FIGURE 2.

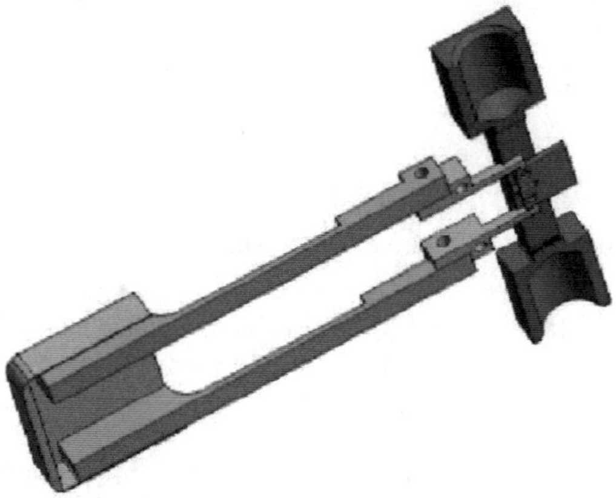

**Figure 2.** Model of the proposed fixture

The assembly diagram of the described device can be seen in FIGURE 3. The fixture for fixing the test specimen comprises an upper 1 and a lower 4 grips with a thread for screwing into rods which are connected to the testing machine. The test specimen 2 is clamped in the jaws by means of special pins with clearance to prevent friction. The pins ensure the correct positioning of the test body in the axis of the machine and pass through the holes in the upper and lower grips 3. Clamping of the test specimen takes place after screwing the upper 1 and lower grips 4 into the rods of the testing machine, the actual clamping of the test specimen 2 is then realized by means of pins through the holes 3 in the upper and lower grips. To ensure friction, there is an H7 / g6 fitting between the pin and the hole in the 0.16T CT test specimen, and there is an H7 / f7 fitting between the hole in the grips and the pin. The special tips 5 of the pressure attachment on the arms of the extensometer 6 are inserted in the groove of the upper and lower grips so that the tips of the extensometer tips fit exactly into the notch of the test 0.16T CT specimen 2. The pressure segment ensures a constant pressure force. This makes the assembly ready for testing.

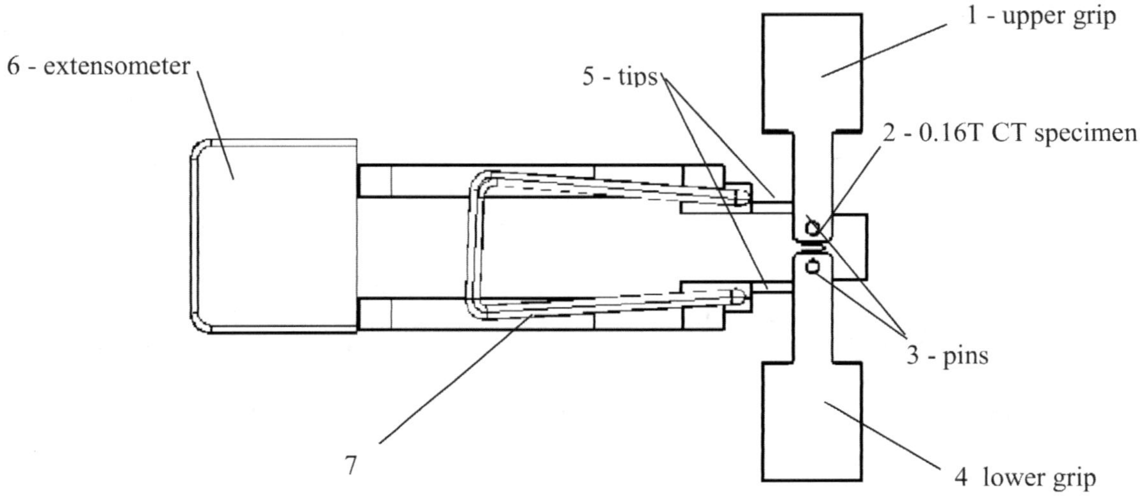


**Figure 3.** Description of the proposed fixture

## 4 Performing tests and evaluation

In order to compare the results of the fracture toughness tests performed using miniaturized and standard specimens, the test specimens were made of 15CH2NMFA material.

A total of six 1T-CT, nine 0.16T CT test specimen and fourteen KLST bodies were tested to determine the reference temperature.

The tests were performed at reduced temperatures in a cooling chamber with a supply of liquid nitrogen. The force-CMOD dependence was measured during the test. The CMOD of the samples was measured during the test with a clip-on gauge COD extensometer, see FIGURE 4. A summary of the determination of the reference temperature $T_0$ is given in Tab. 1. The curve with all measured data is shown in FIGURE 5. It is clear from Table 1 that the reference temperature evaluated using KLST samples and 0.16T-CT samples does not meet the validity criterion $\sum r_i n_i > 1$ and these reference temperatures can be considered as the provisional reference temperature $T_{0Q}$.

**Figure 4.** On the left side – the assembly of the test specimen and the COD extensometer, on the right side – the measurement of the fracture surface

**Table 1** Results for material 15CHNMFA

| Samples | Number of samples | $T_0$ °C | $\sum r_i n_i$ | Difference $T_0$ °C |
|---|---|---|---|---|
| **1T-CT** | 6 | -41.3 | 1.0 | -11.7 |
| **0.16T-CT** | 9 | -62.6 | 1.5 | -9.6 |
| **KLST** | 14 | -51.8 | 0.56 | -1.2 |

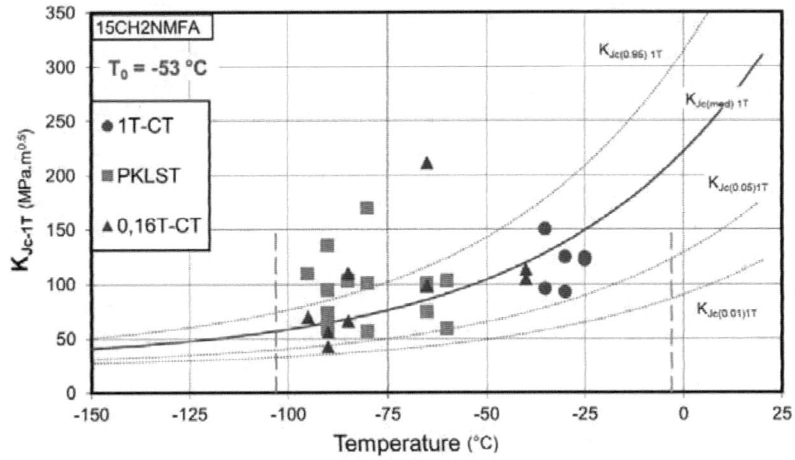


**Figure 5.** Master curve for material 15CH2NMFA

## 5 Discussion of results

The results obtained here point to the fact that there is currently no general solution available to describe the magnitude of the effect in fracture toughness approaches.

To reliably determine the "size independent" value, the material must be examined and the size effect quantified. The lower transition area is well described by the Master Curve approach, including size effects in the evaluation. It is generally considered that in all cases it is not necessary to obtain size-independent values. This can be, for example, an assessment of the properties of components with a small wall thickness, where the normal state of deformation does not in fact prevail. Typical examples of materials made with a small wall thickness that exhibit a high ratio of anisotropy properties are materials and components made by additive manufacturing processes.

## 6 Conclusion

As can be seen in many cases, the values independent of actual size cannot be achieved for the material, and therefore small size techniques are the only way to characterize the properties.

These values apply only to the component and the process under consideration; however, valuable information is provided to enable component design and process optimization. The techniques based on miniaturized test specimens for determining fracture toughness have been shown here as a tool to provide a deeper insight into the fracture behaviour of a material and to understand better the behaviour of the material in cases where a limited number of experiments are available.

**Acknowledgements**

This paper was created by project Service life assessment with the use of mini-samples No.: TH 02020448, financed by the Technology Agency of the Czech Republic.

**References**

[1] Džugan, J.; Procházka R.; Konopík P. Micro-tensile test technique development and application to mechanical property determination. *Small Specimen Test Techniques. 6th Volume. STP 1576.* Mikhail A. Sokolov and Enrico Lucon. Eds.. s. 12–29. ASTM International. West Conshohocken. PA 2014, DOI 10.1520/STP157620140022.

[2] Konopik P Dzugan J and Rund M 2015 Determination of fracture toughness in the upper shelf region using small sample test techniques ,*METAL 2015,* 6(5) pp 710-715

[3] Gotterbarm M, Seifi M, Melzer D, Džugan J, Salem A, Liu Z and Körner C 2020 Small scale testing of IN718 single crystals manufactured by EB-PBF *Addit. Manuf.*, 36, 101449

[4] Melzer D, Džugan J, Koukolíková M, Rzepa S and Vavřík J 2021 Structural integrity and mechanical properties of the functionally graded material based on 316L/IN718 processed by DED technology, *Mater. Sci. Eng. A*, 811, 141038

[5] Sokolov M.A., Development of Mini-Compact Tension Test Method for Determining Fracture Toughness Master Curves for Reactor Pressure Vessel Steels, Materials Science and Technology Division, Oak Ridge National Laboratory, 2017

# Mechanical Behaviour of Type 316L Austenitic Stainless Steel in Low Pressure Hydrogen Steam Environment

**J Janoušek** [1,2]

[1] Research Centre Rez, Hlavní 130, 250 68 Husinec-Řež, Czech Republic

[2] University of West Bohemia, Department of Material Science and Technology, Czech Republic

E-mail: jas@cvrez.cz

**Abstract**. Low pressure superheated $H_2$-steam appears to be an interesting alternative to pressurized water environments, since it is capable of performing accelerated environmentally assisted cracking (EAC) experiments for nickel base and stainless steel alloys. Constant Extension Rate Tensile (CERT) tests were performed with strain rates of $2\times10^{-6}$ or $2\times10^{-8}$ ms$^{-1}$ at 350, 400, 440 and 480 °C on flat tapered specimens of Type 316L austenitic stainless steel. The tapered shape allows the determination of crack initiation over a range of stresses and strains simultaneously on one specimen and therefore the threshold stress value was obtained. The environment was 6 times more oxidizing than the dissociation pressure of NiO. The acquired mechanical properties are summarized and threshold stresses for EAC crack initiation are evaluated.

## 1 Introduction

The requirements for mechanical tests in specific environments are determined especially by the large number of boiling water reactor (BWR) and pressurized water reactor (PWR) power plants used worldwide and the planned supercritical water reactor (SCWR) concept IV Generation. It has been found that the main contributor to degradation in these power plants is stress corrosion cracking (SCC) [1]. The pressure of reactor environments is high, which increases the costs of building test equipment and also increases the safety risk. Therefore, low pressure superheated hydrogen steam appears to be an interesting alternative, since it can accelerate the oxidation and SCC initiation stages whilst respecting the correct electrochemical corrosion potential (*EcP*) in a PWR primary water environment [2,3].

## 2 Investigated Material

An austenitic stainless steel AISI 316L, manufactured by Industeel, Arcelor Group was used in this study. The steel was available in the form of a 15 mm thick sheet, which was hot rolled and subsequently heat treated. Solution annealing was performed at 1050-1100 °C in air. The microstructure consisted of equiaxed grains of austenite and approximately 5% of δ-ferrite longitudinal stringers oriented in the rolling direction. The chemical composition is: Fe ~68 wt. %, Cr ~16 wt. %, Ni ~10 wt. %, Mo ~2 wt. %, C <0.03 wt. % (X2CrNiMo 17-12-2). The steel has a relatively high elongation (~44 % for room temperature and ~26 % at 350 °C). As the temperature rises, the yield strength and tensile strengths drop significantly, which means that cold forming is limited to 20-30 % [4]. The degree of cold working (CW) and yield strength have been found to have a significant effect on crack growth behaviour [5].

Generally a cold-worked metal is more susceptible to corrosion than the same material in an annealed state [6]. Susceptibility to environmentally assisted cracking (EAC) has been investigated in [7] for various surface finishes. It was shown to be associated with high residual stresses as well as with ultrafine-grained and deformed layers that reached up to several microns below the surface. The actual stress needed to initiate the EAC may be the sum of the applied and residual stresses. It has been shown in [8] that an intergranular crack could be initiated after disrupting the oxide or metal-oxide interface. As the temperature rises, a higher EAC rate and shorter initiation times can be expected.

## 3 Specimens

Flat tapered samples were used to study the mechanical behaviour as a function of different strain rates and temperatures in an $H_2$-steam environment. The specimens presented a nominal thickness of 3 mm. The gauge length was 26 mm long, and 6.4 mm wide at the widest part and 4 mm at the narrowest part. This corresponds to a 3° slope that produces a different stress distribution along the gauge length during mechanical loading. In this way, it is possible to determine at what state of stress the cracks still occur and thus to determine the threshold stress. Collecting data in this way has a huge advantage, as there is no need to perform several experiments using constant cross-section samples with different load levels.

The samples were produced by wire cutting using an electro-spark machine tool with the longer side parallel to the plate rolling direction. After machining, the parallel flat surfaces of the samples were subjected to different surface preparations. The first step was hand grinding with abrasive paper with a roughness of P500 in a direction inclined by 25 ± 10° to the load axis. The other side of the sample was ground with abrasive papers with steps P500, P1200, P2000 and P4000. Each grinding step was applied for a minimum of 2 minutes until the grinding marks appeared to be all parallel. The samples were carefully rinsed with deionized water between each step. After the grinding procedures, the P4000 prepared side of the sample was metallographically polished with a 3 μm and 1 μm diamond paste in order to produce a mirror-finish surface. In between these steps, the samples were washed with soapy water and rinsed with ethanol and dried in a stream of hot air. Each sample was sonicated for 15 minutes and then rinsed again with ethanol and finally dried in a stream of hot air.

Back-scattered electron (BSE) analysis revealed a ~1 μm thin layer with a different microstructure near the surface. This deformed layer is created by mechanical processing of the sample during preparation (grinding and polishing) and it is usually thicker (~1-2 μm) for ground surfaces. This may generate different oxidation in the corrosive environment and also different tensile behaviour of the material. Due to this nano-layer, which is mechanically strengthened and therefore more resistant to intergranular corrosion, different material behaviours can be observed on the ground/polished sides.

## 4 Test Equipment, Environment and Methodology

The design of the test apparatus for mechanical testing in a low pressure superheated hydrogen steam environment is based on the oxidation system concept developed by Scenini *et al.* [9]. A similar system is implemented in different independent laboratories [10,11]. A comprehensive study on the thermodynamics of the low pressure $H_2$-steam environment can be found in Volpe *et al.* [12]. This concept was applied to the conversion of a corrosion cell originally designed for testing with heavy liquid metals. The description of the cell was published by the authors [13] in 2018.

The environment was set to perform the EAC experiments in a 6 times more oxidizing environment with respect to the dissociation pressure at the Ni/NiO transition ($p_{O2\ Ni/NiO}$). This value was calculated considering the ratio ($R_{O2}$) between the $p_{O2\ Ni/NiO}$ over the actual partial pressure in the system ($p_{O2}$). The full calculation depends on temperature and $H_2/H_2O$ ratio and can be found in Volpe *et al.* [12]. The environment and hence the *EcP* was created by directly controlling the hydrogen flow rate and water flow rate at room temperature. A gas mixture of 6.5% hydrogen and 93.5% argon was used to create the atmosphere. The gas mixture flow rate was selected to be 50 ml/min. The water flow was then adjusted according to the required temperature as shown in

Table 1. A high-performance liquid chromatography (HPLC) pump was used. Ultra-pure water with a conductivity of 0.055 μS/cm was used which was obtained from reverse osmosis. The temperatures were 350, 400, 440 and 480 °C. The total pressure corresponded to an overpressure of 0.1 bar (10 kPa) and was given by a structural column of water located at the outlet of the corrosion cell. Nickel plates with a purity of 99.5 wt. % and with dimensions approximately 20×20×2 $mm^3$ were placed near the sample to indirectly monitor the *EcP* during the test.

**Table 1.** Overview of the environmental parameters used to generate the low pressure $H_2$-steam environment under oxidizing conditions.

| **Temp. [°C]** | $R_{O2}$ | $F_{gas}$ **[ml/min]** | $F_{H2O}$ **[ml/min]** | $PPR^{-1}$ | $p_{O2}$ **[atm]** | $p_{H2}$ **[atm]** | ***EcP* [mV]** | $[H_2]$ **[ml/kg]** |
|---|---|---|---|---|---|---|---|---|
| 350 | 1/6 | 50 | 2.40 | 981 | $2.47\times10^{-30}$ | 0.078 | 24.05 | 8.05 |
| 400 | | | 2.02 | 826 | $2.06\times10^{-27}$ | 0.103 | 26.02 | 28.86 |
| 440 | | | 1.79 | 732 | $2.26\times10^{-25}$ | 0.127 | 27.60 | 80.14 |
| 480 | | | 1.60 | 655 | $1.49\times10^{-23}$ | 0.152 | 29.05 | 223.46 |

The tapered samples were loaded and strained during constant extension rate tensile (CERT) tests. This technique uses uniaxial tensile testing performed with a very slow constant elongation rate. The displacement rates used were $2\times10^{-6}$ m·s$^{-1}$ (*S1*) and $2\times10^{-8}$ m·s$^{-1}$ (*S2*). The lower strain rate was chosen in order to minimize the effect of the mechanical loading on the EAC behaviour. The tests were interrupted when the maximum stress was reached in order to avoid the full rupture of the sample. In this way, deformation regions are not too distorted and they can be retained. Prior to the start of the test, the sample was exposed to the environmental conditions for 120 hours without any relevant force load (the sample was only loaded with ~100 N).

## 5 Results and Discussion

The CERT tests were performed with the displacement rate *S1* ($2\times10^{-6}$ ms$^{-1}$) at 400, 440 and 480 °C under the $R_{O2}$ = 1/6 $H_2$-steam environment. No significant variations were observed when the results were compared to the 350 °C load-displacement curve which was published in [14]. However, CERT tests performed with the slower displacement rate *S2* ($2\times10^{-8}$ ms$^{-1}$) at 350, 400, 440 and 480 °C showed a more substantial trend. Load-displacement curves are shown in Figure 1. It can be seen from the figure that for temperatures above 400 °C there is a change in behaviour: at higher temperatures, less stress is required to reach the same elongation. This effect is also visible if the maximum values are compared (see Table 2). The curve for 350 °C has a completely different trend. This may be associated with the occurrence of environmentally assisted cracks for higher temperatures. These results are consistent with the other already published results [2,3,14,15], where it was determined that temperature 350 °C did not lead to the initiation of EAC in steam.

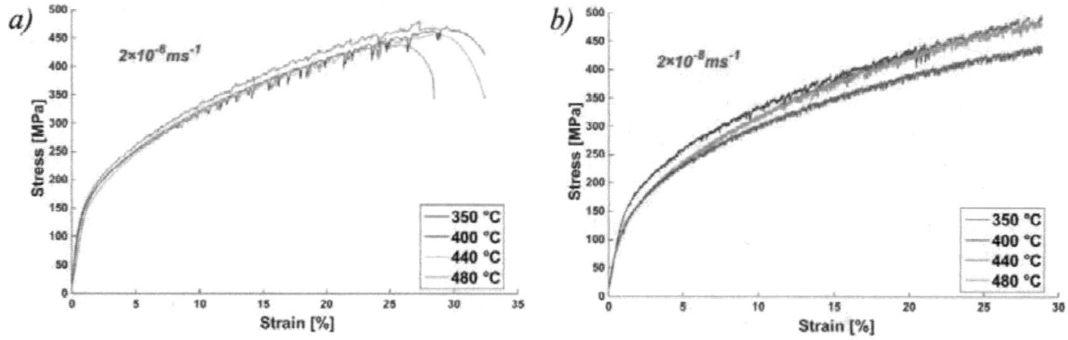


**Figure 1.** Measured CERT load-displacement curves for 316L tapered samples exposed to the $R_{O2}$=1/6 $H_2$-steam environment with the displacement rate a) *S1*: $2\times10^{-6}$ $ms^{-1}$and b) *S2*: $2\times10^{-8}$ $ms^{-1}$.

The boundary between the zone with EAC and the zone without EAC was determined on each side of the specimen using scanning electron microscopy (SEM) LYRA3 GMU (TESCAN company). These results are shown in [15]. EAC was visible near the narrow end of the gauge section. Moving the field of view towards the wide end of the gauge section their density declined. The results of the maximum stresses achieved, including the measured threshold values for the formation of corrosion cracks, are shown in Table 2. The results at 350 °C are not shown in the table because the test did not reach the same elongation as Figure 1 shows. In terms of temperature dependence, it can be stated that a clear trend between threshold values for the ground and polished surfaces was not found. At temperatures above 440 °C EAC cracks were observed on the polished surface further from the smallest cross-section, suggesting that a lower stress level was needed to initiate their formation. At 400 °C the cracks were probably more initiated on the rougher ground surface than on the polished surface. Such phenomena can show that for temperatures above 440 °C the final morphology essentially cannot play a crucial role for a given environment. This finding has to be further reviewed for confirmation.

**Table 2.** Maximum achieved stress with measured threshold values for the formation of corrosion cracks for CERT tests with displacement rate *S2*: $2\times10^{-8}$ $ms^{-1}$.

| **Temperature** *[°C]* | **Maximum stress** *[MPa]* | **Ground surface** threshold stress *[MPa]* | **Polished surface** threshold stress *[MPa]* |
|---|---|---|---|
| 400 | 496 | 386 | 412 |
| 440 | 489 | 406 | 383 |
| 480 | 444 | 397 | 367 |

An interesting change occurred when comparing displacement rates for each temperature separately. As Figure 2 shows, the same conclusions as for temperature 350 °C apply to temperature 440 °C (Figure 2a), namely that a lower displacement rate leads to higher strength values. However, the beginning to the ¼ deformation range does not correspond to this. A completely different situation occurred for the temperature 480 °C (Figure 2b). On the other hand, a higher displacement rate also leads to higher strength characteristics at this temperature. This indicates that another deformation mechanism may occur with rate changes.

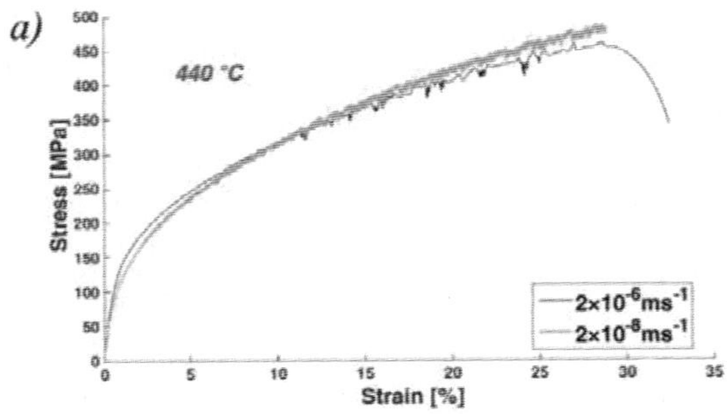


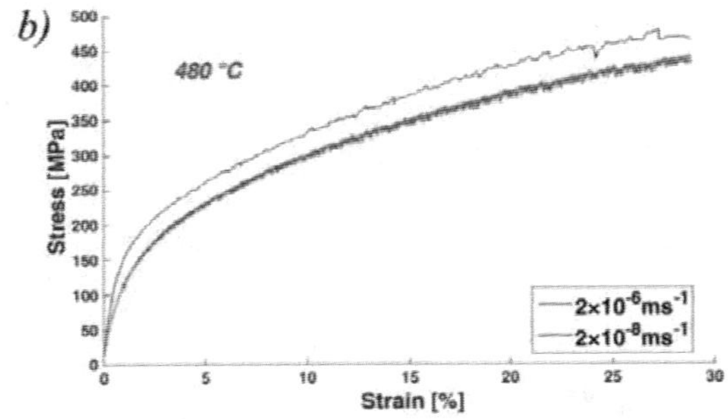


**Figure 2.** Influence of displacement rate on measured CERT load-displacement curves for samples exposed to oxidizing environment with $R_{O2}$ = 1/6 for temperatures a) 440 °C and b) 480 °C.

This is also related to the achieved maximum stress values, which are listed in Table 3. A lower maximum stress was reached at 440 °C for a higher displacement rate (*S1*: 537 MPa, *S2*: 549 MPa), similar to that obtained at 350 °C published in [14]. On the other hand, at a temperature of 480 °C, a significantly higher maximum stress (*S1*: 567 MPa, *S2*: 490 MPa) was achieved for a higher displacement rate. In this case it can be stated that the lower strain rate allows the EAC crack to develop because the mechanical effect is less evident at this strain rate. Regarding the stress thresholds for corrosion cracking, similar findings were obtained as for the lower displacement rate *S2* and the results are shown in Table 3. A marked difference of almost ~100 MPa was observed at 480 °C.

**Table 3.** Maximum achieved stress with measured threshold values for the formation of corrosion cracks for CERT tests with displacement rate *S1*: $2\times10^{-6}$ ms$^{-1}$.

| **Temperature** *[°C]* | **Maximum stress** *[MPa]* | **Ground surface** threshold stress *[MPa]* | **Polished surface** threshold stress *[MPa]* |
|---|---|---|---|
| 400 | 443 | 342 | 358 |
| 440 | 462 | 435 | 430 |
| 480 | 480 | 450 | 351 |

## 6 Summary

The area of interest was environmentally assisted cracking (EAC) in a low pressure superheated hydrogen steam environment 6 times more oxidizing than the Ni/NiO phase transition. It is an approach for accelerating the testing of PWR primary circuit operation conditions. The AISI 316L steel specimen at 350 °C did not lead to the initiation of environmentally assisted cracks, but the occurrence of environmentally assisted cracks was confirmed for temperatures higher than 400 °C. The results obtained from the faster displacement rate $2\times10^{-6}$ ms$^{-1}$ (*S1*) and for 440 °C and 480 °C showed no significant changes in comparison to the 350 °C load-displacement curve. However the temperatures above 400 °C with slower displacement rate $2\times10^{-8}$ ms$^{-1}$ (*S2*) showed a change. The curve for 350 °C has a completely different trend. This was related to the occurrence of corrosion cracks and with reduction in mechanical properties. The more the temperature increases above 400 °C, the less stress is required for the same elongation. At higher temperatures, above 440 °C, EAC cracks were observed on the polished surface further from the smallest cross-section, suggesting that a lower stress level was needed to initiate their formation. It indicates that for temperatures above 440 °C the final morphology essentially cannot play a crucial role for the given environment. This hypothesis was supported by the results for both displacement rates. The lower strain rate allows EAC cracks to develop because the mechanical effect is less evident at this strain rate.

**Acknowledgements**

This work has been undertaken within the SUSEN Project (established in the framework of the European Regional Development Fund (ERDF) in the project CZ.1.05/2.1.00/03.0108 and of the European Strategy Forum on Research Infrastructures (ESFRI) in the project CZ.02.1.01/0.0/0.0/15_008/0000293, which is financially supported by the Ministry of Education, Youth and Sports - project LM2015093 Infrastructure SUSEN. The support of Horizon 2020 MEACTOS GA No. 755151 is also acknowledged.

**References**

[1] Berg H P 2009 Corrosion Mechanisms and Their Consequences For Nuclear Power Plants With Light Water Reactors *R&RATA* vol 2 p 4

[2] Economy G, Jacko R J, Pement F W 1987 IGSCC Behavior of Alloy 600 Steam-Generator Tubing in Water or Steam Tests above 360 C *Corrosion* vol 43 **12** pp 727–73

[3] Capell B M, Was G S 2007 Selective Internal Oxidation as a Mechanism of Intergranular Stress Corrosion cracking of Ni-Cr-Fe alloys *Metallurgical and Materials Transactions A: Physical Metallurgy and Materials Science* 38 **6** pp 1244-1259

[4] Feron D 2012 Nuclear corrosion science and engineering *Woodhead*

[5] Yamazaki S, Lu Z, Ito Y, Takeda Y, Shoji T 2008 The effect of prior deformation on stress corrosion cracking growth rates of Alloy 600 materials in a simulated pressurized water reactor primary water *Corros. Sci.* **50** 835–846

[6] Callister W D, Rethwisch D G 2014 Materials Science and Engineering *Wiley Binder Version* ISBN: 978-1-118-47770-0

[7] Turnbull A, Mingard K, Lord J D, Roebuck B, Tice D R, Mottershead K J, Fairweather N D, Bradbury A K 2011 Sensitivity of stress corrosion cracking of stainless steel to surface machining and grinding procedure *Corrosion Science* **53** pp 3398–3415

[8] Persaud S Y, Korinek A, Huang J, Botton G A, Newman R C 2014 Internal oxidation of Alloy 600 exposed to hydrogenated steam and the beneficial effects of thermal treatment *Corrosion Science* **86** 108–122

[9] Scenini F, Newman R C, Cottis R A, Jacko R J 2005 Alloy 600 oxidation studies related to PWSCC *12th International Symposium on Environmental Degradation of Materials in Nuclear Power System - Water Reactors Lake Tahoe: Minerals, Metals and Materials Society*/AIME pp 891–902

[10] Persaud S Y, Ramamurthy S, Newman R C 2015 Internal oxidation of alloy 690 in hydrogenated steam *Corrosion Science* **90** pp 606–613

[11] Persaud S Y, Ramamurthy S, Korinek A, Botton G A, Newman R C 2016 The influence of the high Fe and Cr contents of Alloy 800 on its inter-and intragranular oxidation tendency in 480 °C hydrogenated steam *Corrosion Science* **106** pp 117–126

[12] Volpe L, Bertali G, Curioni M, Burke M G, Scenini F 2019 Replicating PWR primary water conditions in low pressure H2-steam environment to study alloy 600 oxidation processes *Journal of The Electrochemical Society* 166 **2** C1-C8

[13] Janoušek J, Scenini F, Volpe L, Hojná A, Trojan T 2018 Instrumentation for SCC testing in low pressure superheated hydrogen steam environments, *IOP Conference Series: Materials Science and Engineering* vol 461 Issue 1

[14] Janoušek J, Scenini F, Volpe L, Hojná A, Burke M G 2019 Environmentally-assisted cracking of type 316L austenitic stainless steel in low pressure hydrogen steam environments *Procedia Structural Integrity* vol 17 pp 440-447

[15] Hojná A, Halodova P, Janoušek J, Zimina M 2020 Acceleration of environmentally-assisted cracking initiation of the Type 316L steel in high temperature water and hydrogenated steam vapor environments *CORROSION* 10.5006/3653

IOP Conf. Series: Materials Science and Engineering 1178 (2021) 012025 doi:10.1088/1757-899X/1178/1/012025

# Influence of rolling speed during hot forming using HDQT-R 30-12 machine on the properties of stainless steel 08Ch18N10T

**H Jirkova[1], T Janda[1], M Peković[1], J Říha[2], J Mach[1] and D Krsová[1]**

[1]University of West Bohemia, RTI - Regional Technological Institute, Pilsen, Czech Republic, EU

[2]ŠKODA JS a.s., Orlík 266/15, Pilsen, Czech Republic, EU

E-mail: hstankov@rti.zcu.cz

**Abstract**. Mechanical properties of austenitic stainless steels can be improved by a suitable combination of forming and heat treatment operations. In addition to rolling reduction and speed, the heat treatment parameters are very important as well. Heat treatment typically combines solution and stabilization annealing. In components of WWER nuclear reactors, most of which are made of these steels, it is necessary to achieve the prescribed hot mechanical properties. Bars of 30 mm diameter made of titanium stabilized 08Ch18N10T steel were rolled in an incremental rolling machine. After induction heating to 950°C, different roll speeds were used, from 200 rpm to 25 rpm. The rolled bars were cooled in water and stabilization annealed at 720°C for 10 hours or at 800°C for 1 hour. Mechanical properties were determined by tensile testing at room temperature and at 350°C. Slower rolling speeds led to lower elongation (40%) than the faster ones (50%). Metallographic evaluation focused on the effect of forming on the grain size.

## 1 Introduction

Owing to severe thermal and mechanical stresses which act on WWER nuclear reactor components made of titanium-stabilized 08Ch18N10T austenitic stainless steel, this material is required to meet not only strict microcleanliness criteria but also specifications of mechanical properties for the plant project. Although the steel offers very good corrosion resistance and formability, it also exhibits relatively low yield strength [1, 2]. The yield strength required for use in WWER reactor components is at least 177 MPa at 350°C. The basic means for improving mechanical properties are threefold: substitutional or interstitial strengthening of the austenitic matrix, precipitation hardening and work hardening at reduced forming temperatures [3].

It is difficult to use substitutional elements, such as titanium or aluminium, or interstitial elements C and N for strengthening because their amounts are defined by the material specification [2,4]. Titanium is added to prevent $Cr_{23}C_6$ carbides from forming because the particles deplete the matrix of chromium in their vicinity and make the material more susceptible to intergranular corrosion [2-4]. The strengthening effect is provided by precipitation of carbides and carbonitrides which takes place in the course of heat treatment [3-5]. This heat treatment comprises solution and stabilization annealing. Solution annealing is normally performed in accordance with specifications given in the material data sheet, at 1020-1100°C [5, 6]. Stabilization annealing whose purpose is to improve hot yield strength is carried out at 600-850°C and may take several dozen hours [6, 7].

Work hardening can be induced by cold forming as well as by forming near the lower limit of forming temperatures. Deformation at temperatures no higher than 800°C produces favourable conditions for precipitation-hardening particles to form during subsequent heat treatment. The material develops high dislocation density where the dislocations act as precipitation sites for fine carbonitrides. These precipitates subsequently prevent the movement of dislocations under mechanical load and thus increase the yield strength [3]. By controlling the strain amount and rate along with the forming temperature, one can control dynamic recrystallization of the material which produces dislocation-free equiaxed austenite grains [8].

This steel can be mechanically worked using various processes including forging and rolling. Mechanical properties are governed by the amount and speed of deformation. For these reasons, the present experiment was conducted in HDQT-R 30-12 machine which can be configured to use various roll speeds and heat treatment modules [9].

## 2 Material and Methods

### *2.1 Experimental Material*

The input stock was 08Ch18N10T steel bars with a diameter of 30 mm and length of 90 mm (table 1). It is a titanium-stabilized austenitic stainless steel. Besides inclusion content, the morphology and distribution of carbides and carbonitrides in this steel are controlled. The steel was delivered in the form of rolled peeled stock.

**Table 1.** Chemical composition of 08Ch18N10T steel used for the present experiment (wt. %)

| C | Cr | Ni | Ti | Mn | Si | P | S | Cu | Mo | V | N (ppm) | W | Co | H (ppm) |
|---|---|---|---|---|---|---|---|---|---|---|---|---|---|---|
| 0.05 | 17.75 | 10.05 | 0.43 | 1.8 | 0.52 | 0.024 | 0.014 | 0.1 | 0.08 | 0.11 | 120 | 0.03 | 0.02 | 2.4 |

### *2.2. Incremental Rolling*

Bars of 900 mm were processed in incremental rolling machine HDQT–R30–12 (High Deformation Quenching and Tempering) (figure 1). This rolling mill comprises three conical rolls for helical rolling whose axes are inclined relative to each other at 120°. The rotational movement of the rollers causes the formed bar to be helically drawn into the cylinders. Incremental reduction in the bar diameter is thus applied across its entire length [10]. Prior to forming, the bar stock is heated in inductor coils. The heating temperature is controlled by the inductor power settings. Prior to entry into and exit from the rollers, the temperature of the bars is measured with a pyrometer.

**Figure 1.** Helical rolling mill (HDQT - R 30-12) at Regional Technological Institute [9]

The process routes began with heating the bars to 970°C. Heating was followed by helical rolling. The first pass reduced the stock diameter from 30 mm to 25 mm and the second one produced a diameter of 21 mm. The roll speed was varied from 200 to 25 rpm. The impact of the amount of reduction on grain size was explored by incorporating a route in which the diameter was reduced from 30 mm to 21 mm in a single pass at a roll speed of 100 rpm. The rolled bars were cooled in a water tank at room temperature placed at the end of the roller track. By this operation, the process equivalent to solution annealing was effectively completed.

In order to improved mechanical properties, the products were then stabilization annealed using three different routes. The first one was a standard route, at 720°C for 10 hours. The second one was a combined process: 680°C for 5 hours + 720°C for 10 hours. The last route with an increased temperature and much shorter time, 800°C and 1 hour, respectively, was tested in order to explore the potential for time and energy savings.

### 2.3. *Methods of Evaluation*

Metallographic characterization on longitudinal sections through the bars was performed under an Olympus optical microscope. The evaluation focused on the condition of as-received bars as well as the results of the rolling and heat treatment routes. The grain size near the surface and along the axis of the bar stock was determined using the intercept method according to ASTM A112 [11]. Vickers hardness was measured in the same locations (HV10) according to EN ISO 6507-1:2018 [12]. Tensile tests on test bars 10 mm in diameter and 50 mm in length were carried out at ambient temperature according to EN ISO 6892-1 [13] and at 350°C according to EN ISO 6892-2 [14].

## 3 Results and Discussion

The initial microstructure comprised austenite grains which were distorted by rolling (figure 2a). The grain size was 9.7 μm below the surface of the bar and 10.7 μm along the bar's axis. Slip bands and titanium carbides and carbonitrides were present. The as-received bars had a surface hardness of 247 HV10 and a hardness of 246 HV10 in the centre.

### 3.1. *Effects of roll speed*

Rolling of stock heated to 970°C at various speeds led to a microstructure of fine austenite and typical features including twins, slip bands and titanium precipitates. Rolling at 200 and 100 rpm (R200, R100) led to dynamic recrystallization which produced equiaxed austenite grains (figure 2b). The grain size was between 9.7 and 12 μm (table 2). The expectation that recrystallization led to softening was confirmed by hardness testing. Hardness was lower than in the as-received condition: 145 HV10 near the surface of the bar and 142 HV10 along its axis (table 2). Values were similar for the 100 rpm speed and for the single-pass reduction from 30 to 21 mm (R100-1).

**Table 2.** Rolling parameters, grain size and mechanical properties

| | Roll speed (rpm) | Number of passes | Temperature at exit from the rolls (°C) | Grain size (μm) | | HV10 (-) | |
|---|---|---|---|---|---|---|---|
| | | | | Surface | Axis | Surface | Axis |
| **IS** | - | - | - | 9.7 | 10.7 | 247 | 246 |
| **R200** | 200 | 2 | 1020 | 9.7 | 10.3 | 145 | 142 |
| **R100** | 100 | 2 | 971 | 9.5 | 10.8 | 144 | 141 |
| **R75** | 75 | 2 | 963 | 9.3 | 10.0 | 165 | 163 |
| **R50** | 50 | 2 | 925 | 9.7 | 10.3 | 184 | 184 |
| **R25** | 25 | 2 | 846 | 11.5 | 11.5 | 229 | 200 |
| **R100-1** | 100 | 1 | 1017 | 9.6 | 10.8 | 145 | 143 |

At roll speed of 75 rpm (R75), only partial recrystallization took place, as evidenced by a relatively high hardness: 165 HV10. Slower rolling at 50 rpm (R50) and 25 rpm (R25) failed to support dynamic recrystallization and led to a microstructure of elongated austenite grains (figure 2c). This was reflected in the further increase in hardness to 229 HV10 near the surface of the bar. Slower roll speeds resulted in a slightly larger austenite grain: 11.5 μm (table 2). The higher hardness obtained at the speed of 25 rpm was the result of low finish-rolling temperature due to the long contact between the bar and the water-cooled rolls. At the exit from the rolling mill, the temperature was 846°C. The highest temperature at the exit from the rolls was 1020°C, which was found at roll speed of 200 rpm. Rolling speed had an impact on mechanical properties as well. The ultimate tensile strength at room temperature was 580 MPa for roll speeds of 200 and 100 rpm. Yield strength exceeded 220 MPa and elongation was higher than 50% (figure 3a). Lower elongations, by approximately 10%, were found after routes with slower rolling. The corresponding yield strengths were higher, in excess of 300 MPa, as were the ultimate strengths: more than 610 MPa.

Tensile test at 350°C on materials rolled at higher speeds 100 through 200 rpm revealed ultimate strengths of approx. 460 MPa, yield strength of more than 160 MPa and elongations higher than 36% (figure 3b). From the route with the roll speed of 75 rpm onwards, the elongation levels decreased to 31 and 30% but yield strength increased to 296 MPa at the roll speed of 25 rpm (figure 3b).

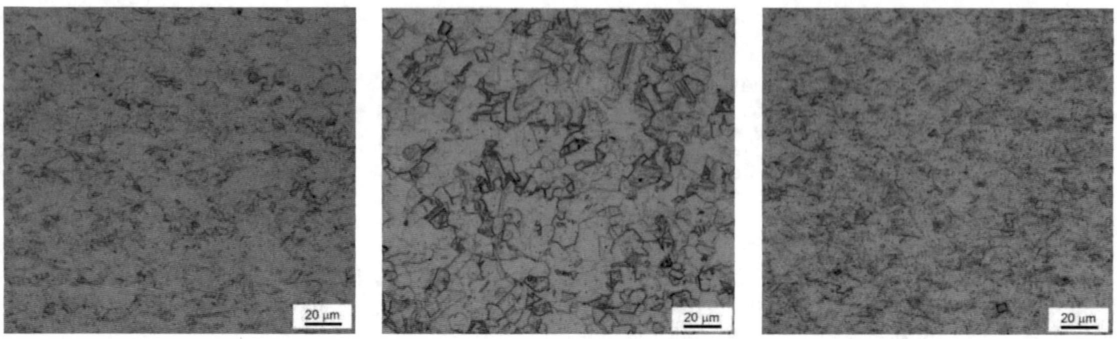


**Figure 2.** Microstructure near the surface of the bar: a) as-received condition, b) after rolling at 200 rpm, c) after rolling at 25 rpm

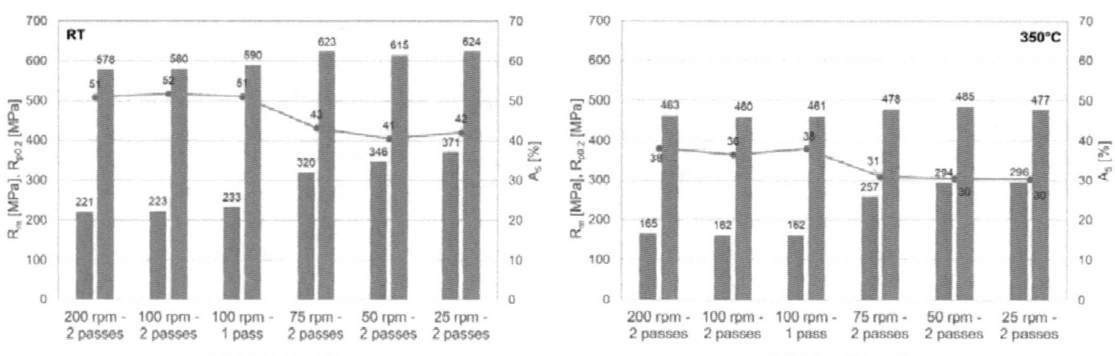


**Figure 3.** Tensile test data for various rolling parameters: a) tensile test at room temperature, b) tensile test at 350°C

## 3.1. *Stabilization annealing*

With a view to improving mechanical properties, namely hot yield strength, the rolling operation was followed by stabilization annealing at 720°C for 10 hours, two-step stabilization annealing at 680°C/5 hours + 720°C/10 hours and an energy-saving annealing cycle at 800°C for a mere hour.

After stabilization annealing, the microstructure of the bars which had been rolled at 200 and 100 rpm speeds consisted primarily of recrystallized grains with precipitates along boundaries and within austenite grains (figure 4a). No microstructural differences were found among the products of the stabilization annealing cycles, including the grain size which was approximately 9 µm in all cases.

After rolling at 75 rpm, recrystallized microstructures were only obtained by combined annealing. After annealing at 720°C/10 hours, the microstructure was not fully recrystallized and contained some regions with very fine grain (figure 4b). A short hold at the higher temperature (800°C) failed to eliminate distortion introduced by rolling. After rolling at the slowest speed, 25 rpm, the distorted grain structure remained unchanged even after stabilization annealing (figure 4c).

**Figure 4.** Microstructure near the bar's surface after stabilization annealing at 720°C/10 hours: a) 200 rpm, b) 75 rpm, c) 25 rpm

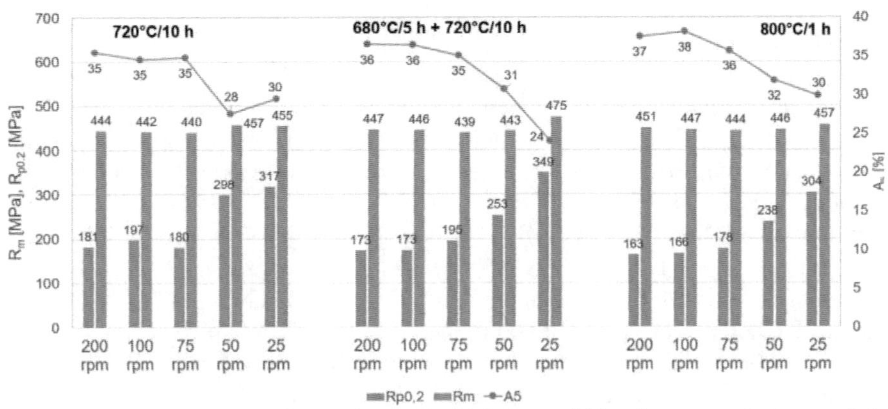


**Figure 5.** Results of tensile testing at 350°C for stabilization-annealed material at different temperatures

The effects of various stabilization annealing procedures were also reflected in mechanical properties measured by tensile test at 350°C. After the standard route with a heating temperature of 720°C and a holding time of 10 hours, the yield strength was above the required value of 177 MPa in all the cases (figure 5). The ultimate strength was above 440 MPa, with elongation at 35%. In bars rolled at 50 or 25 rpm, yield strength was higher, in the range of 298 to 317 MPa. Elongation was, however, lower than before, by approximately 30%.

Rolling at the highest speeds and combined stabilization annealing led to yield strengths of less than the specified 177 MPa. As in the previous case, yield strength increased with decreasing roll speed. At the lowest speed, 25 rpm, it was 349 MPa. This route, however, also led to the lowest elongation level of 24%. The energy and time-saving route with a heating temperature of 800°C and 1-hour holding time produced the lowest yield strength, regardless of the roll speed. The specified yield strength was only achieved at and beyond the speed of 75 rpm. These results show that the conventional stabilization annealing route, 720°C/10 hours, is a reliable way to achieve the prescribed mechanical properties regardless of the speed of the prior rolling process.

## 2 Conclusion

In this experimental study, the effects of roll speed were explored on 08Ch18N10T steel bars processed in an incremental rolling mill. The impact of stabilization annealing on mechanical properties was examined as well.

It was found that at roll speeds of 75 rpm or less, yield strength increased significantly. At 25 rpm, the increase was up to 130 MPa when compared to the highest roll speeds. These results are in agreement with the increase in hardness from 145 HV10 to 229 HV10. The stock passed through the rolls more slowly and was therefore cooled down by the contact with water-cooled rolls. This was reflected not only in mechanical properties but also in the structure of the product which consisted of elongated austenite grains. In all the products, stabilization annealing at 720°C for 10 hours produced hot yield strengths of more than the specified limit of 177 MPa. The values ranged from 181 MPa for roll speed 200 rpm to 317 MPa for roll speed 25 rpm. Combined annealing and the energy-saving stabilization annealing only led to acceptable yield strengths with roll speeds of 75 rpm and less.

**Acknowledgements**
This research was funded by the Technology Agency of the Czech Republic under the project TJ02000274 „Determination of the principles and processes taking place during the stabilization annealing of austenitic stainless steels used in nuclear power ".

**References**

[1] Pardo A, Merino M C, Coy A E, Viejo F, Arrabal R and Matykina E 2008 Effect of Mo and Mn additions on the corrosion behaviour of AISI 304 and 316 stainless steels in H2SO4 *Corrosion Science* **50(3)** 780–794
[2] Rezaei H A, Ghazani M S and Eghbali B 2018 Effect of post deformation annealing on the microstructure and mechanical properties of cold rolled AISI 321 austenitic stainless steel *Materials Science and Engineering A* **736** 364–374
[3] Petrman I. Některé problémy výroby vysokolegovaných ocelí pro komponenty jaderných elektráren typu VVER 440. 1980 *Conference proceedings: Materiálové problémy lehkovodních reaktorů*, Praha: Dům techniky ČSVTS pp. 22-27
[4] Wang J, Su H, Chen K, Du D, Zhang L and Shen Z 2019 Effect of δ-ferrite on the stress corrosion cracking behavior of 321 stainless steel *Corrosion Science* **158** 108079
[5] Moura V, Kina A Y, Taveres S S M, Lima L D and Mainier F B 2008 Influence of stabilization heat treatments on microstructure, hardness and intergranular corrosion resistance of the AISI 321 stainless steel *Journal of Materials Science* **43(2)** 536–540
[6] Kaneko K et al. 2011 Formation of M23C6-type precipitates and chromium-depleted zones in austenite stainless steel *Scripta Materialia* **65(6)** 509–512
[7] Podaný P, Martínek P, Nacházel J and Balcar M Heat Treatment of Reactor Vessel Steel 08Ch18N10T. 2012 *MS&T 2012: Materials Science and Technology Conference and Exhibition* 2012. NY: Materials Science and Technology (MS&T) pp. 1036–43
[8] Peković M, Vrtáček J, Janda T and Volkmanová J Induction hardening of steels with use of the device for incremental forming of round bars HDQT-R 30-12 2020 *IOP Conf. Ser.: Mater. Sci. Eng.* **723** 012025
[9] Ghazani M S and Eghbali B Characterization of the hot deformation microstructure of AISI 321 austenitic stainless steel *Materials Science and Engineering: A* **730** 380–390
[10] Peković M. et al. 2018 Termomechanické zpracování ocelí s použitím zařízení pro inkrementální tváření tyčí HDQT-R 30-12 (Thermomechanical treatment of steels with using the device of incremental bars forming HDQT-R 30-12) *Kovárenství* **65** 26-29
[11] ASTM E112 - Standard Test Methods for Determining Average Grain Size, 2013
[12] EN EN ISO 6507-1 Metallic materials - Vickers hardness test – Part 1: Test method, 2018
[13] EN ISO 6892-1 Metallic materials – Tensile testing – Part 1: Method of test at room temperature, 2020
[14] EN ISO 6892-1 Metallic materials – Tensile testing – Part 1: Method of test at elevated temperature, 2018

# Effect of ageing on the mechanical properties of cold formed S700 rectangular hollow

**A Kaijalainen[1], J Mourujärvi[1], J Tulonen[2], P Steen[2] and J Kömi[1]**

[1] Materials and Mechanical Engineering, CASR, University of Oulu, Oulu, Finland
[2] SSAB Europe Oy, Hämeenlinna, Finland

E-mail: antti.kaijalainen@oulu.fi

**Abstract**. The aim of this work is to study the effect of ageing (250 °C for 1 h) on the mechanical properties of the cold formed S700 rectangular hollow section. The investigated hot rolled steel strip was produced thermomechanical rolling followed by direct quenching. The dimension of the investigated hollow section was 120 x 120 x 10 mm and the corner radii and the other tolerances compliant with EN 10219. Tensile properties and Charpy-V impact toughness were determined for the base material and flat and corner area of the hollow section. The results showed that the tensile strength in the corner was slightly higher in comparison with the flat side, revealing higher cold deformation rate in the corner. Ageing increased the strength level relatively higher than cold forming without losing any elongation properties. The impact energies were at the high level at -40 °C and -60 °C in cold formed and aged materials. Even at -80 °C, the CV results were 118 J/cm$^2$. It is also notable that no difference in CV values between the flat and the corner samples were observed. Thus, the results showed that the flat side specimens testing provides sufficient information of mechanical properties of the cold formed rectangular hollow sections and no need demanding corner sample testing when the structural hollow section is produced by using the thermomechanical controlled and direct-quenched base material. Furthermore, results showed that cold formed S700 is excellent for offshore steels, as steels are used even colder conditions as before.

## 1 Introduction

Formed steel sections are widely used in construction and engineering applications due to their relatively high strength and stiffness properties. Especially, cold-formed welded hollow sections provide cost efficient and environmentally friendly alternatives compared to hot-formed sections [1]. Cold-formed rectangular hollow sections are manufactured from steel strips to final sections in several cold forming stages. Cold deformation causes work hardening of the material, which resulting in enhanced strength, although a corresponding loss of elongation and toughness. When steels are used even colder climate conditions, the requirements especially for toughness properties are increasing. This increases demands for the steel to behave safety in demanding conditions.

Artificial aging is defined in the EN 10225-4 [2] and DNV standard [3]. The technique of artificial aging is often employed to alter the mechanical properties of ferritic and/or martensitic steels. In traditional ferritic steels, strength generally increases while ductility decreases with increasing thermal exposure due to the tempering. Soininen [4] has been reported that, toughness properties decrease gradually from the base material to the flat side and then to the corner of the hollow section, due to ageing at 250 °C for 30 min. Sun and Packer [5] have shown that the impact toughness properties of the corner area are lower compared to the flat side. Likewise, Guo et al. [6] have shown that corner has the highest strength level compared to the flat side, and flat side strengths are higher than raw material. Kaijalainen et al. [7] have been reported that S500 cold formed rectangular hollow section (RHS) produced by the direct-quenched base material has much better toughness properties than conventionally manufactured.

Therefore, aim of this study is to compare microstructure and mechanical properties of direct-quenched S700 in base material, flat and corner area. Additionally, report the effect of artificial ageing on the mechanical properties of the base material and cold formed flat and corner area of RHS.

## 2 Experimental

Microstructures and mechanical properties of S700 cold formed structural rectangular hollow section (120 x 120 x 10 mm) were investigated. The investigated hot rolled base material (in wt.% 0.05C-1.8Mn-0.2Si-0.08Nb-0.1Ti) was produced by a conventional thermomechanical controlled processing and direct quenching (TMCP-DQ). The yield strength in hot rolled stage was 600 MPa, tensile strength 720 MPa and total elongation was 23 %. Microstructure of the base material was ferrite and bainite with small fraction of carbon rich areas, as can be seen in Figure 1. External corner radius of the hollow section was 25 mm, which is in the middle of EN 10219 tolerance range (20-30 mm for 10 mm) [8]. Artificial ageing was carried out at 250 °C for one hour, which is requirement in the EN 10225-4 standard [2].

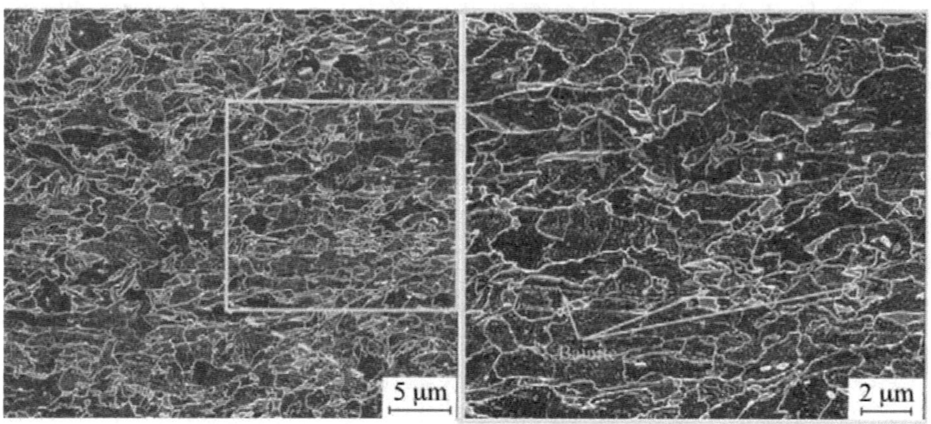


**Figure 1.** Microstructure of the base material.

General characterization of the transformation microstructures was performed with a laser scanning confocal microscope (LSCM) after nital etched specimens. Grain boundary misorientation distribution and grain sizes were measured using Oxford-HKL electron backscatter diffraction (EBSD) system on the Zeiss Ultra plus field emission scanning electron microscope (FESEM). The FESEM for the EBSD measurements was operated at 15 kV and the step size was 0.3 µm. Grain boundaries of low-angle and high-angle (effective grain size) were determined as equivalent circle diameter (ECD) values with the low (>2.5°) and the high boundary misorientation (>15°), respectively.

Samples for the mechanical testing were water-cut and machined suitable for tensile and Charpy-V tests (Figure 2) in the rolling direction. Sample dimension was 10 x 12 x 70 mm. Tensile tests were carried out using Zwick/Roell Z100 at the room temperature in accordance with the standard EN ISO 6892-1:2016. Charpy-V notch (CVN) impact testing was performed in accordance with the standard EN ISO 148-1:2016 at -40 °C, -60 °C and -80 °C (3 specimens / temperature) using longitudinal (L-T) and transversal (T-L) specimens for base and flat side material and longitudinal specimen for corner (L-T). Sub-size impact test specimen was machined to 5 mm x 10 mm x 55 mm size.

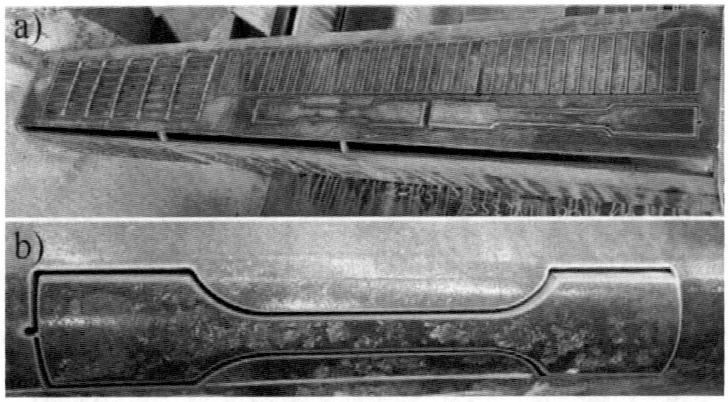


**Figure 2.** Examples of water-cut Charpy-V impact toughness and tensile test samples from a) flat side and b) corner of rectangular hollow section.

## 3 Results and discussion

### *3.1 Microstructure*

The microstructures of the base material, flat side and corner area are presented in Figure 3 and grain sizes respectively in Table 1. Over 2.5° results include both low- and high-angle grain boundaries and over 15° results include only high-angle grain boundaries. From the figures of the corner area it can be seen that the outside area microstructure is stretched, and the inside is compressed.

From the EBSD grain boundary measurements and microstructures it can be seen that the flat side and corner area has smaller grain sizes comparing to the base material. From Table 1 it can be seen that there are no clear differences in the grain sizes in the corner area, except the inner corner. The base material has the smallest fraction of low-angle grain boundaries (red color in Figure 3) and the corner areas have the high fraction low-angle grain boundaries, which indicating higher cold forming level as been seen in Ref. [9].

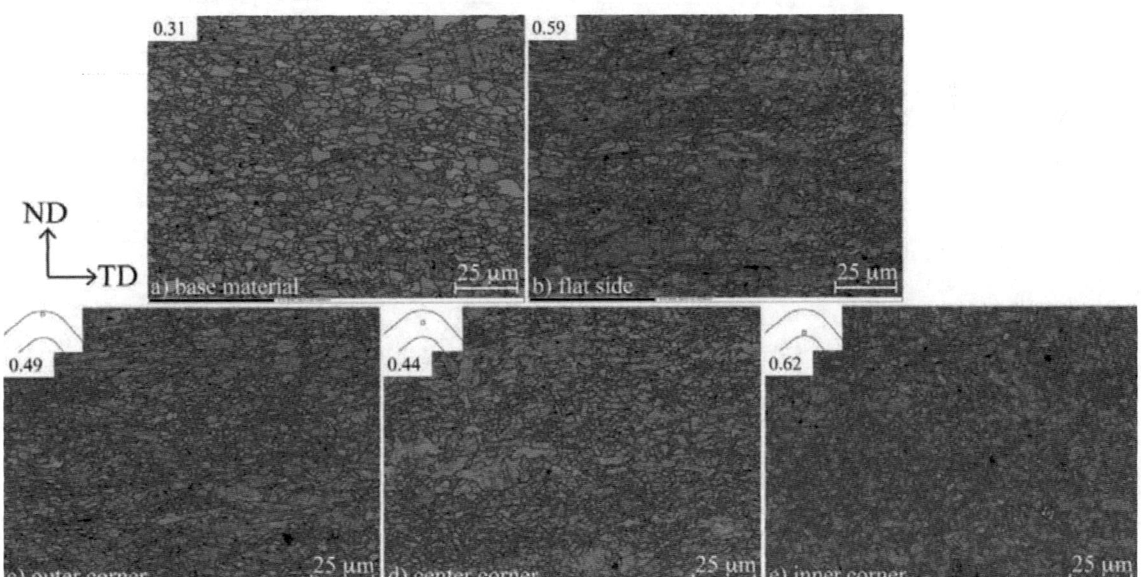


**Figure 3.** EBSD image quality maps with low- (>2.5°, red) and high-angle (15°-62.7°, blue) boundaries. Microstructures of a) base material, b) flat side, c) outer, d) center and e) inner corner area of the rectangular hollow section. Low-angle grain boundary fractions are also presented.

**Table 1.** EBSD grain size (equivalent circle diameter, (ECD)) measurements.

| | base material | flat side | outer corner | center corner | inner corner |
|---|---|---|---|---|---|
| **Low-angle boundaries** | 1.7 µm | 1.1 µm | 1.0 µm | 1.2 µm | 0.8 µm |
| **High-angle boundaries** | 2.6 µm | 2.3 µm | 2.1 µm | 2.2 µm | 1.7 µm |

### *3.2 Mechanical properties*

Table 2 and Figure 4 present the tensile test results carried out from base, flat side and corner specimens. Generally, from the results can be seen that the higher cold deformation increases the strength values and therefore the tensile strength ($R_m$) values in the corner area are the highest. Similar results are observed by Walker and Guo et al. [6,10]. About 115 MPa increasing of the tensile strength was observed from the base material to the flat side of the hollow section and from the flat side to the corner area. When comparing the yield strength, the aged base material specimen have increased yield point, which increase the $R_{p0.2}$ value relatively much comparing the others. Ageing increased the tensile strength level ~40 MPa on the flat and corner area. However, there are no notable differences in elongation values between the corner samples, except of the aged flat side specimen.

**Table 2.** Mean tensile test results of base material, flat side and corner samples. Two sample tested for each material.

| | $R_{p0.2}$ (MPa) | $R_m$ (MPa) | A (%) | $R_{p0.2}/R_m$-ratio |
|---|---|---|---|---|
| **Base material** | 604 | 719 | 23 | 0.84 |
| **Base material Aged** | 670 | 724 | 22 | 0.93 |
| **RHS flat side** | 722 | 838 | 9 | 0.86 |
| **RHS flat side Aged** | 871 | 886 | 14 | 0.98 |
| **RHS corner** | 762 | 948 | 10 | 0.80 |
| **RHS corner Aged** | 785 | 983 | 11 | 0.80 |

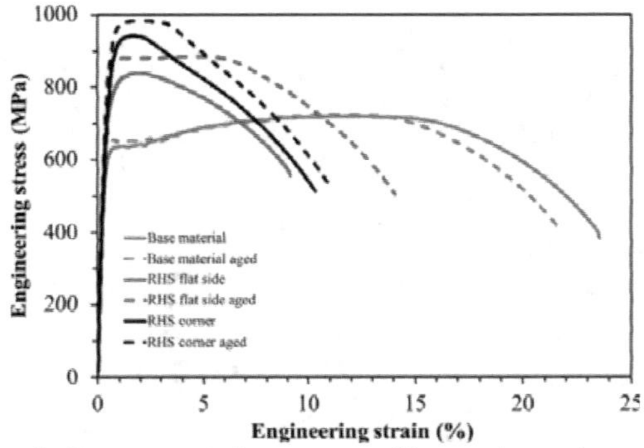


**Figure 4.** Stress – strain curves of investigated materials.

In the Table 3 it is shown individual and average CVN results of the base material, flat side and corner samples at the testing temperatures -40 °C, -60 °C and -80 °C. At -40 °C impact values are same level, and only some exceptions can notice in transversal direction (T-L) with and without aging. It is also notable that there are no significant differences in the CVN values between the flat (142 J/cm$^2$) and the corner areas (153 J/cm$^2$). It can be also seen in Figure 5a, that aging decreases impact energies generally, which corresponding with the study of Soininen [4]. The impact energies are relatively high level at -40 °C and -60 °C when considered strength level in the cold formed and aged materials. Even at -80 °C, the average impact energy is 118 J/cm$^2$ on the flat side, which is exceeding the requirements of EN 10225-4 and EN 10219 standards in the clearly colder testing temperature than required.

**Table 3.** Longitudinal (L-T) and transversal (T-L) Charpy-V impact energies of the base material and the flat and corner areas of the hollow section. The sub-size impact test specimens were used.

| | -40 °C (J/cm$^2$) | | | *average* | -60 °C (J/cm$^2$) | | | *average* | -80 °C (J/cm$^2$) | | | *average* |
|---|---|---|---|---|---|---|---|---|---|---|---|---|
| **Base material L-T** | 185 | 160 | 153 | ***166*** | 183 | 193 | 180 | ***185*** | 163 | 165 | - | ***164*** |
| **Base material L-T Aged** | 155 | 183 | 158 | ***165*** | 158 | 185 | 178 | ***173*** | 180 | 85 | 163 | ***143*** |
| **Base material T-L** | 158 | 133 | 160 | ***150*** | 133 | 98 | 130 | ***120*** | 120 | 120 | - | ***120*** |
| **Base material T-L Aged** | 103 | 133 | 113 | ***116*** | 118 | 128 | 130 | ***125*** | 133 | 65 | - | ***99*** |
| **RHS flat L-T** | 158 | 143 | 125 | ***142*** | 135 | 140 | 133 | ***136*** | 115 | 135 | 105 | ***118*** |
| **RHS flat L-T Aged** | 133 | 153 | 123 | ***136*** | 60 | 128 | - | ***94*** | 80 | 103 | 85 | ***89*** |
| **RHS flat T-L** | 123 | 123 | 118 | ***121*** | 103 | 90 | 45 | ***79*** | 70 | 75 | 73 | ***73*** |
| **RHS flat T-L Aged** | 85 | 88 | 98 | ***90*** | 38 | 38 | 83 | ***53*** | 40 | 50 | 43 | ***44*** |
| **RHS corner L-T** | 153 | 158 | 150 | ***153*** | 145 | 158 | 118 | ***140*** | 93 | 68 | 100 | ***87*** |
| **RHS corner L-T Aged** | 140 | 163 | 140 | ***148*** | 70 | 115 | 100 | ***95*** | - | - | - | - |

In the Figure 5b it is presented a relationship between the tensile strength and impact toughness at -60 °C. From the Figure 5b it can be seen three different phenomenons, (1) aging has a minor effect on the impact energies of the base material when testing direction is same. When changing the CVN testing direction from longitudinal (L-T) to transverse (T-L) drops the base material energy levels by 60 J/cm$^2$. Similar trend can be noticed on the RHS samples as well, where the transversal energies are lower. (2) Artificial ageing increases the strength level, while the toughness values decrease. (3) The corner sample of the hollow section has 110 MPa higher tensile strength compared to the flat side, but still having better impact energy at -60 °C than the flat side. Typically higher strength level decreasing the toughness properties, especially for the corner specimens [5]. However, in this study direct-quenched material improves toughness properties. Similar improvement of the corner impact energies has been found in S500 grade cold formed structural hollow sections, when base material has been direct-quenched [7]. Therefore, it could be concluded that corner sample testing is not necessary, because the flat side specimens testing provides sufficient information of the mechanical properties of the cold formed rectangular hollow sections when using direct-quenched feedstock material.

In the future, in order to improve the knowledge of the mechanical properties of cold-formed structural hollow sections produced by direct-quenched feed stock material, the transition curves with the transition temperature (T28J) or ductile-to-brittle transition temperature (DBTT) must be studied. In addition, supplementary tensile test specimens should be tested to obtain more accurate strength and elongation results.

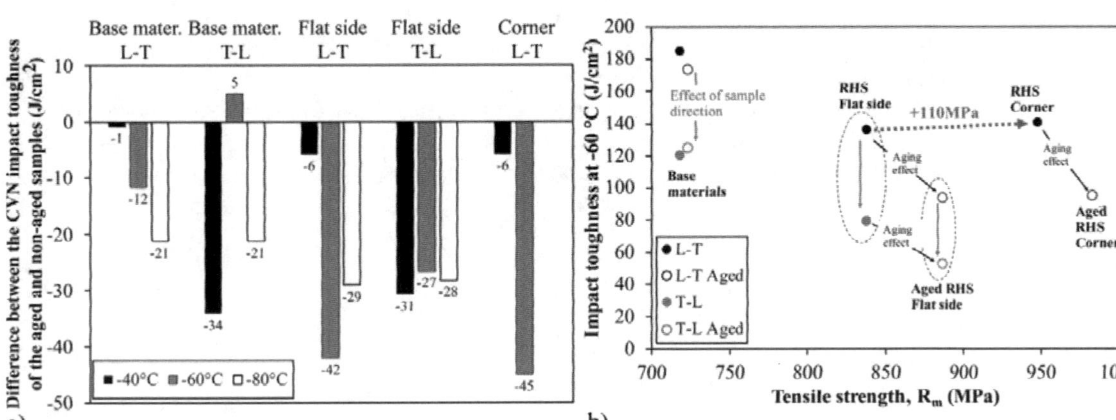


**Figure 5.** a) Effect of the ageing on the impact toughness. b) Relation between the tensile strength and the impact energies at -60 °C of the investigated conditions.

## 4 Summary

Mechanical properties and microstructure of the cold-formed rectangular hollow section manufactured by using the direct-quenched feed stock material was studied. Additionally, the effect of the artificial ageing on the mechanical properties was investigated. Microstructural characterization showed that the inside area of the hollow section corner was compressed, and the outside area was stretched. However, no large differences in the grain sizes between the different corner areas were observed. The corner area had smaller grain sizes comparing to the base material. Higher cold deformation increases the strength values and therefore the strength in the corner areas was the highest. Artificial ageing (250 °C, 60 min) increases the strength level, while the toughness values decrease. The most remarkable result in this study was that the corner area of the hollow section had better impact energy values even at -60 °C than the flat side although having clearly higher tensile strength level. Therefore, it can be concluded that the time consuming corner sample testing is not necessary, and the flat side testing provides the sufficient information of the mechanical properties of the cold formed rectangular hollow sections up to S700, when the cold formed structural hollow section is produced by using the thermomechanical controlled and direct-quenched feed stock material.

**Acknowledgements**
The authors are grateful to Business Finland for financing this work as a part of the research project Steel Ecosystem for Focused Applications (StEFA).

**References**

[1] Puthli R and Packer J A 2013 Structural design using cold-formed hollow sections *Steel Constr.* **6** 150–7
[2] EN 10225-4 2009 Weldable structural steels for fixed offshore structures. Technical delivery conditions, *European Standard, CEN*
[3] DNVGL-CP-0347 2016 Steel hollow sections
[4] Soininen R 1996 Fracture behaviour and assessment of desing requirements against fracture in welded steel structures made of cold formed hollow sections (Lappeenranta University of Technology, Lappeenranta, Finland)
[5] Sun M and Packer J A 2014 Charpy V-notch impact toughness of cold-formed rectangular hollow sections *J. Constr. Steel Res.* **97** 114–26
[6] Guo Y J, Zhu A Z, Pi Y L and Tin-Loi F 2007 Experimental study on compressive strengths of thick-walled cold-formed sections *J. Constr. Steel Res.* **63** 718–23
[7] Kaijalainen A, Mourujärvi J, Tulonen J, Steen P and Kömi J 2020 Effect of direct quenching on the mechanical properties of cold formed S500 rectangular hollow section *Procedia Manuf.* **50** 777–83
[8] 10219 E 2006 *Cold formed welded structural hollow sections of non-alloy and fine grain steels, European Standard, CEN*
[9] Kaijalainen A, Mourujärvi J, Tulonen J, Steen P and Kömi J 2021 Improvement on Impact Toughness of Cold Formed S420 Steel by Direct Quenching *Mater. Sci. Forum* **1016** 648–53
[10] Walker A 1975 *Design and Analysis of Cold-formed Sections* (Wiley)

IOP Conf. Series: Materials Science and Engineering 1178 (2021) 012027 doi:10.1088/1757-899X/1178/1/012027

# Design & Modelling of Double Cantilever structure by Stainless Steel 316L deposited using Additive Manufacturing Directed Energy Deposition Process

**A Kiran[1], J Hodek[1], J Vavřík[1], M Brázda[1], M Urbánek[1] and J Cejpek[2]**

[1] COMTES FHT a.s., Průmyslová 995, 334 41Dobřany, Czech Republic, EU
[2] IDIADA CZ a.s., Pražská třída 320/8, 500 04 Hradec Králové, Czech Republic, EU

E-mail: abhilash.kiran@comtesfht.cz

**Abstract**. Distortion in the metal additive manufacturing due to inherent residual stress during the separation of the deposited part from the base plate is critical. Distortion gets amplified after removing it from the build platform. Internal stress in the deposited sample during the material deposition and cooling phase is crucial to estimate. Residual stress in the structure influences structural distortion. Numerical analysis of residual stress prior to deposition helps to identify inconsistent stress in the built part and helps to take required measures like optimising process parameters, base plate preheat to make it less significant. Residual stress estimation before deposition using numerical calculation plays a major role in the improvement of the built part and establishing technology for a wide application. A numerical model for thermal history and residual stress estimation has been developed for Directed Energy Deposition (DED). A standard double cantilever model was considered for physical deposition and numerical model verification. Thermal history from Finite Element (FE) Analysis was compared with the in-situ measurement. A close agreement was noticed between in situ measurement and simulation results. A double cantilever was built using austenite stainless steel 316L by the DED technique. Internal stress in the deposited part before and after a partial separation from the base plate was compared. Numerical results comparison before and after separation emphasize the impact of residual stress on distortion.

## 1 Introduction

Directed Energy Deposition (DED) is one branch of the metal deposition technique under the Additive Manufacturing (AM) process. A 3D object has been built using a laser/electron beam energy source from a feedstock powder/wire. DED employs a powder blown technique (in the case of powder as feedstock) to supply feedstock directly to the melt pool generated by the concentric heat source. Localized heating and complex thermal cycle during the deposition of solid structure generate residual stress in the built part. Distortion of the built structure after separation from the built platform results in redistribution of internal stress and strain which may lead to dimensional inaccuracy. Internal stress before separation and after separation in the built structure is important to estimate distortion. The residual internal stress calculation provides the information to take measures (such as process parameter optimization, scanning strategy) required to hinder distortion in the built structure.

A numerical calculation for distortion before a deposition can prevent a failure of the part by adopting required design modification or process parameters optimization. Several studies on the process

parameters, distortion, residual stress numerical modeling, and experimental validation have been reported for the Selective Laser Melting (SLM) and DED AM process [1,2,3,4].
The main goal of this work is to develop a Finite Element (FE) model for the DED process simulation to calculate thermal and internal stress in the built part. An FE model was built using 3DExperience Finite Element (FE) Analysis software with an Additive Manufacturing scenario. A double cantilever structure is a popular model for distortion, residual stress, and inherent strain estimation in SLM [5]. The cantilever shape which exhibits distortion at the arms represents the magnitude of residual stress in the built structure. The double cantilever structure can be classified as two simple cantilevers. Distortion/stress measurement for two simple cantilevers can be compared to experimental measurement validation. Thus cantilever distortion measurement is reliable for the numerical model calibration. The double cantilever calibration for the DED numerical model has not been reported to the author's knowledge. The reason to select a double cantilever model is the ease of internal stress estimation, easiness to measure distortion directly and efficiently. In the presented work, thermal validation for FE numerical model was carried out to ensure the efficiency of structural calculation for thermo-mechanical FE simulation. Residual stress in the structure before and after separation from the base plate was evaluated using numerical results. The calibrated numerical model is planned to study different process parameters, material behavior, scanning strategy and, a part design for the DED structure.

### *1.1 Materials*

Austenite stainless steel 316L (Sandvik Osprey LTD, Neath, UK) powder with a 50-150 µm particle size range was used. The base plate 316L steel was heat-treated at 400℃ for four hours to remove potential residual stress before deposition. The chemical properties of steel 316L are listed in table 1.

**Table 1.** Powder and base plate steel 316L chemical composition (wt. %).

| | Fe | Cr | Ni | Mo | Mn | Si |
|---|---|---|---|---|---|---|
| **Powder (316L)** | Bal. | 17.2 | 10.4 | 2.3 | 1.3 | 0.8 |
| **Base Plate (316L)** | Bal. | 16.24 | 10.49 | 2.14 | 1.12 | 0.44 |

The temperature-dependent thermal material properties for numerical calculation were derived from the literature [ [6]].

### *1.2 Experimental setup*

The base plate dimension and thermocouple positions are shown in figure 1 (a). Thermocouples are welded approximately at a distance of 2 mm away from the contour path of the cantilever structure. The thermocouple (TC0) is in the center along the length of the double cantilever. TC1 and TC2 are at a distance of 17mm from the extreme ends of the double cantilever.

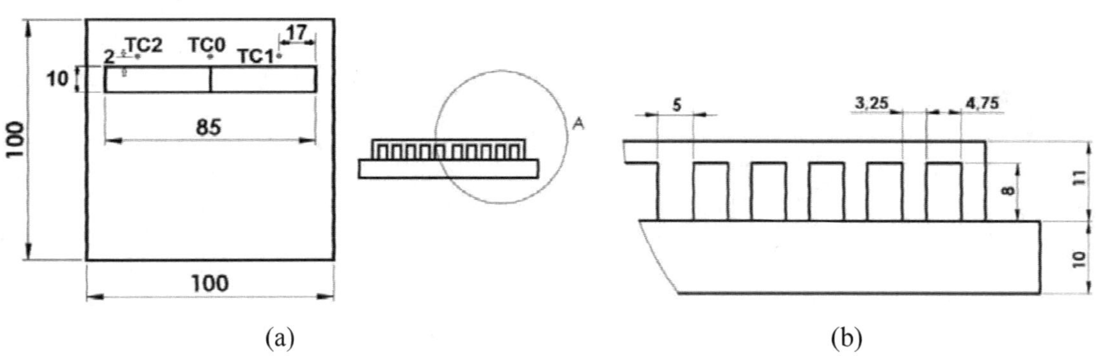


**Figure 1.** (a) Base plate and thermocouples dimensions in mm. (b) Double cantilever dimensions in millimeters

Double cantilever structural dimensions are depicted in figure 1 (b) and designed as per the DED process parameter taking the minimum thickness of the columns, external support for top bar deposition, etc. into account. The double cantilever consists of five columns on either side of the center pillar. The Center column is 5mm thick and the remaining side columns are 3.25mm. Each column is separated at a distance of 4.75mm. All columns are built to the height of 8mm and the top bar is 3mm thick.

InssTek MX-600 metallic deposition system (InssTek, Daejeon, South Korea) is equipped with a DED process technology that was used for the deposition of a double cantilever. The deposition using InssTek MX-600 during the top bar is shown in figure 3. Process parameters are listed in table 2.

**Table 2.** DED process parameters.

| Parameter | Value |
|---|---|
| **Laser power** | 500 W |
| **Laser beam diameter** | 0.8 mm |
| **Scanning speed** | 14 mm/sec |
| **Hatching distance** | 0.5 mm |
| **Powder feed rate** | 3 g/min |
| **Deposition duration** | 130 mins |

The top bar of the double cantilever deposition was supported by placing a 1mm thick 316L sheet cut into a cross-section of 4.75×10mm. The support structure material was the same as the deposition powder to avoid thermal expansion mismatch. Thin supporting sheets were placed between the columns in such a way to provide a continuous flat surface to deposit the top bar. The top bar of the double cantilever measures 3mm thickness and consists of 12 layers. Thin steel 316L sheets were selected to ensure negligible resistance for residual stress redistribution in the top bar and distortion after cutting at the bottom of the columns.

**Figure 3.** Deposition during the top bar of the double cantilever.

A thermal field during the deposition process was measured on the base plate using the thermocouple type 'K' [6]. Thermocouple positions are marked in figure 1. The deposited double cantilever before cutting is shown in Figure 4 (a). Columns of double cantilever were cut using a wire electric discharge machine (diameter of the wire 0.25 mm). The distorted double cantilever model after the cut is shown in figure 4 (b).

(b)

(b)

**Figure 4. (a)** Double cantilever structure built by DED. **(b)** Distortion in the separated columns of the double cantilever.

*1.3 Finite Element (FE) Model*

A Finite Element (FE) model was designed using the 3DExperience (Dassault Systèmes, Vélizy-Villacoublay, France) software. A Thermo-Mechanical approach was implemented [1]. The FE model of double cantilever consists of 43800 elements (0.5mm mesh size) and 53298 nodes. The base plate consists of 28800 elements (1mm mesh size) and 39188 nodes. A hexa dominant with linear order elements were selected. A thermo-mechanical DED process was simulated as two-stage, (1) Thermal simulation with Heat Transfer Coefficient (HTC) of 18 $W/m^2.K$ and emissivity of 0.1 was set to the model [6]. A laser power of 500W was set as per the experimental data with an absorption coefficient of 0.12. A concentrated energy method was adopted for laser energy distribution. The initial temperature of 27℃ was allotted to the structure. Linear heat transfer elements (DC3D8) were selected for the thermal simulation. (2) The mechanical simulation for distortion calculation was driven by thermal output. Linear elements (C3D8) from the 3DExperience library were used for mechanical analysis. The bottom plane of the base plate was clamped and the cutting plane at 2 mm above from the upper surface of the base plate defined in the static step. The meshed model and the cutting plane are shown in figure 5.

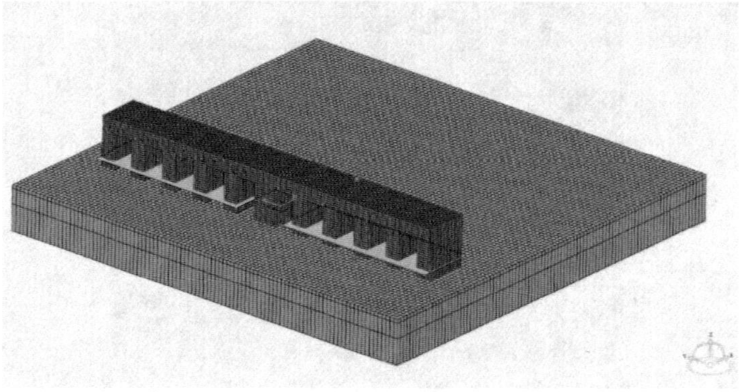

**Figure 5**. FE model and cutting plane.

## 2 Results and Discussion

*2.1 Thermal history*

The nodal temperature was attained from the thermal numerical result. The nodal temperature corresponding to the thermocouple position from thermal analysis closely agrees to the in situ measurement. The average temperature in the structure increases during the deposition of the top bar as shown in figure 6 (a). Whereas an average temperature during the deposition of the columns is below

100°C. A thermal source interaction with the structure during the deposition of the columns was for a short duration. Dwell time as per the scanning strategy and idle time during shifting laser source from one column to another is the reason for the low average temperature. Long and continuous laser energy interaction during the deposition of the top bar develops a higher average temperature on the base plate.

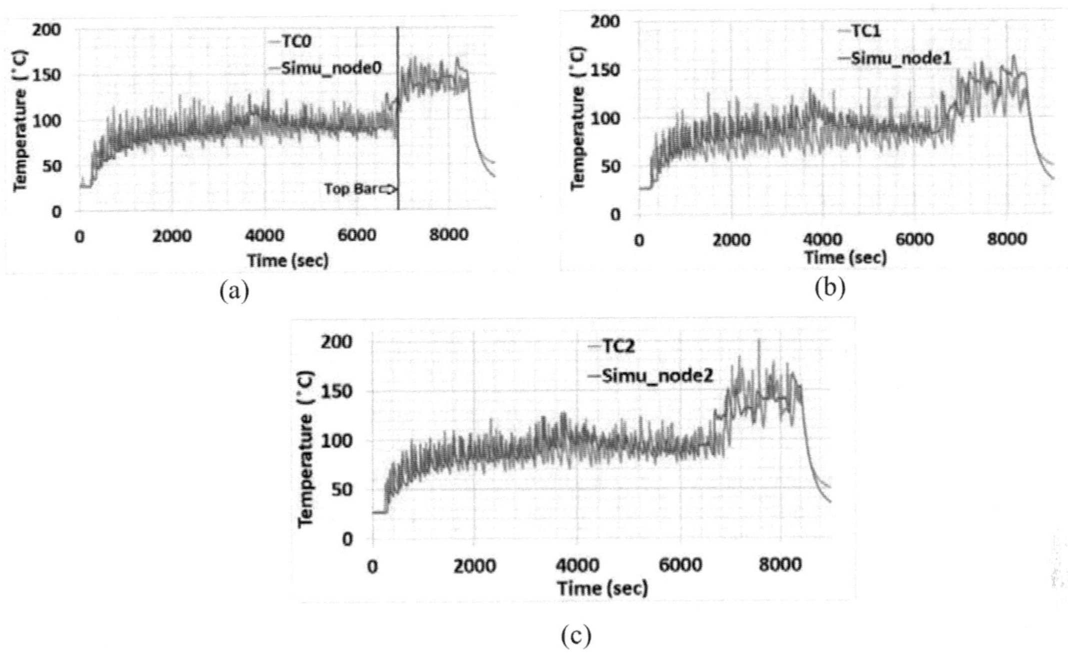


**Figure 6.** Thermal history comparison between nodal temperature from the numerical calculation and thermocouple data from the experiment.(a) Thermocouple TC0 (b) Thermocouple TC1 (b) Thermocouple TC2

## *2.1 Residual Stress*

The structural simulation was derived from thermal results. The structural calculation reveals Von Mises stress in the double cantilever structure distributed uniformly. A stress concentration at the edge of the column in contact with the base plate can be noticed in the before cut model shown in figure 7(a). The structural simulation was conducted until 9000sec. The internal stress development during rapid cooling after the thermal source disconnected from the structure was considered.

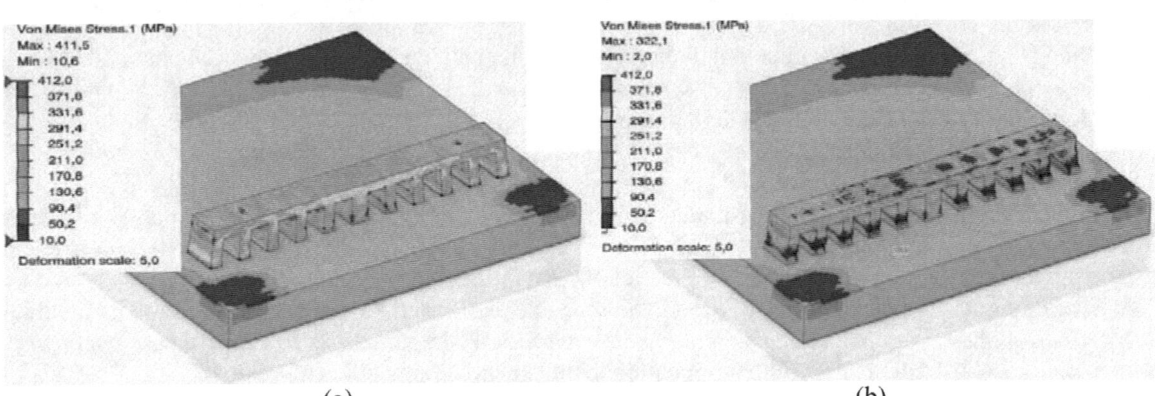

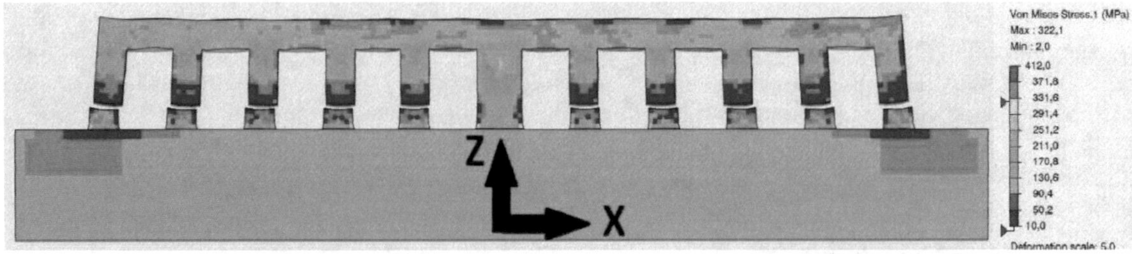


(c)

**Figure 7(a).** Contour map of the Von Mises stress distribution in the before cut sample. **(b)** After the cut sample. **(c)** After cut the sample normal view

Stress relaxation around the cut plane in the column of the double cantilever is depicted in figure 7(b). Stress relaxation on the top surface on the double cantilever forced for the contraction of the top bar. This could be the reason for the two ends of the double cantilever distortion in the positive Z-axis and negative X-axis direction shown in figure 7(c). Distortion can be noticed in the deposited sample shown in figure 4(b).

## 3 Conclusion

The numerical model was built for the Thermo-Mechanical DED simulation to estimate internal stress in the structure. The Double Cantilever structural dimension was designed and succeed in deposition using a DED process condition for 316L. Thermal results from the numerical calculation were validated with experimental results which closely agree. Thermal simulation accuracy is important for reliable structural calculation in Thermo-Mechanical DED simulation. The residual stress calculation illustrates the importance of internal stress on the distortion of the structure. The second stage of the work will be (1) numerical modeling for distortion and validation with the experimental result (2) Evaluating preheating temperature on the base plate to hinder residual stress and distortion.

**Acknowledgments**

The paper was developed with support from the ERDF project Pre-Application Research of Functionally Graduated Materials by Additive Technologies, No. CZ.02.1.01/0.0/0.0/17_048/0007350.

## References

[1] Kiran A; Hodek J; Vavřík J; Urbánek M; Džugan J, *Numerical Simulation Development and Computational Optimization for Directed Energy Deposition Additive Manufacturing Process* Materials, vol. **13**, no. 11, 2020.

[2] Qiang Chen1; Erwan Beauchesne; Francis Arnaudeau; Pierre-Richard Dahoo; Constantin Meis, *Modeling of mechanical behavior in additive manufacturing at part scale*, 8th International Conference on Mathematical Modeling in Physical Science, Bratislava, Slovakia, 2019

[3] J Vavřík and J Hodek, *Residual stress investigation in additively manufactured samples*, IOP Conference Series: Materials Science and Engineering, vol. **723**, 2020

[4] Pedro Alvarez1; Joseba Ecenarro; Iñaki Setien; Maria San Sebastian, *Computationally efficient distortion prediction in Powder Bed Fusion Additive Manufacturing*, International Journal of Engineering Research & Science, vol. **2**, no. 10, 2016

[5] IñakiSetien; MicheleChiumenti; Sjoerdvan der Veen; MariaSan Sebastian; FermínGarciandía; AlbertoEcheverría, *Empirical methodology to determine inherent strains in additive manufacturing*, Computers & Mathematics with Applications, vol. **78**, no. 7, pp. 2282-2295, 2019

[6] Kiran A, Koukolikova M, Vavrik J, Urbanek M and Dzugan J 2021 Base Plate Preheating Effect on Microstructure of 316L Stainless Steel Single Track Deposition by Directed Energy Deposition, *Mater.,* 14(18) 5129

IOP Conf. Series: Materials Science and Engineering 1178 (2021) 012028 doi:10.1088/1757-899X/1178/1/012028

# Preparation of Ti-Al-Si alloys with the various contents of aluminium and silicon

A Knaislová[1], J Linhart[1], P Novák[1], I Szurman[2], T Čegan[2], J Juřica[2], D Vojtěch[1]

[1]Department of Metals and Corrosion Engineering, University of Chemistry and Technology, Technická 5, Prague 6, 166 28, Czech Republic

[2]Department of Non-ferrous Metals, Refining and Recycling, Faculty of Materials Science and Technology, VSB - Technical University of Ostrava, 17. listopadu 15, Ostrava – Poruba, 708 33, Czech Republic

E-mail: knaisloa@vscht.cz

**Abstract**. The intermetallics based on Ti-Al system are a relatively new class of high-temperature construction materials that combine unique physical and mechanical properties for use in aircraft engines, industrial gas turbines, or automotive industry. They are characterized by low density, good mechanical properties at high temperatures and good oxidation resistance. Addition of silicon into the Ti-Al alloys improves oxidation resistance and ultimate tensile strength in compression, but the complicated production limits wider use of these Ti-Al-Si intermetallic alloys. In this work, two main technologies of preparation of intermetallic compounds will be compared - melting and powder metallurgy.

## 1 Introduction

Intermetallic compounds based on the Ti-Al system belong to a relatively new class of high-temperature construction materials that combine unique physical and mechanical properties suitable for application in aircraft engines, industrial gas turbines or automotive industry. They are characterized by low density, which ranges from 3.9 to 4.2 $g \cdot cm^{-3}$, high melting point, good oxidation resistance at 600 – 800 °C and excellent creep resistance. Currently, commercial application is limited by very low ductility at room temperature reaching a maximum value of 1 % [1-5].

In most cases, Ti-Al intermetallics contain other alloying elements in order to improve their properties, including the already mentioned ductility or fracture toughness, as well as maintaining mechanical properties at higher temperatures. Chemical or mechanical properties can be significantly improved by silicon, which has a minimum solubility in the γ-TiAl phase in the solid state. Therefore, Ti-Al-Si intermetallics form titanium silicide $Ti_5Si_3$ particles in the structure, which are extremely hard and thermally stable due to the high melting point and provide resistance to oxidation at high temperatures [6-9].

However, the production of these Ti-Al-based materials is very complicated. Partially modified conventional methods of melting metallurgy are most often used due to their availability, possible high productivity and relatively low production costs. The limitation of melting metallurgy in the preparation of intermetallics based on Ti-Al is extreme reactivity of the melt with materials of crucibles, high melting points and very exothermic reactions during the formation of intermetallic phases. Due to these undesirable properties of intermetallics, a large number of casting defects occur, for example the formation of cracks and pores, which are the cause of the resulting undesirable porous structure. Some shortcomings can be eliminated by modifying the melting technology, which, however, also contributes to increasing the cost of the process. The second alternative for the production of Ti-Al-based alloys is powder metallurgy, which offers more economical processing of raw materials (production of components with minimal waste). Currently, even larger components can be produced by the means of this technology, but it is still a very expensive process due to high energy consumption and the use of raw materials in the form of metal powders [1, 10, 11].

This work deals with intermetallic alloys based on Ti-Al-Si with various content of aluminium and silicon, which are produced by two methods – mechanical alloying combined with Spark Plasma Sintering and centrifugal casting in the induction vacuum furnace Linn Supercast – Titan followed by hot isostatic pressing.

## 2 Experimental

TiAl15Si15 and TiAl35Si5 alloys (wt. %) were prepared by powder metallurgy method, which included mechanical alloying (MA) followed by Spark Plasma Sintering (SPS) method. Powders of pure titanium (particle size < 50 μm, purity 99 %), aluminium (particle size < 50 μm, purity 99.7 %) and silicon (particle size < 50 μm, purity 99.5 %) were used as starting materials. The 5 g powder mixtures, together with 10 grinding balls, were placed in a grinding vessel, which was subsequently closed, filled with inert gas (argon) to prevent oxidation and placed in a Retsch PM 100 ball mill. The milling time of MA was chosen 240 minutes with a change of direction of rotation at 30 minutes intervals. The conditions of mechanical alloying were used as in our previous article [12]. Mechanically alloyed powder mixtures weighing approximately 5 g were compacted in graphite form using the SPS method. The compaction process took place at a temperature of 1100 °C and a pressure of 80 MPa. The total sintering time was 15 minutes.

The reference samples of TiAl15Si15 and TiAl35Si5 alloys were prepared by centrifugal casting (CC) in a vacuum induction furnace Linn Supercast - Titan at VSB - Technical University of Ostrava. A graphite crucible, which was made of isostatically pressed graphite - SGV5-G B 527 XN, was used for the melting of input pure metal powders. The resulting alloys were obtained in the form of a cylinder with a diameter of 20 mm. The selected TiAl35Si5 alloy was then processed by hot isostatic pressing (HIP) at the temperature of 1260 °C and the pressure of 190 MPa, for 4 hours.

To identify the phase composition, mechanically alloyed and cast TiAl15Si15 and TiAl35Si5 alloys were subjected to X-ray diffraction analysis using a diffractometer (PANalytical X´Pert Pro) followed by evaluation in the PANalytical HighScore plus software package using the PDF 2 database. The ground and polished samples were etched with Kroll reagent (5 ml $HNO_3$, 10 ml HF, 85 ml $H_2O$) to highlight the microstructure of the experimental alloys, which was observed with a Nikon Eclipse MA200 light microscope using NIS Elements software and TESCAN Vega 3 LMU. The mechanical properties were performed on compact sample at room temperature. Vickers hardness with the load of 5 kg (HV 5) was measured from 10 indentions in each sample. Fracture toughness was calculated by Anstis equation [13] (1).

$$Kc = 0.016 \cdot \left(\frac{E}{HV}\right)^{\frac{1}{2}} \cdot \left(\frac{F}{c^{\frac{3}{2}}}\right) \quad (1)$$

Where $Kc$ is fracture toughness (MPa·m$^{1/2}$), $E$ is the modulus of elasticity (GPa), $HV$ is the Vickers hardness (HV 1, GPa), $F$ is the load (N) and $c$ is a half of crack length (mm)

Tests of compressive strength were conducted by the means of the universal testing device LabTest 5.250SP1-VM. Values of ultimate tensile strength in compression were determined from the measured loading curves.

## 3 Results and discussion

Phase composition (Table 1) of Ti-Al-Si alloys prepared by powder metallurgy (mechanical alloying and Spark Plasma Sintering) and melting metallurgy (centrifugal casting, hot isostatic pressing) was determined by X-ray diffraction analysis. The TiAl15Si15 and TiAl35Si5 alloys are formed by titanium aluminide TiAl and titanium silicide $Ti_5Si_3$, which has a high thermodynamic stability. The formation of titanium silicide is due to the mutual affinity of silicon and titanium, and is therefore preferentially formed. The remaining titanium reacts with aluminium to form the titanium aluminide matrix. $TiAl_2$ phase is also in these alloys, forming the matrix of the alloys. The TiAl35Si5 alloy has the highest ratio of aluminium to titanium, therefore the $TiAl_2$ and $TiAl_3$ phases are formed, which are very rich in aluminium, and together with the TiAl phase form the matrix of this alloy. All studied cast alloys (TiAl15Si15 and TiAl35Si5) are consist of titanium silicide $Ti_5Si_3$ and titanium aluminide TiAl.

**Table 1.** Phase composition of the Ti-Al-Si alloys

| Alloy | Phases |
|---|---|
| TiAl15Si15 MA+SPS | $Ti_5Si_3$, TiAl, $TiAl_2$ |
| TiAl35Si5 MA+SPS | $Ti_5Si_3$, TiAl, $TiAl_2$, $TiAl_3$ |
| TiAl15Si15 CC | $Ti_5Si_3$, TiAl, C |
| TiAl35Si5 CC | $Ti_5Si_3$, TiAl |
| TiAl35Si5 CC+HIP | $Ti_5Si_3$, TiAl |

Microstructures of Ti-Al-Si alloys prepared by powder and melting metallurgy are shown on Fig. 1. Metallographic samples were formed from the prepared compacted alloys, which were subsequently etched with Kroll's reagent to visualize the structure of the preset titanium aluminides and silicides. These aluminides have reduced chemical resistance, and therefore they are darker areas in the microstructure images. Conversely, silicides are lighter areas. In the electron micrographs, the presence of the titanium silicide $Ti_5Si_3$ (lighter parts) and the titanium aluminide $Ti_xAl_y$ (darker parts) can be distinguished. The compacted alloys excel in very low porosity due to the presence of a fine-grained structure obtained by mechanical alloying followed by SPS. TiAl35Si5 and TiAl15Si15 alloys reach the porosity of 0.95 and 0.4 vol. %, respectively. On the other hand, the TiAl15Si15 alloy prepared by casting consists of large sharp-edged particles of $Ti_5Si_3$ silicide, which are undesirable in terms of mechanical properties. This alloy was contaminated by carbon (in the form of graphite) due to the preparation of the alloy in a graphite crucible. The TiAl35Si5 alloy contains a lower silicon content, and therefore the formation of a hard and brittle phase of the silicide is strictly limited. The porosity of casting of the Ti-Al-Si does not exceed 0.9 vol. %. TiAl35Si5 alloy was then processed by hot isostatic pressing. After HIP, the porosity decreased.

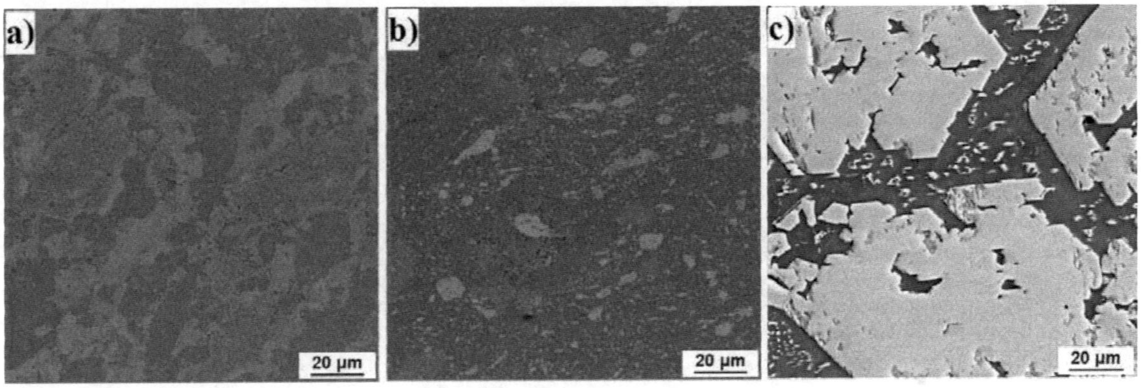

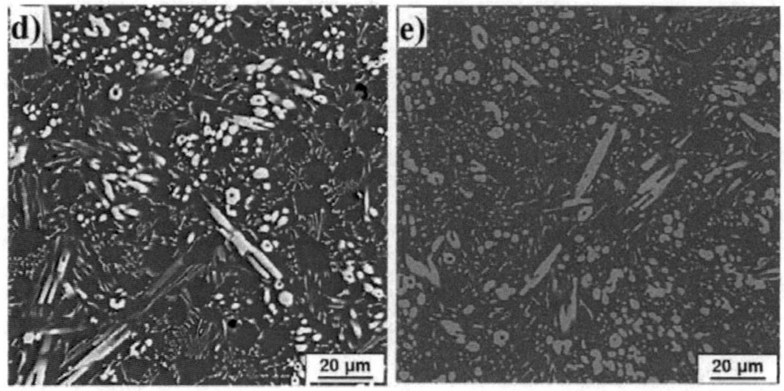


**Figure 1.** Scanning electron micrographs of the Ti-Al-Si alloys: a) TiAl15Si15 (MA+SPS), b) TiAl35Si5 (MA+SPS), c) TiAl15Si15 (CC), d) TiAl35Si5 (CC), e) TiAl35Si5 (CC+HIP)

A comparison of the hardness of the studied alloys is shown in Fig. 2. The hardness of Ti-Al-Si alloys prepared by the conventional centrifugal casting method ranges from 374 HV5 for TiAl35Si5 (CC + HIP) to 459 HV5 for TiAl15Si15 (CC). The measurement results show that the hardness increases with increasing silicon content in the alloys. This allows the formation of a hard phase of titanium silicide $Ti_5Si_3$. Therefore, the TiAl15Si15 alloy achieves higher hardness values compared to the TiAl35Si5 alloys, which contain tougher titanium aluminides in their structure and the content of hard phase $Ti_5Si_3$ is low here. The cast TiAl35Si5 alloy, which was further subjected to hot isostatic pressing (HIP) for comparison, did not show a significant change in hardness. The average measured values of both cast TiAl35Si5 alloys are almost identical. The hardness of reference Ti-Al-Si alloys prepared by mechanical alloying followed by SPS (MA + SPS) takes values of 722 HV5 for TiAl35Si5 and 871 HV5 for TiAl15Si15, which reaches the highest value of all studied alloys. The hardness of Ti-Al-Si alloys prepared by this method is higher than those of alloys prepared by conventional centrifugal casting due to the formation of finer and more homogeneously distributed intermetallic phases that is generally desirable to improve the mechanical properties of alloys.

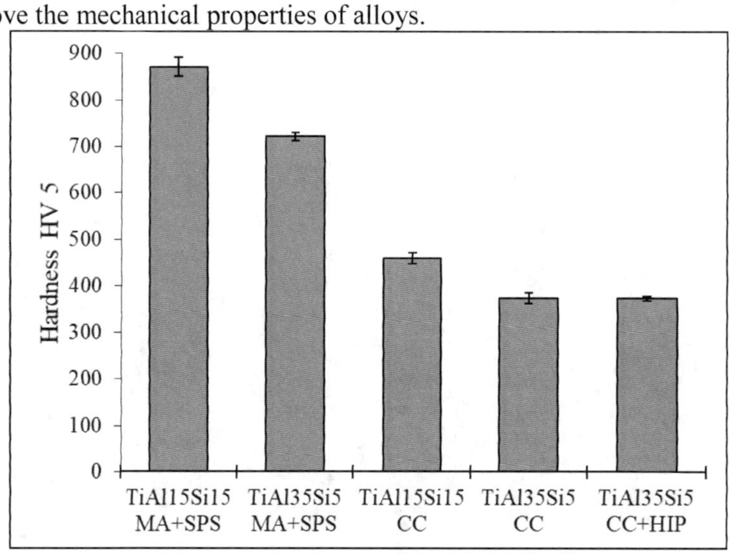


**Figure 2**. Hardness of the Ti-Al-Si alloys

Intermetallics are among the materials that are commonly classified as brittle. For this group of materials, fracture toughness and compressive strength are decisive for potential use. The fracture toughness of Ti-Al-Si alloys was measured by the Anstis method (Equation 1), which, among other parameters, considers crack sizes based on the material after indentation of the hardness indenter (Fig. 3b). However, in the studied cast alloys (Fig. 3a), the fracture toughness was unmeasurable by this method due to the absence of cracks and their propagation, which was most likely in the aluminide phase (Fig. 3). In contrast, the crack propagates through alloys of the same composition, which were prepared by powder metallurgy followed by SPS, and therefore the fracture toughness of these reference alloys could be measured by this method. Figure 3 illustrates the behavior of a TiAl35Si5 prepared by centrifugal casting (a) and TiAl35Si5 alloy prepared by mechanical alloying method followed by SPS (b) after indentation.

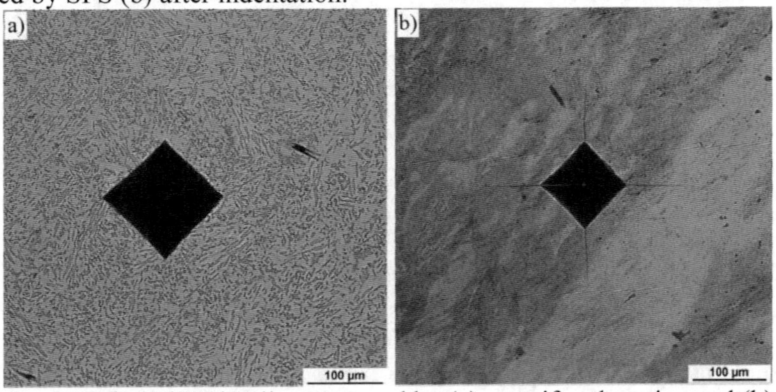


**Figure 3**. Behavior of TiAl35Si5 alloy prepared by (a) centrifugal casting and (b) mechanical alloying method followed by SPS after indentation

Based on the obtained results of mechanical properties and knowledge about the microstructure of the studied alloys, it can be determined that the coarser structure of cast samples led to an increase in fracture toughness. The positive influence of the coarser structure of cast alloys of the Ti-Al-Si system on the resulting fracture toughness is also confirmed and discussed in [11]. On the other hand, reference alloys prepared by mechanical alloying followed by SPS tend to brittle fracture. This statement was confirmed by the Anstis fracture toughness calculation, where the reference alloys showed very low values of 0.32 ± 0.05 MPa·m$^{1/2}$ for TiAl15Si15 (MA+SPS) and 0.60 ± 0.05 MPa· m$^{1/2}$ for TiAl35Si5 (MA+SPS). The higher calculated value for the reference alloy TiAl35Si5 (MA+SPS) is due to the higher content of aluminium and the low content of silicon, when the tougher phases of aluminide predominate in the structure over the brittle phase of titanium silicide $Ti_5Si_3$.

The compressive strength measurements of the alloys were performed three times for each material, and therefore the resulting values of the ultimate compressive strength UCS correspond to the average of the given measurements. The whole process took place at room temperature. Ti-Al-Si alloys produced by the conventional centrifugal casting method increase the strength and ductility limit with increasing aluminium content, and therefore the highest values are shown by the cast alloy TiAl35Si5 (1867 MPa). Due to the higher aluminium content and low silicon content, the resulting alloy structure is largely formed by relatively plastic titanium aluminides, which guarantee higher ductility of the alloy. The formation of a brittle and hard phase of the titanium silicide is limited here due to the lower silicon content, nevertheless its strengthening effect appears to be sufficient on the basis of the obtained results. The higher ultimate compressive strength of the cast TiAl35Si5 alloy was also supported by the chosen centrifugal casting technology due to the gain of a material with a minimum number of cracks and low porosity. In the case of the cast TiAl35Si5 alloy after HIP (1801 MPa), there was an expected decrease in ultimate compressive strength due to the reduction of stress in the whole volume of the material. The TiAl15Si15 (CC) alloy shows lower values of ultimate compressive strength (1205 MPa) due to the higher amount of silicon, which appears to be negative with this processing method and increases the proportion of brittle $Ti_5Si_3$ at the expense of the tougher aluminide phase TiAl. It was the higher amount of hard and brittle silicides, which form large sharp-edged particles in the given alloy (see Fig. 1c), contributed to a significant decrease in the strength and fracture toughness of the given alloy. The reference alloy TiAl15Si15 prepared by mechanical alloying followed by SPS (2055 MPa) showed the highest values of ultimate compression strength of all studied alloys. A balanced ratio of aluminides and silicides combined with a homogeneous and fine-grained structure allowed a rapid increase in the UCS. The reference alloy TiAl35Si5 prepared by the powder metallurgy method (1850 MPa) showed comparable values of compressive strength with the same alloy type prepared by the casting method.

## 4 Conclusion

In this work, TiAl15Si15 and TiAl35Si5 alloys prepared by conventional centrifugal casting in a Linn Supercast - Titan vacuum induction furnace were studied and subsequently compared with reference samples prepared by mechanical alloying method followed by SPS. The microstructure of cast alloys was formed by a mixture of titanium silicide $Ti_5Si_3$ and titanium aluminide TiAl, which formed the matrix of the alloys. Due to the high overheating of the melt, the TiAl15Si15 (CC) alloy caused undesired carbon contamination. For the reference alloys (MA+SPS), the matrix was further enriched with phases richer in aluminium ($TiAl_2$, $TiAl_3$). Based on the obtained results of cast alloys, it was found that there is a growth of hard but brittle phase $Ti_5Si_3$ with increasing silicon content. For this reason, the TiAl15Si15 (CC) alloy was formed by large sharp-edged silicide particles, which significantly contributed with undesired contamination to the deterioration of the resulting properties. The TiAl35Si5 (CC) alloy showed a finer-grained structure without signs of contamination with minimal cracking and low porosity (0.9 vol. %), and therefore a significant improvement in selected mechanical properties was achieved, but at the expense of hardness. Due to the additional hot isostatic pressing of this alloy, the porosity (0.7 vol. %) and the total stress in the material decreased, and thus the ultimate compressive strength decreased. Other measured mechanical properties did not show significant changes. Reference samples prepared by powder metallurgy showed a finer and more homogeneous structure with low porosity up to 0.9 vol. %. The hardness of these alloys reached doubled values compared to the cast samples. The TiAl15Si15 alloy (MA+SPS) showed the best mechanical properties of all studied alloys. However, compared to cast alloys, mechanically alloyed alloys were characterized by the considerable brittleness.

**Acknowledgements**
This work was supported from the grant of Specific university research – grant No A1_FCHT_2020_003 and A2_FCHT_2020_046.

## References

[1] Lapin J 2009 Metal (Hradec nad Moravicí) 1-12
[2] Qu S J, Tang S Q, Feng A H, Feng C, Shen J, Chen D L 2018 *Acta Mater.* **148** 300-310
[3] Kvanin V L, Bilikhina N T, Vadchenko S G, Borovinskaya I P, Sychev A E 2008 *Inorg. Mater.* **44** 1194-1198
[4] Vojtěch D, Lejček P, Kopeček J, Bialasová K 2009 Metal (Hradec nad Moravicí) 1-7
[5] Novák P 2012 *Chem. listy* **106** 884-889
[6] Stoloff N S, Sikka V K 1996 Physical metallurgy and processing of intermetallic compounds (Springer US) p 684
[7] Vojtěch D, Novák P, Macháč P, Morťaniková M, Jurek K 2008 *J. Alloys Compd.* **464** 179-184
[8] Guan Z Q, Pfullmann T, Oehring M, Bormann R 1997 *J. Alloys Compd.* **252** 245-251
[9] Novák P, Kříž J, Průša F, Kubásek J, Marek I, Michalcová A, Voděrová M, Vojtěch D 2013 *Intermetallics* **39** 11-19
[10] Moll J H, McTiernan B J 2000 *Metal Powder Report* **55** 18-22
[11] Knaislová A, Novák P, Kopeček J, Průša F 2019 *Materials* **12** 3084
[12] Knaislová A, Linhart J, Novák P, Průša F, Kopeček J, Laufek F, Vojtěch D 2019 *Powder Metall.* **62** 56-60
[13] Quinn G 2006 *Ceramic Engineering and Science Proceedings* 45-62

IOP Conf. Series: Materials Science and Engineering 1178 (2021) 012029 doi:10.1088/1757-899X/1178/1/012029

# Application of thermal gravimetric analysis and comparison of polyethylene films before and after exposure in various chemical solutions

**T Kojnoková[1], L Markovičová[1] and F Nový[1]**

[1]University of Žilina, Faculty of Mecahnical Engineering, Department of Materials Engineering, Univerzitná 8215/1, 010 26 Žilina, Slovakia

E-mail: tatiana.kojnokova@fstroj.uniza.sk

**Abstract**. In this paper, there are published experimental results of research in the field of polymers. The thermal analysis conveys information about changes in material effected by the melting of the material. The investigated material is low-density polyethylene (LD – PE). The LD – PE films were exposed to aggressive media with different values of potencia hydrogeni (pH). The experiment aims to apply thermal gravimetric analysis (TGA), one of the methods of thermal analysis. TGA monitors the weight of polyethylene films as a function of temperature. Original LD – PE film and the effect of media is evaluated based on TGA and evaluation of the surface morphology (macrostructure) after exposure. The largest weight loss is noticed in the original film without the influence of media. At 200 °C, the wax products are released from the material. During thermal analysis is observed a weight loss of up to 1 mg in the investigated film samples. LD – PE film after the effect of Coca-Cola has the highest decrease in mass and the most changed its macrostructure. LD – PE film after effect of Saponate has the lowest decrease in mass and its macrostructure is almost the same as the original LD – PE film after production technology.

## 1 Introduction

The usage of polymer materials has increased in various engineering fields. Polymer materials can be exposed to aggressive conditions and work in specific conditions. Polyethylene (PE), which is a thermoplastic polymer [1], belongs among the most commonly used polymers. It is produced industrially by polymerization from oil and natural gas. PE can take various forms mainly classified based on molecular weight, density, degree of crystallinity, and chain branching. PE has applications in everyday life due to its simple structure and easy manufacturing. Due to its positive properties, which make them very attractive material, there is a high production of polyethylene products. Polyethylene can be used in various applications and also like wrapping for the transport of liquids. This leads to a major problem of waste disposal. The resulting waste can be recycled for the production of different products or the original application [2].

PE can be classified according to their density which is the result of the degree of crystallinity and type and content of branches. Among commonly accessible types of PE belongs LD – PE, HD – PE, LLD – PE. The LD – PE has a chaotic structure with short and long-chain branches and a lower degree of crystallinity in comparison to HD – PE [3]. Transformation of the polyethylene product to another product is also associated with the analysis of changes in the composition and properties of substances under the influence of defined temperature changes. Thermal gravimetric analysis

IOP Conf. Series: Materials Science and Engineering 1178 (2021) 012029 doi:10.1088/1757-899X/1178/1/012029

monitors the properties of substances as a function of temperature. The result of TGA is a thermogravimetric curve (TGA curve), which shows the dependence of sample weight on temperature or time. During thermal gravimetric analysis, energy is released and absorbed, or magnetic, electrical, and optical properties change [4].

## 2 Material and methodology

The experiment was performed on the original LD – PE films (O LD – PE films) and on the films that were filled with the media at different pH values. The exposure of effected samples was 3 months. The investigated LD – PE films are produced by extrusion and they are of thickness 50 ± 2 μm welded in a longitudinal direction. The exposure period was followed by an evaluation of the TGA curves of LD – PE films and their comparison after treatment with the media. Manufactures pack many drinks and cleansing articles into PET bottles and then produce a lot of polymer waste. Media were chosen from commonly used preparations in the household. Selected media include Savo, Saponate, and Coca-Cola. Savo is an intensive disinfectant with a pH of 13.37. Saponate is a grease removal agent with a pH of 3.35. Coca-Cola is a popular round-flavored soft drink with a pH of 2.23. The LD – PE films were subjected to a thermal load above the melting point. The tabular melting point of LD – PE is the temperature in the range from 100 to 115 °C [5].

### *2.1 Thermal gravimetric analysis*

Thermal gravimetric analysis studies the properties of materials as a function of temperature. The primary use of TGA of polymers is the detection and quantification of the crystalline melting process. This method measures the amount and rate of weight changes in a material as a function of temperature or time in a controlled atmosphere [6]. The used TGA instrument positions the sample vertically. Thermal analysis was performed on a TGA analyzer from company Linseis. The chemical or physical processes that occur when the sample is heated are monitored and measured experimentally. Before measurement, samples are weighed from all investigated films. The sample weight should be in the range from 10 to 25 mg. The weighed sample is placed in a high-temperature crucible on a sample pan. They shall be placed in the holder together with the reference crucible. General conditions and temperature profiles are set in program STA2. Temperatures are selected to capture any changes in the material as a function of temperature. The temperature profile is as follows: heating to 200 °C for 10 min, holding at this temperature for 25 min, and cooling to 30 °C for 50 min. The temperature of 200 °C is chosen based on PE treatment. The processing temperature range for LD – PE is 180 ÷ 240 °C. The measurement is carried out at atmospheric pressure. The TGA curve fits and the obtained gross curves from the loss experiment are further adjusted in the Linseis TA Evaluation program. Measured data are expressed in mg of units as weight change (Δ m). TGA of the material consists of 3 stages as shown in Figure 1: stage I. – melting, stage II. – release of the products from the material and stage III. – solidification.

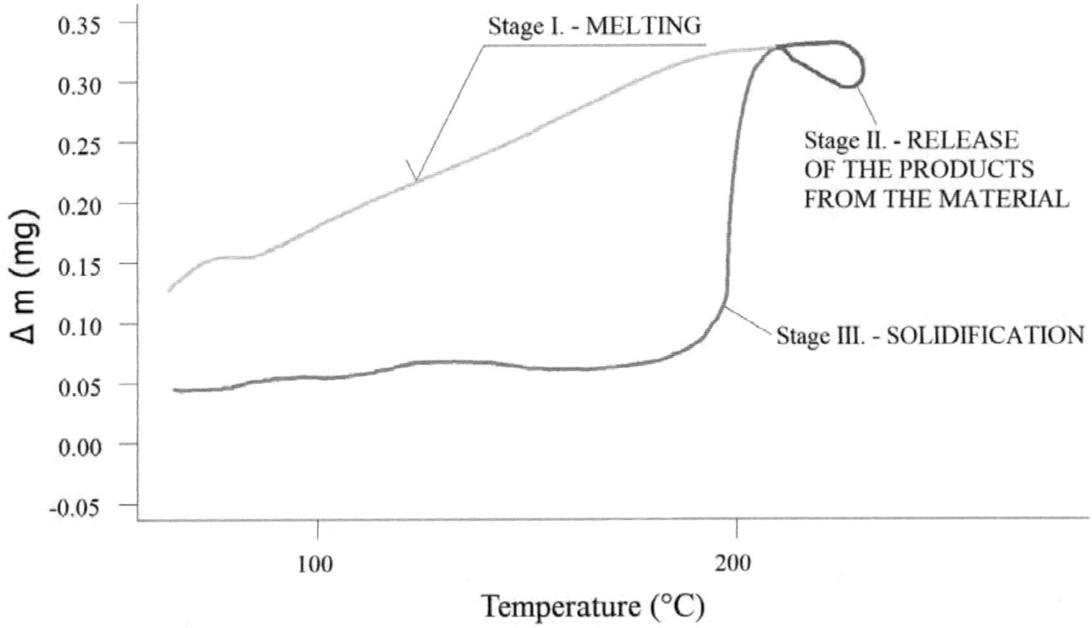


**Figure 1.** Stages of TGA of material

## 3 Results

Thermal gravimetric analysis monitored the change in weight as a function of the temperature of the individual LD – PE films. Figure 2 shows the measured TGA curves obtained by TGA for investigated films. Above 200 °C, energy losses occurred due to the thermal decomposition of the sample.

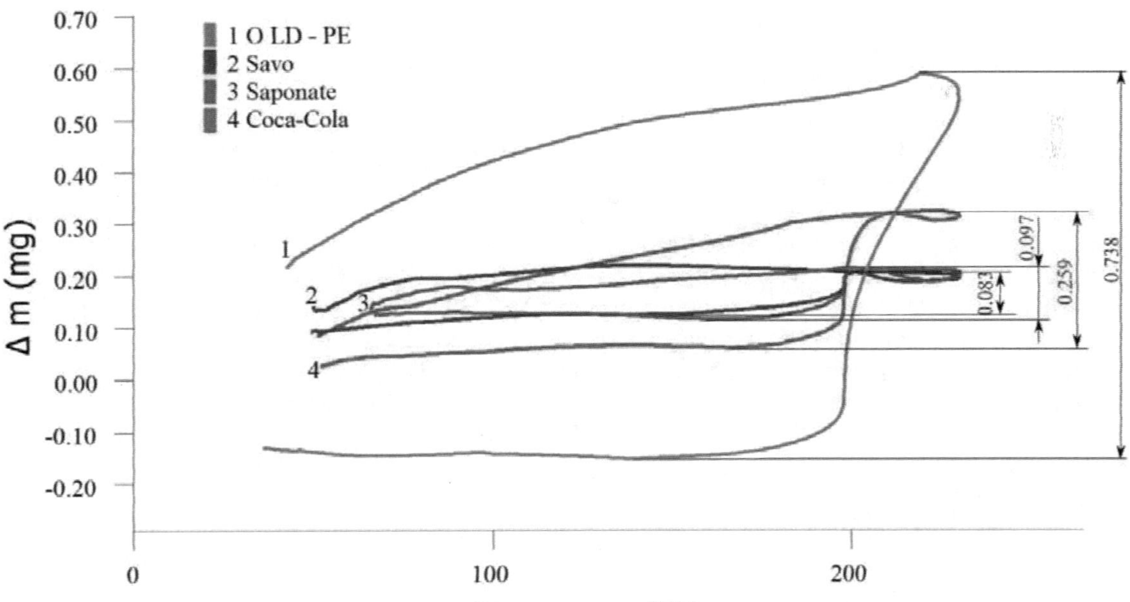


**Figure 2.** TGA curves of investigated LD – PE films

The original (O LD – PE) film without the effect of media has the highest Δ m. The LD – PE films with effect of media has lower Δ m. Used media caused crosslinking and stronger bonds between molecules. The weight loss of the O LD – PE film is therefore the highest and it equals 0.738 ± 0.036 mg. The biggest weight change after 3 months by the media treatment was with Coca-Cola effected LD – PE film, 0.259 ± 0.025 mg. On the other hand, the smallest temperature-dependent change in weight of the Savo-treated LD – PE film was of 0.083 ± 0.006 mg. The change in weight of the LD – PE film treated with Saponate was of 0.097 ± 0.018 mg. The thermal gravimetric analysis leads to processes that are associated with the melting of the material. Weight loss is caused by processes such as e.g. dehydration, sublimation, evaporation, and reduction. The result of the thermal analysis and the shape of the curves obtained by the thermal analysis can be influenced by several factors including sample preparation, carrier size, furnace atmosphere, pressure, humidity, and temperature mode. Change in weight of investigated LD – PE films are compared through a bar chart in Figure 3 [4-6].

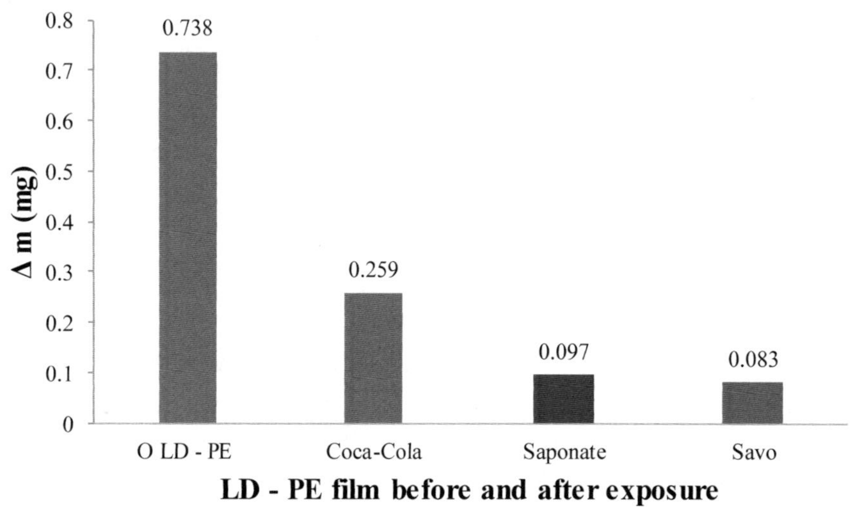


**Figure 3.** Comparison of Δ m for investigated LD – PE films

The effect of media has influence on the macrostructure of LD – PE film. The macrostructure of experimental material can contain 3 types of stripes: a) deep stripes, b) medium-deep stripes, c) shallow very fine strips. The measured area of the samples was of 1 $cm^2$. After exposure – 3 months, the macrostructure of the individual films on the ZEISS illuminated microscope was also evaluated using the AxioVision LE64 program. O LD – PE film (Figure 4a) contains visible deep stripes after technological processing, shallow stripes, and medium-deep stripes. In Figure 4b it is possible to see the surface morphology of the LD – PE film effected by Savo. The surface morphology contains original visible deep strips and medium-deep strips. Shallow strips have been removed from the material. The surface morphology of the LD – PE film effected by the Saponate (Figure 4c) is most similar to the original condition without the presence of shallow strips. The LD – PE film effected by Coca-Cola (Figure 4d) is most changed from the original state of the material. The medium-deep strips and shallow strips have been removed after treatment. Only deep strips remained. It is possible to observe the occasional occurrence of pores.

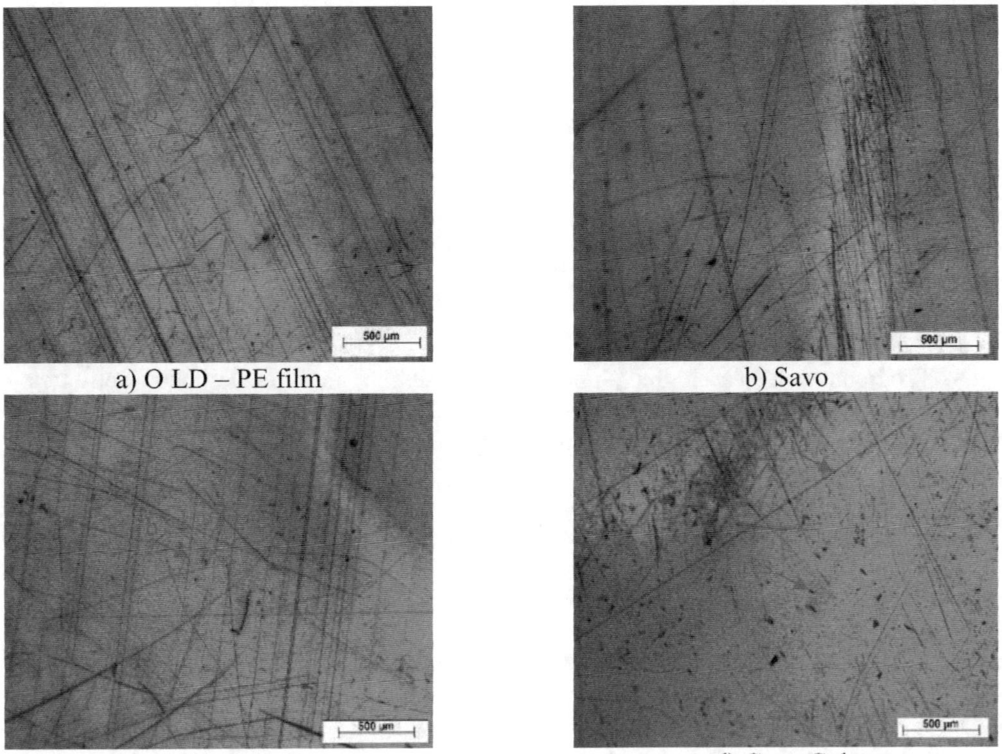

a) O LD – PE film b) Savo

c) Saponate d) Coca-Cola

**Figure 4.** Comparison of surface morphology of investigated LD – PE films

Due to the aggressive environments, polyethylene is degraded, which is manifested by changes in macromolecules, and thus the aging of the material. Due to the increased temperature, polyethylene loses its flexibility and strength. This roughens the material and gives it a greater ability to absorb fluid. Most polymers are attacked by bacteria, which change the structure of the polymer and damage the material [7]. After 3 months of exposure of the individual media to the LD – PE film, the material was degraded. After exposure, Coca-Cola contained mold, which roughened the surface of the material. This was also reflected in the surface morphology changes of the effected films, which contained pores compared to the original film. The occasional occurrence of porosity, which allows the penetration of media into the composition and damage to the LD – PE film. The porous structure is little visible on the LD – PE film by the effect of Coca-Cola. Based on the measurement, it can be stated that the LD – PE film shows the best properties after the application of the Saponate. Its macrostructure is the most similar to the macrostructure of the original LD – PE film [8].

## 4 Conclusions

The aim of this work was to study the impact of the selected media caused a strengthening of chemical bonding and this results in lower weight loss than the original LD – PE film, where the weight loss of tested samples was the largest. Based on obtained results from the thermal gravimetric analysis the following can be concluded:

- TGA monitored the change in weight depending on the temperature.
- In the original LD – PE film, the cleavage of the bonds to the parent chain occurred, which could be due to weak bonds in polyethylene.
- The largest weight loss was in the original LD – PE film without the influence of media.
- Selected media caused crosslinking of structure and reinforcement of the material.
- At 200 °C, the wax products were released from the material.

- During thermal gravimetric analysis was observed a weight loss of up to 1 mg with measurement deviations in the investigated LD – PE film samples.
- LD – PE film after the effect of Coca-Cola has the highest decrease in mass and the most changed its macrostructure.
- LD – PE after effect of Saponate has the lowest decrease in mass and its surface morphology is the most similar to the original LD – PE film in as manufactured condition.
- For further research, it is recommended to study the surface morphology at higher magnifications and a longer time of exposure.

**Acknowledgements**

This research was supported by the projects VEGA 1/0951/17, KEGA 012ŽU-4/2019 and the project of Operational Programme Research and Innovation: Research and development activities of the University of Zilina in the Industry of 21st century in the field of materials and nanotechnologies, No. 313011T426. The project is co-funding by European Regional Development Fund.

**References**

[1] Kloužková A, Zemanová P, Kloužek J and Pabst W 2012 *Termická analýza* (Praha VŠCHT) p 66
[2] Noorunnisa Khanam P and Almaadeed M A 2015 *Processing and characterization of polyethylene-based composites* (Advanced Manufacturing: Polymer & Composites Science) vol 1(2) pp 63-79
[3] Shebani A, Klash A and Elhabishi R 2018 *The Influence of LDPE Content on the Mechanical Properties of HDPE/LDPE Blends* Res Dev Material Sci. vol 7(5)
[4] Uthaman A, Xian G and Thomas S 2020 *Durability of an Epoxy Resin and Its Carbon FiberReinforced Polymer Composite upon Immersion in Water, Acidic, and Alkaline Solutions* (China: Harbin Institute of Technology) p 23
[5] Kojnoková T, Markovičová L and Gaňa D 2020 *Zmeny vlastností LD-PE fólií po expozícii vo vybraných prostrediach* (SEMDOK 2020: 25th International Seminar of Ph.D. Students) pp 64-68
[6] Sepe M P 1997 *Thermal analysis of polymers* Rapra review reports vol 8(11) pp 13-18
[7] Stodghill S P 2010 *Thermal Analysis – A Review of Techniques and Applications in the Pharmaceutical Sciences* American Pharmaceutical Review vol 13(2)
[8] Markovičová L, Zatkalíková V, Kojnoková T, Gaňa D and Liptáková T 2020 *The physical – mechanical properties of low-density polyethylene films* (Development of Materials Science in Research and Education (DMSRE29) IOP Publishing vol 726)

# Effect of high-energy attrition milling and $La_2O_3$ content on the microstructure of Mo-$La_2O_3$ composite powders

**P Komarov [1,*], S Tkachenko [1], M Remešová [1], A Deák [2], D Crawford [3], M Casas-Luna [1], V Bednaříková [1], E Bodoki [4], J Cihlář [1] and L Čelko [1]**

[1] Central European Institute of Technology, Brno University of Technology, Purkyňova 123, 612 00 Brno, Czech Republic

[2] Research Centre for Natural Sciences, Institute of Materials and Environmental Chemistry, Magyar Tudösk Körútja 2, 1117 Budapest, Hungary

[3] School of Chemistry and Biosciences, University of Bradford, Richmond Road, BD7 1DP, Bradford, UK

[4] Department of Analytical Chemistry, Iuliu Haţieganu University of Medicine and Pharmacy, Louis Pasteur Street 4, RO-400349 Cluj-Napoca, Romania

*E-mail: pavel.komarov@ceitec.vutbr.cz

**Abstract.** Mo-$La_2O_3$ composites are potential high-temperature materials for future technology devices operating at temperatures above 1300 °C because of their excellent thermal stability, high mechanical properties and good creep resistance. In this study, we focused on the preparation of Mo-matrix/lanthanum oxide ($La_2O_3$) composite powders using high-energy attrition milling. The effects of rotational milling speed (350 and 800 rpm) and $La_2O_3$ content (2.5 and 10 vol. %) on the microstructural evolution, phase composition, morphology, and distribution of the second phase in the produced composite Mo-$La_2O_3$ powders were investigated in details. The results show that the most interesting composite powder was Mo-10 vol.% $La_2O_3$ produced using a rotational speed of 800 rpm, which exhibited better distribution, smaller particle size and higher amount of ceramic phase introduced in the interiors of the Mo grains.

## 1 Introduction

At present, in connection with the upcoming power and energy engineering technology improvements, a lot of attention is drawn to the development of materials with superior chemical, physical and mechanical properties at elevated temperatures. Molybdenum (Mo) alloys and Mo-based composites are promising high-temperature materials to be applied in aerospace, energy generation processes, electronics, telecommunication, and defense fields. The reason for that is the combination of very high melting point (~2623 °C), good creep strength at elevated temperatures, high thermal conductivity (138 $W \cdot m^{-1} \cdot C^{-1}$), low specific heat (25.1 - 28.4 $J \cdot K^{-1} \cdot mol^{-1}$), comparable density to the Ni-superalloys, etc. [1]. The application of Mo-based alloys and composites in aerospace engines and ground-based stationary gas turbines can be a cost-effective solution to increase their efficiency and to decrease their environmental impact due to reduced amount of $CO_2$ emitted into the atmosphere. However, low oxidation resistance and ductile-to-brittle transition behavior of Mo and its alloys are to be improved to ensure the best performance in the aforementioned applications [2].

Several approaches were employed, including substitutional alloying by silicon and boron as well as the second phase strengthening with oxide ($ZrO_2$, $La_2O_3$, etc.), carbide (TaC, ZrC), or other ($Mo_3Si$, $Mo_5SiB_2$) reinforcements [3-6]. The introduction of rare-earth elements' oxide (REOs) particles, such as $La_2O_3$, into Mo-based composite microstructure leads to their significant strengthening, creep resistance enhancement, and oxidation resistance improvement due to strong metal-REOs-oxygen connection. The REOs particles might play role of the barrier for the grain size growth, increasing the ductility, toughness, strength, and elastic modulus, which in turn reduces the ductile-to-brittle transition temperature (DBTT). The studies on the preparation of the Mo-based composites by high-energy milling are rather scarce [7-8] and still need to be performed systematically. In this context, the attrition milling is an attractive method for production of Mo-$La_2O_3$ composite powders, which then can be potentially used for thermal spray or advanced sintering (e.g., spark plasma sintering). Due to intensive impact and shearing action occurring in this type of milling, effective and homogenous particle refinement results in obtaining pure and ultra-fine powder particles (as small as 20 nm) of solid materials with uniform microstructure and improved properties [9-10].

The present work studies the preparation of the composite Mo-matrix powder with 2.5 and 10 vol. % of lanthanum oxide ($La_2O_3$), using two different rotation speed in rotational high-energy ball mill. Particle morphology, microstructure and phase composition of the fabricated composite powders were examined and analyzed in detail.

## 2 Materials and methods

For the fabrication of the Mo-based experimental powder composition with 2.5 and 10 % vol. of $La_2O_3$, the commercially available 45-90 μm molybdenum (GTV, Germany) and 99.3 % pure lanthanum oxide (Luoyang Golden Egret Ceotools Co., China) powders were utilized. Composite powder mixtures were prepared by measuring of tap density of ceramic powders, considering the volume of 140 g of molybdenum powder. The powder mixtures were processed in the high-energy attrition ball mill Simoloyer CM01 (Zoz Gmbh, Germany) in ambient atmosphere. The ball-to-powder ratio was kept at 100:6, using stainless steel balls with 4.5 mm in diameter. Prior to the powder processing, pre-milling and homogenization of powder mixtures was provided by mixing at 700 rpm for 5 minutes, adding ~0.5 ml of isopropyl alcohol as a lubricant. For the final processing, the low- and the high-speed rotation modes conducted at 350 rpm and 800 rpm, respectively, for 60 minutes were used (Table 1).

**Table 1.** Studied Mo-$La_2O_3$ powder mixtures and parameters of their milling.

| Designation of the powder | Chemical composition | | Milling parameters | |
|---|---|---|---|---|
| | Vol. % | Wt. % | Preliminary homogenization | Final milling |
| Mo-La2.5-350 | Mo – 2.5 vol.% $La_2O_3$ | Mo – 1.05 wt.% $La_2O_3$ | 700 rpm, 5 min | 350 rpm, 60 min |
| Mo-La10-350 | Mo – 10 vol.% $La_2O_3$ | Mo – 4.2 wt.% $La_2O_3$ | 700 rpm, 5 min | 350 rpm, 60 min |
| Mo-La2.5-800 | Mo – 2.5 vol.% $La_2O_3$ | Mo – 1.05 wt.% $La_2O_3$ | 700 rpm, 5 min | 800 rpm, 60 min |
| Mo-La10-800 | Mo – 10 vol.% $La_2O_3$ | Mo – 4.2 wt.% $La_2O_3$ | 700 rpm, 5 min | 800 rpm, 60 min |

The morphology of powder particles in the fabricated composite powders and their elemental analysis were studied by means of scanning electron microscope LYRA 3 (SEM; Tescan, Czech Republic) equipped with X-ray energy dispersive spectroscopy (EDX) detector XFlash 5010 (Bruker, USA). EDX analysis was performed onto 5 areas at least. To exclude electrostatic charging, prior to SEM and EDX study, powder particles were covered by thin carbon layer (~25 nm) using EM ACE 600 coater (Leica Microsystems, Germany). Phase composition of initial and fabricated powders was determined by X-ray diffraction analysis using SmartLab 3kW diffractometer (Rigaku, Japan) by scanning in Bragg-Brentano geometry using Cu Kα radiation (50 kV, 50 mA) for the scanning 2θ range between 10-90° at 3 °/min.

## 3 Results and discussion

*2.1 Initial powders*

XRD analysis of the commercially purchased molybdenum and lanthanum oxide powders showed that they were single-phase materials (Figure 1a). Molybdenum and lanthanum oxide powders showed body-centred cubic (ICSD 65-007-6279) and hexagonal (ICSD 98-015-4586) crystallographic lattices, respectively.

Scanning electron microscopy revealed that Mo powder particles morphology is related to the agglomerated and sintered type with spherical-shaped agglomerates of 30-80 μm composed from fine particles (Figure 1 b). Analysis of the initial $La_2O_3$ powder showed irregularly shaped particles in the range of 4-20 μm (Figure 1 c).

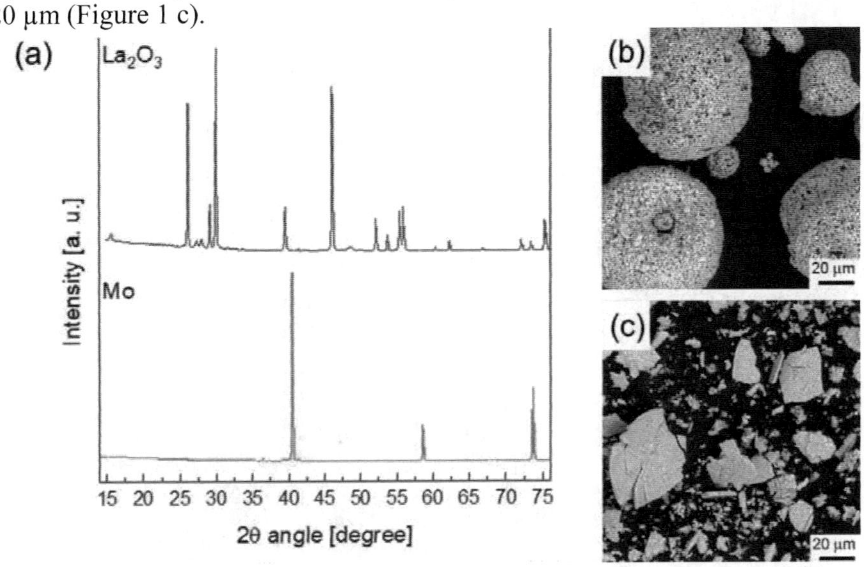


**Figure 1.** Phase composition (a) and morphology of (b) Mo and (c) $La_2O_3$ initial powders.

*2.2 Effect of milling*

The XRD investigation of fabricated composite powders did not detect any intermediate $La_XMo_YO_Z$ compounds after milling in both milling modes, as shown in the example of the composite with 10 vol.% of $La_2O_3$ (Figure 2). It means that no chemical reaction or alloying occurs between Mo and $La_2O_3$ during low- and high-energy milling conditions. At the same time, the appearance of lanthanum hydroxide $La(OH)_3$ shows that the hydroxylation of $La_2O_3$ compound occurs during milling, which is probably due to milling in ambient conditions or use the isopropyl alcohol lubricant. The occurrence of the hydroxide is typical in the chemistry of the lanthanum oxide, moreover, the hydroxide easily decomposes to the oxide by heat treatment around 300°C [11]. Furthermore, after the milling in the high-energy milling mode, a slight half-width peak broadening has been observed along with slight decrease of the peak intensity for all the samples. It can be explained by higher deformation of Mo matrix during milling as well as refinement of $La_2O_3$ particles. As both Mo and $La_2O_3$ are easy mechanically deformable [12], intensive deformation during high-energy milling might provide interaction between ceramic and metallic phases with further improvement of grain refinement of molybdenum matrix.

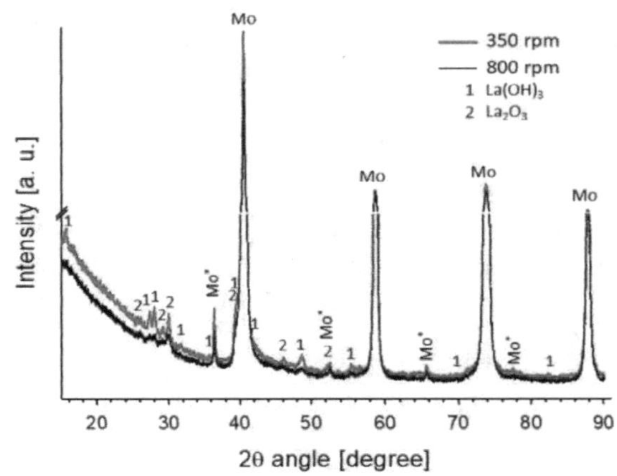


**Figure 2.** Phase composition of Mo-La10-350 (red line) and Mo-La10-800 (black line) powder mixtures after milling.

The effect of the milling processing using 350 rpm and 800 rpm onto the composite powder morphology is shown in Figure 3. The low-energy milling of Mo-La2.5-350 and Mo-La10-350 powder samples resulted mostly in the refinement of Mo and $La_2O_3$ particles, while the formation of larger flake-like powder agglomerates from the deformed and welded Mo particles and crushed $La_2O_3$ did not occur readily. In contrast, during high-energy milling, higher delivered energy to the powder system (samples Mo-La2.5-800 and Mo-La10-800) resulted in the formation of large amount of cold-welded particles agglomerates of ~20-50 μm in size, whose number was the highest in the Mo-La10-800 sample.

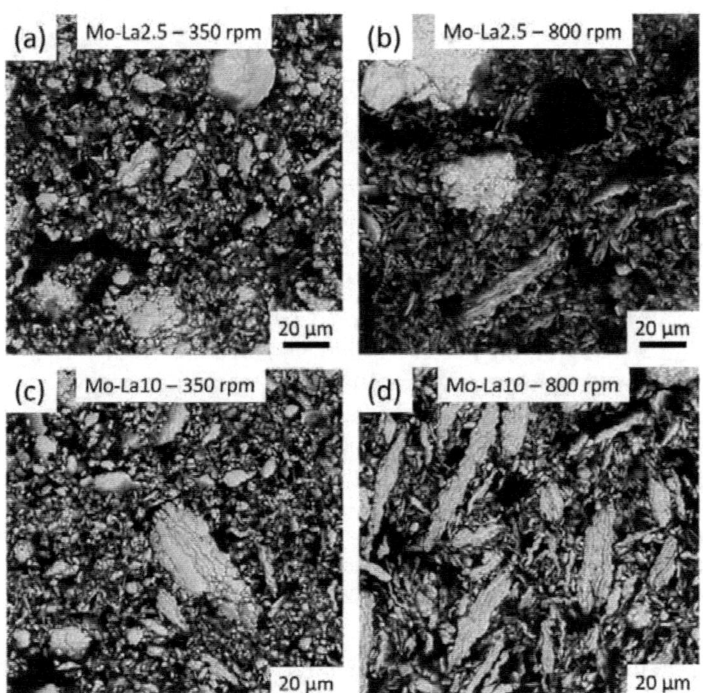


**Figure 3.** Morphology and details of fabricated (a-b) Mo-$La_2O_3$-350 and (c-d) Mo-$La_2O_3$-800 powders.

These agglomerates, which were formed due to intensive plastic deformation, exhibited the formation of flake-like particles with lamellar microstructure (Figure 4). The detailed SEM morphology shows the presence of molybdenum lamellas (light-gray color) with irregular-shaped submicron-sized lanthanum oxide particles located on the grain boundaries (dark-gray color). When comparing the results of the low- and the high-energy milling, it can be seen that the number of flake-shaped particles is higher and the lamellas in the microstructure are thinner or refined with incorporated $La_2O_3$ particles in between. EDX mapping also confirms more intensive incorporation of ceramic particles as well into the agglomerates in high-energy milling mode, while in the low-energy milling mode the most of $La_2O_3$ crushed debris is located in between deformed Mo particles. Furthermore, in the agreement with intensive incorporation of $La_2O_3$ particles into Mo matrix during high-energy milling, EDX elemental analysis showed almost twice higher content of La element in the Mo-La2.5-800 and Mo-La10-800 powders, Table 2.

To summarize, the observed powder microstructure after the mechanical milling processing will definitely impact the mechanical properties of the final sintered product. In this way, powder particles with refined molybdenum matrix produced by high-energy attrition milling are likely to have uniformly distributed lanthanum oxide particles. Besides the reinforcement of the molybdenum, uniformly distributed ceramic particles can reduce the ductile-to-brittle-transition as well that is one of the main aims of the Mo-focused research. However, even though it was possible to produce composite powder particles, further research on formation of spherical shaped composite powder particles is needed, following to the potential application in thermal spraying where the shape of the feedstock powder particles plays significant role.

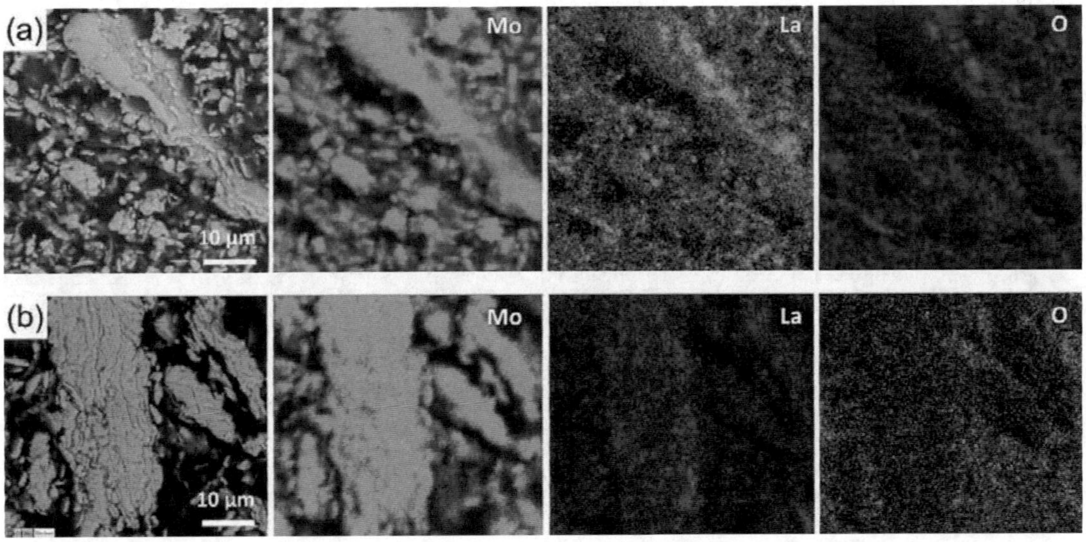


**Figure 4.** EDX elemental mapping of (a) Mo-La10-350 and (b) Mo-La10-800 composite powder particle cross-cuts.

**Table 2.** EDX elemental composition of the Mo-$La_2O_3$ composite powder agglomerates.

| Designation of the powder | Element (wt. %) | | |
|---|---|---|---|
| | Mo | O | La |
| Mo-La2.5-350 | 81.1 | 18.4 | 0.5 |
| Mo-La10-350 | 81.6 | 15.8 | 2.6 |
| Mo-La2.5-800 | 80.1 | 19.1 | 0.8 |
| Mo-La10-800 | 84.1 | 11.0 | 4.9 |

## Conclusions

- In the present work, high-energy attrition ball milling has been employed for the preparation of Mo-$La_2O_3$ composite powders with low (2.5 vol.%) and high (10 vol.%) content of $La_2O_3$ phase. The most interesting powder composition for further powder consolidation experiments was obtained from Mo-10 vol.% $La_2O_3$ mixture using the high-energy ball milling mode at 800 rpm for 60 min, while the use of lower $La_2O_3$ content and low-energy milling resulted mostly only in deformation of Mo and crushing of $La_2O_3$ particles.
- Composite Mo-$La_2O_3$ powder agglomerates show flake-like particles with lamellar microstructure with the uniformly distributed submicron-sized ceramic particles on the grain boundaries, which should beneficially affect the mechanical properties and DBT temperature of the final products sintered from this powder.
- As one of the potential applications of this powder is thermal spraying feedstock, further research on formation of composite powder particles in spherical shape is definitely needed.

## Acknowledgements

Authors acknowledge the Ministry of Education, Youth and Sports of the Czech Republic for funding the project LTC20068 "New materials for emerging energy technologies" under INTER-EXCELLENCE program, sub-program INTER-COST. The work was also carried out with the support of European Union's COST programme within the CA18112 action "Mechanochemistry for Sustainable Industry". CzechNanoLab project LM2018110 funded by MEYS CR is gratefully acknowledged for the financial support of the measurements/sample fabrication at CEITEC Nano Research Infrastructure.

## References

[1] Snead L L, Hoelzer D T, Rieth M and Nemith A A N 2019 Refractory Alloys: Vanadium, Niobium, Molybdenum, Tungsten (Eds. Boston: Elsevier) p 585-640
[2] Parthasarathy T A, Mendiratta M G and Dimiduk D M 2002 *Acta Mat.* **50** p 1857-68
[3] Jéhanno P, Böning M, Kestler H, Heilmaier M, Saage H and Krüger M 2008 *Powd. Met.* **51** p 99-102
[4] Takida T, Kurishita H, Mabuchi M, Igarashi T, Doi Y and Nagae T 2004 *Mat. Trans.* **45** p 143-8
[5] Kurishita H, Kitsunai Y, Shibayama T, Kayano H and Hiraoka Y 1996 *J. Nucl. Mat.* **233-7** p 557-64
[6] Lemberg J A and Ritchie R O 2012 *Adv. Mat.* **24** p 3445-80
[7] Yamauchi A, Yoshimi K, Kurokawa K and Hanada S 2007 *J. Al. Comp.* **434-5** p 420-3
[8] Heilmaier M, Krüger M, Saage H, Rösler J, Mukherji D, Glatzel U, Völkl R, Hüttner R, Eggeler G, Somsen Ch, Depka T, Christ H-J, Gorr B and Burk S 2009 *JOM* **61** p 61-7
[9] Vijatović Petrović M, Bobić J, Radojković A, Banys J., and Stojanović B 2012 *Ceram. Int.* 38-7, p 5347–5354.
[10] Gorrasi G and Sorrentino A 2015 *Green Chemistry 17*-5 p 2610–2625
[11] Fleming P, Farrell R A, Holmes J D, Morris M A 2010 *JACS* **93** p 1187-94
[12] Zhang G, Gang L, Sun Y, Jiang F, Wang L, Wang R and Sun J 2009 *Int. J. of Ref. Met. and H. M.* **27** p 173-6

# New approach to heat treatment of spring steel

**Kotous J Salvetr P Procházka R**

COMTES FHT a.s., Prumyslova 995, 334 41, Dobrany, Czech Republic

jakub.kotous@comtesfht.cz

**Abstract**. In the 54SiCr6 spring steel, a spheroidized structure was achieved by two processing methods: conventional soft annealing above the temperature $A_1$ in a furnace and accelerated ASR annealing by an induction heating equipment. The ASR technology is much faster, even for materials with a worse tendency to the spheroidisation, such as steels with higher silicon content. Thanks to this, a finer structure is obtained. This is the basis of other advantages for the subsequent hardening. Quenching can be done at lower temperatures, higher yield and tensile strength is obtained while maintaining same or improved ductility and contraction. This also results in influencing of fatigue behavior.

## 1 Introduction

Quenching temperature and the time optimizing is an important state to achieving good resultant properties of products. An economic point of view is important too. The effort is to have temperature and time as lowest as possible. In this case, the optimal initial structure is very important [1-3].

The finer the structure, the faster the austenitization. The quenched structures achieved the fastest dissolution into austenite but it isn´t suitable for machining. The soft annealing is maybe the last in effective solution [4]. The different accelerated annealing in furnaces [5][6] are better option but the fastest spheroidisation annealing is achieved by an induction equipment [7]. The Accelerated Spheroidisation and Refinement (ASR) is based on fast-cycling about the temperature $A_1$. It produced very fine globular cementite particles. The ASR treatment is able to achieve by thermo-mechanical processing too [8]. So it´s easy to integrate into hot forming and save more time. It is necessary to achieve at least the same properties after ASR as after conventional soft annealing to ensure applicability in the spring production.

In this article, the influence of soft annealing (SA) and accelerated spheroidizing (ASR) on the spring steel 54SiCr6 grade was studied. Significant study values were resultant properties after hardening which were determined by the tensile tests. Important monitored attribute of spring steel is the fatigue limit therefore the fatigue behavior was also observed.

## 2 Materials and methods

The experimental material was the steel 54SiCr6 grade, which is used commonly for the spring manufacture. The steel was made at COMTES FHT a.s. It was casted in a vacuum induction furnace, forged, hot rolled and air cooled. Then follows normalizing annealing. The chemical composition is given in table 1. The samples were bars with 20 mm in diameter and length of 130 mm.

**Table 1.** The chemical composition of the experimental steel 54SiCr6.

| **Element** | C | Si | Mn | Cr | Mo | Cu | P | S | Fe |
|---|---|---|---|---|---|---|---|---|---|
| **wt. %** | 0.57 | 1.51 | 0.68 | 0.75 | 0.03 | 0.01 | 0.008 | 0.003 | bal. |

**The carbide spheroidisation** was carried out using two different sequences. The process Accelerated Spheroidisation and Refinement (ASR) was completed using medium-frequency induction heating equipment.
Induction heating was carried in a coil with 35 mm in inner diameter, 150 mm length and 14 threads. The temperature was measured with a welded thermocouple. The ASR sequence comprised of three cycles with follows regime: 820 °C/ 15 s + 725 °C/ 5 min/ air cooling to 600 °C. The second cycle was identical. In the third cycle, the hold at 725°C was extended to 10 minutes. The sequence is plotted in figure 1.

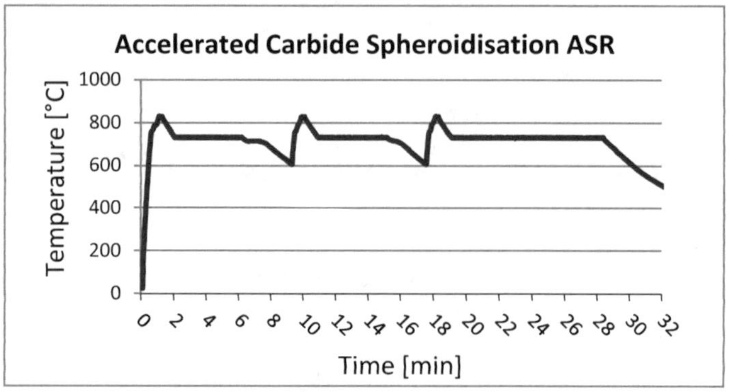


**Figure 1.** The ASR regime in 54SiCr6

The soft annealing (SA) was performed in an electrical air furnace. The material was held just above the $A_{c1}$. The sequence was as follows: heating at 600 °C/hour to 720 °C, then at 15 °C/hour to 770 °C, holding for 5 hours, furnace cooling at 5 °C/hour to 720 °C, then at 25 °C/hour to 650 °C, free cooling in the furnace to 400°C. The total time until the removal of samples from the furnace was 27 hours.
The slow cooling at 5 °C/hour in the SA sequence and the isothermal holds at 725 °C in the ASR sequence were necessary for preventing formation of new cementite lamellae in this high silicon content steel.
**The quenching and tempering** (QT) were carried out in electrical air furnaces. Quenching temperatures were in the range from 810 to 910 °C, the soaking time 20 minutes, medium oil. Tempering was at 400 °C/ 2 hours/ air for all regime.
**The evaluation** of different spheroidisation was mediated by observing microstructures using a scanning electron microscopy (SEM), hardness measuring, tensile testing and fatigue testing of the selected regime.

## 3 Results

### *3.1 Initial state of material*

Both spheroidized microstructures (post-SA and post-ASR) contained globular cementite particles in a ferritic matrix. Small cementite lamellae were rare observed. The ASR process led to densely-dispersed small globular cementite particles (max. 0.5 μm) in the ferritic matrix (figure 2). The soft annealing produced large globular particles up to 2 μm in size (figure 3).

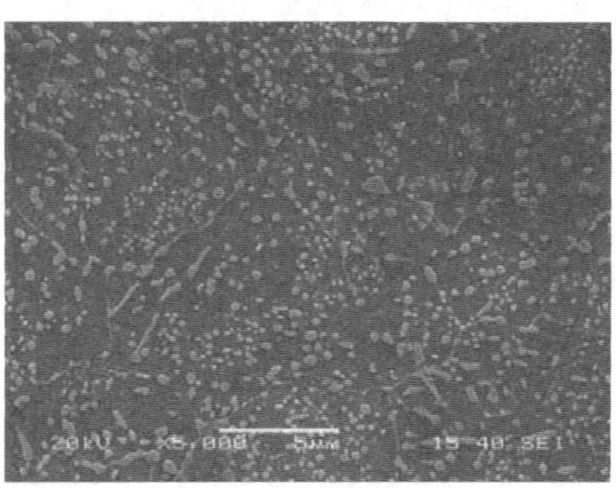

**Figure 2.** The microstructure after accelerated annealing (ASR)

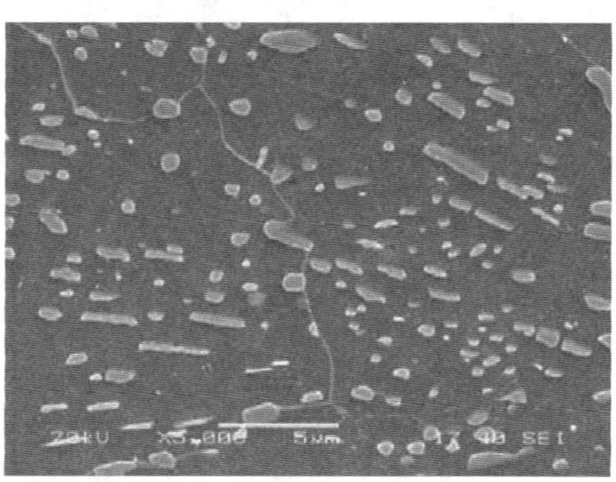

**Figure 3.** The microstructure after soft annealing (SA)

The hardness after the ASR was 224 ± 5 HV10. It was lower for SA - 207 ± 1 HV10. Both values meet the delivery specifications of spheroidized annealing of 54SiCr6 steel.

**Table 2.** The mechanical properties of initial states

| Heat treatment | $R_{p0.2}$ [MPa] | $R_m$ [MPa] | $A_g$ [%] | $A_5$ [%] | Z [%] | HV10 |
|---|---|---|---|---|---|---|
| ASR | 545 ± 11 | 803 ± 5 | 13.1 ± 0.3 | 24.0 ± 0.2 | 62.8 ± 0.1 | 224 ± 5 |
| SA | 356 ± 1 | 688 ± 2 | 16.5 ± 0.2 | 27.9 ± 0.3 | 52.5 ± 0.1 | 207 ± 1 |
| Initial state with lamella cementite | 542 ± 1 | 987 ± 2 | 9.3 ± 0.1 | 16.7 ± 0.2 | 38.8 ± 0.6 | 290 ± 7 |

Big differences were in the results of tensile test, see table 2. The state after ASR had higher yield and tensile test with lower elongation. Surprisingly, the area reduction was better than after SA. The initial state before spheroidisation with lamellar pearlite was also included in results for interest.

### 3.2 *Quenching and tempering*

The quenching was performed in the range of temperatures from 810 to 910 °C. The soft-annealed specimens, which were quenched from 810, 830, and 850 °C, contained ferrite in the microstructure. With higher quenching temperatures, fully martensitic with undissolved cementite was observed (figure 4 and 5). The amount of undissolved cementite decreased with increasing quenching temperatures. Yet, undissolved globular cementite was found in the martensitic matrix after quenching from 910 °C (figure 5). The largest undissolved cementite particles had a size about 1000 nm. They remained in the material even after quenching from the highest temperature because they would need much more time for dissolving.

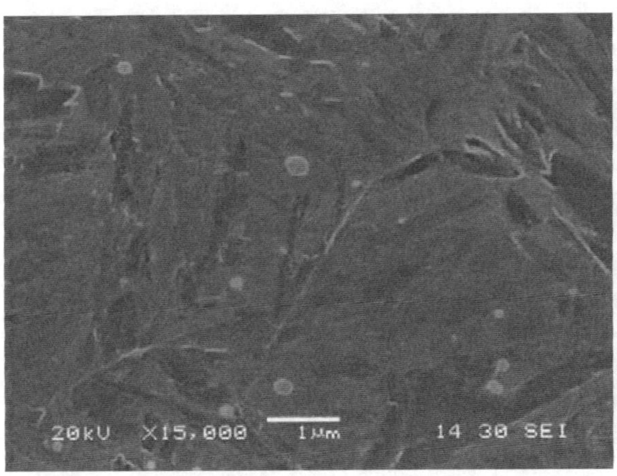


**Figure 4.** The quenched sample ASR 830 °C

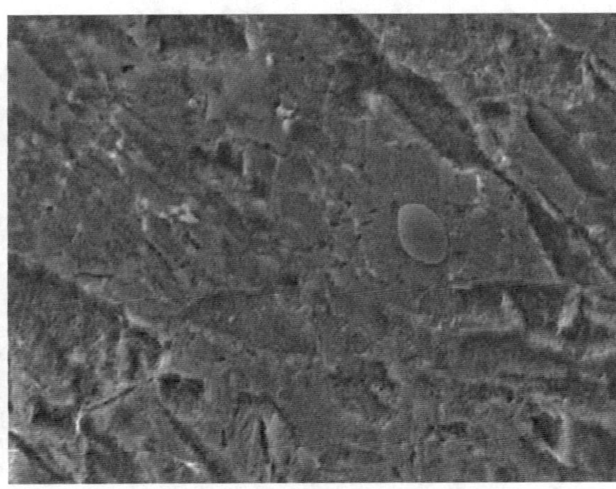

**Figure 5.** The quenched sample SA 910 °C

The ferrite in the quenched ASR specimens was observed only after quenching from 810 °C. Fine globular cementite dissolved quickly. The maximal size of the undissolved cementite particles was 300 nm (figure 4). Following quenching from 870°C, the undissolved cementite was rare to find. During the tempering, cementite precipitated within and along martensitic plates. Apart from the size of undissolved cementite particles, no differences were found between tempered microstructures obtained after the ASR and the SA.

The hardness results are in agreement with the microstructure, see table 3. When the ferrite is completely transformed into austenite, the hardness reached maximum values. It wasn´t increase with dissolving rest of cementite. It was about 750 HV10 after quenching and 550 HV10 after tempering for both initial states. The dissolution of remaining cementite particles has not led to a major increase in overall hardness.

**Table 3.** Hardness HV10

| Heat treatment | | Quenching temperature [°C] | | | | | |
|---|---|---|---|---|---|---|---|
| | | 810 | 830 | 850 | 870 | 890 | 910 |
| Quenched | ASR | 683 ± 2 | 756 ± 4 | 752 ± 5 | 754 ± 6 | 757 ± 4 | - |
| | SA | 600 ± 11 | 667 ± 7 | 702 ± 11 | 742 ± 4 | 745 ± 4 | 746 ± 8 |
| Tempered | ASR | 533 ± 1 | 547 ± 2 | 554 ± 1 | 548 ± 6 | 546 ± 4 | - |
| | SA | 462 ± 2 | 515 ± 3 | 530 ± 4 | 544 ± 2 | 547 ± 4 | 534 ± 5 |

### 3.3 *Tensile tests*

Room-temperature tensile tests were performed in accordance with ČSN EN ISO 6892-1. The gauge sections had a size of Ø10×60 mm. The specimens were tested in tempered condition. Results of the tensile tests are given in table 4.

**Table 4.** Results of tensile tests

| Heat treatment | $R_{p0.2}$ [MPa] | $R_m$ [MPa] | $A_g$ [%] | $A_5$ [%] | Z [%] |
|---|---|---|---|---|---|
| ASR 830 | 1696 ± 19 | 1873 ± 13 | 3,1 ± 0,1 | 8,8 ± 0,1 | 34,9 ± 0,4 |
| ASR 850 | 1686 ± 11 | 1881 ± 1 | 2,9 ± 0,1 | 9,3 ± 0,1 | 38,8 ± 0,9 |
| ASR 870 | 1684 ± 19 | 1863 ± 10 | 3,1 ± 0,1 | 9,7 ± 0,2 | 41,7 ± 1,1 |
| SA 870 °C | 1634 ± 7 | 1832 ± 12 | 3,4 ± 0,1 | 8,9 ± 0,2 | 31,4 ± 0,9 |
| SA 890 °C | 1656 ± 10 | 1847 ± 1 | 3,2 ± 0,1 | 9,6 ± 0,2 | 34,0 ± 1,4 |
| SA 910 °C | 1644 ± 18 | 1835 ± 16 | 3,2 ± 0,1 | 9,9 ± 0,1 | 35,2 ± 1,0 |

The states which were fully austenitized were tested. Although the hardness was the same, the dissolving rest of cementite carbides influences the mechanical properties. The results with both spheroidisation annealing had the same trend in elongation $A_5$ and area reduction Z. The trend was increasing with higher temperature and lowering content of undissolved cementite. The values of elongation were similar. However, the ASR achieved a higher area reduction than the SA. In the point of yield and ultimate strength, the values are similar for the individually initial state in the tracked range of quenching temperatures but all of them are higher after the ASR than the SA. Differences were about 30-40 MPa.

### 3.4 *Fatigue testing*

The high-cycle fatigue testing was carried out on the magneto-resonance machine Vibrophore Rumul at the room temperature. The specimens were tested in the tempered state and there were compared between the ASR quenched from 830 °C and the SA quenched from 870 °C. The gauge sections had a size of Ø5 mm with a radius 64 mm. Loading was alternating with asymmetry cycle R = -1. The results represent the determination of the fatigue limit and the Wohler curves, see figure 6.

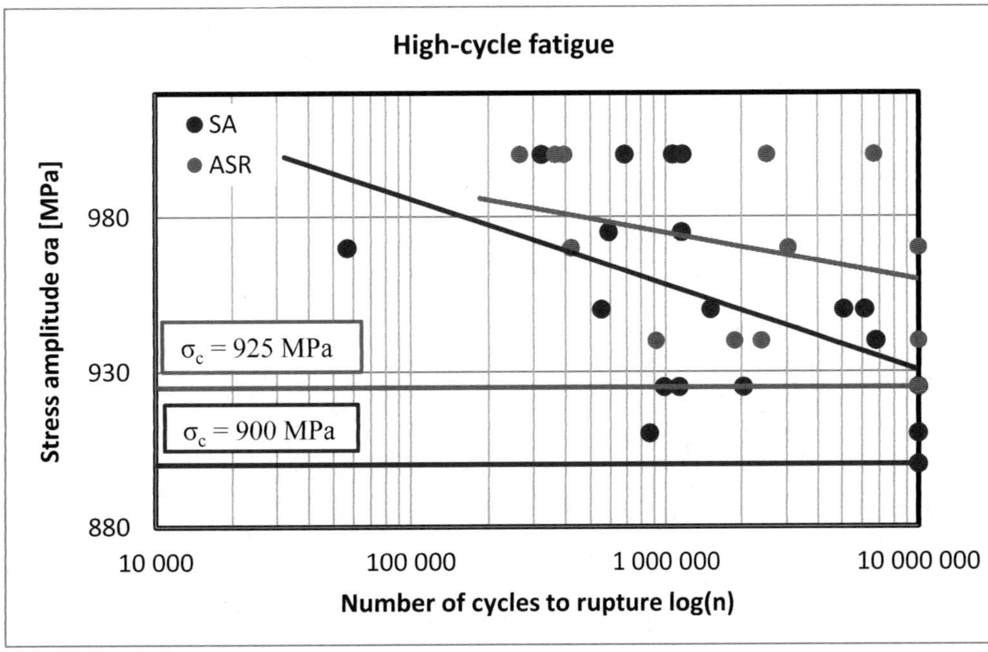


**Figure 6.** The Wohler curves

The results had a large scatter. The fatigue limit of the SA regime was 900 MPa. The ASR fatigue limit was 925 MPa. This result was expected thanks to the fact that ASR achieved higher ultimate strength. The area of timed fatigue limit after the ASR looks a little better too. It last more cycles at the same stress amplitude.

## 4 Discussion

The carbides after the ASR dissolved very fast almost as quickly as lamellar pearlite [4]. In the SA material, the austenitizing was retarded, as cementite was present in the large globules whose neighborhood was depleted of carbon. Hence, slower dissolution of the large particles and the long-distance diffusion of carbon poses greater demands on the austenitizing process before quenching.

The hardened ASR material had probably finer austenite grain, which was reflected in higher strength characteristics with higher area reduction while maintaining good elongation. The grain etching was done without success for quantitative grain size evaluation.

## 5 Conclusion

The effect of different spheroidisation annealing was observed. The article was focused on the influence of this effect on resultant mechanical properties in 54SiCr6 spring steel. Therefore the metallographic analysis and hardness measuring were supplemented by tensile and fatigue testing in the final tempered state.

Two initial states were achieved by the Accelerated Spheroidisation and Refinement (ASR) and the soft annealing (SA). The austenitizing was much faster post-ASR than post-SA. In the 54SiCr6 steel, the difference in optimal quenching temperatures is about 40°C. In the case of the SA, the large undissolved cementite particles were observed at the highest quenching temperature. It also leads to different mechanical properties. The ASR treatment improves the yield and ultimate strength and the area reduction while the elongation and maximal values of hardness remain the same as for that SA samples. The finer carbides after the ASR positively affected resultant fatigue behavior. The fatigue limit was higher by 25 MPa and Wohler curve looked more favorable.

In the conclusion, the ASR annealing combines the advantages of the soft annealing and the state without annealing (lamellar pearlite). The globular morphology of cementite guarantees the reduction in hardness which is the purpose of the spheroidisation annealing. Thanks to the fine carbides, the austenitizing is rapid and lower quenching temperatures are sufficient. In addition, it leads to improving mechanical properties.

**Acknowledgement**
The result was supported from ERDF Research of advanced steels with unique properties, No. CZ02.1.01/0.0/0.0/16_019/0000836.

**References**

[1] Jirková H., Hauserová D., Kučerová L., Mašek B., Energy and Time Saving Low Temperature Thermomechanical Treatment of Low Carbon Plain Steel, Materiali in Technologije/Materials and Technology, 2013, vol. 47/3, pp. 335-9.
[2] Hauserová, D., Dlouhý, J., Nový, Z., Microstructure and Properties of Hardened 100CrMnSi6-4 Bearing Steel after Accelerated Carbide Spheroidisation and Long-duration Annealing. In: Bearing Steel Technologies: 10th Volume, Advances in Steel Technologies for Rolling Bearings, STP 1580, 2015, pp. 389-409, ISSN 2160-2050.
[3] Hauserova D, Dlouhy J and Kotous J 2017 Structure Refinement of Spring Steel 51CrV4 after Accelerated Spheroidisation Arch. of Metall. and Mater. vol 62 p 1473-77.
[4] Kotous, J., Dlouhý, J., Nacházelová, D., Influence of accelerated spheroidisation on quenching process and resultant mechanical properties in spring steel 54SiCr6, Materials Science and Engineering, 2019, vol. 461.
[5] Saha A., Mondal D.K., Biswas K., Maity J., Development of high strength ductile hypereutectoid steel by cyclic heat treatment process, Materials Science and Engineering A 2012, vol. 541, pp 204-15.
[6] Mishra S., Mishra A., Show B. K., Maity J., Simultaneous enhancement of ductility and strength in AISI 1080 steel through a typical cyclic heat treatment, Materials Science and Engineering A, 2017 vol 688, pp 262-71.
[7] Hauserová, D., Dlouhý, J., Nový, Z., Effect of Heating Rate on Accelerated Carbide Spheroidisation (ASR) in 100CrMnSi6-4 Bearing Steel, Archives of Metallurgy and Materials, 2014, Vol. 59, pp. 1199-1203, ISSN 1733-3490.
[8] Hauserová D., Dlouhý J., Nový Z., Zrník J., Duchek M., Forming of C45 Steel at Critical Temperature, Procedia Engineering, 2011, vol. 10, pp. 2955-60.

# Advanced techniques applied to analyse the influence of the heat treatment on the mechanical properties, microstructure and texture of additively produced components from maraging steel

**M Koukolíková[1], V Brunát[1], G Wolf[2], T Simson[3] and J Koch[3]**

[1]COMTES FHT a.s., Průmyslová 995, 334 41 Dobřany

[2]Fraunhofer-Institut für Umwelt-, Sicherheits- und Energietechnik UMSICHT, An der Maxhütte 1, 92237 Sulzbach-Rosenberg, DE

[3]Ostbayerische Technische Hochschule Amberg-Weiden, Kaiser-Wilhelm-Ring 23, 92224 Amberg, DE

Corresponding author: mkoukolikova@comtesfht.cz

**Abstract**. The experimental program Metallic materials in the process chain of additive manufacturing is focused on the characterization of laser powder bed fused materials (L-PBF). Metallographic analysis, electron backscatter diffraction supported by micro tensile testing were used to analyse the material properties of the maraging steel MS1. The main goal of the research was to use advanced techniques to investigate the influence of the additive manufacturing with very fine powders, and subsequent heat treatment on the microstructure, texture and mechanical properties of the components from maraging steel.

## 1 Introduction

In the case of additive manufacturing (AM) production, the component is formed by the gradual application of thin layers of metal powder, which are locally melted by a laser beam. Due to the gradual formation of the component layer by layer and high solidification rates, different microstructures, local residual stresses, or the formation of pores and many other phenomena may occur. Due to the large number of influencing factors and relative novelty of AM technologies based on powder bed fusion, there are numerous areas where the relationships between the parameters of AM processes, materials, and created microstructures are not completely explored.

Maraging steel MS1 (1.2709), officially named LaserForm Maraging Steel (B) by 3D SYSTEMS, is a genuine tool steel fine-tuned for the use with ProX® DMP 200 metal printers. It is easily heat-treatable in a simple age-hardening process resulting in excellent hardness and high - strength, with good weldability and machinability [1]. Martensite in alloys of low-carbon Fe-Ni-C system with high nickel content is plastic and tough. Martensite precipitation hardening is promoted by a fine dispersion of intermetallic phases, which can be simple or complex compounds (e.g. $Ni_3$(Ti, Al, Mo), $Fe_2Mo$, Ni(Al, Fe)). Precipitation hardening takes place at a temperature of 450 to 500 °C for 3 to 6 hours. A strength limit of 1400 to 2140 MPa with high toughness and contraction (50 to 70 %) can be achieved with such processing [2], [3].

## 2 Materials and Methods

Maraging steel 1.2709 supplied by 3D SYSTEMS as a gas-atomized powder was used as the experimental material. Chemical composition supplied by the powder manufacturer is summarized in Table 1. Powder analysis obtained by laser diffraction on a HELOS/BR instrument (Sympatec GmbH) proved probability density function of the particle size distribution $x_{90}$ = 20.6 μm. The parts were manufactured with standard parameters on a ProX® DMP 200 3D SYSTEMS metal printer with laser speed 1200 mm/s at a laser power 135 W and layer thickness 30 μm. This range of 3D metal printers works with very fine powder fractions.

Experimental material was analysed by metallographic analysis and mechanical testing. The experiment was performed in as-built, solution annealed and precipitation hardened states. The heat treatment conditions are summarized in Table 2. Metallographic specimens were prepared by a standard procedure of grinding and subsequent polishing. The microstructures were revealed by etching in Nital solution and photographed using the light microscope NIKON ECLIPSE MA200 equipped with a NIS Elements 5.2 digital image processing and analysis software. Detailed microstructural observation and electron backscatter diffraction (EBSD) analysis were performed on a scanning electron microscope JEOL IT 500 HR. Electron backscatter diffraction, a unique tool for assessing the microstructure and texture of a material, provided detailed information about the additively manufactured samples. Observation was performed under the backscattered electrons with EDAX Hikari Super camera. EBSD analysis was performed at a step size of 0.1 μm, size of the analysed area 128 x 100 μm, acceleration voltage 20 kV, and scanning speed was 90 - 100 diffractograms per second and 4 x 4 binning. EBSD maps were processed in 1000x magnification. Data post-processing (neighbour Confidence Index correlation and grain dilation algorithm) was performed using the OIM software (Version 8.0). Individual data points were grouped into grains if the critical disorientation angle between two adjacent pixels within the grain did not exceed 5 ° and the minimum number of pixels forming the grain was 2. The evaluation of retained austenite fraction were performed using EBSD and X-ray diffraction (BRUKER D8 DISCOVER device with Cu Kα radiation). Diffraction patterns were collected in 2θ range from 30° to 120° with a step size of 0.25° and exposure time 0.5 s/step. The measurement was conducted on the polished surface of samples. The peaks of austenite (200), (202), (311), (222) were used for quantification analysis.

Mechanical properties were analysed by means of the Micro - Tensile Testing procedure (M-TT), which was developed in COMTES FHT based on the demand to measure reliably material characteristics by tensile tests from the minimum amount of the experimental material. The M-TT has proved to be effective in earlier studies [4] - [8]. M-TT samples were produced by means of electro-discharge machining in XZY, 45° and ZXY orientations (Figure 1). The ultimate tensile strength (UTS), yield strength (YS), elongation (El), and cross section reduction (Z) of the specimens were measured from 3 samples in each direction. Hardness HV10 was measured using laboratory hardness tester Struers DuraScan 50 with 3 indents in YZ plane.

**Table 1.** Chemical composition of the maraging steel supplied by 3D SYSTEMS [1].

| Element | Fe | Ni | Co | Mo | Ti | Si | Mn | C |
|---|---|---|---|---|---|---|---|---|
| Weight [%] | Bal. | 17.00 - 19.00 | 9.00 - 11.00 | 4.00 - 6.00 | 0.90 - 1.10 | ≤ 1.00 | ≤ 1.00 | ≤ 0.03 |

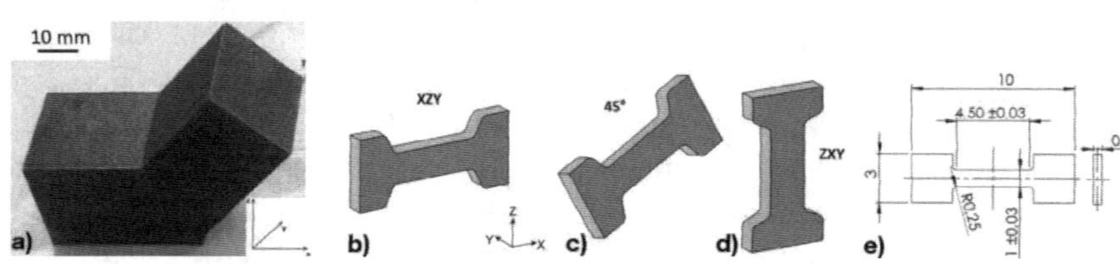


**Figure 1.** Manufactured part (a), M-TT sample XZY orientation (b), M-TT sample 45° orientation (c), M-TT sample ZXY orientation (d), M -TT sample dimensions (e).

**Table 2.** Heat treatment modes for the researched material.

| Sample | Heat treatment condition |
|---|---|
| **0** | As-built |
| **HT1** | Solution annealing 840 °C / 1 h, water quenching |
| **HT3** | Solution annealing 840 °C / 1 h, water quenching, precipitation hardening 480 °C / 6 h, air cooling |

## 3 Results and Discussion

Microstructural analysis performed by means of light (LM) and electron microscopy (SEM) was performed parallel to the building direction. The use of Nital reagent principally revealed the microstructures (Figure 2, Figure 3) of the lath martensite in a low carbon steel.

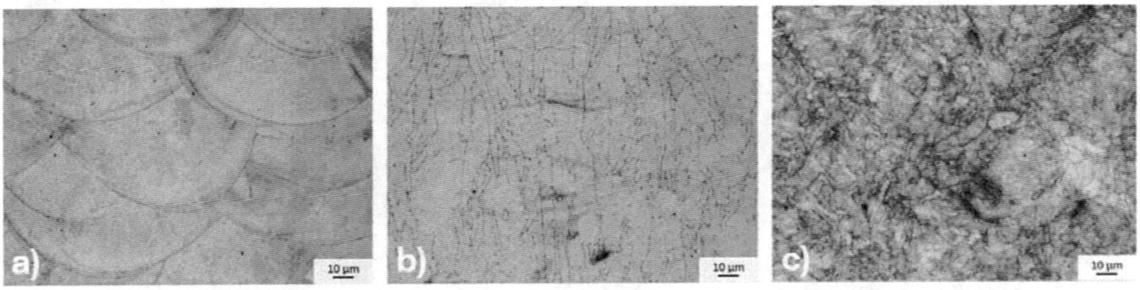


**Figure 2.** LM: a) as-built; b) solution annealing; c) precipitation hardening. Nital.

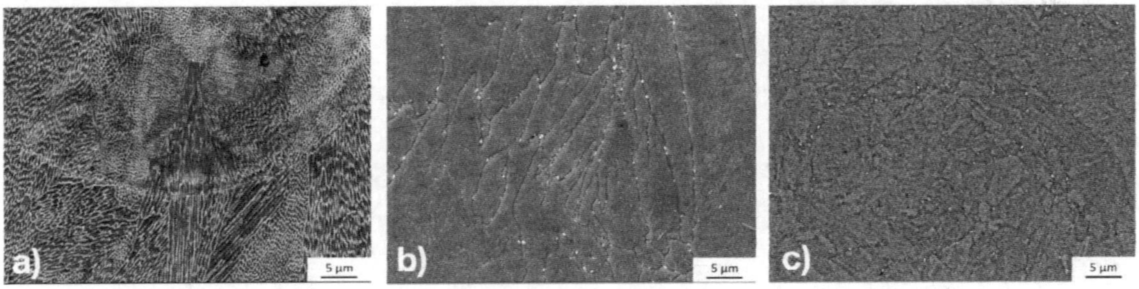


**Figure 3.** SEM: a) as-built; b) solution annealing; c) precipitation hardening. Nital.

The solidification structure of the as-built state (Figure 2a, 3a) with epitaxial grain growth across the individual melt pool boundaries and fine cellular structure was observed by detailed electron microscopy analysis. Solution annealing eliminated cellular microstructure and melts pool boundaries. Undissolved phases (Figure 2b, 3b) in the lath martensitic matrix mainly at the grain boundaries, but also in the matrix of the solid solution state were observed. Aging of the material promoted the formation of intermetallic precipitates within fine martensitic laths (Figure 2c, 3c).

### *3.1 EBSD analysis*

The EBSD analysis provided the information on the development of the microstructure of individual states. The colour Inverse Pole Figure (IPF) maps represent random grain orientations along the building direction of all states (Figure 4). The pole figures which enable the representation of the preferred orientation (texture) in the analysed material (Figure 5) proved that no strong texture was formed during the additive production by the L-PBF method and subsequent heat treatment. The texture in the as-built condition presented in Figure 5a showed that no clear <100> alignment occurred in the building direction during the AM process. After solution annealing at 840 °C, there was a slight grain growth and grain rearrangement, as observed by comparing the IPF and PF shown in Figure 4, 5 a - c. The texture after the aging treatment with prior SA retained almost the same as in the as – built state, as shown in Figure 5c.

The amount of retained austenite detected by EBSD analysis was found to be 1 - 2 % with no significant differences between individual states. The lowest fraction was observed by solution annealed state. These results were verified by X-ray diffraction with the similar results.

**Figure 4.** IPF maps: a) IPF as-built; b) IPF solution annealing; c) IPF precipitation hardening.

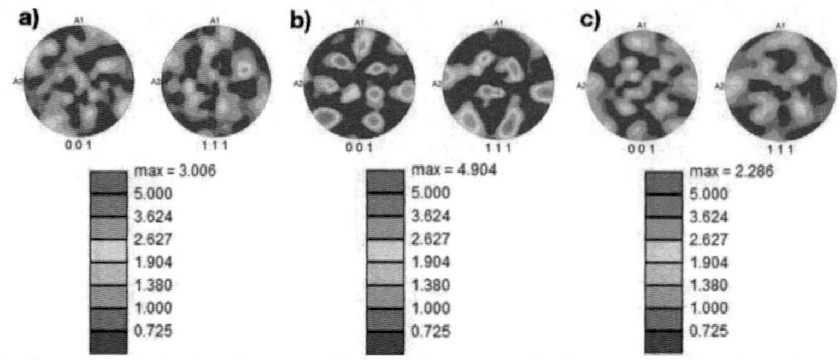


**Figure 5.** Pole figures: a) As-built state; b) solution annealed state; c) precipitation hardened state.

### *3.2 Micro tensile testing (M-TT)*

Micro tensile testing of the supersaturated homogeneous state after solution annealing is characterized by lower strength properties in comparison to the as-built state as the microstructure of the as-built state consisted by a mixture of martensite and tempered martensite, which was formed by further applied layers during the laser powder bed fusion.

The precipitation hardening caused a dramatic increase in the ultimate tensile strength at the cost of decreased elongation at break (El). M-TT testing in XZY, 45° and ZXY directions proved no significant build orientation based anisotrophy of mechanical properties in as-built and heat treated samples (Table 3, Figure 6). The prolonged stress - strain curves were caused by further deformation of the sample after a crack was formed, but at the time of the crack the sample was not completely torn. The decrease in hardness confirmed the trend of decreasing yield strength and ultimate tensile strength after solution annealing, whereas precipitation hardening led to the improvement of the hardness (Table 3).

**Table 3.** Mechanical properties of researched material.

| Specimen | Specimen | YS [MPa] | UTS [MPa] | El [%] | Z [%] | HV10 |
|---|---|---|---|---|---|---|
| As-built (AB) | AB_XZY | 931 ± 17 | 1158 ± 19 | 12 ± 1 | 69 ± 2 | 359 ± 1 |
| | AB_45° | 959 ± 22 | 1163 ± 11 | 11 ± 1 | 63 ± 7 | |
| | AB_ZXY | 981 ± 24 | 1152 ± 11 | 13 ± 1 | 66 ± 0 | |
| Solution annealing (SA) | SA_XZY | 842 ± 21 | 1099 ± 2 | 11 ± 1 | 61 ± 1 | 330 ± 2 |
| | SA_45° | 814 ± 8 | 1078 ± 8 | 11 ± 0 | 59 ± 5 | |
| | SA_ZXY | 801 ± 12 | 1077 ± 13 | 11 ± 2 | 60 ± 10 | |
| Precipitation hardening (PH) | PH_XZY | 2049 ± 89 | 2145 ± 41 | 3 ±1 | 51 ± 7 | 618 ± 2 |
| | PH_45° | 2066 ± 28 | 2157 ± 20 | 4 ±0 | 54 ± 7 | |
| | PH_ZXY | 1989 ± 22 | 2111 ± 67 | 2 ±1 | 50 ± 6 | |

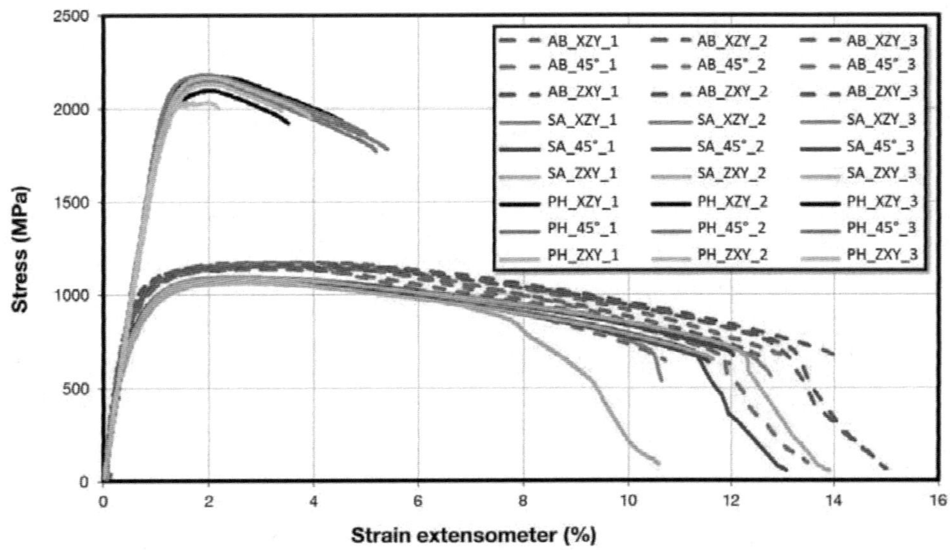


**Figure 6.** Engineering stress-strain curves for XZY, 45° and ZXY specimens of as-built (AB), solution annealing (SA), precipitation hardening (PH) states.

Experience from previous research works [8], [9] monitored the effect of several etchants on revealing different microstructural features. The use of etchants e.g. Adler and Villela – Bain resembled the conventional microstructures with barely revealed traces of 3D process. Nital etching was used to reveal the traces of 3D printing of the researched material. Nital highlighted the laser tracks and cellular microstructure, as it attacks alloy- rich segregations (Ni, Mo, Ti) and solidification structures [8]. The deficiency of texture along building direction of all states is probably caused by changes in the laser paths among layers, which resulted in altering the heat flux direction [2].

The M - TT results are in accordance with theoretical knowledge in [3], that the supersaturated homogeneous state (SA – low carbon lath martensite) is characterized by lower strength properties in contrast to the equilibrium heterogeneous state (AB state - a mixture of martensite and tempered martensite). The decrease in UTS may also be related to a significant reduction in the residual stress found in the material after laser powder bed fusion production and can be mitigated by post heat-treatment [9], [14], [15], i. e. solution annealing.

The effect of subsequent heat treatment on the microstructure and mechanical properties of maraging steel proved that the precipitation hardening caused almost doubling of the values of ultimate tensile strength at the expense of reduced elongation at break. M-TT testing proved no significant build orientation based anisotrophy of mechanical properties in as-built and heat treated samples. The achieved results obtained with the use of the micro-tensile tests samples are in good accordance with the results using the standard specimen sizes declared not only by the powder manufacturer, but also other publications [2], [8] - [14].

## 4 Conclusions

Maraging steel produced by additive manufacturing using the laser powder bed fusion technology was analysed in as-built and two heat treated states by metallographic analysis by means of light, electron microscopy, and electron backscatter diffraction and the results were supported by micro tensile testing.

Metallographic analysis of the as - built state proved the occurrence of a typical line-shaped morphology denoting melt pools. The direction of the grain growth tends to follow the direction of the heat dissipation. Heat treatment of individual states reduced the effect on the melt pool boundaries and fine cell structure. Solution treatment resulted in slight grain growth of the matrix.

The trace amount of retained austenite detected by EBSD and verified by the X-ray diffraction analysis proved no significant differences between individual states. There was also the absence of remarkable texture along building directions.
Micro-tensile tests in the XZY, 45° and ZXY directions reliably measured the material properties. Testing of M - TT is advantageous for additively manufactured materials, which can be tested from a minimum amount of experimental material and thus the costs of AM production can be reduced. Future research will focus on modification of the alloy's chemical composition and its influence on microstructural and mechanical properties.

## Acknowledgements

The research has been realized under the financial support of the Project 3D COVER – Metallic materials in the process chain of additive manufacturing, No.: 185, financed by the ERDF.

## References

[1] LaserForm Maraging Steel (B) Datasheet. https://www.3dsystems.com/materials/maraging-steel
[2] Tan C and Zhou K 2018 Microstructural characterization and properties of selective laser melted maraging steel with different build directions *Science and Technology of Advanced Materials* 19:1 pp 746-758
[3] Kraus V 2000 Tepelné zpracování a slinování (Pilsen: University of West Bohemia) chapter 8.3 pp 130 - 131
[4] Dzugan J and Sibr M 2017 Mechanical properties determination of AM components. *IOP Conference Series: Materials Science and Engineering* 179 012019
[5] Dzugan J and Seifi M 2018 Effects of thickness and orientation on the small scale fracture behaviour of additively manufactured Ti-6Al-4V *Materials Characterization* 143
[6] Podany P and Reardon C 2018 Microstructure, mechanical properties and welding of low Carbon, medium manganese TWIP/TRIP steel *Metals* 8 (4) 263
[7] Pehlivan E and Roudnicka M Effects of build orientation and sample geometry on the mechanical response of miniature CP-Ti Grade 2 strut samples manufactured by laser powder bed fusion 2020 *Additive Manufacturing* 35 101403
[8] Hrbackova T Simson T 2018 The effect of heat treatment on mechanical properties andmicrostructure of additively manufactured components. *IOP Conf. Ser.: Mater. Sci. Eng.* 461 012026
[9] Kucerova L and Zetkova I 2019 Microstructural characterisation and in-situ straining of additive-manufactured X3NiCoMoTi 18-9-5 maraging steel *Materials Science & Engineering* A 750 pp 70–80
[10] Mutua J and Nakata S 2018 Optimization of selective laser melting parameters and influence of post heat treatment on microstructure and mechanical properties of maraging steel *Materials and Design* 139 pp 486–497
[11] Tan C and Zhou K 2017 Microstructural evolution, nanoprecipitation behavior and mechanical properties of selective laser melted high-performance grade 300 maraging steel *Materials & Design* 134 pp 23–34
[12] Simson T and Koch J 2019 Mechanical Properties of 18Ni-300 maraging steel manufactured by LPBF Proc. Int. Conf. on Structural Integrity 17 pp 843–849
[13] Kempen K and Yasa E Microstructure and mechanical properties of Selective Laser Melted 18Ni-300 steel *Physics Procedia 12* pp 255–263
[14] Valenta J 2018 Heat treatment of maraging steel processed by SLM 2018 Barchelor's thesis VUT Brno
[15] Hermann Becker T and Dimitrov D The achievable mechanical properties of SLM produced Maraging Steel 300 components 2016 *Rapid Prototyping Journal* 22(3) pp 487–494

# Microstructure changes occurring in heat-resistant steels during severe plastic deformation and subsequent creep

**P Kral [1,*], J Dvorak [1], V Sklenicka [1], Z Horita [2,3,4], Y Takizawa [5], Y Tang [6], K Kucharova [1], M Kvapilova [1], M Svobodova [7]**

[1] Institute of Physics of Materials, Czech Academy of Sciences, Žižkova 22, 616 62 Brno, Czech Republic

[2] Kyushu Institute of Technology, Kitakyushu, 804-8550 Japan

[3] Magnesium Research Center, Kumamoto University, Kumamoto, 860-8555, Japan

[4] Saga University, Synchrotron Light Application Center, Saga, 840-8502 Japan

[5] Technology Department, Nagano Forging Co., Ltd, Nagano 381-0003, Japan

[6] WPI, International Institute for Carbon-Neutral Energy Research (WPI-I2CNER), Kyushu University, Fukuoka 819-0395, Japan

[7] UJP PRAHA a.s., 156 10 Praha-Zbraslav, Czech Republic

E-mail: pkral@ipm.cz

**Abstract**. This work investigates creep and microstructure characteristics in P92 martensitic and 304L austenitic stainless steel after the imposing of large value of plastic strain. The as-received coarse-grained steels were strongly deformed at room temperature (RT) before tensile creep testing. The flat specimens were tested at 873 K and 923 K under different applied stresses. Microstructures were analysed by means of scanning electron microscope with EBSD (electron back-scatter diffraction) camera and transmission electron microscope. It was observed that application of severe plastic deformation (SPD) techniques at RT significantly reduces grain size and changes phase structure of 304L austenitic steel. The microstructure investigations revealed that the growth of new phases is enhanced in plastically deformed P92 martensitic and 304L austenitic stainless steels compared to their undeformed states. Creep results showed that the mean grain size of SPD-processed steels is still near ultrafine-grained region even after long-term creep testing and the coarsening of microstructure is predominantly influenced by creep time and strain.

## 1 Introduction

The methods of allowing to impose the large deformation into the microstructure provide a unique opportunity to study creep processes in ultrafine-grained or even nano materials [1]. The grain size of these materials is comparable or even finer than stationary subgrain size in their coarse-grained counterparts. For this reason creep mechanisms related to the high-angle grain boundaries (HAGBs) may significantly affect the mechanical properties of materials at high temperatures [2,3]. The grain

boundaries can strengthen the material because they act as strong obstacles to dislocation glide. But they can also reduce strength of materials through the processes associated with grain boundaries, such as grain-boundary sliding [4] or enhanced recovery of dislocations at grain boundaries [5].
The martensitic and austenitic chromium steels are very often used for the high-temperature components of steam power plants. Their creep resistance is significantly influenced by the density of free dislocations and low-energy boundaries, such as low-angle grain boundaries and martensitic lath boundaries whose movement is restricted by precipitates [6,7]. The microstructure of martensitic and austenitic chromium steels change and deteriorate their mechanical properties during long-term annealing and creep exposure at high temperatures. The deterioration of creep properties and degradation of microstructure in these steels is caused by martensite recovery, growth of carbides and formation of new coarse phase or σ phases which reduce the content of W, Mo or Cr in solid solution [6].
It was found that the microstructure refinement through severe plastic deformation leads to the faster precipitation of Laves phase in ultrafine-grained martensitic P92 steel [8]. By contrast, the previous works showed that cold strain imposed into austenitic steels can enhance the precipitation of secondary phases [9-11] which lead to the degradation of properties in stainless steels, in particular an increase in brittleness and a decrease in corrosion resistance [12-14]. It was also shown that plastic strain of many austenitic steels at low temperature may cause the formation of a large amount of deformation-induced martensite [15].
The aim of the present investigation is to evaluate the influence of severe plastic deformation at room temperature (RT) on microstructure and creep in P92 martensitic and 304 austenitic steels.

## 2 Experimental Materials and Methods

The experimental materials used in the present investigation were coarse-grained P92 and 304 L austenitic steels. Their chemical composition (in wt%) and heat treatment can be found elsewhere [8,9].
The specimens with a thickness of 1.1 mm in the form of either disc with 30 mm in diameter or sheets 100 mm long and 10 mm wide were cut from coarse-grained P92 and 304L steels. The discs were deformed by 1 rotation of high-pressure torsion (HPT) at room temperature under the pressure of 6 GPa and rotation speed of 0.1 mm per second. The sheets were processed by high-pressure sliding (HPS) at room temperature with the sliding distance of 5 mm under the pressure of 4 GPa.
The amount of imposed equivalent strain during HPT depends on the radius of the disc and can be estimated by following equation [1]:

$$\varepsilon_{eq} = 2\pi rN/\sqrt{3}t \qquad (1)$$

where $r$ is the distance from the disc centre, $N$ is the number of turns and $t$ is the thickness of disc. The tensile specimen located about 7.5 mm from the disc centre. From this reason the minimum of equivalent strain $\varepsilon_{eq}$ is about 23 and the maximum is about 37 (if we consider the areas near grip parts). The amount of imposed equivalent strain during HPS process can be determined by this equation [16]:

$$\varepsilon_{eq} = x/\sqrt{3}t \qquad (2)$$

where $x$ is the sliding distance of plunger with respect to anvil and $t$ is the thickness of the sample. Thus equivalent strain imposed during HPS was about 7.8. The further details of sample preparation are given elsewhere [17,18]. Constant load tensile creep tests of P92 steel were conducted at 873 K and austenitic 304L steel was tested at 923 K. The experimental materials were tested under different loads using flat specimens of a gauge length of 8 mm, thickness of 1 mm and width 3 mm.

Microstructure investigated by scanning electron microscope (Tescan Lyra 3) equipped with a NordlysNano EBSD detector. The specimens of ultrafine-grained P92 martensitic steel processed by HPT was analysed on the section perpendicular to the rotation axis of the disc. The microstructure of HPT processed states was investigated on areas about 48 μm$^2$ with EBSD step size of 25 nm. The microstructure processed by HPS and after creep exposure was investigated on areas about 3000 μm$^2$ with EBSD step size of 50 nm.
The boundaries with $\theta \geq 15°$ are referred as high-angle grain boundaries (HAGBs) and boundaries with $\theta < 15°$ are referred as low-angle grain boundaries (LAGBs). The mean intercept spacing of LAGBs and HAGBs was performed along test longitudinal and transverse lines.

a)

b)

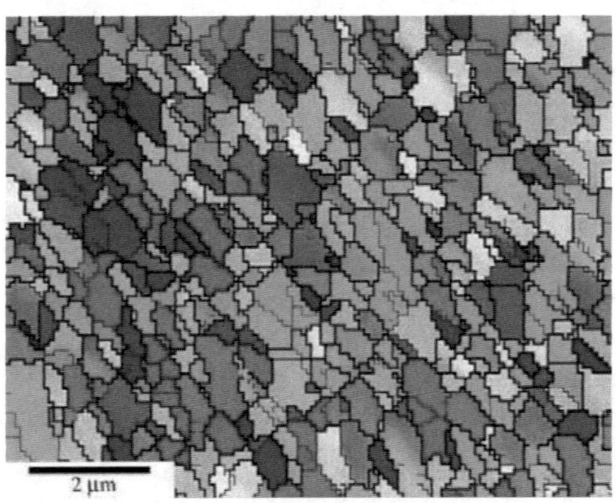

c)

**Figure 1.** Microstructure of SPD-processed steels: a) P92 processed by HPT ( $\varepsilon_{eq} \sim 23$), b) P92 deformed by HPS with $\varepsilon_{eq} \sim 2.6$, c) 304L processed by HPT ($\varepsilon_{eq} \sim 23$).

## 3 Results and discussion

### *3.1 Microstructure processed by HPT*

Figure 1 shows the microstructure of P92 steel processed by HPS and also the microstructures of P92 and 304L steels after one HPT revolution at RT situated about 7 mm from the centre of the HPT disc. The average size of grain in P92 steel in reduced to the value about 0.15 μm after imposed strain $\varepsilon_{eq}$ higher than 20. The microstructure of HPT-processed 304L steel is formed by grains with the mean size about 0.37 μm.

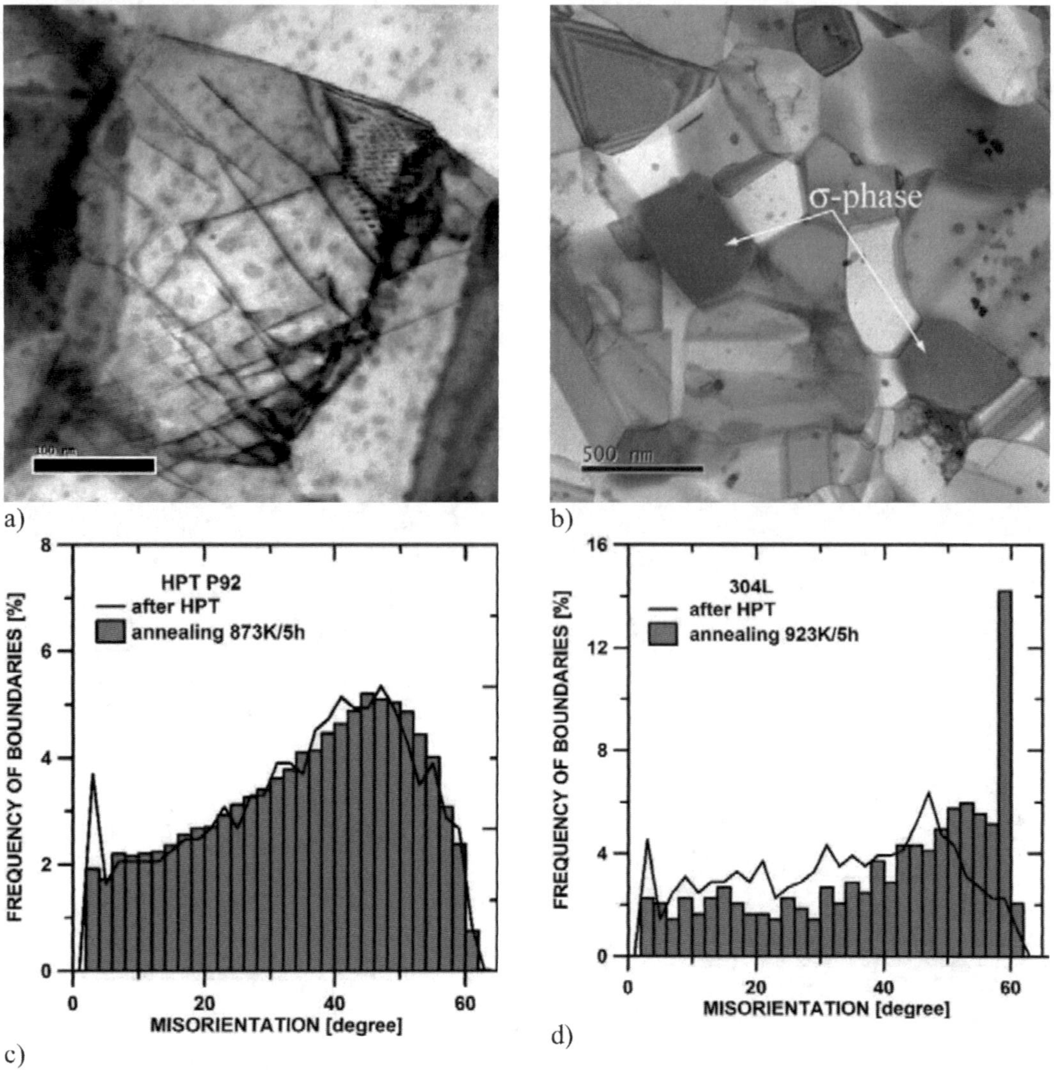


**Figure 2.** Microstructures of HPT-processed steels after short-term annealing for 5 h: a) UFG P92 annealed at 873 K, b) UFG 304L steel after annealing at 923 K. Comparison of misorientation distributions for HPT state and its annealed counterpart c) UFG P92 and d) UFG 304L steel

In the previous studies [1] was shown that the grain refinement by is meaningly affected by imposed strain value and also by processing temperature. It was reported [19] that the largest changes of microstructure characteristics take place during $\varepsilon_{eq}$ up to 10 and $\varepsilon_{eq}$ higher than 20 usually leads to the negligible changes of the mean grain size and number of HAGBs. The saturation of microstructure characteristics after $\varepsilon_{eq}$ higher than 20 was found in several materials. It means that the change in the number of HAGBs and the reduction of average grain size in steels after $\varepsilon_{eq}$ higher than 20 is insignificant. This implies that the saturation process occurring in the microstructure may be significantly affected by dynamic recovery or recrystallization, where boundaries act as sinks and/or sources of dislocations. The previous works [20] revealed that cold SPD may cause not only to the refinement of the microstructure but also phase transformations. This was recently observed also in the HPT-processed 304L austenitic steel [9]. It was found that during HPT deformation occurs the subsequent transformation of FCC into BCC phase. After $\varepsilon_{eq}$ higher than about 20 the microstructure of 304L steel contains only BCC phase [9,21]. This result is valid for the SPD conditions (applied pressure and rotation speed) used in this study because the applied pressure and rotation speed have the influence on the microstructure changes [see 21-23].

### *3.2 Microstructure at the beginning of the creep testing*

This microstructure at the beginning of the creep testing was simulated by annealing at the creep testing temperature for 5 h. This leads to the grain size similar to that in the structure at the beginning of testing. One can see (Fig. 2a) that the fine grains of short-term annealed UFG P92 steel can contain dislocations in their interiors. It means that the dislocations, generated during HPT, are not recovered during annealing at 873 K for 5h. The comparison of misorientation distributions demonstrates that the short-term annealing has neglect effect on the misorientation of boundaries (Fig. 2c). The HPT formed dislocation and LAGBs structure is partially preserved up to higher temperatures and can strengthens the microstructure by deformation hardening. It can be suggested that the influence of possible deformation hardening coming from HPT at room temperature on creep strength can be high at short-term creep test and low creep temperatures. However, significant changes of misorientation distributions occurred in UFG 304L steel after annealing at 923 K/5 h (Fig. 2d). The results demonstrate the occurrence of low energy Σ3 (111/60°) in the microstructure after short-term annealing. The significant changes of misorientation distribution in UFG 304L are related to the reversible transformation of BCC phase into austenite. The annealed microstructure of UFG 304L steel exhibited grain coarsening and the means grain size at the beginning of the creep testing was about 450 nm. In the microstructure of annealed UFG 304L steel was observed σ-phase precipitates (Fig. 2b). The enhanced precipitation of secondary phases in SPD-processed materials was found also in other works [8]. It was concluded that SPD supports the precipitation in UFG microstructure because dislocations act as places for nucleation of precipitates [24].

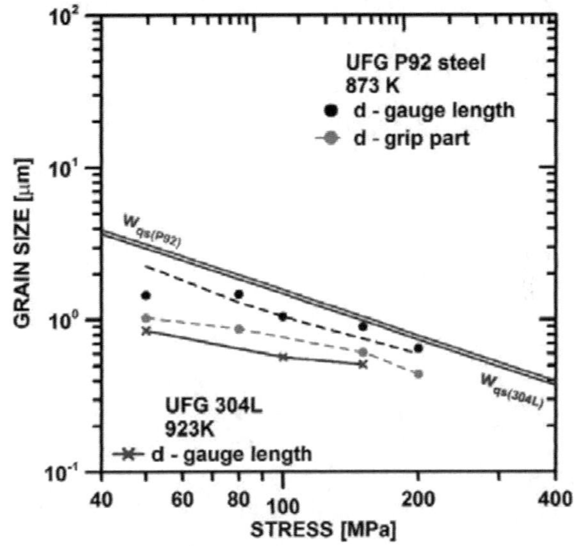


**Figure 3.** Thermal stability of grains in UFG P92 and UFG 304L steels after creep testing under different stresses. The average grain *d* in UFG P92 was measured in the grip part and the gauge length

### *3.3 Microstructure after creep exposure*

Figure 3 shows that grain growth occurs during creep exposure. The grain size was measured after the final fracture of specimens. The comparison of the grain size in the grip (unstressed) part and gauge length of UFG P92 steel tested at 873 K demonstrates that the grain coarsening can be influenced by stress and/or creep strain. One can see that mean grain size of UFG P92 steel is near UFG level even after creep testing at 50 MPa with creep time about 10 500 h. It can be suggested that good thermal stability of microstructure is supported by the presence of carbides and Laves phase precipitates [8] along grain boundaries which restrict their growth. Even if the grain coarsen during creep testing of UFG P92 and 304L steel, the mean grain size of both steels is finer than predicted quasi-stationary subgrain size $w_{qs}$ [5] for their coarse-grained counterparts. This result demonstrates that grains do not contain subgrains. Thus creep processes associated with boundaries, such as grain boundary sliding or enhanced recovery of dislocations, can significantly influence creep in HPT-processed steels.

## 4 Conclusions

Heat-resistant steels P92 and 304L were subjected to severe plastic deformation at room temperature. The microstructure changes after application of SPD, annealing and creep testing were investigated. It was observed that the short-term annealing of UFG P92 caused only minor changes in misorientation distribution and the dislocation formed during HPT at RT are not fully recovered. However, short-term annealing of UFG 304L led to the formation of Σ3 (111/60°) twin boundaries. It was further observed that creep exposure caused the growth of grains but the average grain size was finer than subgrain size predicted for coarse-grained materials.

**Acknowledgements**
The authors are grateful for the support of the Czech Science Foundation (grant No. 19-18725S) and partial support of MEXT, Japan, for scientific research (A) (No. 19H00830). The severe plastic deformation was performed in the International Research Center for Giant Straining for Advanced Materials (IRC-GSAM) at Kyushu University, Fukuoka, Japan.

## References

[1] Valiev R Z, Islamgaliev R K, Alexandrov I V 2000 *Prog. Mater. Sci.* **45** 103
[2] Kral P, Dvorak J, Sklenicka V, Langdon T G 2019 *Mater. Trans.* **60** 1506
[3] Blum W, Dvorak J, Kral P, Eisenlohr P, Sklenicka V 2019 *Metals* **9** 1150
[4] Langdon T G 1994 *Acta Metall. Mater.* **42** 2437
[5] Blum W and Zheng X H 2009 *Acta Mater.* **57** 1966
[6] Abe F, Kern T-U, Viswanathan R 2008 Creep-resistant steels (Woodhead Publishing)
[7] Orlova A, Bursik J, Kucharova K, Sklenicka V 1998 *Mater. Sci. Eng. A* **245** 39
[8] Kral P, Dvorak J, Sklenicka V, Masuda T, Horita Z, Kucharova K, Kvapilova M, Svobodova M 2018 *Mater. Sci. Eng. A* **723** 287
[9] Kral P, Dvorak J, Sklenicka V, Horita Z, Takizawa Y, Tang Y, Kvapilova M, Svobodova M, 2020 *Mater. Sci. Eng.* **785** 139383
[10] Duhaj P, Ivan J, Makovicky E 1968 *J. Iron and Steel Inst.* **206** 1245
[11] Shuro I, Kuo H H, Todaka Y, Umemoto M 2012 *J. Mater. Sci.* **47** 8128
[12] Pohl M, Storz O, Glogowski T 2007 *Mater. Character.* **58** 65
[13] Horváth J, Král P, Janovec J 2016 *Acta Phys. Pol. A* **130** 960
[14] Hosseini V A,, Karlsson L, Wessman S, Fuertes N 2018 *Materials* **11** 933
[15] Padilha A F, Plaut R L, Rios P R 2003 *ISIJ International* **43** 135
[16] Fujioka T, Horita Z 2009 *Mater. Trans.* **50** 930
[17] Sklenicka V, Kral P, Dvorak J, Takizawa Y, Masuda T, Horita Z, Kucharova K, Kvapilova M, Svobodova M 2019 *Adv. Eng. Mater.* 1900448
[18] Kral P, Dvorak J, Sklenicka V, Masuda T, Horita Z, Kucharova K, Kvapilova M, Svobodova M, 2018 *Materials* **11** 787
[19] Renk O, Pippan R 2019 *Mater. Trans.* **60** 1270
[20] Bachmaier A, Pippan R 2019 *Mater. Trans.* **60** 1256
[21] Shuro I, Umemoto M, Todaka Y, Yokoyama S *2010 Mater. Sci. Forum* **654-656** 334
[22] Zehetbauer T, Pippan R, Stüwe H P, Vorhauer A, Schafler E, Kohout J 2003 *Adv. Eng. Mater.* **5** 330
[23] Hebesberger T, Pippan R, Stüwe H P 2002 Proc. of the 2$^{nd}$ Int. Symposium on Ultrafine Grained

Materials, TMS Annual Meeting (Seattle), (Warrendale: USA, The Minerals, Metals and Materials Society) p 133
[24] Gubicza J, Schiller I, Chinh N Q, Illya J, Horita Z, Langdon T G *2007 Mater. Sci. Eng A.* **460-461** 77

# The influence of Zener drag on recrystallization behaviour of twin-roll cast AA8079 alloy after homogenization

R Králík, B Křivská, L Bajtošová, M Cieslar

Charles University, Faculty of Mathematics and Physics, Ke Karlovu 5, 121 16, Prague, Czech Republic

E-mail: rkralik96@gmail.com

**Intermediate annealing is a crucial step of aluminum foil production. The as-cast material hardens during rolling and an annealing step is required at intermediate gauges to soften the material and restore its ductility. Generally, fine recrystallized structure could assure sufficient ductility for further rolling and processing. Recrystallization of the material during intermediate annealing can be controlled by a microstructure tailored during homogenization. Recrystallization behaviour of a twin-roll cast aluminum alloy AA8079 was studied. The as-cast strip was homogenized at 420 °C and 580 °C and rolled down to two different thicknesses. Evolution of microstructure of rolled materials during a model intermediate annealing was studied, and Zener drag and particle stimulated nucleation were determined as mechanisms controlling recrystallization.**

## 1 Introduction

Several properties make aluminum an optimal material for a use as packaging foil stock [1]. The AA8079 alloy is a low alloyed material containing iron and silicon. The equilibrium solubility of both elements in aluminum is low, which leads to a formation of intermetallic particles [2].

Twin-roll casting (TRC) reliably produces strips several millimeters thick [3-6], which is beneficial for the production of foil stock materials, as it reduces the number of required rolling steps during downstream processing when compared to traditional casting methods such as direct chill casting. During TRC the melt is poured on water cooled rolls where it rapidly solidifies [7]. As a result, the as-cast material has a gradient structure along the normal direction of the strip. A homogenization annealing step is required before further processing. This step is traditionally performed at temperatures above 520 °C in order to facilitate phase transformation of present phases of the Al-Fe-Si system into thermodynamically stable Fe-rich phases. This annealing step also changes size and morphology of particles [8]. Distributions of particles and present phases in the as-cast state depend on a specific alloy composition, casting speed and casting gauge [9-12]. Particles of the stable phase grow coarse during homogenization at high temperatures [8], which leads to a promotion of particle stimulated nucleation (PSN). PSN is an effect commonly used to facilitate the formation of a fine grained structure. Deformation zones with high dislocation density form around coarse particles during rolling and serve as nucleation sites for new grains during annealing [13].

However, a lower temperature of homogenization might be of industrial interest due to possible energy savings. Homogenization at lower temperatures does not facilitate noticeable changes in particle size and distribution. Therefore, PSN does not occur and a fine grained structure ought to be reached through an increased influence of Zener drag (ZD) [14]. ZD is a retarding force of grain boundary motion caused by the presence of fine particles. Retarding grain boundary motion allows nucleation of a larger number of grains, which leads to a finer grain structure.

The influence of homogenization temperature on a distribution of intermetallic phases in AA 8079 alloy was studied and mechanisms controlling softening of the material after subsequent rolling and annealing were determined.

## 2 Experimental methods

The experimental AA8079 alloy had a high Fe/Si ratio >> 1 containing 1.06 wt. % Fe and 0.06 wt. % Si. Thickness of the as-cast TRC strip was 7 mm. Samples cut from the strip were homogenized at two different temperatures (420 and 580 °C). Each sample was heated from room temperature for 8 h, soaked at the selected temperature for 8 h and then furnace cooled back to room temperature for another 8 h. Afterwards, the samples were cold rolled to 600 μm or 60 μm gauges and subjected to a model intermediate isochronal annealing. The isochronal heating was performed with an effective heating rate of 1 K/min. The annealed samples were then water quenched to room temperature.

Light optical microscope (LOM) observations were performed using Zeiss Axio Observer 7 in a bright field and polarized light regimes. Scanning electron microscopy (SEM) observations were done in a FEI Quanta 200F electron microscope. Microhardness was measured with the Qness Q10 device. Samples for all these measurements were prepared using a combination of mechanical polishing with SiC papers and electro-chemical polishing and etching methods. Samples for LOM and SEM particle observations were polished at room temperature in a 0.5 % solution of HF in ethanol. Samples for grain observations were polished at -15 °C in Barker's reagent. All observations and measurements were performed in a transversal direction of studied strips.

## 3 Results and discussion

### *3.1 Recrystallization behavior*

Evolutions of microhardness of rolled materials homogenized at different temperatures are shown in figure 1. The thicker material (600 μm) fully recrystallizes around 300 °C regardless of the homogenization temperature, and microhardness remains constant within an experimental error at higher annealing temperatures. The thinner material (60 μm) has higher initial microhardness due to a higher degree of plastic deformation during rolling. However, the thinner material softens over an extended temperature range. This range is smaller for the material with higher homogenization temperature. The influence of homogenization temperature on a grain structure in these materials is strong. The microstructure of specimens annealed isochronally to 350 °C is shown in figure 2. The microstructure differs significantly in both materials at this temperature. The thicker material (600 μm) homogenized at 420 °C coarsens significantly at 350 °C. The recrystallized grains are elongated in the rolling direction. Their size is approximately (50 ± 7) μm in the normal and (250 ± 50) μm in the rolling direction. The thicker material homogenized at 580 °C contains a bimodal grain structure. Coarse elongated grains appear close to the surface of the strip and fine equiaxed grains are located in the center of the strip. The coarser grains are approximately (111 ± 6) μm long in the rolling direction and approximately (52 ± 7) μm long in the normal direction. The finer grains in the strip interior have a diameter of (45 ± 5) μm.

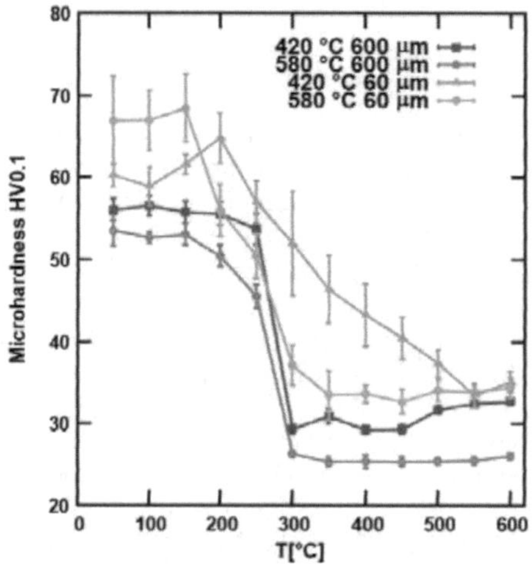


**Figure 1.** Evolution of microhardness during isochronal annealing after two different temepratures of homogenization and two different reductions.

The behavior of materials rolled to 60 µm is inversed. The material homogenized at 580 °C has coarser grains (slightly elongated, approx. (33 ± 8) µm in the ND and (42 ± 9) µm in the RD) while only partially recovered and partially recrystallized fine-grained structure was observed in the material homogenized at 420 °C.

**Figure 2.** Microstructure of the TRC material after homogenization at 420 or 580 °C, rolling with two reductions and isochronal annealing up to 350 °C.

### 3.2 *Influence of particle distribution*

The driving force of recrystallization is a result of a high stored deformation energy. Particles and other material defects retard recrystallization. The influence of Zener drag of particles was quantified through microscopic observations. A simplified equation (1) can be used to quantify this retarding force [14]:

$$P_Z = \frac{3}{4}\frac{f\gamma_{GB}}{R} \ . \tag{1}$$

In this equation $f$ is the volume fraction of particles, $R$ is their diameter and $\gamma_{GB}$ is grain boundary energy. Grain boundary energy depends on a specific type of boundary and material. We assume a value of 0.3 J/m$^2$ for certain high angle grain boundaries in aluminum [15]. Size and volume fraction of particles is affected by a selected homogenization temperature. LOM images are used to determine the volume fraction of particles (figure 3) as they offer better statistics from a larger area of the sample. SEM images are more suitable for the determination of the particle size due to a better resolution (figure 4).

LOM images show particles distributed in eutectic colonies in material homogenized at 420 °C. These particles are fine and contribute to the control of recrystallization by ZD (figure 3 a), 4 a)). The material homogenized at 580 °C contains bimodal particle distribution with a combination of finer and coarser particles (figure 3 b), 4 b)). The coarse particles serve predominantly as potential nucleation sites after rolling and contribute to recrystallization controlled by PSN. Summary of measured particle volume fractions, the size of particles and an estimate of Zener drag according to equation (1) is shown in table 1.

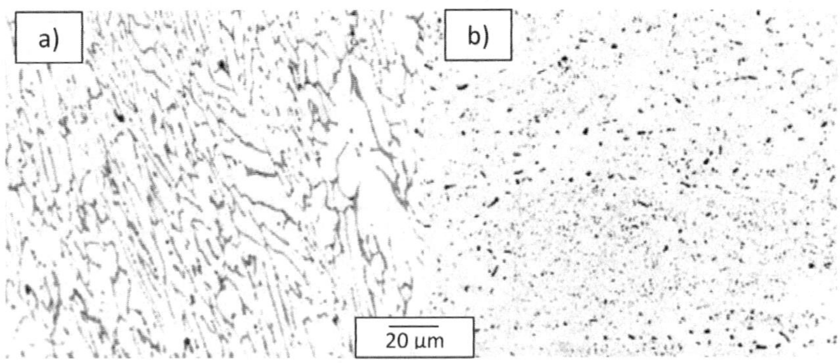


**Figure 3.** LOM image of a material homogenized at: a) 420 °C and b) 580 °C.

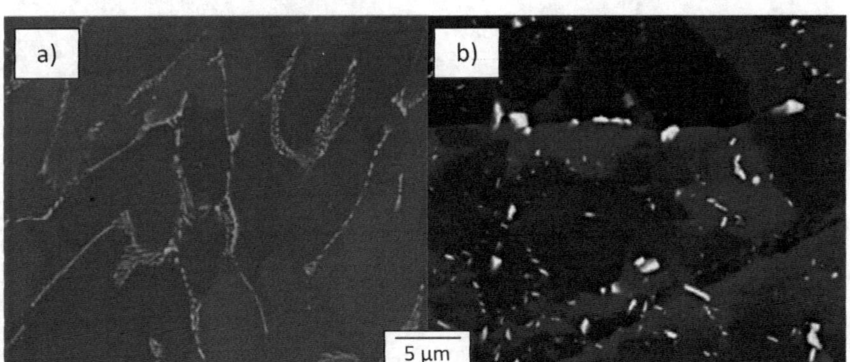


**Figure 4.** SEM image of a material homogenized at: a) 420 °C and b) 580 °C.

**Table 1.** Characterization of particles and estimation of Zener drag

| | fraction f [%] | Diameter R [µm] | Zener drag Z [kPa] |
|---|---|---|---|
| 420 °C | 8 ± 1 | 0.3 ± 0.1 | 43 ± 11 |
| 580 °C | 6 ± 1 | 0.5 ± 0.1 | 24 ± 5 |

The cold rolling process does not affect volume fraction or size of particles. However, the simplified equation (1) only applies for materials with homogenous distribution of particles and uniform particle sizes. This is not the case of the studied material. Thresholding of images is not exact either. As a result, the values of Zener drag are rather apparent and they affected by a large experimental scatter (table 1).

Fenomenological description of the effect of inhomogeneous particle distributions on recrystallization is difficult. Nes claims that a grain boundary moving through parallel layers of particles of thickness $d$ and spacing $l$ experiences effective Zener drag $P_Z$' in these layers given by equation [14]:

$$P_{Z'} = \frac{l}{d} P_Z \,. \tag{2}$$

This effective Zener drag is higher than Zener drag in a material with random particle distribution, but two major cases have to be considered – a boundary that is normal or parallel to the rolling direction. The normal boundary is retarded by the Zener drag given by equation (1) if it stretches over multiple particle layers or is not retarded at all when constrained within one interdendritic space. The parallel boundary is not affected by Zener drag in between dendrites and is pinned by Zener drag according to equation (2) after it reaches a layer with particles. This means that the retarding forces are weaker in the direction of eutectic dendrites (RD) than retarding forces in the normal direction. As a result, recrystallized grains can grow faster in the rolling direction, which results in elongated grains. Those were observed in the thicker material (600 µm) homogenized at 420 °C. This material coarsened significantly despite the higher apparent Zener drag value. Birol [8] shows, that this is a result of a preferred growth of several selected. The thinner material homogenized at 420 °C shows the smallest grain size after partial recrystallization. This could be a result of changes in particle distribution caused by rolling. SEM observations show that interdendritic spacing of particles decreases with increasing reduction (figure 5 a), b)), and the value of ZD is closer to the one given by the equation (1). This implies a higher value of ZD in the rolling direction, which prevents preferential growth in this direction. The above-mentioned facts are confirmed by SEM (figure 5) in materials annealed up to 300 °C – a temperature at which recrystallization starts. The thicker material (figure 5 b)) contains grains already coarsened in the rolling direction (highlighted by the white oval), while no such grains are observed in the thinner material (figure 5 a)).

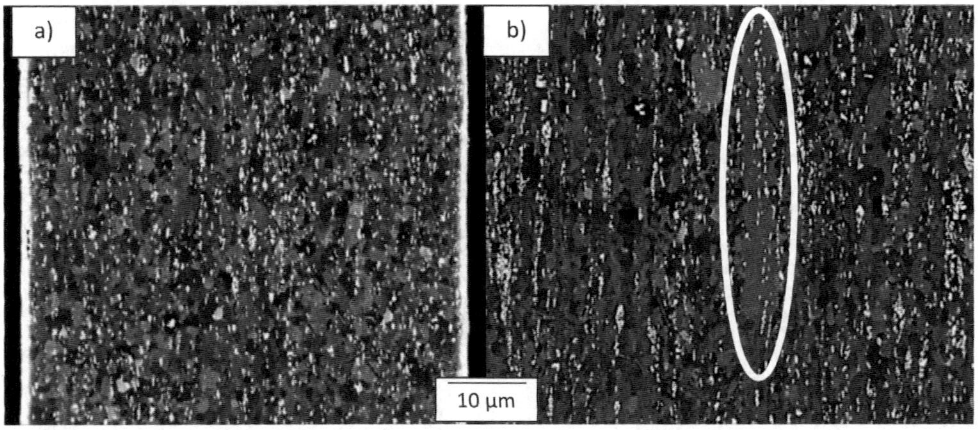


**Figure 5.** Particle distribution in TRC materials homogenized at 420 °C, and annealed up to 300 °C after rolling to: a) 60 µm, b) 600 µm.

The Zener drag of particles is weaker in materials homogenized at 580 °C. However, the increased nucleation through PSN results in a formation of fine recrystallized grains. Grain sizes after recrystallization are similar for both material thicknesses. However, grains in the thinner material (60 µm) are slightly elongated.

## 4 Summary

The studied material was homogenized at two different temperatures (420 and 580 °C) in order to tailor processes controlling recrystallization. Both materials homogenized at 580 °C behave similarly regardless of thickness. Recrystallization of these materials results in a coarser structure with grain diameters between 30 and 50 µm in the central part of the strip thanks to the influence of PSN.

The Zener drag of particles in the material homogenized at 420 °C is higher. Behavior of the thicker material is in accordance with previously published results, and softening controlled through Zener drag yields coarse grains at 350 °C. However, softening process differs significantly after rolling to 60 µm. The material retains partially recrystallized fine grained structure at 350 °C. This grain structure is significantly finer than the one controlled by PSN in materials with higher homogenization temperatures.

**Acknowledgement**
The financial support of the Charles University Grant Agency project 704119 and project TRIO FV 20337 of the Czech Ministry of Industry and Trade are highly acknowledged.

**References**

1 Kerry J 2012 Aluminum foil packaging *Packaging Technology, Fundamentals, Materials and Processes* 163-77
2 Hatch J E 1984 *Aluminum: Properties and physical metallurgy* American Society for Metals Metals Park, OH ISBN 0-87170-176-6
3 Berg B S, Hansen V, Zagierski P T, Nedreberg M L, Olsen A and Gjonnes J 1995 Gauge reduction in twin-roll casting of an AA5052 aluminium alloy: The effects of microstructure *J Mater Process Tech* **53** 65-74
4 Yun M, Lokyer S and Hunt J D 2000 Twin-roll casting of aluminum alloys *Prog mater sci* A **280** 116-23
5 Cook R, Grocock P G, Thomas P M, Edmonds D V and Hunt J D 1995 *J Mater Process Tech* **55** 76-84
6 Lee Y S, Kim H W and Cho J H 2014 Effect of casting parameters on roll separation force during twin-roll casting *Procedia Engineer* **81** 1547-52
7 Song R, Harada Y, and Kumai S 2017 Twin roll casting of aluminum alloy strips *J Mater Process Tech* **153-154** 42-7
8 Birol Y 2008 Thermomechanical processing of a twin-roll cast Al-1Fe-0.2Si alloy *J Mater Process Tech* **202** 564-8
9 Birol Y 2009 Analysis of macro segregation in twin-roll cast aluminum strips via solidification curves *J Alloy Compd* **468** 168-72
10 Spathis D and Spiros J 2013 The influence of casting speed in the as-cast strip mechanical properties of 8079 and 8006 alloys *Light Met 2013* 305-9
11 Engler O, Aegerter J and Calmer D 2020 Control of texture and earing in aluminum alloy AA8011A-H14 closure stock *Mater Sci Eng*: A **775** 69-80
12 Lentz M, Laptyeva G, Engler O 2016 Characterization of second-phase particles in two aluminum alloy foil alloys *J Alloy Compd* **660** 276-88
13 Humphreys F J 1977 The nucleation of recrystallization at second phase particles in deformed aluminum *Acta Metall Mater* **25** 1323-44
14 Nes E, Ryum N and Hunderi O 1985 On the Zener drag *Acta Metall Mater* **33**11-22
15 Zheng H, Li X, Tran R, Chen Ch, Horton M, Winston D, Persson K A and Ong S P 2020 Grain boundary properties of elemental metals *Acta Metall Mater* **186** 40-9

# Resistivity and Formation of Intermetallic Layer in Aluminum-Steel Clad Strip

**B Křivská[1], M Šlapáková[1], R Králík[1], L Bajtošová[1], M Cieslar[1], O Grydin[2], M Stolbchenko[2], M Schaper[2]**

[1]Charles University, Faculty of Mathematics and Physics, Ke Karlovu 5, 121 16 Prague, Czech Republic

[2]Paderborn University, Chair of Materials Science, Warburger Str. 100, 33098 Paderborn, Germany

E-mail: krivska.barbora@seznam.cz

**Abstract**. Cladding of two dissimilar materials such as aluminum and austenitic steel can bring outstanding material characteristics beyond the ones of each conventional monomaterials. The study focused on development of an Al-Fe-rich intermetallic phase formed on the aluminum/steel interface during heat treatment. Electrical resistivity measurements were performed to monitor diffusion processes and phase transformations during isochronal and isothermal annealing. Resistivity measurement results were supplemented by microscopic methods revealing the interfacial microstructure.

## 1 Introduction

Al-steel clad sheets manufacturing enables to get a material combining steel's high strength with a low weight of aluminum, making this material very attractive for structural applications in shipbuilding, aircraft constructions or car bodies [1,2].

Clad materials represent joints of two or more layers of different metals. They can be produced in various forms e.g. strips, foils, tubes, rods or wires. Nowadays, more than 90 % of the clad strips production is comprised by a cold or hot roll bonding process [3]. However, these methods suffer from severe drawbacks resulting in a high cost of the final product. The issue is that the substrate's preparation consists of several steps, including multi-pass hot rolling and complicated surface pretreatments.

Twin-roll casting used as a cladding process [4,5] is a novel method, which poses a solution overcoming such drawbacks since clad strips are prepared directly from a melt of one metal and a solid sheet of the other one that are fed simultaneously between the rolls of a twin-roll caster. Demands on the surface preparation are then partially relinquished.

The clad material could be subjected to a thermal exposure during manufacturing of a final product, which could activate diffusion between the metals leading to undesirable formation of brittle intermetallic compounds (IMC), influencing the final mechanical properties of the clad material [6,7]. Akramifard et al. [8] investigated microstructure and mechanical properties of a three-layered composite AA1050/304L/AA1050 prepared by hot rolling. During a subsequent heat treatment, growth of an IMC layer was observed deteriorating the joint's quality. Cracks propagated along the Al–IMC interface and caused an even-by-hand delamination of the bond. $Al_5Fe_2$ and $Al_{13}Fe_4$ phases were found as the main constituents of the IMC layer formed on the interface [8–11].

The aim of this paper is to investigate the evolution of the interface between Al and steel in twin-roll cast Al-steel clad sheet at elevated temperatures using resistivity measurements and observation of the resulting structure by light optical microscopy (LOM) and scanning electron microscopy (SEM).

## 2 Experimental

### *2.1 Material*

A melt of technically pure aluminum EN AW-1070 was clad on a solid strip of austenitic steel type 1.4301. The casting process was realized by a laboratory twin-roll caster equipped with two water-cooled rolls with a diameter of 370 mm and a length of 200 mm [12]. A vertical operation plane was chosen. No mold release coating was applied on the rolls' surface either before or during the casting process. The final thickness of the composite strip was 2.5 mm with the aluminum-to-steel thickness layer ratio of 4:1.

Since a subsequent heat treatment could influence the bond strength between the steel and aluminum layer, isochronal and isothermal annealing were performed using a laboratory air furnace. The isochronal step-by-step annealing scheme with step 50 °C / 25 min was applied with a water-quenching between every two steps. The isothermal annealing was done at 500 °C with a logarithmically changing time-step length.

### *2.2 Methods*

Electrical resistivity measurements (four-point probe at the temperature of liquid nitrogen) were carried out after each annealing step and solid solution decomposition and phase transformations were investigated. Light optical microscopy (LOM, Zeiss Axio Observer metallographic microscope) and scanning electron microscopy (SEM, Quanta 200F) observations in backscattered electrons (BSE) were used for the characterization of microstructural changes.

## 3 Results and discussion

### *3.1 Initial state*

Both LOM and SEM show colonies of primary phases arranged in interdendritic chains obliquely orientated in the casting direction of both outer sides of the aluminum layer (figure 1, 2a).

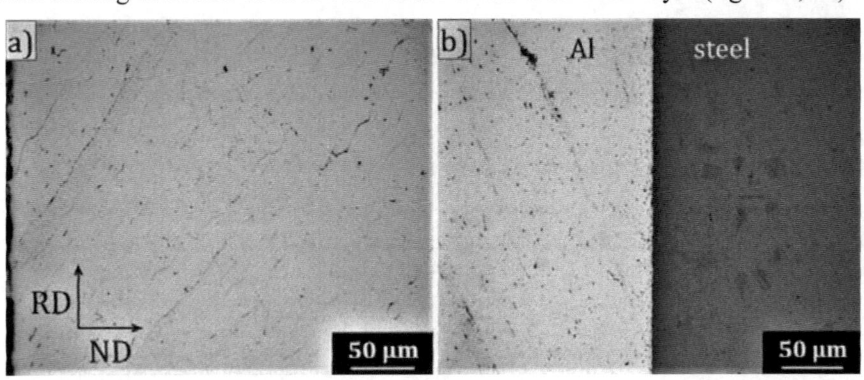


**Figure 1.** Light optical micrograph of a) Al layer surface without steel coating and b) Al/steel interface.

A slight difference in the phase distribution on the opposite sides of the aluminum layer – with and without the adjacent steel substrate – could be noticed due to different cooling rates caused by the added steel sheet [13,14]. The steel substrate acts as an additional heat barrier so the aluminum solidifies from that side at a considerably slower rate and a position of the kissing point [13] is shifted towards the steel layer.

The presence of the IMC layer at the Al-steel interface in the as-cast state was not confirmed by LOM or SEM (figure 1b, 2b). The observation is in contradiction with results given in [2,15], where a formation of the IMC during the casting was reported. It should be noted that clad strip casting conditions were different for the compositions used in our study and in the referenced works. In work of Grydin[2], a thicker Al-layer, a thinner steel layer and a higher Al pouring temperature were used. In work of Stolbchenko [15], the same clad parameters were used (Al layer 2.0 mm, steel layer 0.5 mm) but with higher casting temperature and casting rate. Thus, a higher strip outlet temperature could result in the IMC growth without any additional annealing.

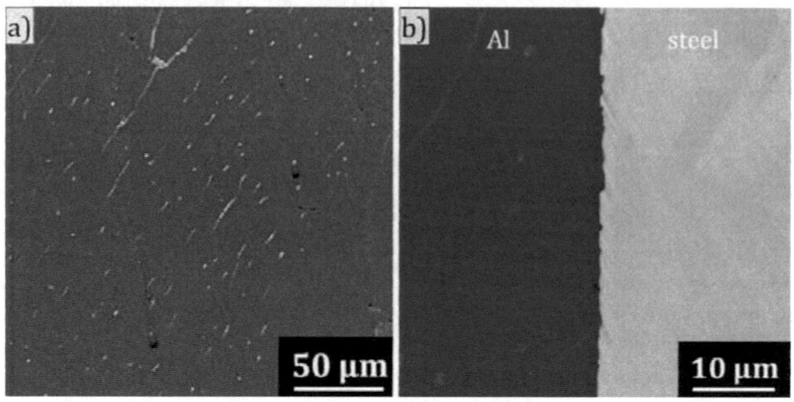


**Figure 2.** BSE images a) of aluminum and b) interface region.

### *3.2 Annealing*

Figure 3a shows electrical resistivity annealing curves of the composite and separate Al and steel layers during isochronal annealing. A pronounced resistivity drop was measured in the composite and Al sample between 200 °C and 450 °C. Since very similar shapes of the resistivity curves were observed for Al and the composite sample, phase transformation ongoing in Al layer has a controlling role in the resistivity curves evolution. The decrease of resistivity is connected with a higher purity of the matrix and a depletion of the solid solution. Contrarily, a backward increase above 450 °C could be a sign of particle dissolution.

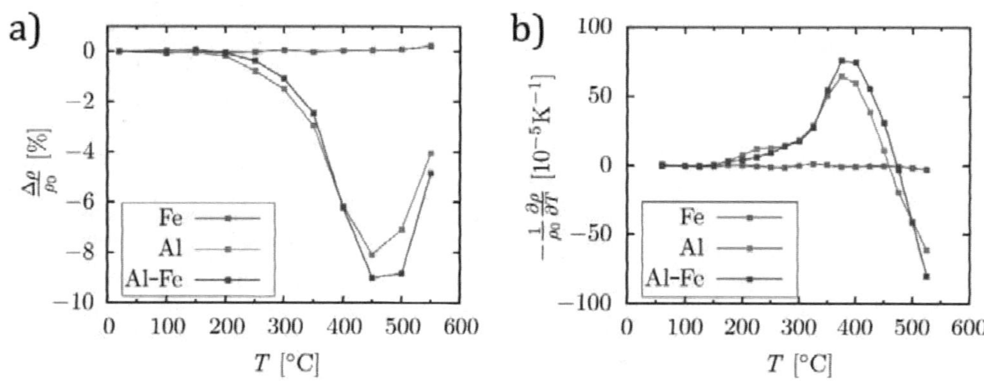


**Figure 3.** a) Relative changes of electrical resistivity during isochronal annealing, b) resistivity annealing spectra calculated from resistivity data.

The calculated negative differential curves $-(d\rho/dT)/\rho_0$, referred to as resistivity annealing spectra, exhibit a pronounced peak at 400 °C for both aluminum and composite samples (figure 3b).

Since the derivative directly corresponds with an intensity of resistivity changes, peaks in positive values are related to precipitation, coarsening of the precipitates and solid solution depletion. In contrast, negative values measured above 450 °C are connected with solid solution enrichment and dissolution of the particles. A small peak present below 250 °C in Al sample is often observed in Al alloys and it is related to a redistribution of Si atoms [16]. The effect is very weak and it can be overlapped by other transformations ongoing at higher temperatures. Partial suppression of the peak in the composite could be the result of a different heat loss due to the added steel layer.

Noticeable coarsening of existing particles corresponding with the peak in resistivity annealing spectra is confirmed between 350 °C and 450 °C by LOM micrographs (figures 4a, b, c). Above 450 °C, predicted particle dissolution was also observed (figure 4d). The steel layer remained unchanged during all annealing experiments and the resistivity of the steel sample is rather constant during the whole annealing process.

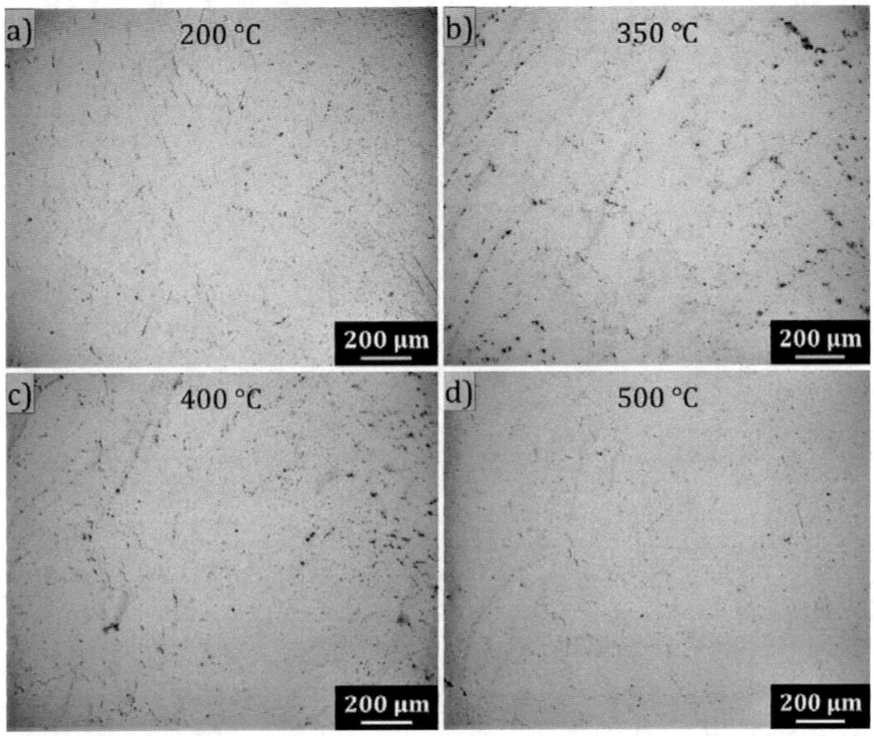


**Figure 4.** LOM images of the particles distribution in Al during isochronal annealing up to: a) 200 °C, b) 350 °C, c) 400 °C, d) 500 °C.

The development of the interface region was studied by SEM (figure 5). Growth of an IMC layer at the Al-steel interface was observed above 500 °C. Its influence on the resistivity curves remains rather disputable. However, the slope of the resistivity annealing spectra above 450 °C was slightly steeper in the case of the composite sample (figure 3b).

Considering the convincing development of the interface region above 500 °C, the isothermal annealing was performed at 500 °C. The resistivity curves are shown in the figure 6a. A similar evolution of the Al and composite resistivity was obtained, indicating the main influence of the phase transformations in the Al layer. In contrast to the isochronal annealing, the resistivity curves of the Al and composite samples during isothermal annealing do not contain the backward increase suggesting that 500 °C annealing leads only to a depletion of the matrix and coarsening of existing particles. However, a larger drop of resistivity observed in the clad material indicates an influence of formation of an IMC layer at the Al-steel interface. Similarly, as observed by several researches [9–11], the interfacial phase layer grows according to the parabolic law (figure 6b and figure 7).

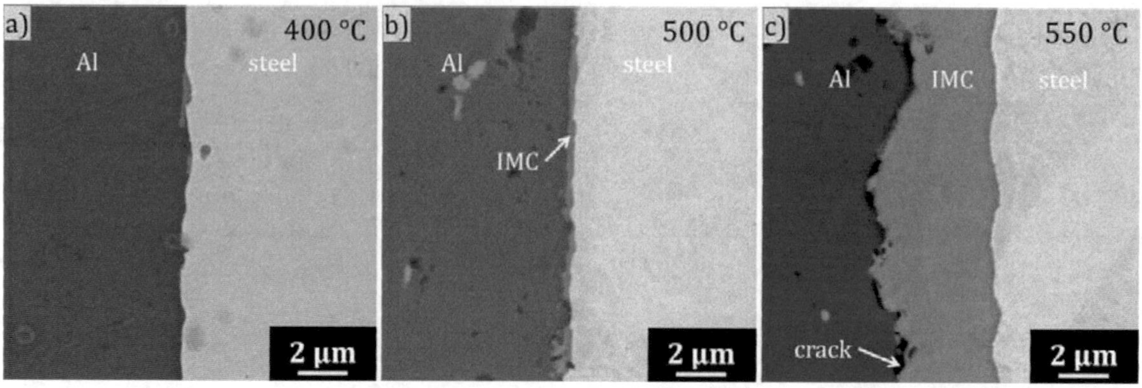


**Figure 5.** BSE images of the interfacial phase evolution during isochronal annealing up to: a) 400 °C, b) 500 °C, c) 550 °C.

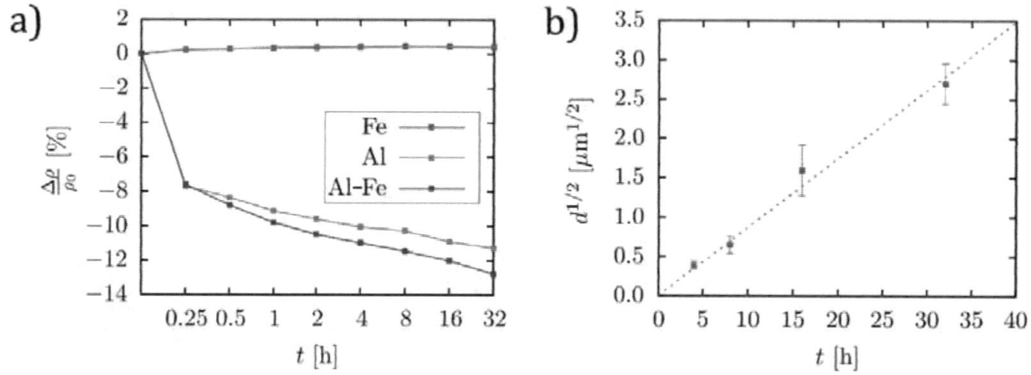


**Figure 6.** a) Relative changes of electrical resistivity, b) time evolution of the IMC layer thickness $d$ during isothermal annealing.

Formation of the thick IMC layer could result in formation of cracks between Al and IMC and deterioration of the bond [8,15]. Isochronal annealing up to 600 °C leads to full delamination of the aluminum layer from the steel sheet. Isothermal annealing at 500 °C for 32 h results only in crack formation in Al along the IMC layer (figure 7c). EDX analysis confirms formation of Al-Fe rich phases [17] observed in Al-steel clad sheets [8–11].

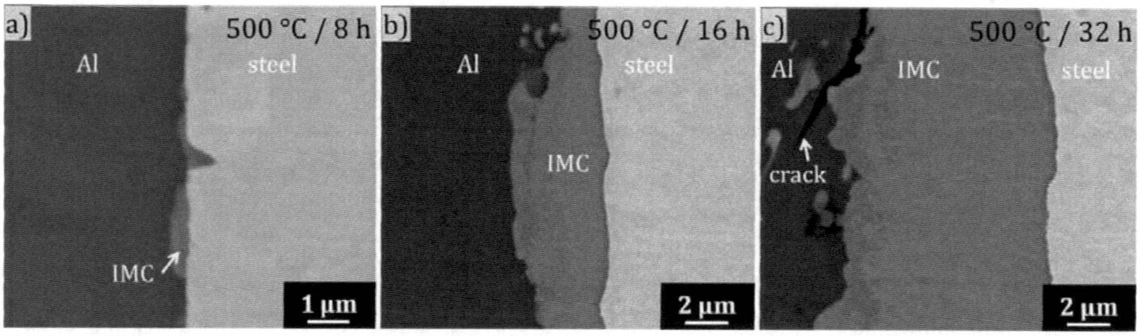


**Figure 7.** BSE images of the interfacial region after annealing at 500 °C for: a) 8 h, b) 16 h, c) 32 h.

## 4 Summary

Evolution of microstructure, resistivity and IMC layer in twin-roll cast Al-steel clad material were studied during annealing. Significant coarsening of precipitates was observed in the Al strip below 450 °C during isochronal annealing. Corresponding maximum in the resistivity spectrum was detected at 400 °C. Partial dissolution of coarser particles occurs above 450 °C, resulting in presence of negative values in the resistivity spectrum. A similar evolution of resistivity was also observed in the composite material. The influence of the IMC layer growth on the resistivity of the clad is rather small, however, evinced by larger decrease of the composite resistivity at 500 °C. The interfacial IMC phases grow parabolically with time.

**Acknowledgements**

The financial support of GAČR project 20-00355Y, SVV-260582 and German Research Foundation (DFG) in regard to the project SCHA 1484/21-1 are highly acknowledged.

**References**

[1] Groche P, Wohletz S, Brenneis M, Pabst C, Resch F 2014 Joining by forming – A review on joint mechanisms, applications and future trends, *J. Mater. Process.Technol.*, **214**, p 1972–1994

[2] Grydin O, Gerstein G, Nürnberger F, Schaper M and Danchenko V 2013 Twin-roll casting of aluminum–steel clad strips. *J. Manuf. Process.* **15** p 501–507

[3] Khaledi K, Rezaei S, Wulfinghoff S and Reese S 2018 Modeling of joining by plastic deformation using a bonding interface finite element. *Int. J. Solids Struct.* **160** p 68–79

[4] Hudson E B 1936 Direct casting of sheetlike metal structures. US Patent No. 2128941

[5] Papich K S 1997 Method of continuously casting composite strip. US Patent No. 5669436

[6] Prasada Rao A K, Kim K H, Bae J H, Bae G T, Shin D H and Kim N J 2009 Twin-Roll Cast Al-Clad Magnesium Alloy. *Mater. Sci. Forum* **618–619** p 467–471

[7] Bae J H, Prasada Rao A K, Kim K H and Kim N J 2011 Cladding of Mg alloy with Al by twin-roll casting. *Scr. Mater.* **64** p 836–839

[8] Akramifard H R, Mirzadeh H and Parsa M H 2014 Cladding of aluminum on AISI 304L stainless steel by cold roll bonding: Mechanism, microstructure, and mechanical properties. *Mater. Sci. Eng. A* **613** p 232–239

[9] An J, Liu Y, Lu Y and Sun D 2001 Hot roll bonding of Al–Pb-bearing alloy strips and steel sheets using an aluminized interlayer. *Mater. Charact.* **47** p 291–297

[10] Springer H, Kostka A, Payton E J, Raabe D, Kaysser-Pyzalla A and Eggeler G 2011 On the formation and growth of intermetallic phases during interdiffusion between low-carbon steel and aluminum alloys. *Acta Mater.* **59** p 1586–1600

[11] Li R, Yuan T, Liu X and Zhou K 2016 Enhanced atomic diffusion of Fe–Al diffusion couple during spark plasma sintering. *Scr. Mater.* **110** p 105–108

[12] Grydin O, Ogins'kyy Y, Danchenko V and Bach F-W 2010 Experimental twin-roll casting equipment for production of thin strips. Metallurgical and Mining Industry **2** p 348–354

[13] Stolbchenko M, Grydin O, Samsonenko A, Khvist V and Schaper M 2014 Numerical analysis of the twin-roll casting of thin aluminium–steel clad strips. *Forschung im Ingenieurwesen* **78** p 121–130

[14] Chen G, Li J and Xu G 2017 Bonding process and interfacial reaction in horizontal twin-roll casting of steel/aluminum clad sheet. *J. Mater.* **246** p 1–12

[15] Stolbchenko M, Grydin O and Schaper M 2019 Manufacturing and Characterization of Twin-Roll Cast Aluminum–Steel Clad Strips. *Adv. Eng. Mater.* **21** p 1800454

[16] Cieslar M, Slámová M, Uhlíř J, Coupeau Ch, Bonnevile J 2007 Effect of composition and work hardening on solid solution decomposition in twin-roll cast Al-Mn sheets *Kovove Mater.* **45** p 91-97

[17] Křivská B, Šlapáková M, Grydin O, Cieslar M 2020 Aluminium-steel clad material prepared by twin-roll casting. *Metal 2020 Conference Proceedings* Ostrava (Tanger Ltd.) p 989–993

# Evaluation of the usefulness of rotors for aluminium refining

**K. Kuglin[1] and D Kalisz[1]**

[1] AGH University of Science and Technology, Faculty of Foundry Engineering, 23 Reymonta Street, 30-059 Kraków, Poland

E-mail: dak@agh.edu.pl

**Abstract**. Refining of liquid aluminium by blowing with an inert gas is one of the most effective methods of purifying liquid metal from hydrogen and non-metallic inclusions. In order to improve the efficiency of this process, it is necessary to develop a rotor design. The study compares rotors designed and their usefulness. For this purpose, the theoretical number of nozzles, the amount of gas per nozzle and the gas bubble radius were calculated.

## 1 Introduction

The process of refining with neutral gas plays important role in recycling processes and use of secondary materials [1-12]. The type of applied neutral gas depends on the involved contaminations: argon and nitrogen are used for hydrogen removal, a mixture of chlorine and argon or nitrogen is used for removing metallic contaminants.

In industrial practice this process is realized in reactors equipped with a whirling head introducing neutral gas, either continuously or cyclically [1, 5]. The efficiency of the refining process is determined by a number of parameters, but the design of the impeller as most important and decisive for the production of properly dispersed neutral gas bubbles in the liquid metal. This finally guarantees that the liquid metal is thoroughly mixed in the ladle [1-10]. The barbotage refining technology with a whirling head has a number of design variations depending on the number and location of nozzles in the rotor moving with a specific speed. As compared to the traditional refining procedure, the introduction of additional centrifugal force causes the formation of a whirling stream of fine-dispersive neutral gas bubbles in the volume of a metal bath. This phenomenon favors the removal of nonmetallic inclusions to the gaseous bubbles [1-12]. Although numerous research works [5-10] on this issue were conducted, the practical implementation of this technology to the structure of a given foundry still encounters many problems connected with the implementation of such key production parameters as: intensity of mixing and consequently the dispersion of gaseous bubbles in metal, and their connection with the average size of the gas bubble and mass transfer coefficient. The efficiency of the refining process is determined by the type and design of the nozzle, design of the refining system and also surface of the refining gas/metal interface. The design of the whirling head decides about the intake ability and transfer of gas to the system and about the bubble size, though the design of the whirling head and the nozzle is one of the decisive parameter.

## 2 Comparison of rotor operation

The 3D technique has been used for making whirling heads. As presented in figure 1, the designed whirling heads are equipped with 4, 8 and 12 nozzle outlets, respectively.

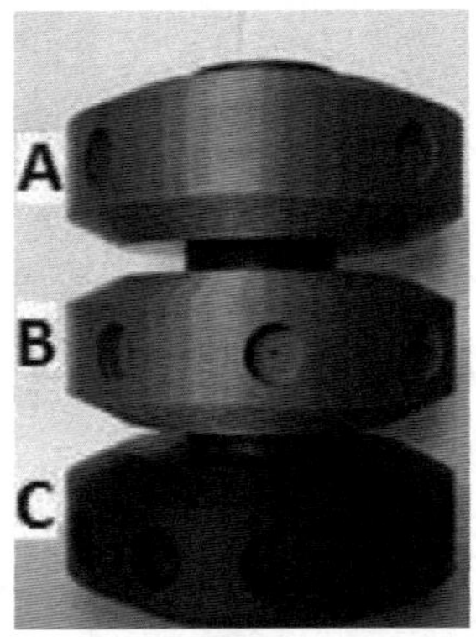


**Figure 1.** Whirling head, a)- 4 nozzle outlets, b)- 8 nozzle outlets, c)-12 nozzle outlets

The parameters of the designed impeller are presented in tables 1-3. Table 1 presents characteristic parameters of applied refining system.

**Table 1**. Characteristic parameters of applied refining system

| No. | Parameters | | |
|---|---|---|---|
| 1 | **Nozzle type** | | Pompa |
| | | | Propeller |
| 2 | **Nozzle design** | Number of outlets | 4 - 24 |
| | | Diameter of outlets | 1 [mm] |
| | | Shape of outlets | Round |
| | | Shape of nozzle | Cut cylinder |
| 3 | **Refining system design** | Shape of ladle | Cylinder |
| | | Distance between nozzle and bottom of ladle | 110 [mm] |
| | | Capacity | 225 [$dm^3$] |
| | | Number of compartments in system | 1 |
| | | Shape of bottom | round |
| | | Thickness of shaft | 56 [mm] |
| | | Immersion depth of shaft | 590 [mm] |

Table 2 presents characteristic of gas the basic variants of the tests carried out at the gas flow rate in refining process and impeller revolutions.

**Table 2**. The basic variants of the tests the gas flow rate and impeller head operations during refining process

| | Unit amount of gas to be refined | |
|---|---|---|
| **No.** | [l/min] | [cm$^3$/s] |
| **1** | 10 | 166.67 |
| **2** | 20 | 333.33 |
| **3** | 30 | 500.00 |
| | [r/min] | [obr/s] |
| **Number of impeller head operations** | 200 | 3.34 |
| | 300 | 5.00 |
| | 400 | 6.67 |
| | 500 | 8.34 |

As part of the work, it was decided to conduct comparative tests of the calculated mass transfer area for the designed rotors. The following equation was used to calculate the surface area, where : k is number of holes in the head; n is rotary impeller speed [obr/s]; $Vc_j$ is unit amount of gas to be refined [cm$^3$/s]:

- Calculation of theoretical number of outlets ( $\mathbf{k_0}$ ) : $k_0 = k \cdot n$ (1)
- Calculation of gas flowing through the i-th outlet ( $\mathbf{V_i}$ ) : $V_i = \frac{V_{c_i}}{k_0}$ (2)
- Calculation of unit radius of a bubble at a temperature of 20°C ( $\mathbf{r_i}$ [cm]) : $r_i = \sqrt[3]{\frac{3 \cdot V_i}{4 \cdot \pi}}$ (3)
- Calculation of the gas contact surface ( **A** [m$^2$]) : $A = \frac{3 \cdot V_c}{r_i}$ (4)

Considering the change of volume of all bubbles of identical diameter with height h measured from the bottom of the ladle bottom, its value increased in relation to the summaric input surface :

$$A = A_0 \cdot \frac{1}{L} \cdot \int_0^L \left( \frac{P_0 + \rho_M \cdot g \cdot L}{P_0 + \rho_M \cdot g \cdot (L-h)} \right)^{\frac{2}{3}} \cdot dh \tag{5}$$

$A_0$ denotes a summaric real surface of all bubbles with initial diameter $d_p$, produced at a given gas flow value $Q$. $A$ denotes a summaric surface of bubbles, taking into account their increase. To calculate the real surface $A_0$, i.e. number of bubbles with diameter $d_p$, the compression of the introduced gas to pressure $P_{\max}$ should be accounted for: $Q_{rz} = \frac{P_0}{P_{\max}} \cdot Q$ (6)

The number of bubbles with diameter $d_p$ introduced in 1 second: $n = \frac{6 \cdot Q_{rz}}{\pi \cdot d_p^3}$ (7)

Hence the total initial surface of bubbles introduced in 1 second equals to: $A_0 = n \cdot \pi \cdot d_p^2$ (8)

If the average flow rate of a bubble is $u$, then its outflow time equals to: $t_w = \frac{L}{u}$ (9)

With this parameter we can calculate the number of bubbles present simultaneously in the metal, and therefore the reaction surface: $A_r = t_w \cdot A$ (10)

The hydrogen removal rate, expressed by mass, is calculated from the diffusion control formula:

$$\dot{m}H = \frac{A_r \cdot \rho_M}{1400} \cdot k_N \cdot \{[\%H]-[\%H]_{gr}\} \quad (11)$$

[% H], $[\% H]_{gr}$ - hydrogen in liquid aluminium and metal-argon bubble interface.

## 3 Results of calculations

One of the most important parameters determining impeller usability is the degree to which the gaseous bubbles are dispersed in a volume of liquid metal in the ladle, and which is connected with the contact surface of a gaseous bubble with radius ($r_i$) and liquid metal (A). The results of calculations of contact surface for gas A are presented in figures 2-5, results of calculation hydrogen removal $\dot{m}_H$ in kg/s for [%H] = 0.001 are presented in table 3.

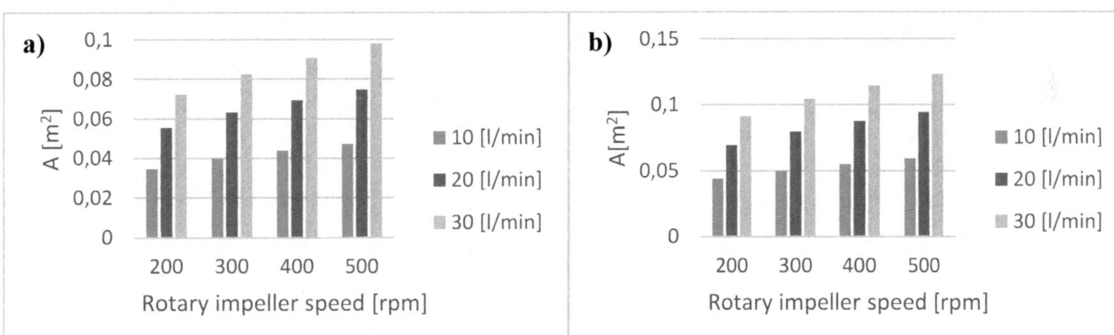


**Figure 2.** Calculated contact surface A [$m^2$] for a a)4, b)8 nozzle rotor with a flow rate equal to 10, 20 and 30 [l/min]

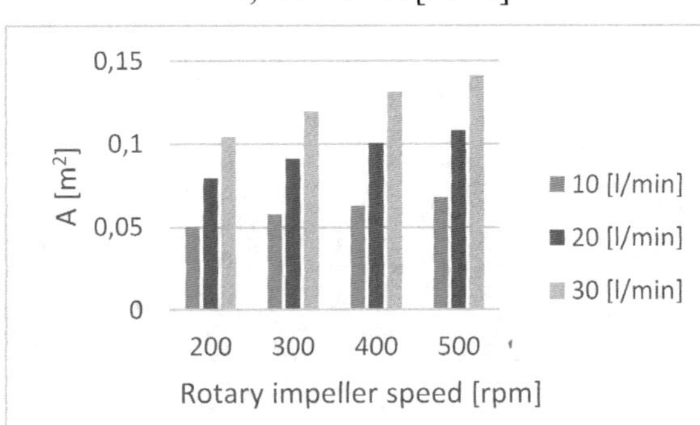


**Figure 3.** Calculated contact surface A [$m^2$] for an 12 nozzle rotor with a flow rate equal to 10, 20 and 30 [l/min]

Next the contact surface A was determined in a function of rpm for rotors equipped with 4, 8, 12, 16, 20 and 24 nozzles, and neutral gas flow rate 10, 20 and 30 [l/min].

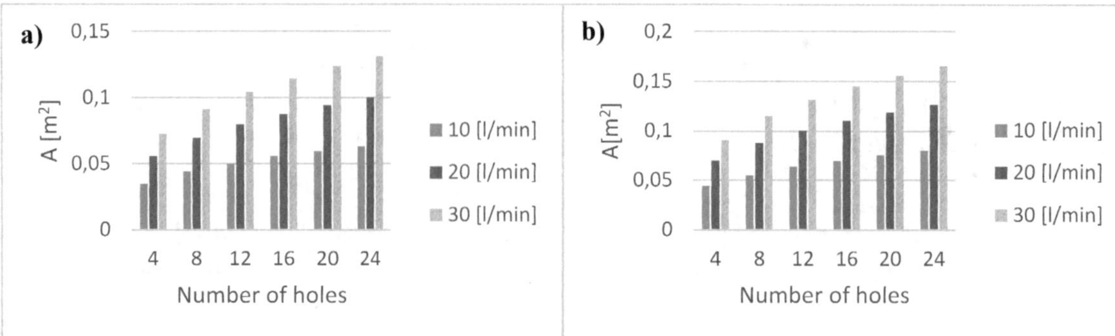


**Figure 4.** Calculated contact surface A [$m^2$] at a flow rate equal to 10, 20, 30 [l/min] and a)200, b)400 rpm

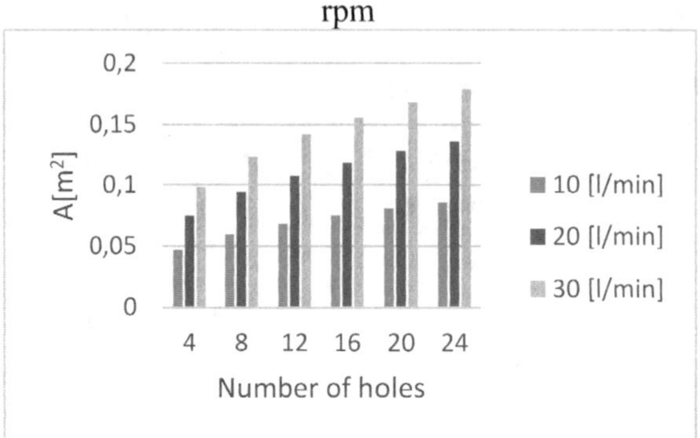


**Figure 5.** Calculated contact surface A [$m^2$] at a flow rate equal to 10, 20, 30 [l/min] and 500 rpm

The calculations reveal that a 12-nozzle rotor is most efficient for all neutral gas flow rates (10, 20, 30 [l/min]). The dynamics of rotations is another equally important parameter. It was observed that the same refining efficiency can be obtained for a smaller number of nozzles and increased number of rotations. For instance, surface A is similar for an 8-nozzle rotor with a flow rate equal to 30 l/min and 500 [rpm] to that for a 12-nozzle rotor with the same flow rate conditions and 300 [rpm].

**Table 3**. Calculation of rate of nitrogen removal $\dot{m}_H$ in kg/s for [%H] = 0.001.

| **Bubble diameter** | **Volumetric argon flow rate $Q$ ($Nm^3/s$)** | | | | |
|---|---|---|---|---|---|
| $d_p$ $(m)$ | **0.001** | **0.005** | **0.01** | **0.05** | **0.1** |
| 0.01 | 2.67E-07 | 1.33E-06 | 2.67E-06 | 1.33E-05 | 2.65E-03 |
| 0.03 | 2.96E-08 | 1.48E-07 | 2.96E-07 | 1.48E-06 | 9.75E-05 |
| 0.05 | 1.07E-08 | 5.33E-08 | 1.07E-07 | 5.33E-07 | 2.09E-05 |
| 0.08 | 4.16E-09 | 2.08E-08 | 4.16E-08 | 2.08E-07 | 5.04E-06 |

The results of calculation show how important is the diameter of gas bubble. It only depends on geometry of whirling head. These results relate to the case where the rate determining step of the process is the diffusive transport of the hydrogen in aluminium. If this step turned out to be the reaction on the surface of the bubbles, then the kinetic equation of the second-order reaction should be considered.

## 4 Conclusions

The design of the rotor decides about the potential of neutral gas transfer to refined liquid metal or alloy. The design and the number of rotations gives information whether or not gas can be introduced to metal; the smaller is the gaseous bubble, the bigger is the contact surface. The calculation of contact surface (A) was based on the assumption of uniform distribution and dynamics of neutral gas flow as well as an ideal round shape of the bubble with a constant predefined radius. This assumption was very simplified as the flowing bubbles collide and coalesce, so the calculated contact surface A [$m^2$] produces overestimated values. According to empirical calculations, the refining efficiency defined by surface A can be modified by using a specific number of rotor revolutions, flow and number of nozzles. In rotors equipped with a bigger number of nozzles the high degree of gaseous bubbles dispersion can be achieved for a smaller number of rotations or gas flow. The same refining efficiency can be achieved by using a constant gas flow with fewer nozzles and by varying the number of rotations of the rotor. For example, for a gas flow of 30 l / min, the same refining efficiency is obtained for 8 and 12 nozzles if the speed is changed from 500 to 300 [rpm].

**Acknowledgements**
The authors acknowledgement the AGH University of Science and Technology for its financial support within POWR.03.05.00-00-z307/17

**References**

[1] Bonderek S and Rzadkosz S 1999 Problemy uszlachetniania ciekłych stopów aluminium [in Polish] Metals and Alloys (Krzepnięcie Metali i Stopów) Vol 1(41).
[2] Saternus M 2015 Zanieczyszczenia ciekłego aluminium-metody ich usuwania [in Polish] Obróbka Metali Vol. XXVI(2) pp 115-132
[3] Ransley C E, Talbot D E J and Barlow H C 1957-1958 An instrument for measuring the gas content of aluminium alloys during melting and casting. Inst. Metals Vol 86 pp 212-219
[4] Adamski C and Misiag M 1954 Gazy w metalach nieżelaznych [in Polish] PWT (Warszawa)
[5] Saternus M and Merder T 2018 Physical modeling of aluminium refining process conducted in bath reactor with rotary impeller Metals Vol 8(9)
[6] Saternus M 2018 Determination of RTD curves for aluminium refining process conducted in batch reactor- Physical modeling Arch. Metall. Mater. Vol 63(4) pp 1909-1914
[7] Saternus M 2011 Rafinacja aluminium i jego stopów przez przedmuchiwanie argonem [in Polish] (Gliwice: Wyd. Politechnika Śląska)
[8] Saternus M, Merder T and Warzecha P 2011 Numerical and physical modeling of aluminium barbotage process Solid State Phenomena Vol 176 pp 1-10
[9] Saternus M and Botor J 2003 Modelling of the continuous refining process of AK-64 alloy Archives of Metallurgy Vol 48(3) pp 321-341
[10] Michalek K, Tkadlečková M, Socha L, Gryc K, Saternus M, Pieprzyca J and Merder T 2018 Physical modelling of degassing process by blowing of inert gas Arch. Metall. Mater. Vol 63(2) pp 987-992
[11] Żak P L, Kalisz D, Lelito J, Szucki M, Gracz B and Suchy J S 2015 Modelling of non-metallic particles motion process in foundry alloys Metalurgija Vol 54(2) pp 357-360 .
[12] Gaye H, Huin D, Riboud P 2000 Nitrogen Alloying of Carbon and Stainless Steels by Gas Injection Met. Mater. Trans. B 31B pp 905 – 912.

# Investigation of mechanical properties of secondary AlSi7Mg0.3 cast alloys before and after corrosion

**L Kucharikova[1], E Tillová[1], I Švecová[1] and T. Liptáková[1]**

University of Žilina, Faculty of Mechanical Engineering, Department of Materials Engineering, Univerzitná 8215/1, 010 26 Žilina, Slovakia

Corresponding author's e-mail address: lenka.kucharikova@stroj.uniza.sk

**Abstract**. The casting industry currently use about 35% of secondary (recycled) aluminum and about 65% of primary aluminum to meet their needs. The manufacturer wants to use the secondary aluminum alloys more and more, but these materials have the higher amount of the undesirable elements. The most negative element in such alloys is Fe. Although the corrosion current increased due to the presence of alloying elements in the matrix of aluminum alloys, the higher content of Fe (which led to increasing amount of Fe-rich phases) could affect the corrosion resistance. The basic metallography assessment of experimental materials microstructure confirm the higher amount of Fe-rich phases in secondary alloys. The results of mechanical properties shows that the higher content of Fe (also Fe-rich phases) does not lead to decreasing the mechanical properties. The similar results of mechanical properties were reached after corrosion attack with 3.5 wt. % NaCl. The assessment of corrosion attack confirmed the presence of pitting corrosion on experimental samples, which increase with increasing Fe content.

**Keywords:** aluminum alloys; recycling; iron; corrosion behavior; mechanical properties;

## 1 Introduction

Eutectic and near eutectic Al-Si alloys are widely used in casting industry due to superior abrasion and corrosion resistance, low thermal expansion coefficient and high strength/weight ratio [1]. The AlSi7Mg0.3 cast alloys belongs to near eutectic alloys which are used for engine constructions, vehicle and aero plane constructions, shipbuilding, electrical engineering and engine constructions, car wheels and so on [2].

Automotive and aerospace industries have interest in decreasing weight, costs and $CO_2$ emissions. However, most of aluminum alloy scrap for recycling is generated from the automotive and beverage industries, the application of aluminum alloys especially recycled is required. The main impurities that exist in recycled Al-Si cast alloys are Fe, Mn, Cu and Zn [1,3]. Iron is most critical because Al has limited solubility of iron and hence forms intermetallic compounds during solidification of the alloy. Also iron leads to formation of casting defects and eventually deteriorates the mechanical properties of alloys [1,4]. The main Fe-rich intermetallic phases formed in Al-Si alloys are α-$Al_{15}(Mn,Fe)_3Si_2$ (Chinese script type structure), α- $Al_8Fe_2Si$ (phase is hexagonal shaped) and β-$Al_5FeSi$ (phase has a monoclinic/ orthorhombic structure [4-6]. The other β phases identified in Al-Si alloys are $Al_3FeSi_2$, $Al_4FeSi$, $Al_9Fe2Si_2$, b-$Al_5FeSi$, b-$A_{4.5}FeSi$ and $Fe_2SiAl_5$ [4]. All of β phases formation in plate like form (needles in 2D) decreasing mechanical properties of materials especially ductility. Their formation can by affected with cooling rate but also with neutralizers (Mn, Be, V, Mo, etc.) [7-9]. The Mn is the best neutralizers which transforms coarse $Al_5FeSi$ needles to a finely dispersed constituent $Al_{15}(Mn,Fe)_3Si_2$ exhibiting a Chinese script type structure in the present [10]. 2006). The condition is the Mn/ Fe ratio of 0.5 [6]. Also was researched that if an iron amount exceeds a value of the mass fraction w = 0.45 %, the recommended addition of Mn should not be lower than half of the iron content [11-13]. The critical iron content (in wt. %) for an alloy can be calculated using the following relationship (1) [8,9]:

$$Fe_{crit} \approx 0.075 \cdot [wt.\ \%\ Si] - 0.05.* \qquad (1)$$

*at 5 wt. % of Si, the critical iron content is ~0.35%, at 7 wt. % of Si it rises to ~0.5, at 9 wt. % it is ~0.6 and by 11 wt. % it reaches ~0.75% [14].

Previous works shows that mechanical properties of commonly used Al-Si casting alloys do not only depend on the chemical composition. However, dendritic α-Al(Si) morphology and other intermetallic in the microstructures have important effects on the microstructural properties [1,4,5,8,14-16]. The Ambant [17] in their work state that the presence of Fe-rich intermetallic particles has a detrimental effect on corrosion resistance. Iron rich phases are catalytic sites for cathodic reactions, and at the same time are sites for nucleation of pits. Furthermore, iron is a noble element, and so it is possible, that its presence in solid solution will affect anodic dissolution kinetics [17]. One can see that aluminum is stable in the pH range from 4 to approximately 9 due to the presence of aluminum oxide film, which makes its use possible for many applications where corrosion resistance is required. However, in acidic and alkali environments, the aluminum corrodes with the formation of $Al^{3+}$ and $AlO_2^-$ ionic species, respectively. It should also be noted that the passivity of aluminum in neutral solutions could be overruled, if in the solution there are species like chlorides that can cause the breakdown of the oxide film and promote corrosion [16]. Also the phase $Mg_2Si$ phase is seen to be 'active' compared to the Al-alloy matrix [18].

From these reasons were examined mechanical properties of secondary (recycled) AlSi7Mg0.3 cast alloys considerably affected from microstructure and corrosion attack in this study. The materials were casted into the sand mold in Uneko Ltd. company, Zátor, Czech Republic in form of bars 20 x 300 mm.

## 2 Experimental methods

The experimental materials were casted as primary (set marking P3 with 0.123 wt. % of Fe) and secondary (sets marking as S4 with 0.454 wt. % of Fe and S5 with 0.655 wt. % of Fe) in order to reflect the effect of secondary aluminum alloys chemical composition to mechanical properties (table 1). Since the material is used for casting, which are exposed to corrosive environment the effect of corrosion to mechanical properties, were also studied.

**Table 1** The chemical compositions of experimental materials (in wt. %).

| Material | $N_o$. of set | Si | Fe | Cu | Mn | Mg | Zn | Ti | Al |
|---|---|---|---|---|---|---|---|---|---|
| primary | **P3** | 7.028 | **0.123** | 0.013 | 0.009 | 0.354 | 0.036 | 0.123 | 92.253 |
| secondary | **S4** | 7.340 | **0.454** | 0.021 | 0.009 | 0.302 | 0.020 | 0.118 | 91.673 |
| secondary | **S5** | 7.315 | **0.655** | 0.030 | 0.010 | 0.292 | 0.028 | 0.120 | 91.486 |

The samples for measurements of mechanical characteristic were prepared with turning and milling operation. The sets of experimental materials were divided to half in order to examine properties before and after corrosion attack. According to the fact [16] that presence the chlorides in solution can cause the breakdown of the oxide film (promote corrosion) and that Fe rich phases are catalytic sites for cathodic reactions, and at the same time are sites for nucleation of pits [17] the 3.5 wt.% NaCl solution was used as corrosion environment. The corrosion tests were performed according to rules of immersion tests. The temperature during testing was 20 °C for three weeks. The solution of 3.5 % was used because these content correlate to the average salinity level of the sea and these results could by compared to results of potentiodynamics tests. The mechanical tests were performed immediately after degreasing samples in ethanol and drying. The equipment INSTRON Model 5985 was used for tensile tests, which took place according to the standard ISO 6892-1:2009. Test rates and control were set according to Method A recommended range. The calculated results include ultimate tensile strength (UTS) and ductility (A). Hardness measurement was performed by a Brinell hardness tester with a load of 2451.66 N (250 Kp), 5 mm diameter ball and a dwell time of 15s. The bending impact test was performed on Charphy tested hammer with nominal energy 300 J. The results of such tests include resistance of the material to impact stress (Impact energy in J). The mechanical characteristic value at each state was obtained by an average of at least six measurements.

The observation of corrosion attack of specimens for tensile tests was performed on the stereo microscope. The assessment of the effect of higher content of the Fe to microstructure was evaluated with using an optical microscope Neophot 32 and quantitative analysis software NIS Elements. Samples were prepared by standards metallographic procedures (wet ground, DP polished with diamond pastes and etched by Dix-Keller). The assessment of corrosion attack were performed on mechanical specimens before testing (stereo microscopy study) and on metallography samples prepared after testing (non-etched).

## 3 Research results and discussions

### *3.1 The basic microstructure of experimental materials*

The microstructure of primary alloys (P3) correlate with binary diagram and knowledge about basic microstructural features [5,6,8,14] and consist of the: matrix ($\alpha$ = Al(Si)), eutectic (mechanical mixture of eutectic Si and matrix), Mg-rich intermetallic phases ($Mg_2Si$) and Fe-rich intermetallic phases ($Al_5FeSi$, $Al_{15}(FeMg)_2Si_2$) (figure 1). The other set`s (secondary alloys S4 and S5) microstructure is creating from the same microstructural features, but the percentage of the Fe-rich phases - $Al_5FeSi$ grows and the Fe-rich phases $Al_{15}(FeMg)_2Si_2$ is not formed (in very small volume) and $Mg_2Si$ is also in small volume (figure 2).

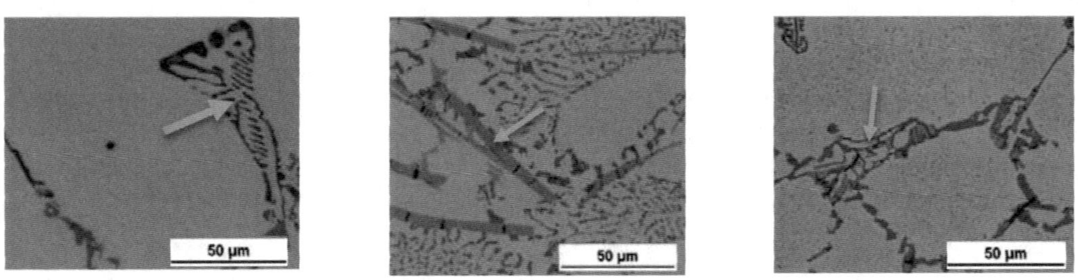


a) b) c)

**Figure 1** The intermetallic phases in microstructure of experimental materials, etch. Dix-Keller a) $Mg_2Si$; b) $Al_5FeSi$; c) $Al_{15}(FeMg)_2Si_2$.

The calculation (1) was used for assessment of $Fe_{crit}$ according to Taylor [14]. The results are: for set P3 = 0.4771, for set S4 = 0.5005 and for set S5 = 0.498625. Experimental materials has fulfilled the condition at 7 wt. % of Si is the $Fe_{crit}$ ~ 0.5. Therefore, were confirmed the ratios Mn/Fe for each material. The results show that in primary experimental alloy is ratio 0.07, for set S4 is 0.0198 and for set S5 is 0.0153. The results demonstrated that ratios Mn/Fe [11-13] are not fulfilled. These results point to the fact that even if the content of Fe does not reach the $Fe_{crit}$ is important to observe the specified content of Mn for the lower formation of the $Al_5FeSi$ Fe-rich phases. When it is no so the formation of such phases increase (figure 2). The quantitative analysis shows that the average length of the Fe-rich phases increase (for set P3 = 30.44 $\mu m^2$, for set S4 = 59.77 $\mu m^2$ and for set S5 = 52.57

μm²) and correlate with results of the ratio, because the highest was on samples S4. Moreover, Esquuivel [20] reported that increasing the Mn/Fe ratio would result in increased corrosion resistance in various cast alloys containing Mn, Fe, Si and minor additions of Mg. Therefore, there is assumptions that secondary experimental alloys will have lover corrosion resistance, which caused decreasing of the mechanical properties.

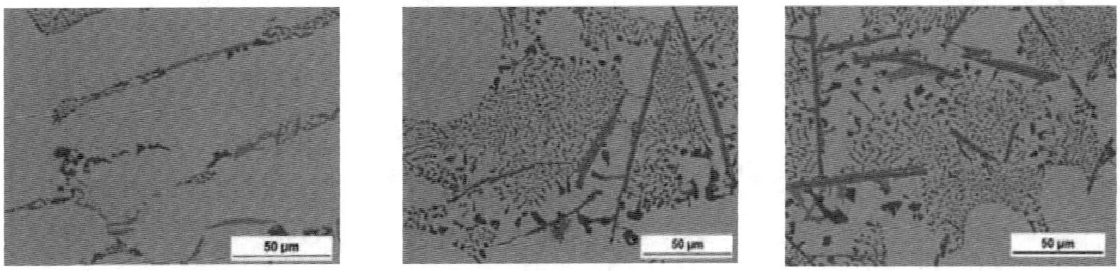


a) b) c)

**Figure 2** The changes in microstructure caused with higher the Fe content, etch. Dix-Keller a) set P3; b) set S4; c) set S5.

### 3.2 *The corrosion attack*

The assessment of corrosion attack shows presence the pitting corrosion on each experimental samples. The presence of such corrosion correlate with corrosion environment and also with chemical composition [16,17]. With increasing Fe content has been observed increasing density and scale of corrosion pit (figure 3). These changes correlate with research work of Samuel [19], that the higher content of Fe lead to the increasing pitting corrosion of Al-alloys. Also, Esquuivel [20] and Ambant [17] demonstrated that the chemical composition, size, number and distribution of the intermetallics in the matrix govern the corrosion of Al alloys.

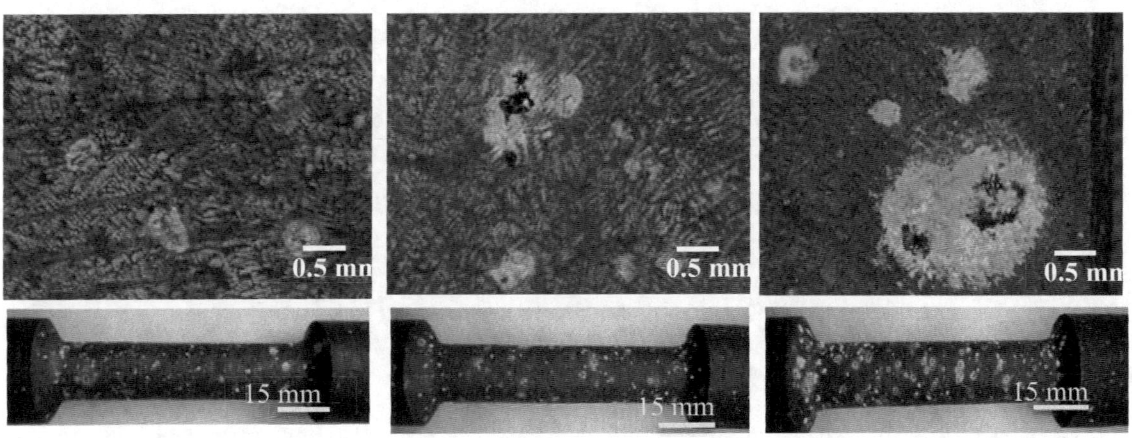


a) b) c)

**Figure 3** The pitting corrosion of experimental alloys a) set P3; b) set S4, c) set S5.

### 3.3 *The changes in mechanical properties after and before corrosion*

The results of mechanical properties of experimental materials shows that combination of the corrosion in 3.5 % NaCl and Fe content influence these properties (figure 4). The basic mechanical properties in materials before corrosion attack are more or less the same. The differences were very small (table 2, figure 4). These results confirm the knowledge [1,4,14-16] that the chemical composition affecting the mechanical properties, but also confirm that the presence of higher amount of $Al_5FeSi$ Fe-rich phases make these materials cleavages.

The results of basic mechanical properties after corrosion shows that the Fe content have a great influence on the properties, because the UTS and HBW increase with the Fe content. The impact energy and ductility decreasing with increasing Fe content in experimental materials after corrosion (table 2, figure 4). The corrosion effect on the mechanical properties is not clearly, because the results are not according to knowledge's in research works of [16-19]. The important is the note that with increasing content of Fe increase strength properties but decreasing tension properties in samples after corrosion.

**Table 2** The mechanical properties of experimental materials

| | Before corrosion | | | After corrosion | | |
|---|---|---|---|---|---|---|
| **$N_o$. of experimental sets** | **P3** | **S4** | **S5** | **P3** | **S4** | **S5** |
| UTS [MPa] | 144 | 150 | 142 | 130 | 140 | 144 |
| HBW 5/250/15 | 52 | 55 | 54 | 58 | 59 | 59 |
| Ductility [%] | 1.45 | 1.91 | 1.58 | 1.7 | 1.5 | 1.3 |
| Impact energy [J] | 14 | 12 | 10.5 | 10.5 | 8 | 7.8 |

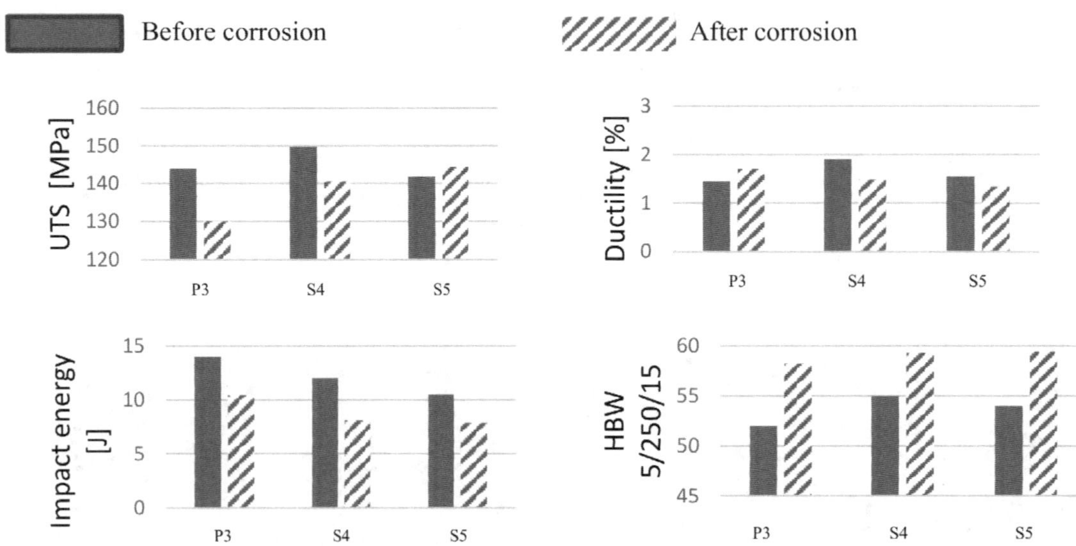


**Figure 4** The results of mechanical properties of each experimental material before and after corrosion.

## 4 Conclusions

The secondary experimental materials (S4 and A5) have increasing amount of the $Al_5FeSi$ Fe-rich phases, because the ration Mn/Fe is not sufficient.

The negative effect of the $Al_5FeSi$ phases on to basic mechanical properties was also not confirmed (the primary alloy have similar properties as secondary). It supposes that these phases have not the critical length (more than 500 μm [14]) and their amount can act as barriers to the propagation of the fracture.

The negative effect of the corrosion on the properties was also not confirmed. Decreasing only UTS and Impact energy. Moreover, the properties of secondary alloys after corrosion have similar, small, or less high value as primary alloys.

These results lead to creating new research works for declaration that secondary alloys with higher content of Fe have not so negative properties and are completely useful for casting the automotive casts.

**Acknowledgement**
This work has been supported by the grant projects N° 016ŽU-4/2020,, N° 1/0398/19, N° 012ŽU-4/2019. The authors also thank for cooperation with casting samples to Uneko Ltd. Company.

## References

[1] Irizalp S G and Saklakoglu N 2014 Effect of Fe-rich intermetallic on the microstructure and mechanical properties of thixoformed A380 aluminium alloy. *Engineering Science and Technology, an International Journal.* 17/2(2014) p. 58-62.

[2] Bangyikhan, K 2005 Effects of Oxide film, Fe-rich phase, Porosity and their Interactions on Tensile Properties of Cast Al-Si-Mg Alloys. Ph.D. thesis, University of Birmingham.

[3] Moustafa M A 2009 Effect of iron content on the formation of ß-$Al_5FeSi$ and porosity in Al-Si eutectic alloys. *Journal of Materials Processing Technology.* 209(2009) p. 605-610.

[4] Mathew J, Remy G, Williams M A, Tang F, and Srirangam P 2019 Effect of Fe Intermetallics on Microstructure and Properties of Al-7Si Alloys. *The Journal of the Minerals, Metals & Materials Society.* JOM71/12(2019) p. 4362–4369

[5] Závodská D, Tillová E, Švecová I, Chalupová M, Kuchariková L, Belan J 2019 The effect of iron content on microstructure and porosity of secondary AlSi7Mg0.3 cast alloy. *Periodica Polytechnica Transportation Engineering.* 47/4(2019) p. 283-289.

[6] Ji S, Yang W, Gao F, Watson D, and Fan Z 2013 Effect of iron on the microstructure and mechanical property of Al–Mg–Si–Mn and Al–Mg–Si die cast alloys. *Materials Science & Engineering.* A564(2013) p. 130–139

[7] Ashtari P, Tezuka H, and Sato T 2003 Influence of Sr and Mn additions on intermetallic compound morphologies in Al-Si-Cu-Fe cast alloys. *Materials Transactions.* 44/12(2003) p. 2611-2616.

[8] Hurtalová L, Tillová E, Chalupová M, Belan J, Vaško A 2014 Microstructure control of secondary A321 cast alloy used in automotive industry. *Manufacturing technology.* 14/3(2014) p. 326-333.

[9] Rana R S, Purohit R, and Das S 2012 Reviews on the Influences of Alloying elements on the microstructure and mechanical properties of aluminum alloys and aluminum alloy composites. *International Journal of Scientific and Research Publications.* 2/6(2012) p.1-7.

[10] Kim H Y, Park T Y, Han S W, and Lee H M 2006 Effects of Mn on the crystal structure of α-Al(Mn,Fe)Si particles in A356 alloys. *Journal of Crystal Growth.* 291/1(2006) p. 207-211.

[11] Fortini A, Merlin M, Fabbri E, Perletti S, and Garagnami L G 2016 On influence of Mn and Mg additions on tensile properties, microstructure and quality index of the A356 aluminum foundry alloy. 21st European Conference on Fracture. *Procedia Structural Integrity.* 2(2016) p. 2238-2245.

[12] Elsharkawi E A, Samuel E, Samuel A M, and Samuel F H 2010 Effects of Mg, Fe, Be additions and solutions heat treatment on the π-AlMgFeSi iron intermetallic phase in Al7Si-Mg alloys. *Journal of Materials Science.* 45(2010) p. 1528-1539.

[13] Podprocká R., Bolibruchová D, and Chalupová M 2017 Reducing the Negative of the Iron in the Alloy Based on Al-Si-Mg by Manganese. *Archives of foundry engineering.* 17/2(2017) p. 212-213

[14] Taylor J A 2004 The effect of iron in Al-Si casting alloys, In 35th Australian Foundry Institute National Conference, Adelaide, South Australia, p. 148-157.

[15] Seifeddine S 2007 The influence of Fe on the microstructure and mechanical properties of cast Al-Si alloys. "Literature review - Vilmer project". Jönköping University, Sweden.

[16] Yasakau K A, Zheludkevich M L, and Ferreira M G S 2018 Role of intermetallics in corrosion of aluminum alloys. Smart corrosion protection. Intermetallic Matrix Composites. Properties and applications. Chapter 15(2018) p. 425 – 462.

[17] Ambat R, Davenport A J, Scamans G M, and Afseth 2006 A Effect of iron-containing intermetallic particles on the corrosion behaviour of aluminium. *Corrosion Science.* 48(2006) p. 3455–3471

[18] Liang W J, Rometsch P A, Cao L F, and Birbilis N 2013 General aspects related to the corrosion of 6xxx series aluminium alloys: Exploring the influence of Mg/Si ratio and Cu. *Corrosion Science.* 76(2013) p. 119-128.

[19] Samuel A M, Samuel F H, Doty H W, and Valtierra S 2018 Beta $Al_5FeSi$ phase platelets-porosity formation relationship in A319.2 type alloys. *International Journal of Metal casting.* 12/1(2018) p. 55-70.

[20] Esquiel J, Gupta R K 2016 Simultaneous improvement of mechanical and corrosion properties of aluminium alloys. *Light metals* 2016. Chapter 26. p. 151-156.

# Influence of surface structures on wettability

**L Kvitek[1], D Saran[1], L Vaclavek[2], J Kapusta[3] and P Kozmin[4]**

[1]Department of Physical Chemistry, Faculty of Science, Palacky University in Olomouc, 17. listopadu 12, 77146 Olomouc, Czech Republic
[2]Joint Laboratory of Optics, Faculty of Science, Palacky University in Olomouc, 17. listopadu 50A, 77207 Olomouc, Czech Republic
[3]Department of Geology, Faculty of Science, Palacky University in Olomouc, 17. listopadu 12, 77146 Olomouc, Czech Republic
[4]Hofmeister s.r.o., Mezi Ploty 12, 326 00 Plzen, Czech Republic

E-mail: libor.kvitek@upol.cz

**Abstract**. Wettability of solid surfaces is a key parameter for many industrial applications of solid materials. The wettability is influenced not only by the nature of the material but also by the existence of surface structures. Crucial influence of the oriented surface structures on the shape of liquid drop was observed in this study. Observed differences in value of contact angle measured on the metal surface at two perpendicular directions were 19° for water and 23° for diiodomethane. These surface structures can significantly influence liquid and heat transfer on structured surfaces.

## 1 Introduction

The surface structuring of metallic materials significantly affects their application properties. Research aimed at formation of these structures is thus at the forefront of interest not only in the field of materials technology, but also in the field of basic materials research. Recently, methods using intense laser radiation [1] have been investigated, which are useful for structuring of the surface of metallic materials preparing wide scale structures from micrometric to nanometric. Since the first research bringing the method of laser surface texturing (LST), number of specific methods have been developed, differing mainly in the type of laser used, where the basic parameters are the wavelength of radiation as well as a pulse time. The last parameter significantly affects the dimensions of the resulting surface structures. Probably the most commonly used for surface structuring are femto-second lasers, with which surface structures with dimensions from tenths up to tens of micrometers can be achieved by a combination of power and the number of laser's pulses repetitions - laser induced periodic surface structuring (LIPSS) [2]. With picosecond lasers, up to submicrometric surface structure dimensions were achieved by the method commonly named as picosecond direct laser interference patterning (ps-DLIP) [3]. On the other hand, with nanosecond lasers, structures up to hundreds of micrometers in size can be created [4].

The effect of surface texturing is manifested mainly in the change of wettability of surfaces [5] and associated effects such as self-cleaning [6], adhesion of microorganisms [7], corrosion resistance [8], improving of adhesive bonding [9] and wear resistance [10], reduction of friction [11] etc. Another interesting effect relates to specific geometry of surface structures in the case of their linear arrangement to a line-pattern. This is manifested in the case of wetting with biaxial drop shaping [12], or in the so-called wetting anisotropy [13]. This phenomenon causes interesting effects especially in the transport of heat [14] and liquids [16] over such a structured surface.

Presented study is focused on the evaluation of the influence of laser-generated microstructures on the process of wetting metal surfaces. The wetting process is monitored using goniometric measurements of static contact angles using standard liquids with polar (water) and non-polar (diiodomethane) character. The aim of this study is to evaluate the influence of linearly arranged surface structures on the wetting process in terms of the wetting anisotropy.

## 2 Materials and methods

The metal surfaces used for this study, designated as VUTS and C1, were prepared at company HOFMEISTER s.r.o. They are plates with dimensions of 9 x 9 x 3 mm, which surfaces are modified by laser method forming linear surface structures. The basic substrate (steel) is covered by layer of sintered carbide based on WC + Co (approx. 8%), and by the subsequent layer of 4th generation called $nAlCrO_4$ with a thickness of about 3 μm composed from nanocrystalline AlCrN grains, which are embedded in an amorphous $Si_3N_4$ matrix. The final layer consists of a nc-AlCrN / a-$Si_3N_4$ nanocomposite coating, forming a quaternary (four-component) nanocomposite. Silicon is involved in the formation of the nanocomposite structure, the addition of which to CrAlN can significantly affect grain size, phase composition, and mechanical properties of the coating, and generally also serves to improve nitride coatings. In general, quaternary nanocomposite coatings have better properties than ternary coatings, of the TiAlN or CrAlN type, due to higher hardness, toughness, corrosion resistance as well as higher temperature resistance up to 1100 °C.

To measure the contact angle, a KRÜSS GmbH device marked KRÜSS DSA30 was used in conjunction with the ADVANCE software (version 1.4.1.2) of the company of the same name, which was used to evaluate the shape of the drop and thus the contact angle. The non-polar liquid chosen for determining the contact angle values was diiodomethane (Sigma-Aldrich s.r.o .; 99%), due to its zero value of the polar contribution to the surface tension, and the polar liquid was ultrapure water (conductivity 18.2 MΩ · cm at 25 °C). The geometry of the surface treatment was evaluated using optical microscopy (Olympus BX-50p polarizing microscope). The Olympus Lext OLS 5000 confocal laser scanning microscope and the Taylor Talson Form Talysurf Series 2 contact profilometer were used to obtain the exact geometry of the surface structures of the VUTS sample.

Measurements on a droplet shape analyser were preceded by cleaning of the plate surface. The plates were washed with ethanol and then dried in an oven at 60 ° C for fifteen minutes, and then stored in a desiccator until were taken for measurement. Drops with a volume of 4-5 μl were placed manually on the plate surface using a Hamilton microsyringe. The measurement of the droplet shape was repeated 6 times, the results show the average values from these six measurements.

## 3 Results and discussion

### *3.1 Geometry of surface structures*

The geometry of the studied surface structures is documented by optical microscopy images (figure 1) and geometric models derived from these images (figure 2). The surface of the VUTS type plates is arranged in mutually parallel grooves with a fine transition, resembling the shape of waves, which are 118 μm apart. The grooves are also parallel to the diagonal of the plate. The surface structure of C1 type plates has a similar treatment as VUTS, but in this case some difference can be observed. The sample again resembles the shape of waves spaced 107.5 μm apart, which are parallel to each other. However, lines are parallel to the edge of the plate and the lower part of the pattern is intermittent and resembles the shape of welds that are on average 58 μm apart.

Accurate evaluation of the surface treatment of the VUTS type plate was performed using confocal laser scanning microscopy. The profile recorded from the center of the surface structures is shown in figure 3, from which can be seen that the exact distance of the wave peaks is 81.5 μm and the profile depth is 18.06 μm. This geometry was verified by measurements on a Form Talysurf Series 2 contact profilometer, which showed very similar dimensions obtained by confocal laser scanning microscopy (figure 4).

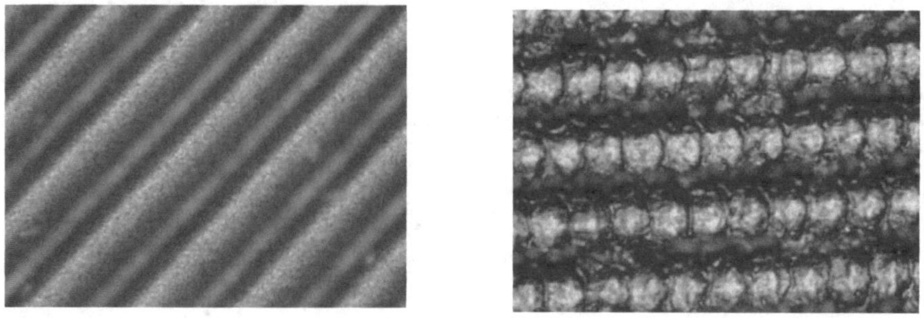

**Figure 1.** Microscopic images of the surface of a) VUTS type and b) C1 type plates. Magnification 200x.

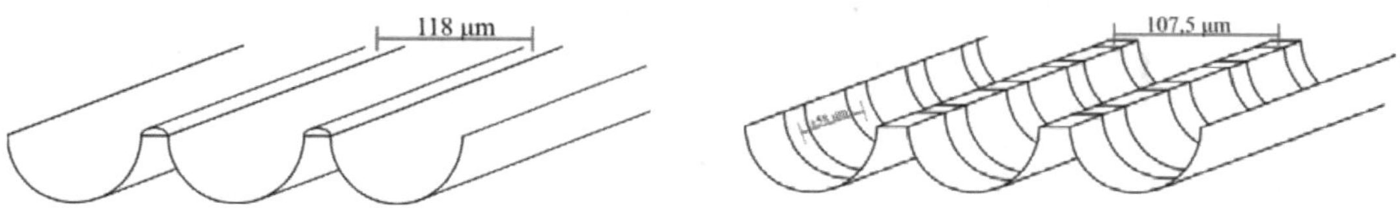


**Figure 2.** Models of geometry of surface structures of plates of a) VUTS type and b) C1 type.

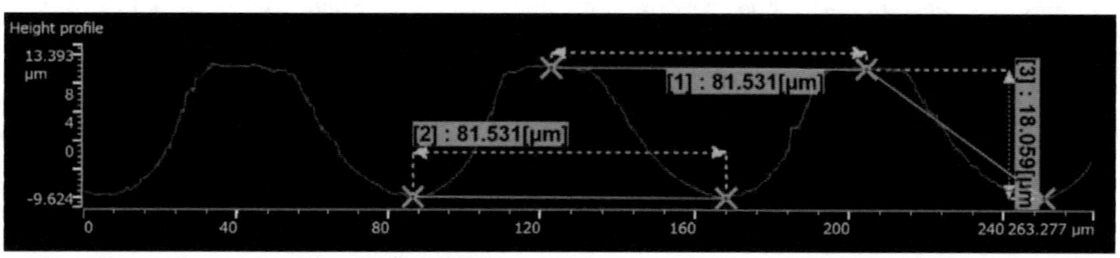


**Figure 3.** Image of the surface of VUTS type plate from a confocal laser scanning microscope.

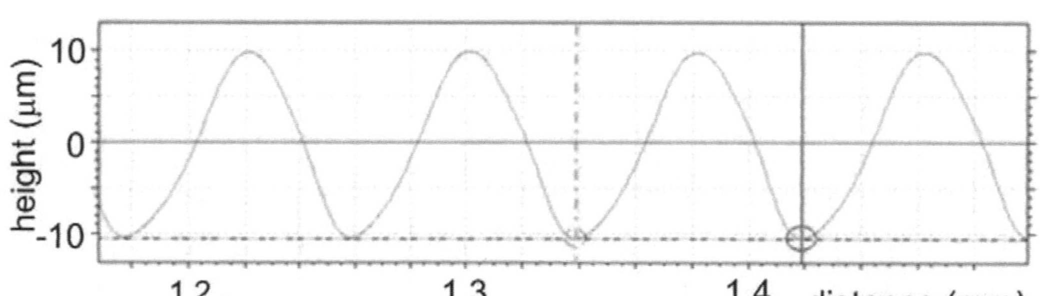


**Figure 4.** Surface geometry of VUTS type plate from Form Talysurf Series 2 profilometer. Legend: Aktuální bod=Actual point; Referenční bod=Reference point; Rozdíl=Difference

### *3.2 Wetting of surface structures*

The wettability of the surface-structured plates was measured by the goniometric method on a KRÜSS DSA30 instrument. Water and diiodomethane were chosen as test liquids. Diiodomethane is a standard liquid used in the evaluation of the surface energy of solids as a non-polar solvent with a high value of surface tension. Therefore, it possible to study non-polar interactions on number of surfaces, where common organic liquids provide perfect wetting. The two studied surface structures did not differ significantly in their geometry, but it was essential for the study itself that in one case the linear surface structure was oriented parallel to the edge of the test plate, while the other was oriented diagonally. Due to the parallel orientation of linear surface structures it is possible to study the shape of the drop in two directions perpendicular to each other with respect to the optical axis of the measuring system. The measured values of contact angles on studied samples and for different orientations are given in table 1 as averages from measurements on 6 separate drops. Typical test liquid drop profiles are shown for water in figure 5 and for the diiodomethane in figure 6.

**Table 1.** Contact angles measured on the studied samples with linear surface structures.

| Liquid | Contact angle (°) | | |
|---|---|---|---|
| | VUTS | C1 ⊥ | C1 ‖ |
| Water | 86±1.5 | 85±3.2 | 104±0.5 |
| Diiodomethane | 41±1.0 | 48±2.7 | 71±1.6 |

Legend: ⊥ the optical axis is perpendicular to the surface structure
‖ the optical axis is parallel to the surface structure

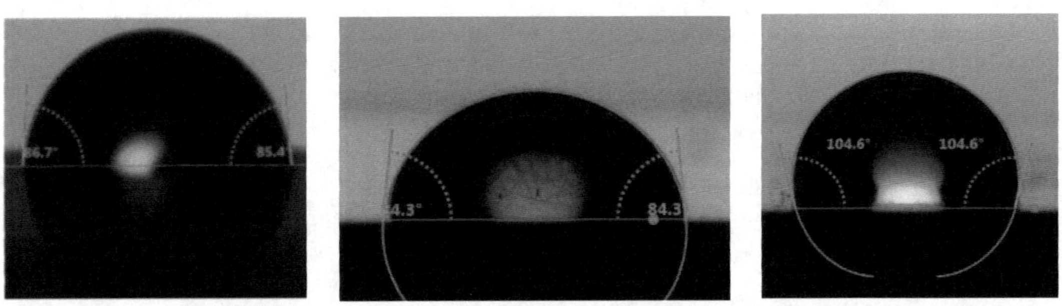


**Figure 5.** Typical profiles of water droplets on surfaces a) VUTS, b) C1⊥ and c) C1‖.

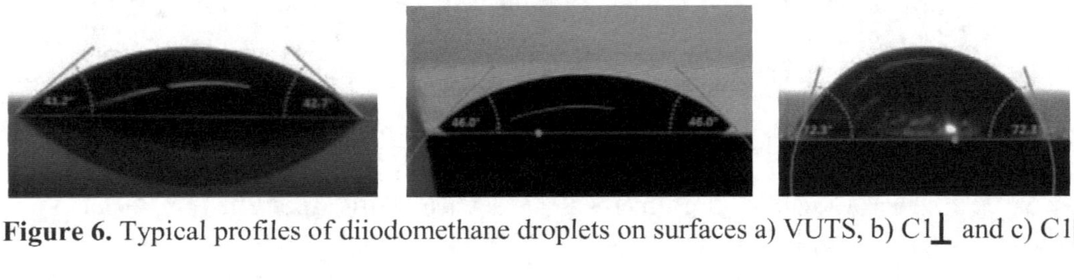

**Figure 6.** Typical profiles of diiodomethane droplets on surfaces a) VUTS, b) C1⊥ and c) C1‖.

The obtained results are in accordance with previously published studies describing the so-called wetting anisotropy [13]. In the case of different orientation of the optical axis of the measuring system with respect to the line of surface structures, the values of the contact angle differed for water by 19°, for diiodomethane even by 23°. These values are very similar to previously published results of studies with similarly structured metal surfaces [16]. However, it is interesting that in the case of the diagonal orientation of the linear surface structures (angle 45 ° to the optical measurement axis), the contact angles are closer to the situation for the perpendicular direction of the surface structure lines to the optical measurement axis. The differences in the magnitude of the contact angles observed in mutually perpendicular directions are explained by the existence of capillary forces which cause a drop of test liquid to flow in a direction parallel to the lines of the surface structures [17]. This is also a very probable reason for the transport of liquids over the surface in this direction in dynamic systems, which are closer to the situation of real practice in the interaction of such a modified surface of the machining tool with the cutting fluid. In the same direction, heat transfer is facilitated, which, with a suitable orientation of the surface structures, can help to more efficiently cool a part of the tool in contact with the material to be machined. The study of similarly structured surfaces can thus bring important knowledge for improving the service life of machine tools in mechanical engineering.

## 4 Conclusions

The performed study proved the fundamental influence of linear surface structures on the wetting of such treated surfaces from the point of view of asymmetry of the wetting angle observed on the droplet in two mutually perpendicular directions. Linear surface structures can thus directionally affect the interaction of liquids with metal surfaces, which to a large extent affects the important properties of cutting tools used in mechanical engineering, in particular the wear resistance of the tool surface.

**Acknowledgments**

The authors acknowledge financial support by Internal Grants of Palacky University in Olomouc (IGA_PrF_2020_034 and IGA_PrF_2020_007).

## References

[1] Kietzig AM, Hatzikiriakos SG and Englezos P 2009 *Langmuir* **25** 4821
[2] Ling EJY, Saïd J, Brodusch N, Gauvin R, Servio P and Kietzig AM 2015 *Appl. Surf. Sci.* **353** 512
[3] Aguilar-Morales AI, Alamri S, Voisiat B, Kunze T and Lasagni AF 2019 *Materials* **12** 2737
[4] Ma Q, Tong Z, Wang W, Dong G 2018 *Appl. Surf. Sci.* **455** 748
[5] Wanga Y, Zhaoa X, Kea C, Yua J, Wang R 2020 *Colloid Interface Sci. Comm.* **35** 100256
[6] Grynyov R, Bormashenko E, Whyman G, Bormashenko Y, Musin A, Pogreb R, Starostin A, Valtsifer V, Strelnikov V, Schechter A and Kolagatla S 2016 *Langmuir* **32** 4134
[7] Patel DS, Singh A, Balani K and Ramkumara J 2018 *Surf. Coatings Technol.* **349** 816
[8] Lian Z, Xu J, Yu Z, Yu P and Yu H 2019 *J. Alloys Compounds* **793** 326
[9] Cheng J, Hu X and Li S 2020 *J. Mat. Sci. Mat. in Electronics* **31** 15086
[10] Bhosale DG and Rathod WS 2020 *Surf. Coatings Technol.* **390** 125679
[11] Yilbas BS, Akhtar SS, Ali H, Karatas C and Al-Qahtani H 2020 *J. Manufact. Proces.* **56** 1230
[12] Wang D, Liu Y, Sridhar S, Li Y, McHale G, Lu H, Yu Z, Wang S and Xu BB 2020 *Adv. Mater. Interfaces* 2001199
[13] Orazi L, Gnilitskyi I, Pavlov I, Serro AP, Ilday S and Ilday FO 2015 *CIRP Annals - Manufact. Technol.* **64** 193
[14] Qi B, Zhou J, Wei J and Li X 2018 *Experimental Thermal Fluid Sci.* **90** 28
[15] Holmes HR and Böhringer KF 2015 *Microsyst. Nanoeng.* **1** 15022
[16] Hans M, Müller F, Grandthyll S, Hüfner S and Mücklich F 2012 *Appl. Surf. Sci.* **263** 416
[17] Zhang Q, He L, Zhang X, Tian D and Jiang L 2020 *ACS Nano* **14** 1436

# Comparison of high temperature corrosion resistance of NiCr-based coatings applied by High Velocity Oxygen Fuel and Twin Arc Spray processes

**K. Lencová, R. Mušálek*, F. Lukáč*, M. Vostřák, Z. Česánek**

Research and Testing Institute
Pilsen, Tylova 1581/46, 301 00 Pilsen, Czech Republic
lencova@vzuplzen.cz

*Institute of Plasma Physics CAS, v.v.i.
Za Slovankou 3, 182 00 Prague, Czech Republic

**Abstract**
High temperature corrosion in a molten sodium sulfate salt environment is a serious problem of coal-fired boiler components. The promising possibility for surface protection is to apply thermally sprayed coatings. However, the key to success is to choose the optimal material and the proper spraying technology for a given application. Therefore, this study is aimed at a comparison of the high temperature behaviour of two protecting NiCr-based coatings sprayed by different technologies on 1.4903 (W.nr.) chrome-molybdenum steel. NiCrMoWFe was deposited by High Velocity Oxygen Fuel technology (HVOF). Twin wire arc spray technology (TWAS) was used to deposit multi-layered “multicoat” consisting of NiCrMoNbTa and NiCrTi with ceramic overlay Armaguard coated by brush. Chrome-molybdenum steel 1.7715 (W.nr.) commonly used in coal-fired boiler was also exposed to the high temperature corrosion test to evaluate the potential benefit of the coating. Coated as well as the uncoated specimens were exposed to the molten salt environment 18% $Na_2SO_4$ and 82% $Fe_2(SO_4)_3$ at 690 °C under cyclic condition. The thermogravimetric approach was used to determine kinetics of corrosion. The X-ray diffraction (XRD), scanning electron microscope (SEM) and energy dispersive spectroscopy (EDS) techniques were used to analyze the corrosion products and determine the corrosion mechanism. The results of high temperature corrosion resistance show that NiCrMoWFe coated by HVOF and multicoat sprayed by TWAS was comparable.

## 1 Introduction

High temperature corrosion is accelerated corrosion that occurs when metals and alloys are covered with a thin film of molten salt. High temperature corrosion of coal-fired boilers is caused by burning low grade fuel in combination with the low emission combustion process. Low grade fuel contains a high amount of impurities such as sulphur, sodium, vanadium, chlorine, and incombustible material which produce ash during combustion. Fly ash can deposit on boiler components and forms the molten salt film. This complex salt deposit attacks the protective oxide scale formed on the surface of the material which leads to severe high temperature corrosion [1, 2, 3, and 4].

Thermal sprayed coatings are applied to the surface of coal-fired boiler components to allow them the function under extreme conditions and extend their service life. Protective coating serves as an effective diffusion barrier between the corrosion environment and substrate material to provide corrosion and oxidation resistance. The coating reacts with the environment and produces the protective oxide scale which should result in surface passivation [5, 6 and 7].

The objective of this work is to compare two thermal coating technologies to apply such coatings. High Pressure/High Velocity Oxygen Fuel (HP/HVOF) and Twin Arc spray technology (TWAS). HVOF technology allows applying coatings with low porosity (about 1%) and low oxide particles content with no significant thermal influence and mechanical alterations of the substrate material. On the other hand, TWAS technology is intended for on-site coating/repairs and lower financial costs.

## 2 High temperature corrosion test

A high temperature corrosion test was performed on Ni-based coatings deposited by HP/HVOF and TWAS technology as well as uncoated steels samples. NiCrMoWFe coating was deposited by HVOF technology. Twin wire arc spray technology was used to deposit Multicoat NiCrMoNbTa and NiCrTi coatings with Armaguard (ceramic overlay coated by brush). All of these coatings were deposited on high temperature corrosion resistant chrome-molybdenum steel 1.4903 (W.nr.). Uncoated chrome-molybdenum steel 1.7715 (W.nr.) commonly used in coal-fired boiler and chrome-molybdenum steel 1.4903 were included in the high temperature corrosion test for comparison.

The specimens were exposed to a molten salt environment 12% $Na_2SO_4$ + 82% $Fe_2(SO_4)_3$ at 690 °C under cyclic condition, simulating the working environment in low-emission boilers of coal-fired power plants. For the high temperature test, samples of size 20 x 20 x 5 mm were used. The exposed surface of the tested specimens was polished down to Ra = 1 µm. Salt deposit (3-5 mg/cm$^2$) of 12% $Na_2SO_4$ + 82% $Fe_2(SO_4)_3$ with uniform thickness was applied with a brush on the preheated samples. Then the specimens were placed to ceramic crucible. Each cycle of the test included 1 hour of heating at 690°C in a furnace followed by 20 min cooling at room temperature for 50 cycles. Weight change measurements were performed after each testing cycle to determine the kinetics of corrosion. The samples were weighted with the ceramic crucible.

The results of the high temperature corrosion test were evaluated using the thermogravimetric method to determine the kinetics of corrosion products. X-ray diffraction (XRD), scanning electron microscope (SEM) and energy dispersive spectroscopy (EDS) techniques were used to analyze the corrosion products and determine the corrosion mechanism [8].

## 3 Results

### 3.1 *Corrosion kinetic*

The weight gain versus the number of cycles plots for tested specimens is shown in Figure 1. The first 10 testing cycles of the test were not displayed in the weight gain graph because the corrosion behavior of the samples was affected by the formation of the passivation oxide layer and the stabilization of the whole test.

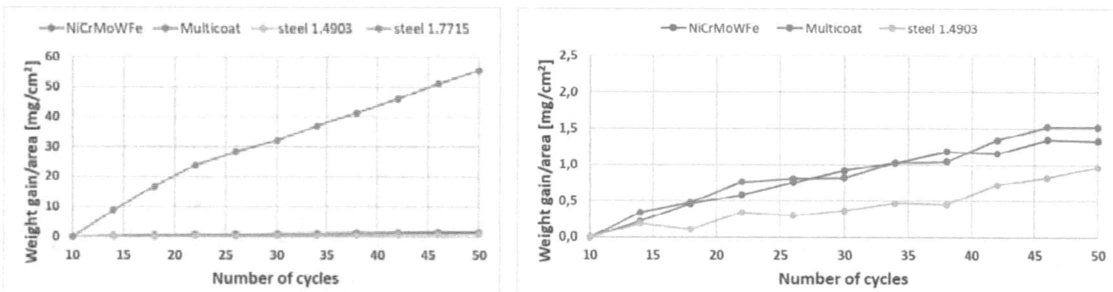


**Figure 1**. Weight gain versus the number of test cycles graph for tested samples after the exposure to the 12% $Na_2SO_4$ + 82% $Fe_2(SO_4)_3$ corrosion environment

Weight gains that were observed during the high temperature corrosion test were caused by the formation of corrosion products. Therefore, the best corrosion resistance shows the sample with the smallest weight gain. Both NiCr based coatings show similar behavior during the test. Although uncoated steel 1.4903 shows the smallest weight gain during the entire period of the test, peeling of the scale was observed during the test (Figure 2 c). The uncoated 1.7715 steel showed significant weight gain and extensive peeling (Figure 2 d). The weight gain of the both uncoated steel wasn't affected by peeling of the scale (Figure 2 a, b), because of all parts of the scale were captured in the crucible and were weighted with the samples.

IOP Conf. Series: Materials Science and Engineering 1178 (2021) 012039 doi:10.1088/1757-899X/1178/1/012039

a) NiCrMoWFe b) Multicoat c) 1.4903 d) 1.7715

**Figure 2**. Macroscopic photography of tested samples after exposure to 12% $Na_2SO_4$ + 82% $Fe_2(SO_4)_3$ corrosion environment

### 3.2 *SEM and EDS cross-section analysis*

SEM and EDS techniques were used to analyze the corrosion products and determine the corrosion mechanism. SEM and EDS micrograph in Figure 3-6 depict the cross-section of coated and uncoated samples after exposure to the 12% $Na_2SO_4$ + 82% $Fe_2(SO_4)_3$ molten salt environment.

After the molten salt corrosion, a continuous oxide scale was formed on the surface of NiCrMoWFe coating (Figure 3). The oxide scale consists mainly of chromium and molybdenum. A chromium depletion area can be seen below the surface of the coating. Also, small amounts of tungsten and iron are observed in the volume of scale and nickel on the top of the oxide scale. The substrate steel seems to be unaffected. Please note that molybdenum and sulfur peaks in EDS spectrum overlap, so they could not be distinguished.

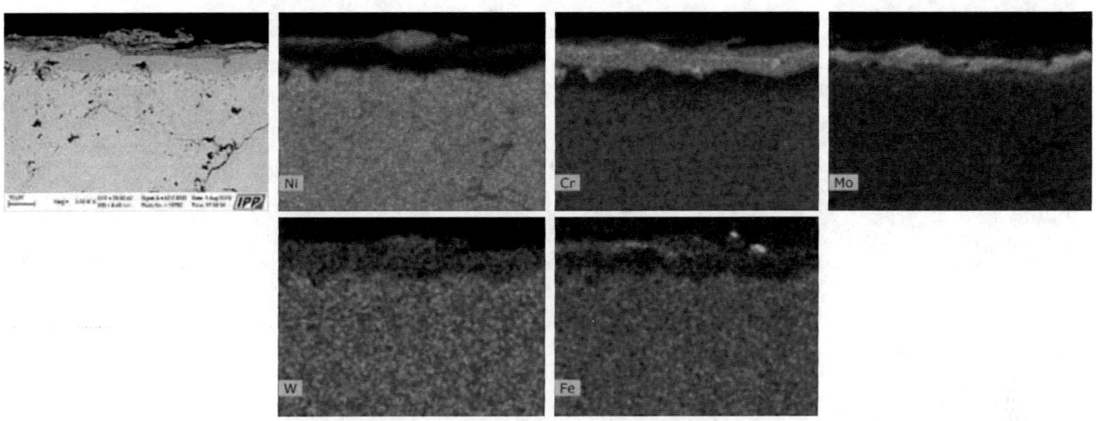


a) Elements of the coating

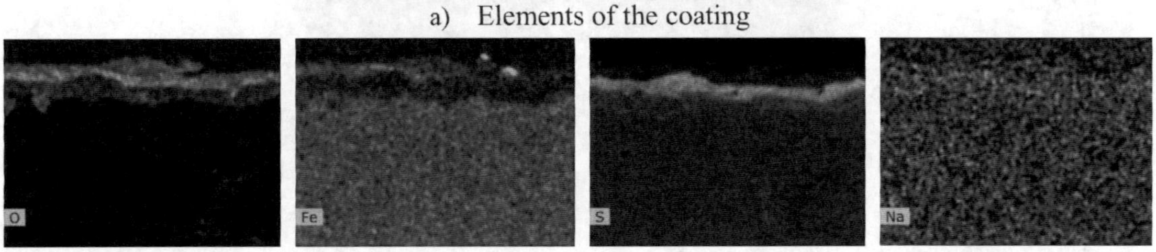


b) Elements of corrosive environment

**Figure 3**. SEM photograph and EDS map of the cross-section of HP/HVOF sprayed NiCrMoWFe coating after the exposure to the 12% $Na_2SO_4$ + 82% $Fe_2(SO_4)_3$ corrosion environment

The cross-sectional analysis SEM/EDS of Multicoat sample after exposure to the molten salt environment is shown in Figure 4. Armaguard - ceramic overlay coating is mainly formed by zirconium and silicone oxides. The oxide protective scale is not formed. A thin area of iron is observed on the top of Armaguard overlay coating and interface of the upper coating NiCrTi, which came from corrosion environment. The oxides contained in the upper coating NiCrTi were mostly formed during TWAS process. NiCrMoNbTa lower coating and substrate steel seem to be unaffected.

The oxide scale on uncoated steel 1.4903 consists mainly of chromium with a small amount of vanadium and manganese, see Figure 5. On the interface of oxide scale and steel, there is an area rich in chromium. A thin chromium depletion area can be seen underneath. On the top of steel, under the oxide scale, is the higher content of silicone. Again, molybdenum and sulfur overlapped. It can be assumed that the thin area on the interface of the oxide layer and the substrate is sulfur because of the content of molybdenum in steel 1.4903 is just about 1%.

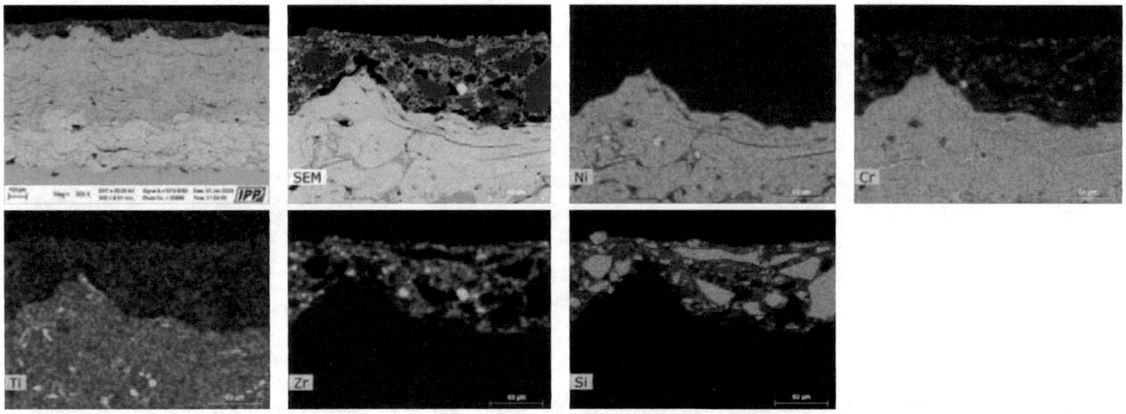


a) Elements of the coating

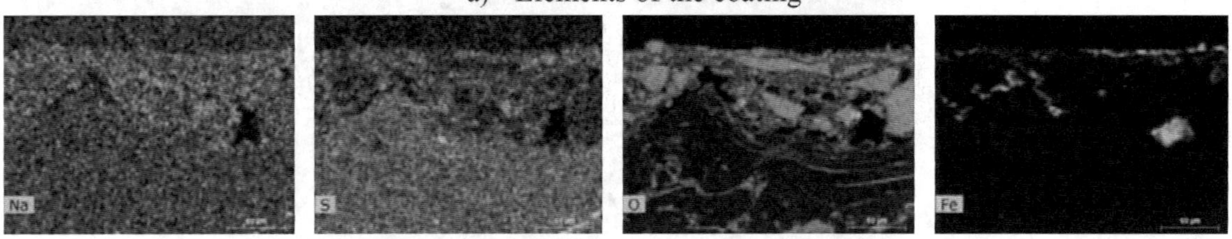


*b)* Elements of corrosive environment

**Figure 4**. SEM photograph and EDS map of the cross-section of TWAS sprayed Multicoat after the exposure to the 12% $Na_2SO_4$ + 82% $Fe_2(SO_4)_3$ corrosion environment

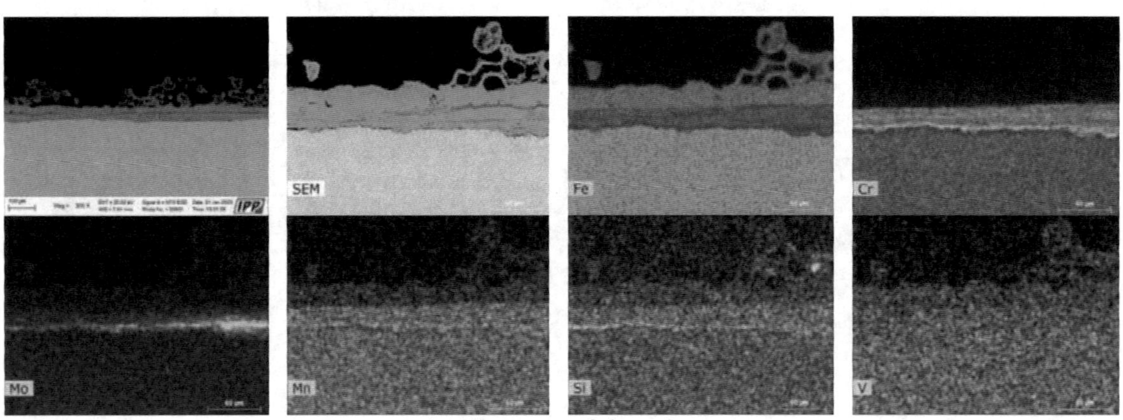


a) Elements of the substrate

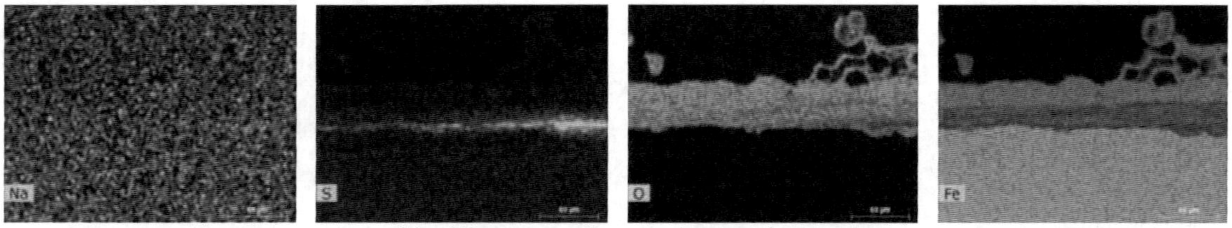


b) Elements of corrosive environment

**Figure 5**. SEM photograph and EDS map of the cross-section of uncoated steel 1.4903 after the exposure to the 12 % $Na_2SO_4$ + 82% $Fe_2(SO_4)_3$ corrosion environment

The SEM micrograph (Figure 6) for high temperature corroded steel 1.7715 reveals the thick delaminating oxide scale formed mainly by iron. It can be seen a thin layer of the silicon on the top of the oxide scale. Also, there can be observed a small amount of chromium in the upper part of the oxide scale.

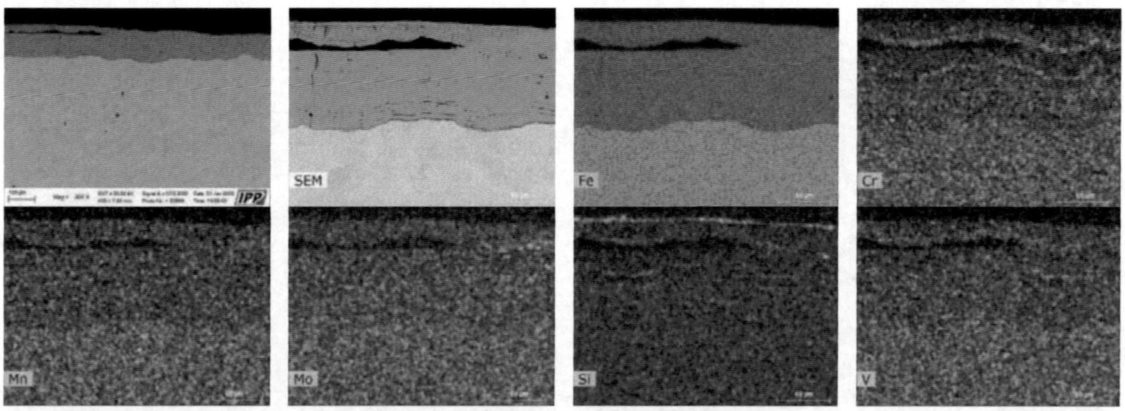


a) Elements of the substrate

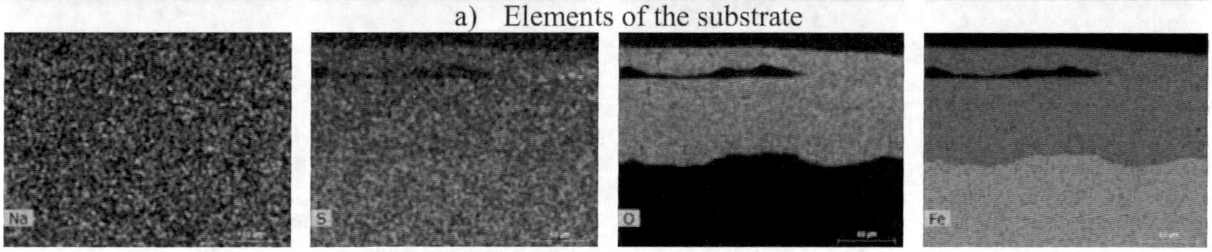


b) Elements of corrosive environment

**Figure 6**. SEM photograph and EDS map of cross-section of uncoated steel 1.7715 after the exposure to the 12% $Na_2SO_4$ + 82% $Fe_2(SO_4)_3$ corrosion environment

### 3.2 *XRD Analysis*

**Table 1.** XRD results for tested samples after the exposure to the 12% $Na_2SO_4$ + 82% $Fe_2(SO_4)_3$

| Sample | Majority phases | Minority phases |
|---|---|---|
| **NiCrMoWFe** | NiO, $Cr_2O_3$, $NiCr_2O_4$, | $FeMoO_4$, $NiMoO_4$,$Fe_2O_3$ |
| **Multicoat** | $Fe_2O_3$, $ZrO_2$, $SiO_2$ | $Cr_2O_3$, $Y_2O_3$ |
| **Uncoated steel 1.7715** | $Fe_2O_4$, $Fe_3O_4$ | - |
| **Uncoated steel 1.4903** | $Fe_2O_4$, $Fe_3O_4$ | $Na_2SO_4$ |

The XRD results for the coated and uncoated samples after exposure to the 12% $Na_2SO_4$ + 82% $Fe_2(SO_4)_3$ molten salt environment are reported in Table 1. NiCrMoWFe coating was found to have the nickel and chromium oxides and spinel $NiCr_2O_4$ as the majority phase. $FeMoO_4$, $NiMoO_4$, $Fe_2O_3$ was observed as minority phase. The XRD analysis results of Multicoat indicate that the scale is mainly consisting iron, zirconium, and silicon oxides. Also, there can be seen yttrium oxide. The oxide scale of each uncoated steel is mainly formed by iron. The residual amount of corrosion environment was found in the case of steel 1.4903.

## 3 Conclusion

In the current investigation are compared NiCr-based coatings applied by two different types of thermal spray technologies. The specimens were exposed to the 12% $Na_2SO_4$ + 82% $Fe_2(SO_4)_3$ molten salt environment at 690 °C under cyclic condition, simulating the working environment in low-emission boilers of coal-fired power plants.

NiCrMoWFe HP/HVOF sprayed coating shows high corrosion resistance in the given environment. The passivation oxide scale formed by NiO, $Cr_2O_3$, $NiCr_2O_4$ contributed to the corrosion resistance and have shown the tendency to act as a diffusion barrier to the oxidizing species. Multicoat applied using by TWAS technology also shows high corrosion resistance due to protective ceramic overlay Armaguard. Outer ceramic overlay was mainly formed by zirconium and silicone. Although uncoated steel 1.4903 shows the smallest weight gain during the high temperature test. On the surface of 1.4903 steel was created thick oxide scale formed mainly by iron with tendency to peel. Uncoated steel 1.7715 suffered severe corrosion attack. Intensive peeling was observed during the high temperature corrosion test. The results show a high oxidation rate during exposure to the corrosion environment and the presence of large amounts of iron oxides.

Both protective coating showed high corrosion resistance during test. Results of High temperature corrosion test showed that the specially designed Multicoat applied with TAWS technology can fully replace more expensive HP/HVOF NiCrMoWFe coating in case of coal fired boiler protection.

ACKNOWLEDGEMENTS

The paper has originated in the framework of institutional support for the long-time conception development of the research institution provided by the Ministry of Industry and Trade of the Czech Republic.

## References

[1] Sidhu T S Malik A Prakash S et al. 2007 Oxidation and Hot Corrosion Resistance of HVOF WC-NiCrFeSiB Coating on Ni- and Fe-based Superalloys at 800 °C Journal of Materials Engineering and Performance Vol. 16 pp 844–849

[2] Sidhu T S Prakash S and Agrawal R D 2006 Hot corrosion and performance of nickel-based coatings, *Tribology International* Vol. 90 pp. 41-47

[3] Topolska S Labanowski J 2010 Corrosion of evaporator tubes in low emission steam, *Archives of Materials Science and engineering* Vol. 42 pp. 85-92.

[4] Ishigai S 1999 Steam Power Engineering: Thermal and Hydraulic Design Principles 1rd ed *Cambridge University Press* p.189

[5] Singh H Kaur M Prakash S 2016 High-temperature Exposure Studies of HVOF-Sprayed $Cr_3C_2$-25(NiCr)/(WC-Co) Coating *Journal of Thermal Spray Technology* Vol. 25 pp 1192-1207

[6] Kaur M Singh H Prakash S 2010 Role of detonation gun spray $Cr_3C_2$–NiCr coating in improving high temperature corrosion resistance of SAE–213–T22 and SAE–347H steel in presence of $Na_2SO_4$–82%$Fe_2(SO_4)_3$ salt deposits *Surface Engineering* Vol. 26 pp 428-439

[7] Kaur M Singh H Prakash S 2009 High-Temperature Corrosion Studies of HVOF-Sprayed Cr3C2-NiCr Coating on SAE-347H Boiler Steel *Journal of Thermal Spray Technology* Vol. 18 pp 619–632

[8] Lencová K Česánek Z Houdková Š Lukáč F Mušálek R 2019 High temperature corrosion studies of HVOF sprayed coatings in molten sulphate salts environment *Conference proceedings - metal 2019* pp 1079-1085

# Theoretical study of the process of balls rolling with a diameter of 40 mm on a ball rolling mill

**A Naizabekov [1], B Mukhametkaliyev [2], S Lezhnev [1] and E Panin [3]**

[1] Rudny Industrial Institute, Rudny, Kazakhstan; [2] JSC "Sokolov-Sarybai Mining Production Association", Rudny, Kazakhstan; [3] Karaganda Industrial University, Temirtau, Kazakhstan

E-mail: sergey_legnev@mail.ru

**Abstract**. The following calibration scheme for rolling round steel is used for producing thermally hardened grinding balls with a diameter of 40 mm at JSC SSGPO for 320 variety rolling mill: a round billet with a diameter of 40 mm is produced from the initial billet of 120×120 mm. In the last developed scheme the square billet 150x150 mm is used as an initial billet for whole mill where the round steel with diameter of 40 mm is obtained from the above mentioned billet and then this round steel is used for production of balls with a diameter of 40 mm. Computer simulation of the ball rolling process in the Simufact Forming program was performed. The results of strain state distribution showed that the maximum processing level in the surface layers of balls is observed, while the stress state distribution showed that in the surface layers prevail high compressive stresses up to -400 MPa. Such stress-strain state leads to formation of gradient structure in the ball cross-section, where the surface layers will have more high mechanical properties. That is the most optimal result for details worked in condition of high values of friction.

## 1 Introduction

The worldwide coronavirus pandemic has a negative impact on almost all sectors of the economy, including the extractive industry. Therefore, declining prices for ore raw materials are forcing subsurface users to look for new innovative ways to extract and enrich ore raw materials with reduced costs. At the initial stage, the process of mining and processing different ores has a lot in common, regardless of the metal extracted from the ore. So after extraction from the ground, before enrichment, any ore must be crushed and grinded. Grinding, as the final stage of ore preparation before enrichment, is carried out in mills of various designs using in most cases grinding balls. During this process, due to the abrasive action of the milled material on the grinding balls, they wear out and are irretrievably consumed in the process of grinding raw materials (wear on average is 0.5-2 kg/t). According to various estimates, the cost of purchasing grinding balls is from 15 to 35% of the total technological costs of the respective industries. As a result, rapid wear of grinding balls leads to a significant increase in the total cost of crushing and grinding, which is about 60-70% of the cost of processing raw materials.

Often arising losses are associated not only with increased consumption of grinding balls and decrease in indicators of the efficiency of the milling equipment due to frequent outages, and deterioration of product quality due to the increase of iron content in the final product. The same problem is also relevant when using mills for grinding raw materials in the production of cement and energy. Thus, the problem of increasing the resource of steel grinding balls is very relevant. Therefore, the solution to this problem is promising not only for Kazakhstan, which is among the world leaders in the production of ores of various metals, but it is actively developed in many developed countries of the West and the Asia-Pacific region, which also produce minerals. It is the introduction of innovations in the extraction and processing of ores that will allow mining companies to maintain the profitability of their enterprises even in the conditions of crisis caused by various factors, including the coronavirus pandemic.

One of the most promising technologies for producing steel grinding balls is the screw rolling technology, where steel balls are produced on a screw rolling mill [1-3]. According to this technology, the rolling process of balls is carried out from a round bar billet using two rolls with screw calibers. For single-pass calibration, one ball is rolled for each revolution of the rolls. For multi-pass calibration, the number of balls coming out of the rolls per revolution is equal to the number of passes of the screw gauge. When leaving the rolls, the balls are intensely cooled in water and hardened, which ensures sufficient wear resistance of the balls in mills for grinding ore at processing plants. This technology is currently used for the production of grinding balls at the metal rolling plant of JSC "Sokolov-Sarybai Mining Production Association" (SSGPO). For these purposes, the SHPS 30-60 ball rolling mill is used, which is focused on the production of grinding balls with a diameter of 30 to 60 mm made of carbon, alloy and spring steel. The billet for rolling is a round hot-rolled steel rod with a length of 3 to 5 m, obtained by hot rolling at the variety rolling mill 320 of the same enterprise.

At present time, the following calibration scheme for rolling round steel is used for producing thermally hardened grinding balls with a diameter of 40 mm at JSC SSGPO for the 320 long-range rolling mill: a round billet with a diameter of 40 mm is produced from the initial billet of 120×120 mm [4]. But when analyzing the technical and economic indicators of this rolling scheme at the metal rolling plant of JSC "SSGPO", it was revealed that in this case there will be an undeveloped potential for the production capacity of the long-range rolling mill 320.

Therefore, we have developed a new calibration of rolls for rolling round steel No. 40 from a billet with a cross section of 150 x 150 mm in the conditions of mill 320 of JSC "SSGPO" [5]. This increased the total number of passes from 9 to 11, and evaluation of the appropriateness of the emerging values of the power parameters of permissible values (for the exception of equipment failure), and also conformity assessment of the shape and size of the final profile plant requirements (to prevent defects of the geometry of the rental). The research carried out in [5] proved that the proposed calibration for the production of round steel with a diameter of 40 mm fully meets the technical parameters of the long mill 320.

## 2 Scheme of balls rolling

This work is devoted to the development and calculation of calibration rolls for rolling balls with a diameter of 40 mm on a ball rolling mill in the conditions of JSC "SSGPO", obtained from the original cylindrical billet with a diameter of 40 mm, as well as computer modeling of this rolling process.

Rolling on a ball rolling mill is carried out between two rolls that rotate in one direction, and on the barrels of which screw calibers are cut. The axes of the rolling rolls of this mill are inclined at a slight angle to the axis of the rolled billet, which ensures the axial feed of the cylindrical billet in the rolls. In the rolls, the rotating initial billet is compressed by the gauge edges and takes the form of a ball connected by a jumper to the rest of the billet. With further progress in the rolls, the ball is calibrated and completely separated from the original cylindrical billet.

In this process the deformation zone has two main sections [6]:

- forming section, where the workpiece is compressed with a change in the shape and size of the screw edge;
- finishing section, where the shape and size of the edges remain unchanged, and compression is carried out by ovalizing the workpiece.

Ball forming is carried out by the edges of the rolls, the height of which gradually increases. To simplify the calibration calculation and roll manufacturing, it is assumed that the height of the gauge edge changes according to the law of a straight line. For a normal rolling process, the profile and dimensions of the forming section of the gauge are calculated in such a way that the following two main provisions are observed during the compression of the workpiece [7]:

1) the volume of metal compressed in the gauge must remain constant throughout the forming process;
2) changing the profile and size of the gauge edge must correspond to the stretching of the compressed workpiece.

According to the first provision, it is necessary that the volume of some part of the billet captured by the rolls remains unchanged as it passes through the remaining sections of the caliber. In this case, there will be no excess of metal at any time of rolling. When an excess of metal appears, the geometric shape of the ball is distorted and voids may form in the axial zone of the workpiece. The presence of a small excess of metal is allowed only at the beginning of the caliber, when the edge is still relatively low and does not prevent the displacement of metal from the caliber.

The second provision is that the normal formation of the bowl shape, the stretching of compressed zone should conform to the changing shape and sizes of caliber edge. In the case of balls rolling, the length of the compressed jumper must be equal to the width of the straight section of the caliber edge. If the change in the width of the caliber edge is less than the draw of the compressed billet, the metal will move away from the edge, and a roll will form on the surface of the billet, which will roll out in the film when the

billet is further compressed. If the change in the width of the caliber edge is greater than the workpiece stretching, then axial tensile stresses occur in the compressed jumper, which can lead to a break in the jumper. Thus, to fulfill both the first and second conditions, the edge on different sections of the caliber must have a strictly defined thickness. In this regard, the forming section of the caliber has a variable cutting step.
When developing the roll calibration, we decided to use the calculation algorithm, which is described in detail in [8-9]. In this case, the initial data for calibration of rolls are the technical characteristics of the rolling mill, the size of the rolls, and the diameter of the resulting grinding ball. The main dimensions of the rolls of the rolling mill are shown in figure 1.

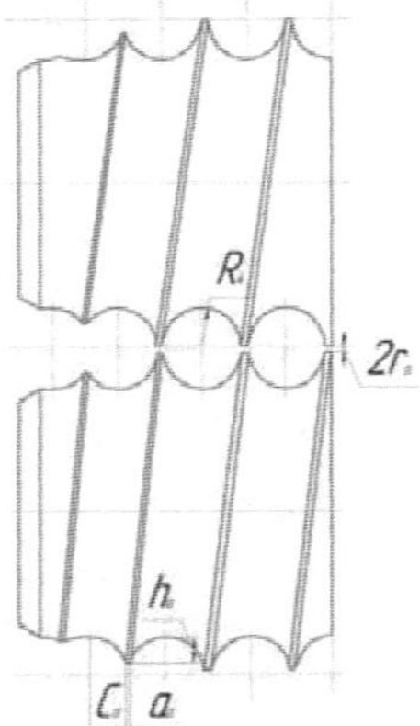


**Figure 1.** Basic elements of rolls calibration of for balls rolling: $h_\alpha$ and $a_\alpha$ – height and width of the edge; $r_\alpha$ and $R_K$ – the radii of the jumper and caliber; $C_\alpha$ and $S_\alpha$ – width of the spherical and cylindrical sections of the caliber

Ball calibration calculation is performed for the left roll, since the calibration of the right roll is almost the same. The only difference is that after cutting the edges of the right roll, it is cut in height by a certain amount. The initial data for calculating the calibration of the rolls of the SHPS 30-60 rolling mill are: ball diameter $d_b$ = 40 mm; rolls diameter $d_r$ = 300 mm. At the same time, when calculating and constructing the calibration of rolls, it is necessary to follow the design recommendations given in table 1.

**Table 1.** Initial data for calibration calculation

| $d_B$, mm | $d_r$, mm | Length of caliber, deg | | 2r, mm | Edge size, mm | | |
|---|---|---|---|---|---|---|---|
| | | $\alpha_{total}$ | $\alpha_{fin}$ | | height on the capture | width on the finishing section | width on the section 270° of capture |
| 40 - 50 | 180 - 300 | 900 - 1080 | 540 - 720 | 3,0 – 3,2 | 2,5 - 3,8 | 3,7 - 5.4 | 2,0 - 2,4 |
| 60 - 80 | 280 - 460 | 1080 - 1350 | 630 - 810 | 3,6 - 5,0 | 4,4 - 4,5 | 5,8 - 8,4 | 3,0 |
| 100 - 125 | 550 - 690 | 1260 - 1350 | 630 - 810 | 6,0 | 5,4 | 9,0 - 11,0 | 3,5 - 4,0 |

According to GOST 7524-2015 "Hot-rolled steel balls for ball mills" we choose the nominal diameter of the ball for subsequent calibration of the rolls, taking into account the permissible deviations. We accept $d_B$ = 40 mm, the permissible deviation of the diameter is ±2 mm. Then the diameter of the workpiece is $d_W = d_B / \eta_0 = 40 / 1{,}03 = 38{,}8$ mm, where $\eta_0$ is the coefficient that takes into account the radial growth of the ball size during rolling (we take $\eta_0 = 1.03$).
According to the regulations, the balls are rolled from a 40 mm blank. Then the diameter of the ball will be equal to:

$$d_B = d_W / \eta_0 = 40 \cdot 1{,}03 = 41{,}2 \text{ mm}. \tag{1}$$

Diameter of caliber

$$d_C = d_B \cdot \eta_t = 41{,}2 \cdot 1{,}013 = 41{,}7 \text{ mm}. \tag{2}$$

where $\eta_t$ = 1,013 - coefficient that takes into account the temperature expansion of the metal.
Minimum width of the edge

$$a = 0{,}04R_C + 1{,}3 = 0{,}04 \cdot 20{,}85 + 1{,}3 = 2{,}1 \text{ mm}, \quad (3)$$

where $R_C = d_C/2$ - radius of caliber, mm;
Initial height of the edge

$$h_0 = 1{,}5 + 0{,}07(R_C - 10) = 1{,}5 + 0{,}07(20{,}85 - 10) = 2{,}3 \text{ mm}; \quad (4)$$

Minimum radius of the jumper

$$r = 1 + 0{,}04(R_C - 10) = 1 + 0{,}04(20{,}85 - 10) = 1{,}4 \text{ mm}; \quad (5)$$

Width of the edge before cutting

$$a_{cut} = 4 + 0{,}14(R_C - 10) = 4 + 0{,}14(20{,}85 - 10) = 5{,}5 \text{ mm}. \quad (6)$$

The length of the caliber (the number of turns) is conventionally measured in degrees of sweep of the helix. For the beginning of the caliber (0°), the center of the radius of the caliber is taken, located on a straight line passing along the end of the shaft from the exit side of the finished ball. When rolling balls, we follow the recommendations of table 1 and set the total length of the caliber $\alpha_{TOTAL}$ = 900°, the length of the forming section $\alpha_F$ = 270° and the dividing section $\alpha_{FIN} = \alpha_{TOTAL} - \alpha_F$ = 900 - 270 = 630°.We divide the length of the gauge into equal sections; for preliminary calculation, the split angle $\Delta\alpha$ = 90°is sufficient. The radius of the jumper at the end of the forming at $\alpha = \alpha_F - \Delta\alpha$ is determined from the ratio

$$r_{cut} = r + \frac{R_C - h_0 - r}{\alpha_f}\Delta\alpha = 1{,}4 + \frac{20{,}85 - 2{,}3 - 1{,}4}{270}90 = 7{,}1 \text{ mm} \quad (7)$$

Changing the height of the edge when the roll is rotated by 90°:

$$\Delta h = r_{cut} - r = 7{,}1 - 1{,}4 = 5{,}7 \text{ mm}. \quad (8)$$

For each angle of rotation of the rolls from $\alpha_{TOTAL}$ to $\alpha_F$ at interval 90° the height of the edge:

$$h_{\alpha-90} = h_\alpha + \Delta h = h_\alpha + 5{,}7, \quad (9)$$

where $h_\alpha$ - current height of the edge, mm.
While the jumper size on the finishing section $\alpha_{FIN}$ will be r = 1.4 mm, the radius of the bridge on the molding section is determined by the formula

$$r_{\alpha+90} = r_\alpha + \Delta h = r_\alpha + 5{,}7. \quad (10)$$

Determine the width of the spherical section of the caliber from the expression

$$C_\alpha = \sqrt{R_C^2 - r_\alpha^2} = \sqrt{20{,}85^2 - r_\alpha^2}. \quad (11)$$

So cut the jumper begins during the ball forming, then take the width of the spherical segments plot it on the angle of rotation of the roll $\alpha_{FIN+90}$ pre-finishing area of the caliber, i.e. $C_{CUT} = C_{FIN+90}$. The main calibration calculations are shown in table 2.

**Table 2.** Calculation data for rolls calibration of ball rolling with a diameter of 40 mm

| $\alpha$, ° | 900 | 810 | 720 | 630 | 540 | 450 | 360 | 270 |
|---|---|---|---|---|---|---|---|---|
| $h_{\alpha-90}$, mm | 2,3 | 8,0 | 13,7 | 19,4 | 19,4 | 19,4 | 19,4 | 19,4 |
| $r_{\alpha+90}$, mm | 18,5 | 12,8 | 7,1 | 1,4 | 1,4 | 1,4 | 1,4 | 1,4 |
| $C_\alpha$, mm | 9,6 | 16,4 | 19,6 | 20,8 | - | - | - | - |

## 3 FEM simulation

Based on the above calibration calculations of the SHPS 30-60 ball rolling mill, a computer simulation of balls rolling process with a diameter of 40 mm was carried out in the Simufact Forming program. The original billet had a diameter of 40 mm and a length of 150 mm. When modeling in the Simufact Forming software package, the following conditions and assumptions were accepted:

IOP Conf. Series: Materials Science and Engineering 1178 (2021) 012040 doi:10.1088/1757-899X/1178/1/012040

- material of the workpiece in the initial state (before deformation) was isotropic and there were no initial stresses and deformations;
- grid of 250,000 hex-type finite elements was created on the workpiece; the average length of the element face was 1.3 mm;
- material used in the simulation – AISI 1065 steel corresponding to 65G steel, the hardening curves were taken from the Simufact Forming library;
- initial temperature of the workpiece was assumed to be equal to 1150 °C, there was deformation heating and heat transfer between the workpiece, the tool and the environment;
- material model of the initial billet was assumed to be elastic-plastic;
- friction coefficients between the tool and the workpiece were selected based on recommendations from [6] and assumed to be equal to 0.5 at the contact of the workpiece with the rolls;
- roll rotation speeds were taken from the characteristics of the 320 mill.

In addition to these conditions, which are common to many metal forming processes modeling programs, the key features of the Simufact Forming complex were also used. In particular, the option of plotting coons surfaces was activated to improve the construction of a finite element grid on the curved surface of a round billet. Figure 2 clearly shows the difference between grid construction in the programs Deform and Simufact Forming. The grid in Deform is rougher, and the surface is not a circle, but a polyhedron (figure 2a). In Simufact Forming, by calculating the geometry, the final elements are curved to take the shape of an arc (figure 2b).

a b

**Figure 2.** Building a hex grid in Deform and Simufact Forming

As a result, a model of balls rolling from a round billet was obtained (figure 3 a), in which the first ball billet is not fully formed due to tightening (figure 3b). The rest of the blanks are solid balls with a diameter of 40 mm. To study the stress-strain state (SSS), the parameters "effective plastic deformation" and "average normal stress" were considered.

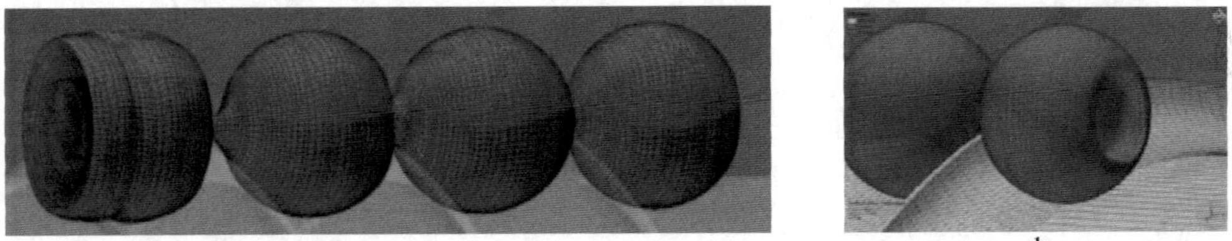

a b

**Figure 3.** Model of ball rolling with a diameter of 40 mm

When studying the strain state, its extremely uneven distribution over the section of the workpiece was noted (figure 4a). The central layers of the workpiece (0÷20% of the radius) receive almost no increase in deformation. Here the level of strain reaches 2.3÷2.5. In the intermediate layers (20÷70% of the radius), the effective plastic deformation is much higher, the level of strain reaches 3÷10. This is due to the fact that the metal in this zone, in contrast to the center of the workpiece, moves along a circular trajectory of the ball gauge more intensively. In the surface layers (70÷100% of the radius), the most intensive development of effective plastic deformation is observed, and at a distance of 95÷100% of the radius, the values of strain are the maximum, about 30÷33. This is due to the fact that the surface layers are subjected not only to the most intense deformation during the formation of the ball, but also to subsequent surface processing, after the ball is formed.

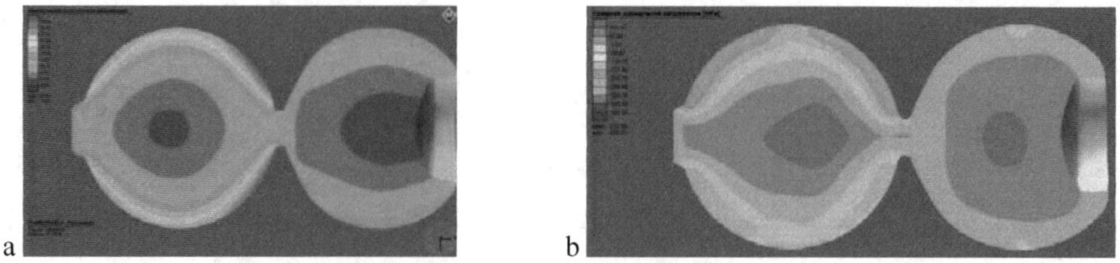

a b

**Figure 4.** SSS Parameters: a) strain state; b) stress state

When studying the stress state, it was noted that the uneven distribution of the average normal stress across the section of the workpiece has a certain tendency (figure 4b). The central layers of the workpiece (0÷20% of the radius) are subjected to tensile stresses, but their value is small, about 0÷55 MPa. In the intermediate layers (20÷70% of the radius), there are no tensile stresses, only compressive stresses act here, their values are at the level of 0÷-116 MPa. This is due to the fact that the metal in this zone, moving along the trajectory of the gauge, is more exposed to normal stresses. In the surface layers (70÷100% of the radius), only compressive stresses act, their values reach -460 MPa. This sharp increase in normal stress is a consequence of the fact that the surface layers are most affected by normal stresses from the deforming caliber.

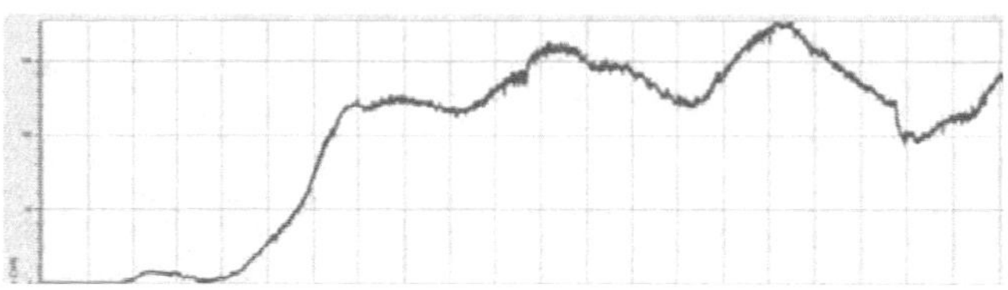

**Figure 5.** Graph of rolling force

Analysis of the resulting force during balls rolling showed that the force graph is non-monotonic. The peaks that correspond to the rolling of individual balls are clearly visible, with the maximum force value in each peak corresponding to the moment when the jumper is cut. At the same time, the general nature of the graph is increasing, which is directly related to the filling of ball gauges with deformable metal. Thus, figure 5 shows three peaks: the first peak is the forming of 1$^{st}$ ball in the caliber (the force value is about 120 kN), the second peak is the forming of 2$^{nd}$ ball (force is approximately 160 kN), the third peak is the forming of 3$^{rd}$ ball (force is about 180 kN), the moment is shown in figure 3a.

## 4 Conclusions

The results of theoretical studies of the balls rolling process with a diameter of 40 mm at the SHPS 30-60 ball rolling mill in the conditions of the Sokolov-Sarybai Mining Production Association, including the results of computer modeling in the Simufact Forming software complex, have allowed to more fully study the production technology of grinding balls using screw rolling.

## References

[1] Tomczak J, Pater Z and Bulzak T The Effect of Process Parameters in Helical Rolling of Balls on the Quality of Products and the Forming Process. Materials. – 2018. –Vol. 11. - № 2125

[2] Pater Z New methods of steel ball rolling. International journal of materials & product technology. – 2013. – Vol. 47. – Iss. 1-4. – P. 12-22

[3] Pater Z Multi-Wedge Cross Rolling of Balls. Journal of iron and steel research international. – 2013. – Vol. 20. – Iss. 10. – P. 46-50

[4] Metal rolling complex of RMZ JSC "SSGPO" 2010 Volume 1 p 147

[5] Naizabekov A B, Mukhametkaliev B S, Lezhnev S N, Panin E A and Andreev E V 2017 Proceedings of the 11th international Congress of Rolling Production (Magnitogorsk) pp 182-8

[6] Afonkin M and Zvyagin V 2007 Production of blanks in mechanical engineering (Politechnika) p 254

[7] Tselikov A I, Barbarich M V, Vasilchikov M V, Granovsky S P and Zhukevich-Stosha E A 1971 Special rolling mills (Moscow: metallurgy) p 236

[8] Peretyatko V N, Klimov A S and Filippova M V Roller grooving in ball-rolling mills. Part 1. Steel in Translation. - 2013. - V. 43. no. 4. - p. 168-70

[9] Peretyatko V N, Klimov A S and Filippova M V Roller grooving in ball-rolling mills. Part 2. News of higher educational institutions. Ferrous metallurgy. - 2013. - № 6. - P. 16-20

# Temelin Irradiated Cladding Project – TIRCLAD

**Stanislav Linhart[1], Josef Běláč[1], Martin Ševeček[1,2], Radomír Řeháček[1], Vladislav Starý[1]**

[1]ALVEL a.s., Stefanikova 110/41, 602 00 Brno - Ponava, Czech Republic
[2]Czech Technical University in Prague, Faculty of Nuclear Sciences and Physical Engineering, V Holešovičkách 2, 180 00 Praha 8, Czech Republic

E-mail: stanislav.linhart@alvel.eu

**Abstract.** It is widely known, that zirconium-based alloys have been used for nuclear fuel cladding fabrication for decades [1], [2]. Therefore, it is crucial for fuel vendors, utilities as well as safety authorities to understand their behaviour during reactor operation with strong fast neutron and gamma irradiation. To complement the information about material properties for current as well as prospective advanced zirconium-based alloys, a long term research project implemented by ALVEL has been carried out, in cooperation with its partners (ČEZ, Škoda Nuclear Machinery, UJV Rez, Research Centre Rez) [3], [4]. The project is implemented on the background of a large material research program sponsored by JSC TVEL and ČEZ (this program started in 2012 and it is planned to be finished in 2024). It is fully independently implemented in the Czech Republic.

Irradiation of material samples started at Temelin NPP in 2014 and finished in spring 2020. The first three batches were already transported to UJV Rez and Research Centre Rez for post-irradiation evaluation. After finishing testing of reference non-irradiated samples testing in 2018 (determination of initial state, mechanical properties and as-received microstructure), six batches of irradiated material have been analysed sequentially with their increasing neutron fluence. The final irradiation damage of the last sixth batch corresponds approximately to fuel burnup of 80 MWd/kgU.

This paper presents the developed methodologies and the expected experimental and other important outcome. Several material properties are being studied on different scales including high resolution microstructural and chemical analysis, creep and mechanical behaviour, nanoindentation, oxide layer morphology. The main objectives of the project include evaluation of materials' microstructural and bulk properties and the derivation of dose- and temperature-dependent correlations that can be implemented into FEM models and fuel performance codes such as FRAPCON/FRAPTRAN [5], [6] or Transuranus [7]. These models can be then used to support licensing of new designs influenced by fuel cladding behaviour during long term dry storage, justification of dry storage for an extended period of time or justification of new safety criteria that can lead to enhanced operation of VVER reactors and spent nuclear fuel storage [8].

Keywords: fuel cladding, irradiation, fuel performance, post-irradiation examination, FEM models

## 1 Introduction.

The presented project is realized on the background of a large material research program sponsored by JSC TVEL and ČEZ. It is fully independently implemented in the Czech Republic. The project has started in 2012 and it is scheduled to finish in 2024. Figure 1 shows the timeline of the project.

For this purpose, a set of six Material Cluster Assemblies (MCA) was designed by experts from OKBM, Nizhniy Novgorod and Bochvar institute, Moscow, and then manufactured in Mashinostroitelny zavod (MSZ) in Elektrostal, Moscow region. Their main components from the project point of view are the ampoules made of E110 alloy cladding material containing three types of neutron fluence monitors and material samples. All the MCAs were loaded into the reactor core of Temelin NPP Unit 1 in 2014. After completion of 1 to 5 irradiation cycles (expected neutron fluence after 5th irradiation cycle is 5,3 . 1026 neutrons / m2 according to core design models = VVER fuel burnup of approx. 80 MWd/kgU), ampoules with their content are separated from the cluster assembly in the unique POMA device, an equipment specially designed and manufactured for their separation by Skoda Nuclear Machinery in Pilsen. Separated ampoules with samples are transported to the hot-cell facilities in Rez for material analyses, testing and further investigation. Before transport to the Czech Republic and their loading into the reactor core of Temelin NPP Unit 1, a licensing process at the State Office for Nuclear Safety (SONS) was successfully concluded. The design of MCAs is presented in Figure 2 and their main parameters are presented in Table 1. Reference non-irradiated materials are studied first to determine initial states and as-received microstructure [9], [10]. Six batches of samples will be studied over time with increasing neutron dose of every batch. The neutron dose and neutron spectra of every batch are evaluated both experimentally and by simulation [11]. The neutron dose and spectra are calculated using sophisticated models and they are also evaluated based on activation of three types of neutron fluence monitors that are placed inside the assemblies, in the middle of ampoules.

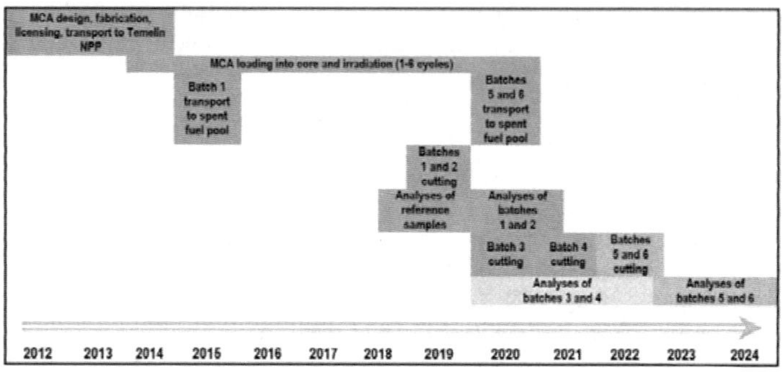


**Figure 1.** Timeline of the project

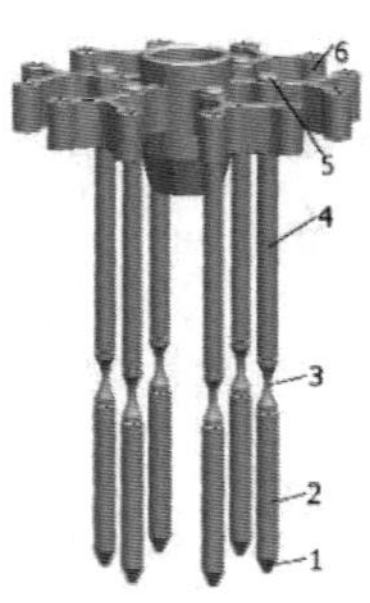


**Figure 2.** Schematics of the MCA irradiated at the Temelin core (not in scale). 1 – MCA end plug; 2 – Ampoules with samples to be separated and analysed; 3 – Joint with the thinned region for separation; 4 – Spacer rodlet (finger); 5 – Screws; 6 – MCA (RCCA) Spider.

| Parameter | Value |
|---|---|
| Number of ampoules in each MCA | 6 pieces |
| Length of the ampoule (separated part) | ~ 300 mm |
| Outer diameter of cladding | 9,1 mm |
| Cladding Material | E110 |
| Mass of one MCA | 8 kg |
| MCA design irradiation time | 6 years |

**Table 1.** Main parameters of MCAs

## 2 AIM OF THE PROJECT AND SCOPE OF ANALYSES.

### *2.1 Material tests and characterization*

The main objectives of the project are to study cladding performance during normal operation of reactors and during wet and dry storage of the spent nuclear fuel. Therefore, a set of experiments were defined to support the development of new models and correlations that can be used in (fuel performance) codes. These models that are directly derived from experiments with irradiated materials can be later used for licensing of the new design of fuels, an extension of dry storage periods and for support of current or new operational limits and technical specifications. The geometry of samples and ampoules are being measured to estimate irradiation growth including anisotropic effects before destructive mechanical testing and characterization [12]. Material analyses of samples placed inside the irradiated ampoules are the part of the above mentioned research program sponsored by TVEL and ČEZ. These analyses include PDE measurement and evaluation, irradiation induced growth measurement, metallographic analysis, TEM analysis and neutron fluence evaluation.

*2.1.1 Mechanical Testing.* The experimental matrix of mechanical tests includes axial and ring tensile specimens. The shape and geometry of the specimens were optimized based on the FEM modelling and testing with non-irradiated materials in order to produce reliable relevant data. The objectives of geometry optimization are to produce as many samples as possible taking into account limited volume of irradiated alloy but still produce reliable results comparable to data produced elsewhere. The temperature for mechanical tests ranges from room temperature up to 400°C with the main focus on operational and storage temperatures between 300-350°C. Standard tensile tests and creep tests, were defined, for example, in [13] and [15]. The deformation rates and stress amplitudes are defined based on FEM modelling and expected magnitudes during normal operation, wet storage, drying of storage casks and dry spent fuel storage. The mechanical tests are supplemented with fractographic analysis of the damage surfaces with quantification of brittle/ductile fractures. Additionally, nanoindentation tests were defined in the same temperature range with a goal of measuring dose- and temperature-dependent properties of a virgin, corroded and hydrided Zr-based alloys [11]. Based on the evaluation of hydrogen pickup and precipitation, the models can account for different hydrogen-related failure mechanisms.

*2.1.2 Material Characterization and Microstructure Analysis.* The fundamental mechanical tests summarized in the previous section are accompanied by several characterization methods and techniques. The main microstructural tool to be used is a High-resolution Transmission Electron Microscopy (HRTEM) including detailed microstructure analysis of the deformed (after mechanical tests) and as-irradiated samples. Scanning electron microscopy (SEM) is also used to analyse the TEM foils mainly with EDX and EBSD detectors [12], [15]. The standard metallographic analysis is done according to the established ISO and ASTM standards and includes morphological analysis of the oxide layers, quantification of precipitates and hydrides and determination of the present phases [16] – [18]. Microhardness measurements are done on the metallographic samples in both longitudinal and transverse directions.

*2.2. Modelling and simulations.* FEM-based models with material properties and models currently used for licensing purposes have been utilized to optimize the geometry of samples for mechanical testing. Sample geometries originating in the French program PROMETRA and Hungarian institute MTA EK were used as a reference and further optimized [13], [19], [20]. A finite element model of axial tensile test with results of Von-Mises is shown in Figure 3. Based on the experimental testing, change of materials' microstructural and bulk properties are evaluated and new correlations derived. They will include dependence on temperature as well as neutron dose. They are then implemented into codes used for nuclear fuel licensing such as ABAQUS, Frapcon and Fraptran [5], [6], [21]. The models account also for an anisotropic behaviour of Zr-based alloys thanks to the testing in the two main directions. The models can be later implemented also to integral and system codes and more precise behaviour of the fuel system can be then simulated.

### *2.3 TESTING OF THE REFERENCE NON-IRRADIATED MATERIAL*

The testing started with non-irradiated (reference) samples serving as a reference for future comparison between irradiated cladding samples and non-irradiated. The methods used were developed based on the testing with reference material and the testing parameters were optimized in order to focus on normal operation and spent fuel storage conditions. As-received Zr1Nb material in its initial state, pre-oxidized as well as hydrided samples were included to study reference material in various expected states. The geometry of samples for mechanical testing was optimized and the final sample shapes of axial and ring samples are shown in Figures 3 and 4. The microstructure has been studied using TEM, SEM and LOM. The EBSD map of the as-received material from the SEM of the 100 x 100 μm area is shown in Figure 5. The step size is 0.1 μm and the resolution 10,000 x10,000 pixels.

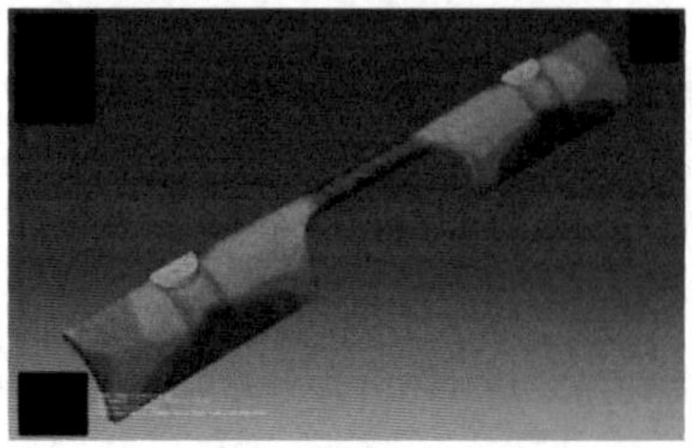

**Figure 3.** Calculation of Von-Mises stresses during the tensile test with the axial tensile sample using the FEM model

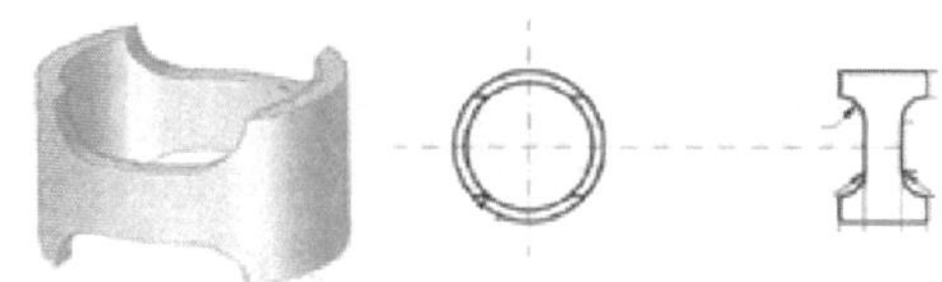

**Figure 4.** Ring sample used for tensile testing and creep tests¨

**Figure 5.** EBSD map of the as-received non-deformed reference sample. Grain size distribution, misorientation and the levels of recrystallization were calculated using software tools from the maps. Strong preferential orientation of the α-Zr grains is obvious.

The project itself and its outcomes will support the implementation of new nuclear fuel designs and in particular their licensing. Dose and temperature dependent material/physical properties will be determined, the tests to be performed are discussed in the following subsection. Additionally, the data measured will be compared to the large database of results obtained during past decades in various research reactors including the LVR-15 reactor operated in the Czech Republic. In comparison with commercial power reactors, the research reactors might operate with slightly different parameters such as neutron spectra, temperature, thermal loading or water chemistry. Therefore, based on the comparison of the old data and newly produced results, the methodologies used in research reactors can be validated or modified.

These activities are included in the In-pile Creep Studies of ATF Claddings (INCA) project which is a new joint experimental program included in the Framework for Irradiation Experiments (FIDES) coordinated by the OECD/NEA [22]. All of the relevant data produced during the project are summarized in an extensive TIRCLAD database. The structure and content of the TIRCLAD database are shown in Figure 6. The database is generally based on the IFPE format defined and extensively used by OECD/NEA for collecting of nuclear fuel performance-related data [23], [24].

## 3 PROJECT STATUS

First four batches of ampoules and material samples were separated from the whole MCA in POMA equipment at Temelin NPP in April 2019 (first two batches), April 2020 (third batch) and April 2021 (fourth batch). All the material samples were extracted from the ampoules were transferred to the semi-hot lab for evaluation of their geometry and other material analyses specified above. The remaining tube-formed ampoules (claddings) were transferred to CVR hot cell for evaluation of their microstructure, mechanical and creep properties. The POMA equipment for ampoule separation is presented in Figure 7, and some moments of ampoule separation are presented in Figure 8. A transport of separated ampoule in a transport cask is presented in Figure 9.

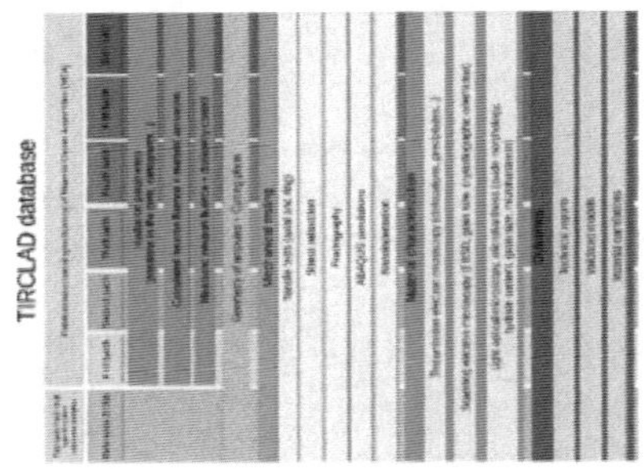


**Figure 6.** Overview of the structure of the TIRCLAD database to be produced according to the IFPE format requirements defined by OECD/NEA.

**Figure 7.** POMA equipment for ampoule separation in the transport container shaft

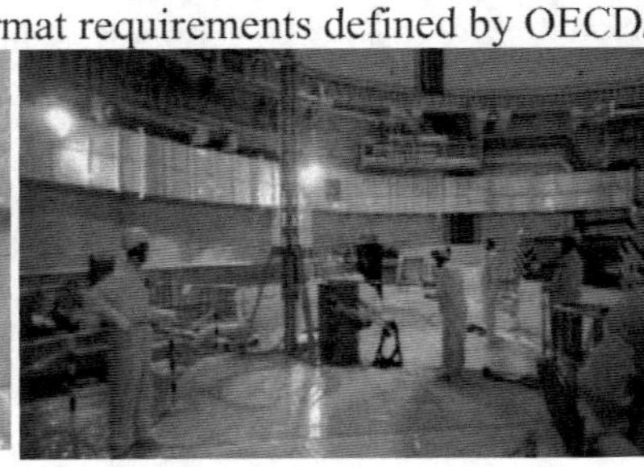

**Figure 8.** Separation of the first batch of irradiated material.

**Figure 9.** Transport of the cask with ampoules from Temelin NPP to UJV Rez.

## 4 DISCUSSION, CONCLUSION AND FUTURE PLANS

An extensive material testing project called TIRCLAD has been initiated in the Czech Republic in 2012. Material assemblies have been first designed to be compatible with VVER fuel design and fuel handling machine [8], [25]. They were then approved by the utility, licensed at SONS and manufactured. Six material assemblies were loaded into the Unit 1 of Temelin NPP. In April 2019 the first and second assemblies were successfully placed into the POMA device, cut and the irradiated material ampoules were transported to hot cells for PIE. Fourth batch of ampoules was already cut in the POMA device and transported to hot cells this April. The remaining two assemblies will follow in the next outages with maximal radiation damage corresponding to fuel cladding irradiated up to 80 GWd/tU. Meanwhile, methodologies for testing were developed on non-irradiated samples. The geometry of samples for mechanical testing was optimized by UJV Řež using detailed FEM models and new test rigs were designed and build in the hot cells in the Research Centre Rez. PIE of non-irradiated and six batches of irradiated samples are focused on material evaluation to support licensing and new designs related to spent fuel behaviour during the long term wet and dry storage. New reliable models will be derived and a database called TIRCLAD will be produced as one of the outcomes of the project as well.

Nowadays, first three batches of irradiated material after first three cycles of irradiation have been cut and transported to UJV Rez and Research centre Rez facilities for testing and evaluation. On-going extension of the project is under preparation. The MCA-ATF program is to be focused on Advanced Technology and /Accident Tolerant Fuel claddings with irradiation starting in 2022 [26].

## 5 REFERENCES

[1] D. R. Olander, "Fundamental aspects of nuclear reactor fuel elements: solutions to problems," California Univ., Berkeley (USA). Dept. of Nuclear Engineering, 1976.

[2] R. B. Adamson and A. H. G. Rickover, "The Decision to Use Zirconium in Nuclear Reactors," in Zirconium Production and Technology: The Kroll Medal Papers 1975–2010, R. B. Adamson,

Ed. 100 Barr Harbor Drive, PO Box C700, West Conshohocken, PA 19428-2959: ASTM International, 2010, pp. 9-9–11.

[3] M. Ševeček, J. Běláč, S. Linhart, R. Řeháček, M. Malá, and O. Srba, "Introduction of the temelin irradiated cladding Project—TIRCLAD 1," American Nuclear Society, Sep. 2019.

[4] S. Linhart et al., "Material research of advanced zirconium-based alloys irradiated at temelin NPP reactor core," VVER conference proceedings, Prague, Nov. 2019.

[5] K. Geelhood, W. Luscher, P. Raynaud, and I. Porter, "FRAPCON-4.0: A Computer Code for the Calculation of Steady-State, Thermal-Mechanical Behavior of Oxide Fuel Rods for High Burnup," PNNL-19418, Vol.1 Rev.2, 2015.

[6] K. Geelhood, W. Luscher, J. Cuta, and I. Porter, FRAPTRAN-2.0: A Computer Code for the Transient Analysis of Oxide Fuel Rods. 2016.

[7] V. Di Marcello, A. Schubert, J. van de Laar, and P. Van Uffelen, "The TRANSURANUS mechanical model for large strain analysis," Nucl. Eng. Des., vol. 276, pp. 19–29, 2014.

[8] V. V. Novikov, V. I. Kuznetsov, and A. V. Salatov, "Modern Approaches to VVER Fuel-Element Design," At. Energy, 119 (2016), 5, 332, doi: 10.1007/s10512-016-0068-0.

[9] P. Gávelová et al., "Microstructure of Zirconium Fuel Claddings: TEM and EBSD Studies of As-Re-ceived and Neutron-Irradiated Materials."

[10] O. Libera, P. Halodová, P. Gávelová, and J. Krejčí, "Nanoindentation of hydrogen enriched zr-1nb zirconium alloy nuclear fuel claddings," Acta Polytech. CTU Proc., 27 (2020), 155.

[11] ALVEL a.s., "Stanovení fluence neutronů v místech vzorků MCA ve vnitřních vodících trubkách palivového souboru," 12-20A-003-02, 2018.

[12] Committee B10, Ed., Deformation Mechanisms, Texture, and Anisotropy in Zirconium and Zircaloy. 100 Barr Harbor Dr., PO Box C700, West Conshohocken, PA 19428-2959: ASTM Int-l, 1988.

[13] M. Király, D. M. Antók, L. Horváth, and Z. Hózer, "Evaluation of axial and tangential ultimate tensile strength of zirconium cladding tubes," Nucl. Eng. Technol., 50 (2018), 3, 425,

[14] D. Franklin, G. Lucas, and A. Bement, Eds., Creep of Zirconium Alloys in Nuclear Reactors. 100 Barr Harbor Drive, PO Box C700, West Conshohocken, PA 19428-2959: ASTM Int-l, 1983.

[15] E04 Committee, "Test Methods for Determining Average Grain Size," ASTM Int-l, 2013.

[16] ASTM E407-07(2015)e1, "Standard Practice for Microetching Metals and Alloys," ASTM International, West Conshohocken, PA, 2015. doi: 10.1520/E0407-07R15E01.

[17] ASTM International, West Conshohocken, PA, "ASTM B614-16 Standard Practice for Descaling and Cleaning Zirconium and Zirconium Alloy Surfaces," 2016.

[18] J. Schemel, Ed., ASTM Manual on Zirconium and Hafnium. 100 Barr Harbor Drive, PO Box C700, West Conshohocken, PA 19428-2959: ASTM International, 1977.

[19] M. Király, Z. Hózer, M. Horváth, T. Novotny, E. Perez-Feró, and N. Vér, "Impact of thermal and chemical treatment on the mechanical properties of E110 and E110G cladding tubes," Nucl. Eng. Technol., vol. 51, no. 2, pp. 518–525, Apr. 2019, doi: 10.1016/j.net.2018.11.002.

[20] B. Cazalis et al., "The PROMETRA Program: Fuel Cladding Mechanical Behavior under High Strain Rate," Nucl. Technol., 157 (2007), 3, 215, doi: 10.13182/NT07-A3814.

[21] D. S. Simulia, "Abaqus," Anal. User's Man. Version, vol. 6, 2010.

[22] J. Klouzal, M. Mikloš, M. Ševeček, and M. Dostál, "In-pile creep studies of ATF claddings: INCA," Sep. 2019.

[23] P. Menut, E. Sartori, and J. A. Turnbull, "The public domain database on nuclear fuel performance experiments (ifpe) for the purpose of code development and validation," p. 14.

[24] P. M. Chantoin, E. Sartori, and J. A. Turnbull, "The compilation of a public domain database on nuclear fuel performance for the purpose of code development and validation," p. 21.

[25] H. Weidinger, "PWR and WWER fuel performance. A comparison of major characteristics," 2005, Accessed: Jan. 31, 2018. [Online]. Available: http://inis.iaea.org/Search/search.aspx?orig_q=RN:37098303.

[26] J. Krejčí et al., "Development and testing of multicomponent fuel cladding with enhanced accidental performance," Nucl. Eng. Technol., p. S1738573319303432, Aug. 2019, doi: 10.1016/j.net.2019.08.015.

# Mechanical and physical properties of Mg alloys prepared by SPD methods

**P Lukáč[1,2,3] Z Trojanová[1,2] J Džugan[2] and K Halmešová[2]**

[1]Charles University, Faculty of Mathematics and Physics, Ke Karlovu 3, 121 16 Praha 2, Czech Republic

[2]COMTES FHT, Průmyslová 995, 223 41 Dobřany, Czech Republic

[3]Corresponding author's E-mail: Pavel.Lukac@mff.cuni.cz

**Abstract**. Magnesium alloys were prepared by severe plastic deformation (SPD). Three methods were used. The mechanical and thermal properties were estimated. The increase in strength could be attributed to the grain refinement and increased dislocation density. The thermal conductivity and expansivity was measured for samples prepared with different methods. Some examples are shown and discussed.

## 1 Introduction

In recent decades, many papers describing the deformation behavior of Mg alloys under different conditions have been published. It was shown that the yield strength of Mg alloy polycrystal $\sigma_y$ depends on the grain size d according the following Hall-Petch relation

$$\sigma_y = \sigma_0 + k_y d^{-1/2} \quad . \tag{1}$$

Here $k_y$ is a constant and $\sigma_0$ is a friction stress. According to this relation, ultra-fine-grained materials (UFG) can exhibit a considerable higher strength. The micrometer or nanometer sized material may be produced by different methods.

Over the last decades, a significant effort has been devoted to obtaining UFG materials using several plastic deformation (SPD) methods. The processing mechanisms and technological parameters are especially important for the materials microstructures. The dislocation density, grain size, and texture are produced. Recently, Estrin and Vinogradov [1] have presented a review on extreme grain refinement by SPD techniques. The principal processing scheme and mechanisms were described. The samples processed by SPD have enhanced mechanical properties.

Besides the mechanical strength, also the physical properties of samples after SPD are influenced. This is important for applications [2]. In the present work, some examples concerning the effect of SPD on the mechanical and thermal properties of Mg alloys will be presented.

## 2 Materials and methods

Two magnesium alloys AZ31 (3wt%Al-1wt%Zn-balance Mg) and AX52 (5wt%Al-2wt%Ca-balance Mg) were processed by SPD. Among the severe plastic deformation techniques, three processes were used: a) Equal-channel angular pressing (ECAP) [3-5] b) accumulative roll bonding (ARB) [6,7] and c) Rotary swaging (RSW) [8]. The microstructure was analyzed by light microscopy and by scanning electron microscopy (SEM). Mechanical properties and thermal properties were estimated in a wide temperature range from RT to 350 °C. Samples were deformed in tension and/or compression.

Thermal diffusivity, $a$, and specific heat capacity at the constant pressure, $c_p$, were measured using the laser flash method in a wide temperature range of 20 and 350 °C. The details can be found in references [4, 6]. Thermal conductivity, $\kappa$, was calculated according the following relationship

$$\kappa = a\rho c_p \ , \quad (2)$$

where ρ is the density estimated by the hydrostatic method. Thermal expansion was measured using LINSEIS L75PT-1600 dilatometer with a horizontal dual sample arrangement. Two temperature cycles (heating up to 400 °C and cooling to room temperature) were performed for each sample using heating and cooling rates of 1 K/min in a resistance furnace in an argon protective atmosphere. Thermal expansion coefficient, α, was estimated as an extrapolated value of the temperature dependence to room temperature [6].

## 3 Experimental results

### *3.1 Mechanical properties*

The grain size of AX52 magnesium alloy processed by ECAP in the route A was substantially refined. The grain size decreased after 8 passes from 60 μm to 3.6 μm (see Table 1). The tensile yield strength (TYS) obtained at two different strain rates $10^{-3}s^{-1}$ and 50 $s^{-1}$ increased with the increasing number of passes for both strain rates as it is shown in Table 1. Although the tensile tests were performed at room temperature, the sensitivity of the TYS to the strain rate indicates the presence of the thermally activated process/es [3]. The strength of ECAPed AZ31 samples is dependent also on the route used [5]. The TYS and UTS (ultimate tensile stress) of AZ31 samples estimated using ECAP in routes A, BC, and C after different passes are listed in Table 2, The strain rate used was in the order $10^{-3}s^{-1}$. Equivalent strain achieved after one ECAP pass is 1.15.

**Table 1**. The grain size, tensile yield stress, ultimate tensile strength, and thermal conductivity of the AX52 alloy depending on the number of ECAP passes.

| Nr of passes | Grain size μm | TYS($10^{-3}s^{-1}$) MPa | TYS($50s^{-1}$) MPa | κ(20 °C) W/m.K | κ(250°C) W/m.K |
|---|---|---|---|---|---|
| 0 | 63 | 77.1 | 85.2 | 70.2 | 68.4 |
| 1 | 28 | 137.3 | 313.2 | 87.0 | 84.9 |
| 2 | 21 | 170.2 | 346..0 | 79.1 | 68.3 |
| 4 | 4 | 197.8 | 358.9 | 73.9 | 61.3 |
| 8 | 3.4 | 205.9 | 358.8 | 75.5 | 71.0 |

Mechanical properties of AZ31 magnesium alloys were also investigated using samples processed by rotary swaging. The yield strength and ultimate strength determined in tensile (TYS, UTS) and compression tests (CYS, CPS) at room temperature are presented in Table 3 depending on the number of the swaging steps. The yield strength values estimated in tension are higher than the values found in compression tests; the tensile compression anisotropy is observed. Plastic anisotropy was also observed on AZ31 sheets submitted to accumulative roll bonding [7]. Values of the TYS measured in the rolling direction (L) are higher compared to the TYS in the transversal direction (T) (see Table 4). The deformation stresses increased with the number of passes in the rolling mill for both samples of orientation L or T.

**Table 2**. The stress characteristics of AZ31 alloy estimated for the three processing ECAP routes: A, $B_C$, C.

| **Nr of passes** | **TYS (A) MPa** | **UTS (A) MPa** | **TYS (B) MPa** | **UTS (B) MPa** | **TYS(C) MPa** | **UTS (C) MPa** |
|---|---|---|---|---|---|---|
| 1 | 149 | 252 | 149 | 252 | 149 | 252 |
| 4 | 144 | 253 | 154 | 271 | 149 | 236 |
| 8 | 124 | 231 | | | 148 | 249 |

**Table 3**: The stress characteristics estimated in tension and compression (strain rate of $10^{-3}s^{-1}$) for AZ31 alloy submitted to rotary swaging.

| **Nr of RSW** | **0** | **1** | **2** | **3** | **4** | **5** |
|---|---|---|---|---|---|---|
| Grain size (μm) | 20-300 | 60.5±55.2 | 41.5±28.6 | 40.3±26.8 | 26.3±13.9 | 16.2±12.6 |
| TYS (MPa) | 102 | 105 | 120 | 123 | 106 | 154 |
| UTS (MPa) | 341 | 331 | 343 | 343 | 342 | 329 |
| CYS (MPa) | 98 | 103 | 123 | 131 | 130 | 186 |
| CPS (MPa) | 374 | 374 | 382 | 390 | 378 | 428 |
| True strain | 0 | 0.54 | 0.83 | 1.16 | 1.39 | 1.83 |

**Table 4**. Tensile yield strength and thermal expansion coefficient estimated for AZ31 alloy after ARB in the rolling direction (L) and transversal direction (T).

| **Nr of passes** | **TYS(L) (MPa)** | **TYS(T) (MPa)** | **UTS(L) (MPa)** | **UTS(T) (MPa)** | **CTE(L) $\times 10^{-6}K^{-1}$** | **CTE(T) $\times 10^{-6}K^{-1}$** | **Equivalent strain** |
|---|---|---|---|---|---|---|---|
| **0** | 162.5 | 85.0 | 307.0 | 297.0 | 25.92 | 27.16 | 0 |
| **1** | 223.0 | 172.4 | 344.9 | 365.6 | 26.05 | 26.49 | 1.86 |
| **2** | 227.0 | 212.0 | 356.6 | 370.0 | 25.04 | 26.41 | 3.72 |

### *3.2 Thermal properties*

Thermal properties of magnesium alloys are especially important for their industrial applications. In the following, we would like briefly to summarize the effect of SPD processing, as mentioned in subsection 3.1, on the thermal conductivity of AX52 and AZ31 alloys. Table 1 lists the thermal conductivity of AX52 magnesium alloy samples after ECAP using the route A after 1, 2, 4, and 8 passes estimated at 20 and 250 °C. The thermal conductivity at a higher temperature is lower than that at room temperature.

Table 5: Thermal conductivity of AZ31 alloy submitted to accumulative roll bonding.

| **κ ($Wm^{-1}K^{-1}$)** | **20 °C** | **100 °C** | **200 °C** | **300 °C** |
|---|---|---|---|---|
| 0 pass | 76.0 | 85.7 | 92.4 | 99.5 |
| 1 pass | 80.1 | 89.4 | 96.9 | 102.9 |
| 2 passes | 87.1 | 96.8 | 105.5 | 112.0 |

The differences were found at all passes (at all grain sizes). The effect of temperature on the thermal conductivity of AZ31 specimens processed by accumulative roll bonding is summarized in Table 5.

Figure 1 shows the coefficient of thermal expansion (CTE) extrapolated to room temperature of AZ31 alloy after ECAP at different routes (A, BC, and C) as a function of the route number. The permanent strain estimated after the temperature cycle (heating to 400 °C and cooling to 20°C) is shown in Figure 2.

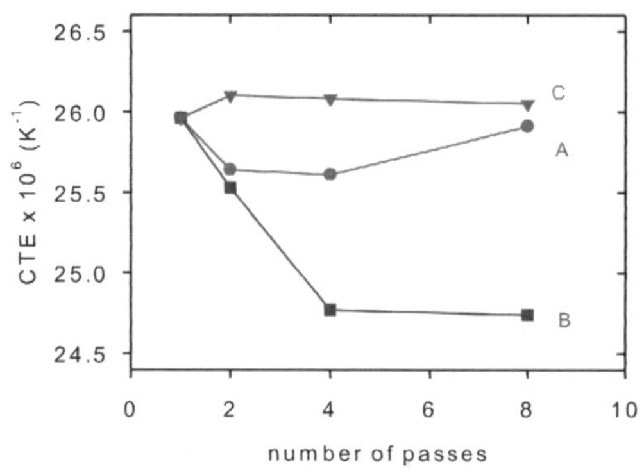


Figure 1. Thermal expansion coefficient of AZ31 alloy estimated for A, BC, and C processing routes.

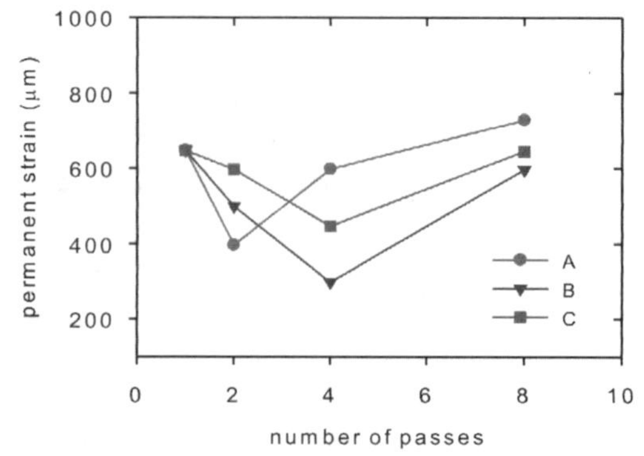


Figure 2. Permanent strain in AZ31 samples after heating to 400 °C and cooling to RT estimated for A, BC, and C processing routes.

Thermal expansion characteristics of magnesium alloys depend on the temperature range, chemical composition, and processing conditions.
The thermal conductivity of AZ31 sheets increases with increasing temperature. Similar results were found by Lee et al. [9]. The thermal conductivity consists of two components: an electron component and a lattice (phonon) component. It is important to stress that the electron component strongly depends on the crystal defects.

## 5 Discussion

The yield strength and deformation mechanisms of magnesium alloys are extremely sensitive to the microstructure (grain size, dislocation density, texture) achieved by severe plastic deformation. Magnesium materials submitted to severe plastic deformation exhibit the pronounced texture which depends on the SPD type. This texture development is due to the fact that the magnesium contains only two easy slip systems. To fulfill the von Mises criterion, five independent slip systems must cooperate to secure the compatible deformation [10]. Besides of basal slip, non-basal prismatic or pyramidal slip systems should be activated. The further possibility for compatible deformation is mechanical twinning, which is usually observed at the beginning of the plastic deformation at lower temperatures. The texture observed in ECAPed samples depends on the processing route. In the AZ31, three different textures were developed for A, BC, and C routes. In the samples extruded via A and $B_C$ the pyramidal planes $\{10\bar{1}1\}$ and $\{10\bar{1}2\}$ tend to be aligned nearly parallel to the extrusion direction (ED). Such orientation is favourable for motion of 1/3 $<11\bar{2}0>$ dislocations in the basal or prismatic planes. After the processing by route C, the c-axis is nearly oriented to ED than the deformation process starts with the twinning. Generally, it is possible to conclude that the strengthening effect of the ECAP and the TYS increase are saturated in the 4$^{th}$ pass and the changes estimated after the 8$^{th}$ pass were found to be marginal. Comparing the TYS estimated for cast AZ31 alloy (TYS(0)=57 MPa), we may conclude that the strengthening effect of the ECAP is significant already after the first pass; further passes increase the deformation stress only marginally even decrease. A significant increase of TYS was observed in the AX52 alloy after the ECAP using the route A. The highest effect of ECA extrusion was observed just after the first pass.

Observed texture in the AX52 alloy was mostly developed after the first pass via route A. The basal planes are parallel to the extrusion direction, while the prismatic planes rotate along the <c> axis. A part of the basal planes, oriented 45° to the ED, lay near the maximum shear stress plane. The observed

IOP Conf. Series: Materials Science and Engineering 1178 (2021) 012042 doi:10.1088/1757-899X/1178/1/012042

increase of the yield strength corresponds to a decrease in the grain size. It indicates that the main reinforcing mechanism is the Hall-Petch strengthening. The main thermally activated mechanism is very probably the motion of <a> dislocations in the prismatic planes [3]. Pronounced planar anisotropy was observed in the AZ31 sheets submitted to the ARB. The yield strength of L-samples where the tension axis was parallel to the ED was higher than that measured in T-samples with the axis perpendicular to the ED [7]. This anisotropy decreased with the increasing number of passes and increasing deformation temperature. The rolled sheets of the AZ31 alloy after the ARB exhibited a typical fibre texture with the splitting of the basal poles in the transversal direction. The plastic anisotropy consists in the different deformation mechanisms occurring during straining. While the deformation in L samples occurs mostly by the dislocation mechanism, the beginning of straining in the T samples was realised by the mechanical twinning. Formation of twins during the deformation of T samples was confirmed by the acoustic emission detection. The twinning activity decreases with decreasing grain size [11]. The tension – compression asymmetry was observed in the rotary swaged AZ31 samples. The reason for this asymmetry consists again in the different deformation mechanisms in tension and compression. The fibre texture in the swaged samples oriented basal planes parallel to the ED. Straining of samples in the <c> axis of the hexagonal magnesium cell is possible by the twinning mechanism. While in tension tests six twins' variants may be formed, in compression are only two.

Similarly, the thermal properties depend on the microstructure and defects in the crystal structure. The thermal conductivity of the AX52 alloy increased after the first ECAP pass and then again decreased up to the 4th pass, eventually again slightly increased in the 8$^{th}$ pass. The thermal conductivity of the AZ31 sheets after the ARB increased with the measurement temperature and the number of passes in the rolling mill. The thermal conductivity of metallic materials, $\kappa$, may be written as a sum of two components - the electronic part, $\kappa_e$, and phonon part, $\kappa_{ph}$:

$$\kappa=\kappa_e + \kappa_{ph} \ . \tag{3}$$

The thermal conductivity may be affected by various microstructural factors such as texture, grain size, solute atoms, precipitates, and dislocations [12]. The electron conductivity is proportional to the electrical conductivity, σ, via the Wiedemann-Franz law

$$\kappa_e = \sigma L T, \tag{4}$$

where the constant L is the Lorenz number. Thermal conductivity of alloys can be described with the Smith-Palmer equation [13]

$$\kappa = C\frac{LT}{\rho_e} + D\,, \tag{5}$$

where C and D are constants and $\rho_e$ the electrical resistance. The Smith-Palmer equation assumes a temperature-independent phonon thermal conductivity via the constant D. Bass measured electrical resistivity in magnesium single crystals [14]. He estimated that the electrical conductivity along the <c> axis is higher than that in the <a> direction i.e. $\kappa(c)/\kappa(a)>1$. The ECAP process oriented the basal planes in AX52 samples parallel to the ED and the channel walls. The conductivity is measured mostly in the <a> direction. Similar effect may be observed in the AZ31 alloy submitted to the ARB. In this case, the conductivity is measured perpendicular to the sheet surface and owing to the texture parallel to the <c> axis. Improved texture after the first and second rolling pass increases the conductivity. On the other hand, the conductivity decreases with increasing concentration of crystal defects (dislocations, point defects). Severe plastic deformation by the ECAP may increase the dislocation density and so simultaneously decrease the conductivity, which is also observed in the AX52 samples.

The planar anisotropy of the thermal expansion coefficient in the AZ31 sheets after the ARB is very probably connected with the elastic anisotropy of the hard and soft direction in the basal plane [15].

Note that the samples after the ECAP processing exhibited significant thermal stresses (see Figure 2). These stresses may be different for various processing routes. Shortening of the AZ31 samples (originally 20 mm in length) after the first pass achieved ~ 600 μm, then decreased, and again increased in the 8$^{th}$ pass. The permanent deformation is different for the samples prepared with diverse processing routes. Observed results are a consequence of the different texture developed in various processing routes and elastic anisotropy of the hexagonal magnesium cell.

## 6 Conclusions

Sever plastic deformation may substantially refine the microstructure of Mg alloys. On the other hand, the significant texture developed in materials depends on the processing technique and composition of alloys. This texture is reason for the plastic anisotropy, tension-compression asymmetry, and anisotropy of thermal properties. It is necessary to bear in mind these findings in various structural applications of magnesium alloys submitted to severe plastic deformation.

**Acknowledgements**. The result was achieved with the use of support from the Ministry of Industry and Trade of the Czech Republic in the form of institutional funding.

**References**

[1] Estrin Y and Vinogradov A 2013 *Acta Mater.* **61** 782
[2] Li S, Yang X, Hou J and Du W 2020 *J. Magnes. Alloys* **8** 78
[3] Trojanová Z, Halmešová K, Džugan J, Palček P, Minárik P. Lukáč P 2018 *Letters on Materials* **8** 278
[4] Trojanová Z, Drozd Z, Máthis K, Kövér M, Džugan J, Lukáč P, Halmešová K 2019 *Adv. Mater. Lett.* **10** 887
[5] Trojanová Z, Halmešová K, Drozd Z, Šíma V, Lukáč P, Džugan J, Minárik P 2018 *Crystals* **8** 278
[6] Trojanová Z, Halmešová K, Džugan J, Drozd Z, Minárik P, Lukáč P 2020 *Crystals 10* 497
[7] Trojanová Z, Džugan J, Halmešová K, Németh G, Lukáč P, Minárik P, Bohlen J *Materials* 2018, **11**(1) 73
[8] Trojanová Z, Drozd Z, Škraban T, Minárik P, Džugan J,Halmešová K, Németh G, Lukáč P, Chmelík F 2020 *Adv. Eng. Mater.* **22** 1900596
[9] Lee S, Ham H J, Kwon S Y, Kim S W, Suh C M 2013 *Int. J. Thermophys.* **34** 2343
[10] von Mises R 1928 *Z. Anwew. Math. Mech.* **8** 161
[11] Máthis K, Čapek J, Zdražilová Z, Trojanová Z *Mater. Sci. Eng.* A **2011** 528 5904
[12] Lumley RN, Polmear IJ, Groot H, Ferrier J *Scripta Mater.* **58**, 2008, 1006
[13] Uher C. Thermal conductivity of metals. In: Thermal Conductivity. Theory, Properties and Applications. Tritt, M. Ed. Kuwen Academic/Plenum Publishers, New York, 2004
[14] Bass J, Fisher KH Landolt-Börnstein database 1982 New Series III/15a Springer
[15] Drozd Z, Trojanová Z, Halmešová K, Džugan J, Lukáč, P, Minárik P *Acta Physica Polonica A* 2018 **134** 820

IOP Conf. Series: Materials Science and Engineering 1178 (2021) 012043 doi:10.1088/1757-899X/1178/1/012043

# Environmental impact on the life of a polymeric composite with polyamide matrix and glass fibres

**L. Markovičová[1], V. Zatkalíková[1], T. Kojnoková[1]**

[1] Department of Material Engineering, Faculty of Mechanical Engineering, University of Žilina, Univerzitná 8215/1, 01026 Žilina, Slovakia

lenka.markovicova@fstroj.uniza.sk

**Abstract.** Polymer construction materials in composite design achieve high strength, have high chemical and thermal resistance as well as resistance to UV radiation. Due to the fact that polymer composites are often used for a long time outdoors, they are exposed to the weather and living organisms. The polymer composite PA66 + GF30 was used in the experiment. The composite consists of a polyamide matrix (PA) and a short glass fiber. Samples of the composite were subjected to accelerated aging in a UV chamber under precisely defined conditions. After UV degradation, the samples were still exposed in wet soil. After exposure, the samples were evaluated for a change in weight and a change in absorption in boiling water. Changes in the surface topography of the samples were evaluated using a light microscope.

## 1. Introduction

Composites are extensively used as materials in making aircraft, electronic packaging to medical equipment, and space vehicle to home building [1]. Composites consist of two or more components, insoluble in each other, which are combined to form a useful engineering material possessing certain properties not found in the constituents [2]. Polymer composites generally consist of thermoplastic or thermoset matrix with organic (e.g., wood flour, chicken feather) or inorganic (mineral or glass materials) fillers (particulates or fibers) [3]. Thermoplastic polymers in composites usually consist of polyolefins such as polyethylene (PE), polypropylene (PP), polystyrene (PS), and polyvinyl chlorine (PVC), or of polyesters such as PLA and polyhydroxyalcanoates (PHA), of which PHB is the most common. Thermosets include acrylics, polyesters/vinyl esters, epoxies, polyurethanes, amines, furans, phenolics. Glass Fibre Reinforced Polymers (GFRPs) is a fibre reinforced polymer made of a plastic matrix reinforced by fine fibres of glass. Fibre glass is a lightweight, strong, and robust material used in different industries due to their excellent properties. Although strength properties are somewhat lower than carbon fibre and it is less stiff, the material is typically far less brittle, and the raw materials are much less expensive [4]. Its bulk strength and weight properties are very favorable when compared to metals, and it can be easily formed using moulding processes [5].

Polyamides have been selected due to the balance of physical and chemical properties against cost. Injection moulding, using mould flow, is well suited for mass production of components with complex geometry. Polyamide 66 (PA66) has interesting properties, including heat resistance, a high strength, toughness, and good wear resistance, and is therefore widely used for sliding parts in automobiles or industrial machines, such as gears and bearing retainers. Unreinforced PA66, which does not contain reinforcement fibers, is used for sliding parts. In addition, reinforcement fibers such as glass fibers (GFs) or carbon fibers (CFs) may be added to PA66 to increase its strength and stiffness, and thus meet the requirements for a downsizing of parts under severe conditions. To use such parts under a high-temperature environment, such as inside an engine compartment of an automobile, it is necessary to understand the temperature dependence of the tribological properties [6 - 8].

Polymers used for outdoor applications deteriorate due to the combined effects of photolysis and pursuant photo-oxidative reactions, more commonly known as UV degradation. The rate of degradation is dependent on exposure conditions such as sunlight intensity, temperature and relative humidity, as well as the structure of the polymer itself and associated impurities. UV degradation manifests itself as a change in the physical characteristics of the bulk polymer. Weathered polymers show yellowing, embrittlement, cracking and hazing. In order to minimize UV degradation, UV screeners are added to the bulk polymer during processing. In general, these protect by absorbing UV light themselves thereby reducing the amount of UV light that can be absorbed by the chromophores present in the polymer [9 - 11].

## 2. Experimental material and methods

A commercial polymer composite, polyamide 66 (trade name PA66 + GF30 or Nylon 6/6 glass fiber reinforced) was used as the experimental material. The matrix of the composites is made of polyamide 66 (PA66) and the reinforcement is made of chopped glass fibers (GF). The fiber length is 4 mm and the fiber diameter is 10 μm. The content of glass fibers in the matrix was 30 %.

Polyamide type PA66 with the addition of 30 % glass fibers is a material that is used for the production of abrasion-resistant and dimensionally stable components, pulleys, but also for the production of highly loaded structural elements, etc. Among other things, this material is used in the electrical, automotive and engineering industries. It excels in high strength, abrasion resistance, high dimensional stability and relatively low absorbency.

Degradation testing of polymeric materials is one of the most important tests in terms of life estimation. Aging tests can be performed under real conditions of polymer use or artificial conditions of accelerated aging are used, which provide faster results. The artificial aging test was performed on a Cofomegra Solarbox 1500e. The device guarantees accelerated artificial aging according to the specified parameters in accordance with the standard. The parameters that were set at the beginning of the test are given in table 1.

**Table 1** Cofomegra Solarbox parameters

| Exposure time (hrs.) | Dry phase (min.) | Wet phase (min.) | Black panel temperature (°C) | Radiation (W/m$^2$) |
|---|---|---|---|---|
| 500 | 40 | 5 | 65 | 250 |

The extent of damage to plastic by the action of a biologically active environment can be determined by visual evaluation of the appearance of the test body, a change in weight, a change in physical properties. These methods are suitable for materials with a smooth surface, porous samples such as foam products are unsuitable. Test specimens, which had been weighed in advance, were placed in a beaked and completely covered with soil, which was then flooded with distilled water. The exposure time was 2 weeks. At the end of this time, soil sample were taken, cleaned and weighed.

The water absorption test for plastics is carried out in cold (method A) and in boiling water (method C) (EN ISO 62). The method consists in determining the weight gain of test pieces immersed in cold and boiling water over a prescribed time. The test specimens should have the same shape and dimensions.

At least 3 immersion test bodies shall be tested the prescribed time into the water container. After the time has elapsed, they are removed from the water, dried and weighed to the nearest 1 mg. The amount of water absorbed by the test specimen in mg is calculated using method A ($X_1$) and C ($X_2$) according to formula 1:

$$X_{1,2} = m_2 - m_1 \quad (mg) \tag{1}$$

Amount of water absorbed by the test piece per unit of surface in mg. cm$^{-2}$ is calculated using method A ($X_5$) and method C ($X_6$) using formula 2:

$$X_{5,6} = \frac{m_2 - m_1}{A} \quad (mg.cm^{-2}) \tag{2}$$

The amount of water absorbed by the test specimen in% is calculated using method A ($X_9$) and method C ($X_{10}$), using formula 3 [12]:

$$X_{9,10} = \frac{m_2 - m_1}{m_1} \cdot 100 \quad (\%) \tag{3}$$

The topography of the samples was also evaluated on a light microscope.

## 3. Experimental results and discussion

All test specimens were weighed before the start of the experiment. The first set of samples was placed in a beaker with wet soil for 2 weeks. The second set of samples was first exposed UV radiation (500 hrs.) and then placed in a beaker and wet soil for 2 weeks. At the end of the exposure period, soil samples were taken, cleaned and prepared for weighing. The values of the resulting masses are recorded in table 2. The value of $m_1$ is the mass recorded before the exposure of the sample to the soil, the value of $m_2$ is the mass of the sample after its exposure in the soil and the last row of the table indicates the difference between these values ($\Delta m$).

**Table 2.** Change in weight of samples before and after UV degradation in wet soil

| | **PA6.6 + 30 GF** | **PA6.6 + 30 GF + 500 UV** |
|---|---|---|
| $m_1$ (g) | 1.9852 | 1.9314 |
| $m_2$ (g) | 2.0924 | 2.6011 |
| $\Delta m$ (g) | 0.0442 | 0.1745 |

Further testing was testing for absorbency and changes in absorbency due to UV radiation. The test specimens were weighed at the beginning of the absorbency test. Absorbency in the boiling water was realized. The exposure time in boiling water was 30 minutes, followed by cooling in cold water for 15 minutes and the samples were weighed again. The change in weight (absorbency) and the amount of water absorbed by the surface of the test sample were evaluated. The results of the measurements are given in table 3 and 4.

**Table 3.** Absorbency of samples before and after UV radiation

| | **PA6.6 + 30 GF** | **PA6.6 + 30 GF + 500 UV** |
|---|---|---|
| $m_1$ (g) | 2.4115 | 2.3976 |
| $m_2$ (g) | 2.4360 | 2.4430 |

| | | |
|---|---|---|
| Δm (g) | 0.0245 | 0.0454 |

**Table 4.** Amount of water absorbed by the surface of the test specimens

| | PA6.6 + 30 GF | PA6.6 + 30 GF + 500 UV |
|---|---|---|
| m (%) | 1.0160 | 1.8936 |

From the measured values, it can be said that for the PA6.6 + 30 GF composite, the change in weight after exposure in soil was also recorded in the evaluation of water absorption. After exposure of the samples to UV radiation, this change in weight was even more pronounced in both tests.
Changes in the surface of the samples due to UV radiation and microorganisms were also confirmed by observing the surface changes using a light microscope. The surface of the samples was observed in reflected light on an Axio Imager A1m microscope from Zeiss. In figure 1 we can see surface of the sample without degradation effect of the environment. The surface of the sample is homogeneous, see traces of injection and solidification of the material in the mold.

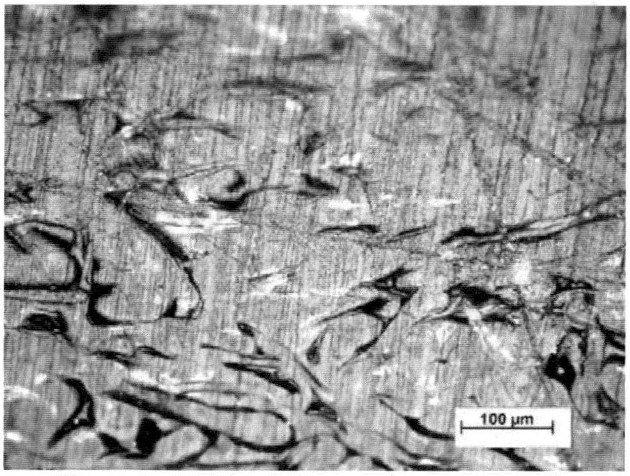

**Figure 1.** PA66 + GF30 - surface, 0 UV, 200x

Samples under the influence of UV radiation (250 hrs.) changed the character of the surface (figure 2). We can identify emerging cracks on the surface of the samples. Cracks are preferably formed at the locations of the protrusion that have formed during the process of injecting the samples into the mold.

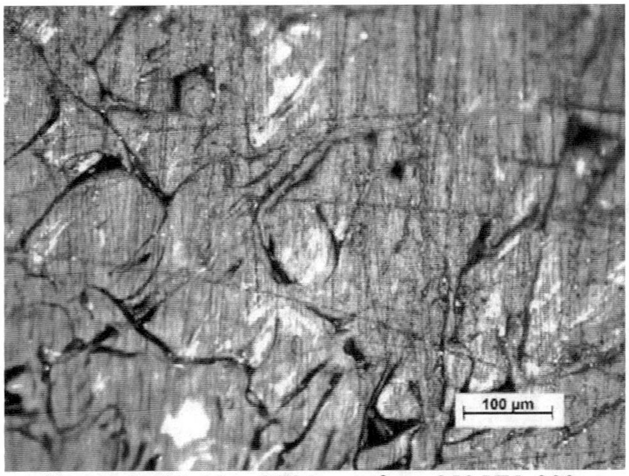


**Figure 2.** PA66 + GF30 - surface, 250 UV, 200x

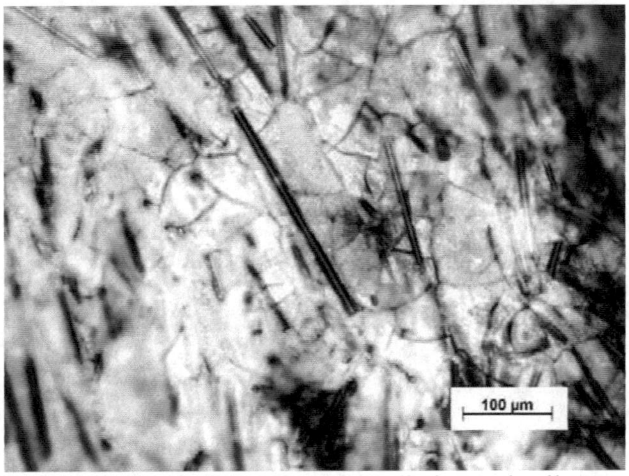


**Figure 3.** PA66 + GF30 - surface, after UV radiation and soil exposition, 200x

After exposure in the UV chamber, the test specimens were placed in wet soil for 2 weeks. These samples showed the most significant change in the character of the surface compared to the original samples (figure 3). The density of surface cracks increased. The individual cracks joined together to form a network for the failure of the samples. In figure 3 we can see that the character of the matrix has also changed.

## 4. Conclusions

Based on experiments performed on the polymer composite PA66 + GF30 after exposure to UV radiation and subsequently in wet soil, we can conclude:

- After exposure of the polymer composite in wet soil, we can state increasing in weight in both cases. In the case of samples without UV radiation, the weight gain was 5% compared to the original weight.
- In the case of samples that were first exposed to UV radiation and then exposure to soil, we observed a change in weight of 35%.
- In the boiling water absorbency test, absorbency was recorded for samples without UV radiation of 1% and for samples after UV radiation of 2%.

- Polyamides are characterized by absorbency, which depends on the concentration of amide groups. This property was confirmed by storing the samples in wet soil as well as in the absorbency test. At the same time, UV radiation caused changes in the topography of the surface, the formation of cracks. Cracks allow easier entry of the liquid into the matrix and thus changes in properties.

**Acknowledgement**
The research was supported partially by Scientific Grant Agency of Ministry of Education, Science and Sport of Slovak Republic and Slovak Academy of Science, grant KEGA No. 049ŽU-4/2017 and grant KEGA No. 013ŽU-4/2019.

**References**
[1] Shaw A, Sriramula S, Gosling P D, Chryssanthopoulo M K 2010 Composites Part B, 41, 446 – 453.
[2] ASTM. Standard Terminology Relating to Plastics, ASTM: West Conshohocken, PA, USA, 2012, p. D883
[3] Brebu M 2020 Environmental Degradation of Plastic Composites with Natural Fillers – A Review, Polymers 2020, 12(1), 166, https://doi.org/10.3390/polym12010166
[4] Jarukumjorn K, Suppakarn N 2009 Effect of glass fiber hybridization on properties of sisal fiber polypropylene composites, Compos: Part B, 40 (2009), pp. 623 - 627
[5] Dwivedi U K, Navin chand 2009 Influence of MA – g – PP on abrasive wear behaviour of chopped fiber reinforced polypropylene composites, J Mater Process Technol, 209 (2009), pp. 5371 - 5375
[6] Lates M T, Velicu R, Gavrila C C 2019 Temperature, pressure, and velocity influence on the tribological properties of PA66 and PA46 Polyamides, Materials, 12 (20) (2019), p. 3452
[7] Habrman M, Ráž K, Kalina T 2019 Effect of the cascade injection molding on the filling behaviour, Manufacturing Technology, volume 19, issue 6, 2019, pp. 936-940
[8] Chval Z, Raz K, Sedlacek F. 2019 Dimension Stability of Thin-Walled Parts from 3D Printed Composite Materials, Structural Integrity, vol. 5, 2019, pp. 91-92.
[9] Ray B C, Prusty R K, Rathore D K 2018 Fibrous polymeric composites: environmental degradation and demage, CRC Press, ISBN 978-1-4987-8401-6
[10] Ray B C, Rathore D 2014 Durability and integrity studies of environmentally conditioned interfaces in fibrous polymeric composites: Critical concepts and comments, Advances in Colloid and Interface Science, vol. 209, p. 68-83
[11] Mahato K K, Dutta K, Ray B C 2018 Static and Dynamic Behavior of Fibrous Polymeric Composite Materials at Different Environmental Conditions, Journal of Polymers and the Environment 26, p. 1024-1050
[12] Standards EN ISO 62 (640112) Plastics. Determination of water absorption in water

# Study of the deformation process in an equal-channel step matrix of a new design based on computer simulation

**A Naizabekov [1], E Panin [2], S Lezhnev [1] and I Volokitina [1]**

[1] Rudny Industrial Institute, Rudny, Kazakhstan; [2] Karaganda Industrial University, Temirtau, Kazakhstan

E-mail: naizabekov57@mail.ru

**Abstract**. The paper presents the results of computer simulation of the deformation process in equal-channel step matrix, the distinctive feature of which is the location of intermediate channel at two angles to the input and output channels. Models with values of the inclination angle in the longitudinal direction (angle α) of 135 degrees and inclination angles in the transverse direction (angle β) of 15, 25 and 35 degrees are considered. To study the stress-strain state, the following parameters were considered: equivalent strain (total strain intensity); equivalent stress (stress intensity), and the resulting deformation force. It was found that with an increase in the value of the angle β, all the characteristics of the stress-strain state increase in their absolute value. When the angle β increases from 0 to 35 degrees, the equivalent strain increases by an average of 22÷25%. The analysis of the forces showed that when the value of the angle β increases, the deformation force increases at both stages, the nature of the increase is non-monotonic. The increase in force occurs sequentially by 10, 12 and 23 kN, which indicates the exponential nature of increase in force depending on the value of the angle β. The maximum force value of 257 kN that occurs when using a matrix with angle of β=35° exceeds the standard value for β=0° by only 22%. Therefore, the best option for performing equal-channel step pressing with the flow of metal in 3 planes can be considered a matrix with angle α=135° and angle β=35°.

## 1 Introduction

Equal-channel angular pressing (ECAP) is one of the most well-known methods of severe plastic deformation, which allows obtaining bulk blanks with an ultrafine-grained (UFG) structure. This method has been known for more than 40 years [1] and is still being actively investigated [2-5]. Despite the fact that a number of new devices and technologies have already been developed [6-9], which are based on the principle of ECAP, the vast majority of these ideas have not been widely applied in industry. The reason for this is the disadvantages of ECAP inherent in most of the developed technologies: the relatively small size of the resulting blanks, the discreteness of the deformation process, and the need to implement a large number of deformation cycles to form the UFG structure. And if the first two disadvantages are successfully coped with by developing so-called "combined" processes [10-13], where the principle of continuous deformation of long samples operates, then the last disadvantage is quite difficult to overcome. Its essence lies in the fact that in order to obtain the UFG structure in the deformable material, it is necessary to ensure the implementation of high degrees of deformation for grain grinding. Based on this, to reduce the required number of deformation cycles, it is necessary to ensure an increase in the level of accumulated deformation in one pass. To do this, it is necessary to improve the design of the deforming tool.

Another effective way to increase the metal processing during ECAP is to turn the sample. In [14], this factor was thoroughly investigated: it was found that the most optimal method of edging is the BC route (90° edging along the longitudinal axis of the sample, i.e. changing the horizontal direction of deformation in the cross section to the vertical one).

The purpose of this work is to study the deformation process in an equal-channel step matrix of a new design, the distinctive feature of which is the location of the intermediate channel at two angles to the input and output channels, i.e. the deformation occurs not in the horizontal or vertical directions, but in both simultaneously (diagonal direction). In fact, the deformation in the matrix of this design is a conventional ECAP with partial edging of the sample. The diagonal movement of the metal in this matrix is shown in figure 1a.

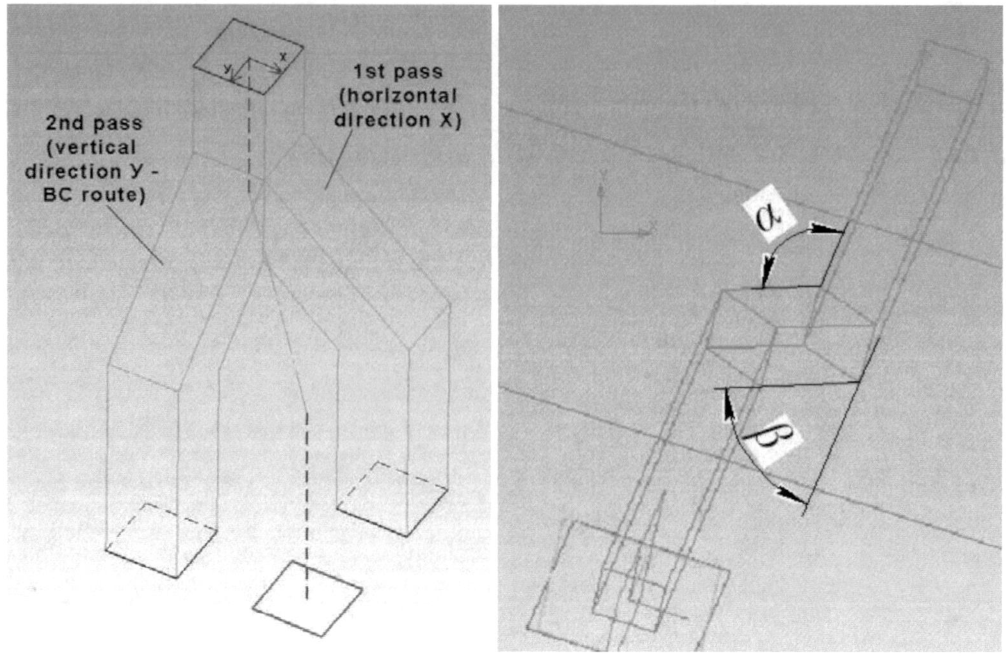


**Figure 1.** Scheme of the channels design in the matrix with a metal flow in 3 planes

## 2 Models for research

4 matrix models were constructed in the KOMPAS program to simulate equal-channel step pressing using matrices with a metal flow in 3 planes. The first matrix had a "classical" equal-channel step structure, in which the direction of metal flow changed only in 2 planes. The other 3 matrices had a similar design to the first one, with the only difference that in them the direction of metal flow is carried out not in two, but in three planes. The channel junction angle in the longitudinal direction (angle $\alpha$) in all matrices was assumed to be equal to 135 degrees, as the most optimal for equal-channel step pressing [15]. The angle of the channel junction in the transverse direction (angle $\beta$) in the first matrix was zero, in the other three matrices it was equal to 15, 25 and 35 degrees, respectively (figure 1b).

The initial blank had dimensions of 20 x 20 x 100 mm. A grid of finite elements evenly distributed over the entire volume of the workpiece was applied to this model. The average size of the final element was 1.3 mm, the number of nodes was 30699, and the number of elements was 96024. Steel AISI1035, heated to a temperature of 1100°C, was chosen as the material of the billet. The rheological properties of the material were taken from the Deform database. A non-isothermal type of calculation was chosen, i.e. in addition to the heat output to the tool, the workpiece also gave heat to the environment, the temperature of which was assumed to be 20°C. In an earlier study of equal-channel step pressing, it was found that the value of the friction coefficient in the matrix will depend on the required force required for the stable flow of the deformation process. To minimize its value, the value of the friction coefficient equal to 0.1 was taken.

To study the stress-strain state, the following parameters were considered: equivalent strain (strain intensity) and equivalent stress (stress intensity).

## 3 FEM simulation

The accumulation of equivalent strain occurs when the workpiece moves in the second and third channels of the matrix. In a matrix with an angle $\beta = 0°$, the distribution of equivalent strain, as in the first stage, is uneven (figure 2), zones of different strain values are clearly visible, the value of which varies in the range of 0.7÷1. As the value of the angle $\beta$ increases, the distribution of the equivalent strain along the length of the workpiece becomes more uniform, and its numerical value increases. In a matrix with an angle $\beta = 15°$, the value of the equivalent strain at the second stage is 0.75÷1.07; in a matrix with an angle $\beta = 25°$, it is 0.85÷1.1; in a matrix with an angle $\beta = 35°$, it is 0.92÷1.22.

IOP Conf. Series: Materials Science and Engineering 1178 (2021) 012044 doi:10.1088/1757-899X/1178/1/012044

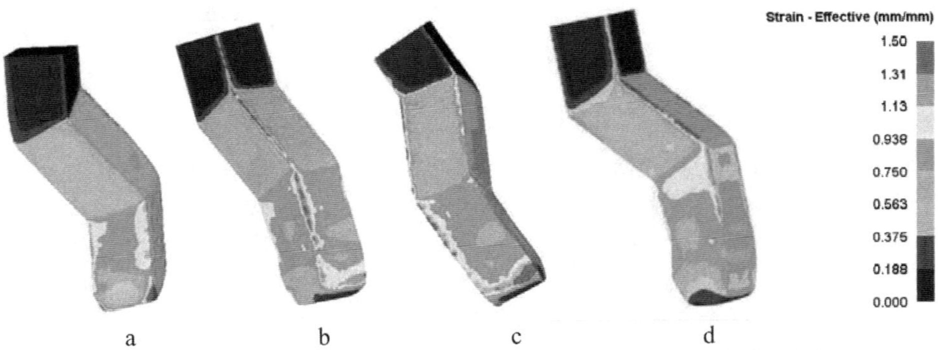


a - angle $\beta = 0°$; b - angle $\beta = 15°$; c - angle $\beta = 25°$; d - angle $\beta = 35°$

**Figure 2.** Equivalent strain distribution

The equivalent stress distribution is located in two distinct zones: in the junction zone of the first and second channels and in the junction zone of the second and third channels (figure 3). In a matrix with an angle $\beta = 0°$ in the docking zone, the value of the equivalent stress varies in the range of 180÷200 MPa. In the area of the channel junction, the value of the equivalent stress increases and is in the range of 420÷450 MPa. As the value of the angle β increases, the numerical value of the equivalent stress increases. In a matrix with an angle $\beta = 15°$, the value of the equivalent stress in the docking zone is 220÷250 MPa; in the area of the channel junction - 430÷470 MPa. In a matrix with an angle $\beta = 25°$, the value of the equivalent stress in the docking zone is 260÷320 MPa; in the channel junction zone-460÷490 MPa. In a matrix with an angle $\beta = 35°$, the value of the equivalent stress in the docking zone is 290÷340 MPa; in the channel junction zone-480÷500 MPa.

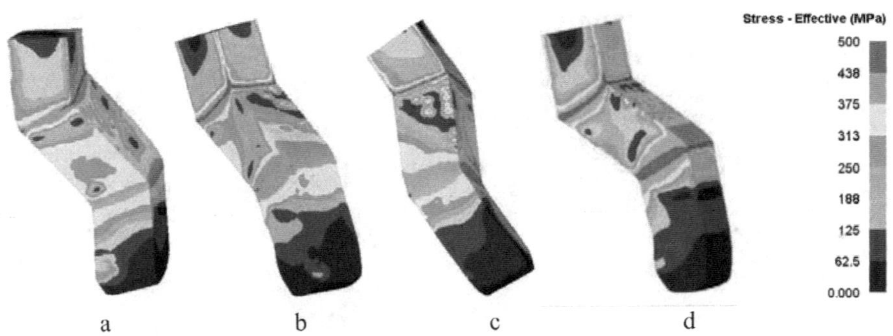


a - angle $\beta = 0°$; b - angle $\beta = 15°$; c - angle $\beta = 25°$; d - angle $\beta = 35°$

**Figure 3.** Equivalent stress distribution

To complete the study, it is also necessary to evaluate the resulting deformation force, since an excessively large value of this parameter entails large energy costs and can lead to equipment failure. Graphs of the resulting force were constructed, one of which is shown in figure 4, the force values for the first and second stages of deformation were recorded on them. For convenience, the effort values were presented as a summary chart (figure 5).

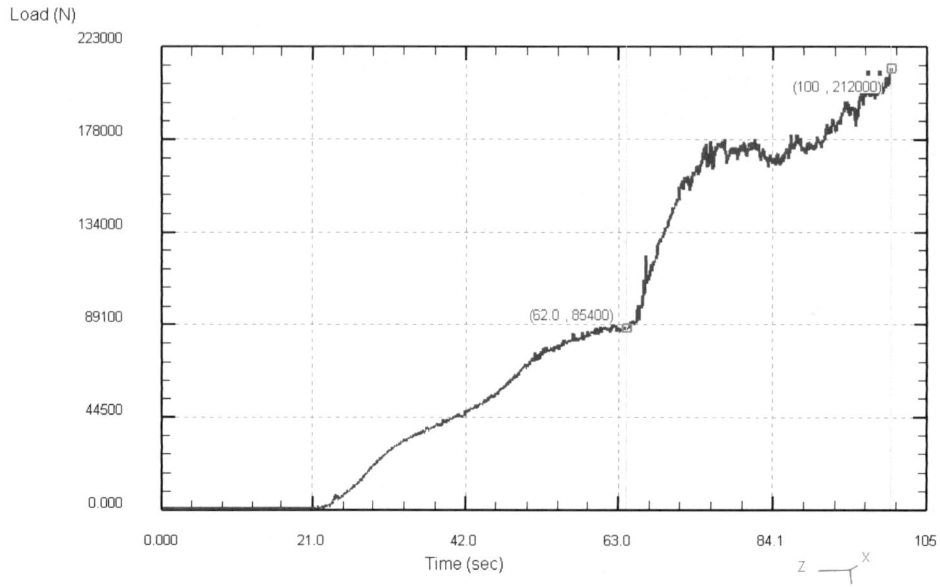


**Figure 4.** Deformation force in a matrix with an angle β=0°

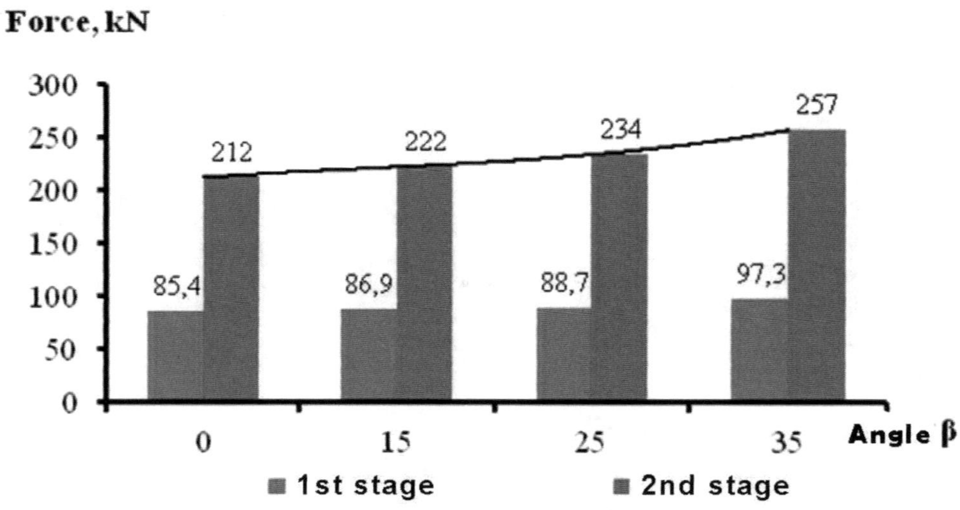


**Figure 5.** Summary diagram of deformation forces

## 4 Discussion of the obtained results

As can be seen from the data presented in figures 2-3, with an increase in the angle β, all the characteristics of the stress-strain state increase in their absolute value.

Due to the absence of compression in ECA matrices as such, the growth of equivalent strain in this process occurs solely due to the implementation of the shear scheme on inclined planes connecting adjacent channels. Therefore, the key factor determining the resulting level of strain in a single pass is the value of the angle (in our case, the angles) of the inclination of these planes to the input and output vertical channels. When the angle β increases from 0 to 35 degrees, the equivalent strain increases by an average of 22-25%.

Equivalent stresses have a similar distribution in both zones – with an increase in the angle β, this parameter increases. However, the change in the channel junction zones is smoother than in the docking zones. This is due to the fact that the channel junction zones, where shear deformations are actually implemented, have a rather small length compared to the lengths of the matrix channels. As a result, the stress concentration in these zones occurs in strictly defined intervals, which is not the case for docking zones, the length of which can reach up to half the length of the channels.

As can be seen from the diagram, when the value of the angle β increases, the deformation force increases at both stages. The increase in force occurs sequentially by 10, 12, and 23 kN, which indicates the exponential nature of the increase in force depending on the value of the angle β. The maximum force value of 257 kN that occurs when using a matrix with an angle β=35° exceeds the standard value for β=0° by only 22%. Therefore, a matrix with an angle α=135° and an angle β=35°can be considered the best option for performing equal-channel step pressing using metal in 3 planes.

Based on this, we can conclude that the most optimal deformation scheme occurs when the angle β increases, i.e. in a matrix with an angle β=35°.

## 5 Conclusion

The paper presents the results of computer simulation of the deformation process in an equal-channel step matrix, the distinctive feature of which is the location of the intermediate channel at two angles to the input and output channels. Models with values of the angle of inclination in the longitudinal direction (angle α) of 135 degrees and angles of inclination in the transverse direction (angle β) of 15, 25 and 35 degrees are considered. Analysis of the parameters of the stress-strain state and the resulting deformation force showed that the most optimal deformation scheme occurs when the angle β increases, i.e. in a matrix with an angle β=35°.

## References

[1] Segal V M 1977 Method of pressing. Patent USSR 575892.

[2] Shaeri M H, Salehi M T, Seyyedein S H, Abutalebi M R and Park J K 2014 Microstructure and mechanical properties of Al-7075 alloy processed by equal channel angular pressing combined with aging treatment *Materials & Design* **57** 250-7.

[3] Zhang X, Liu X, Wang J and Cheng Y 2016 Effect of route on tensile anisotropy in equal channel angular pressing *Materials Science and Engineering A* **676** 65-72.

[4] Wei W, Wang S L, Wei K X, Alexandrov I V, Du Q B and Hu J 2016 Microstructure and tensile properties of Cu-Al alloys processed by ECAP and rolling at cryogenic temperature *Journal of Alloys And Compounds* **678** 506-10.

[5] Mostaed E, Fabrizi A, Bonollo F and Vedani M 2015 Microstructural, texture, plastic anisotropy and superplasticity development of ZK60 alloy during equal channel angular extrusion processing *Metallurgia Italiana* **11-12** 5-12.

[6] Zhao X, Chen N and Zhao N 2012 Numerical Simulation of equal channel angular pressing for multi-pass in different routes *Applied Mechanics and Materials* **268-270** 373-7.

[7] Gzyl M, Rosochowski A, Yakushina E, Wood P and Olejnik L 2013 Route Effects in I-ECAP of AZ31B Magnesium Alloy *Key Engineering Materials* **554-557** 876-84.

[8] Krivtsova O, Talmazan V, Arbuz A and Sivyakova G 2014 Study the process of equal-channel angular pressing with quasi-ultra-small angles of joint channels using computer modeling in program complex DEFORM *Advanced Materials Research* **1030-1032** 1337-41.

[9] Tavakkoli V, Afrasiab M, Faraji G and Mashhadi M M 2015 Severe mechanical anisotropy of high-strength ultrafine grained Cu–Zn tubes processed by parallel tubular channel angular pressing (PTCAP) *Materials Science and Engineering A* **625** 50–5.

[10] Raab G I, Fakhretdinova E I, Valiev R Z, Trifonenkov L P and Frolov V F 2016 Computer study of the effect of tooling geometry on deformation parameters in the plastic shaping of aluminum wire rod by Multi-ECAP-Conform *Metallurgist* **59** 1007-14.

[11] Chembarisova R G and Aleksandrov I V 2016 Simulation of the Elastoplastic Behavior of Grade-4 Ti in the ECAP-C Process *Metal Science and Heat Treatment* **58** 236-44.

[12] Korchunov A G, Chukin M V, Polyakova M A and Emaleeva D G 2011 Principles of designing a continuous method for producing steel wire with an ultra-fine-grained structure *Bulletin of Magnitogorsk state technical University named after G. I. Nosov* **1** 43-6.

[13] Fastykovskiy A R 2010 Experimental study of the rolling-pressing process *Forging and stamping production* **11** 11-4.

[14] Valiev R Z and Langdon T G 2006 Principles of equal-channel angular pressing as a processing tool for grain refinement *Progress in Materials Science* **51** 881–981.

[15] Naizabekov A B, Ashkeev Zh A, Lezhnev S N and Toleuova A R 2005 Investigation of the process of deformation of billets in an equal-channel step matrix *Izv. vuzov. Ferrous metallurgy* **2** 16-8.

IOP Conf. Series: Materials Science and Engineering 1178 (2021) 012045 doi:10.1088/1757-899X/1178/1/012045

# Analysis of forming capacity of HCT490X and DX57D depending on strain rate

**V Novák[1], M Valeš[1], F Tatíček[1], J Šanovec[1], L Chrášťanský[1]**

[1]Faculty of Mechanical Engineering, Czech Technical University in Prague

E-mail: Vit.Novak@fs.cvut.cz

**Abstract**. To evaluate the forming capacity in sheet metal forming, experimentally determined Forming Limit Curves (FLC) are often used. These curves represent the forming limits of a material. The most widely used approach to experimentally determine the onset of necking of sheet metals is formulated in EN ISO 12004-2. The first goal of this work is to investigate the limit strains for dual-phase steel HCT490X (commonly referred to as DP500) with a nominal thickness of 0.6 mm and deep drawing steel DX57D with a nominal thickness of 0.7 mm. The second goal is to evaluate the influence of punch speed (resulting in strain rate) on the position of forming limit curve for materials HCT490X and DX57D. The FLC curve is measured with the Nakajima test on a universal testing machine BUP 600. The test setup and the shape of the specimens follow EN ISO 12004-2. The stochastic, isotropic, and high contrast pattern was applied on the specimens and was measured using a 3D photogrammetric system ARAMIS by GOM. The punch movement velocity was gradually changed, and the experimentally measured results are compared with numerical simulation in AutoForm Forming R8.

## 1 Introduction

The goal of the article is to observe the behavior of the materials at different strain rates and thus determine what effect the strain rate has on the formability of the material. To do this, it is necessary to construct FLC curves for different punch speeds. It is necessary to deal with this issue in order to refine numerical simulations. At present, numerical simulations solve the problem of strain rate only in relation to the strain hardening curve (uniaxial deformation) and not in the whole range of deformations from uniaxial deformation to biaxial deformation. In sheet metal forming, formability is an important property. A material can undergo plastic deformation to a given shape without defects. Many factors contribute to formability, which can be divided into four basic groups. These are the conditions of physical metallurgy of steel, thermomechanical properties, stress state, and technological conditions of processing. One of these factors is the strain rate, which can be expressed as the change in strain (deformation) over the change in time. Thus, the strain rate is not the same as the work tool speed, but there is a link between them. Various methods are used to evaluate formability. One of them is the estimation of the stamping characteristics of sheet metal materials using the Forming Limit Diagram (FLD), which is divided into two regions separated by the Forming Limit Curve (FLC). The FLC curve is part of the forming limit diagram (FLD). The FLC curve contains information about the major and minor strain of a product, thus describing the formability from uniaxial deformation to biaxial deformation. The region below the FLC corresponds to a safe strain state and the region above represents the failure strain state. Factors influencing the position of the FLC curve include the thickness of sheets, the strain hardening, the anisotropy of the sheet metals, and yield strength. [1 – 5]

## 2 Material and experimental procedures

The FLC curve is measured with the Nakajima tests were performed on a Zwick/Roell BUP 600 testing machine, see **Figure 1.C** It is a device for determining the formability of sheet metal with a maximum test load of 600 kN. The flow of deformation was recorded using two DIC cameras and evaluated using the ARAMIS Professional 2019 software. During the experiment, the BUP 600 machine was specifically equipped with a sensor ARAMIS 4M, which works on the digital image correlation principle. ARAMIS Professional 2019 software was used to evaluate the measurements. [8] The parameters for the Nakajima test are prescribed by the ISO 12004-2:2008. The standardized punch speed is 1.5±0.5 mm/s and the test temperature is set at 23±5 ° C. During the test, it is necessary to ensure an almost linear flow of deformation. This can be achieved by a suitable clamping force that keeps the specimen in its place. In order to determine the strain rate effect on the FLC, the test was performed at a punch speed of 2 mm/s, which corresponds to the standard, and at speeds of 10 mm/s, and 15 mm/s for material DX57D and speeds of 2 mm/s and 12 mm/s for material HCT490X. The clamping force was set at 300 kN, which should prevent the material from being pulled in. Another important parameter was the setting of the frame rate of the two DIC cameras. The faster the punch speed, the higher the frame rate was needed to sufficiently record the test. For a punch speed of 2 mm/s, a frame rate of at least 10 Hz was required, for 10 mm/s, it was 35 Hz, for 12 mm/s it was 40Hz, and for 15 mm/s at least 50 Hz. [5, 8] It is essential to use all measured specimens that must have the same tribological system (same amount of lubricant) to ensure the same initial conditions for all performed samples. This regulation was done by a three-layer system of PTFE foil, PE foil, and lubricant.

The materials used in this study were DX57D and HCT490X steel. DX57D is a mild, galvanized steel grade for cold forming with extreme deep drawing quality designed even for the most demanding deep drawing operations. HCT490X is a hot-dip galvanized (5 – 12 μm) dual-phase steel. The structure consists of a hard martensitic phase dispersed in the form of islands and a ductile ferritic matrix. [7 - 9] The mechanical properties of the steels can be seen in **Table 1** and the chemical composition in **Table 2**.

**Table 1.** Mechanical properties of HCT490X and DX57D steels according to ČSN EN 10346 [6]

| Material designation | Proof strength (MPa) | Tensile strength (MPa) | Elongation after fracture (wt%) | Plastic strain ratio (-) | Strain hardening exponent (-) | Bake hardening (MPa) |
|---|---|---|---|---|---|---|
| | $R_{p0,2}$ | $R_m$ | $A_8$ | $r_{m/20}$ | $n_{10\text{-}20/Ag}$ | BH |
| **HCT490X** | 290 - 380 | ≥ 490 | ≥ 24 | ≈*1.0* | ≥ 0.15 | ≥ 30 |
| **DX57D** | 110-170 | 260-330 | ≥ 41 | ≥ 1.8 | ≥ 0.22 | - |

**Table 2.** Chemical composition of HCT490X and DX57D steels according to ČSN EN 10346 [6]

| Element (wt. %) | C | Si | Mn | P | S | Al | Ti | Cr, Mo |
|---|---|---|---|---|---|---|---|---|
| **HCT490X** | ≤ 0.14 | ≤ 0.75 | ≤ 2.00 | ≤ 0.08 | ≤ 0.015 | 0.015-1.0 | ≤ 0.15 | - |
| **DX57D** | ≤ 0.02 | ≤ 0.50 | ≤ 0.30 | ≤ 0.02 | ≤ 0.02 | ≥ 0.01 | ≤ 0.30 | ≤ 1.0 |

To cover the whole FLC curve, five geometries with a diameter of 200 mm were used. These needed to be designed so that there were two geometries of specimens in the left part (uniaxial deformation) of the graph, one in the area of zero minor deformation and two geometries in the right part (Biaxial deformation) of the graph. [1, 2, 5, 8]

The remaining blank width of the specimen used for material DX57D (punch speed 2, 10, and 15 mm/s) in the experiment was G1= 20 mm, G2= 65 mm, G3= 105 mm, G4= 120 mm, and G5= 140 mm and for material HCT490X (punch speed 2 and 12 mm/s) was 20, 60, 100, 125 and 150mm see **Figure 1.B**. For FLC created under standard conditions (2 mm/s), three additional geometries characterized by the remaining blank widths of 50, 80, and 110 mm were used to better describe the curve for material DX57D. For both materials, the shaft length was 35 mm, and the radius was set at 20 mm, see **Figure 1. B**. The specimens were made with DX57D of a thickness of 0.7 mm and for material HCT490X thickness of 0.6 mm.

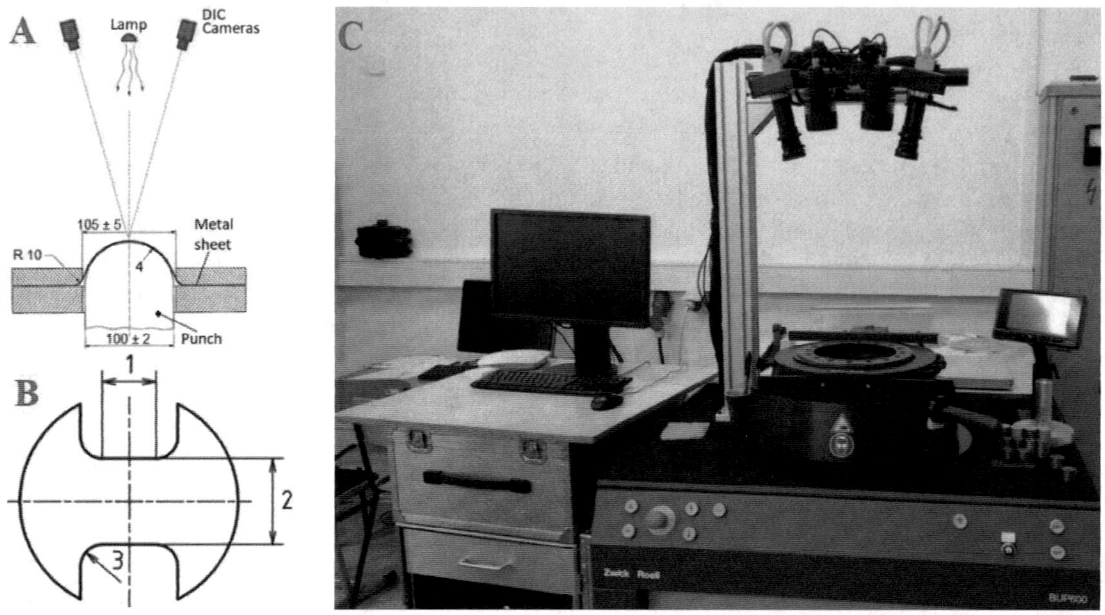


**Figure 1. A, C)** Scheme of the Nakajima and Universal metal testing machine BUP600 with ARAMIS system, **B)** dimensions of the test sample (1=shaft length, 2=remaining blank width, 3=radius, 4=lubrication layer) [8]

For each point on the Forming Limit Curve (FLC), three valid measurements were performed, which are prescribed by the ISO 12004-2: 2008 standard. [5] The values obtained for a certain geometry and the punch speed were then averaged to create the final point in the FLC. By repeating the test for individual geometries and velocities, it was possible to obtain the resulting FLC curves for materials HCT490X and DX57D, see **Figure 2** and **Figure 2** showing that the higher value of the major strain, the more sensitive the material is to the strain rate. The effect of strain rate on the FLC position is shown in **Table 3**. The values of the strain rate obtained for a certain geometry and the punch speed were averaged from three values in the area of the crack, see **Table 3** and **Figure 3.** Problems with measuring the specific pattern at high strain rates: with increasing strain rate, it is necessary to increase the frequency of image capture during observation, but with this is the need to increase the intensity of exposure. This requirement can be reduced to some extent if the entire area of the test specimen is not monitored, but only a certain area where failure is expected to occur is observed. The actual observations also affected the cohesion of the network (with the increasing rate of deformation, the importance of this problem increased), so for some samples (material HCT490X), the observation was limited to the speed of movement of the tool to 12 mm / s. The second sample (material DX57D) was more favorable for observations and allowed observations to be made at higher punch speeds.

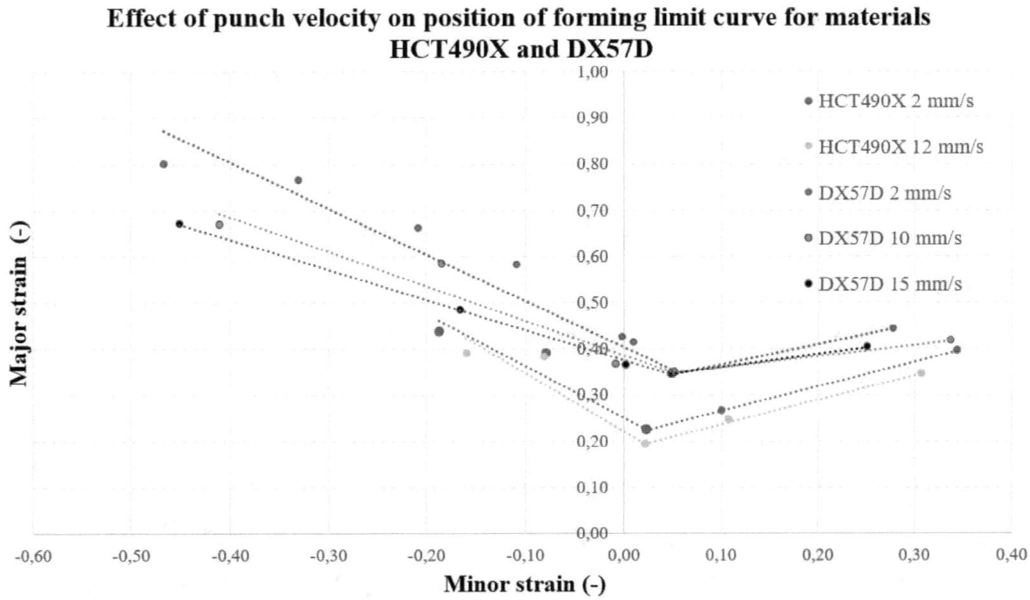


**Figure 2.** Effect of punch speed on the position of FLC for materials HCT490X and DX57D

**Table 3.** Influence of punch velocity on strain rate for material DX57D

| Punch velocity (mm/s) | Major strain rate for different geometries ($s^{-1}$) | | | | |
|---|---|---|---|---|---|
| | G1 | G2 | G3 | G4 | G5 |
| **2** | 0.40 | 0.354 | 0.34 | 0.56 | 0.30 |
| **10** | 2.23 | 1.79 | 1.57 | 2.13 | 1.85 |
| **15** | 3.43 | 1.99 | 1.83 | 2.40 | 2.16 |

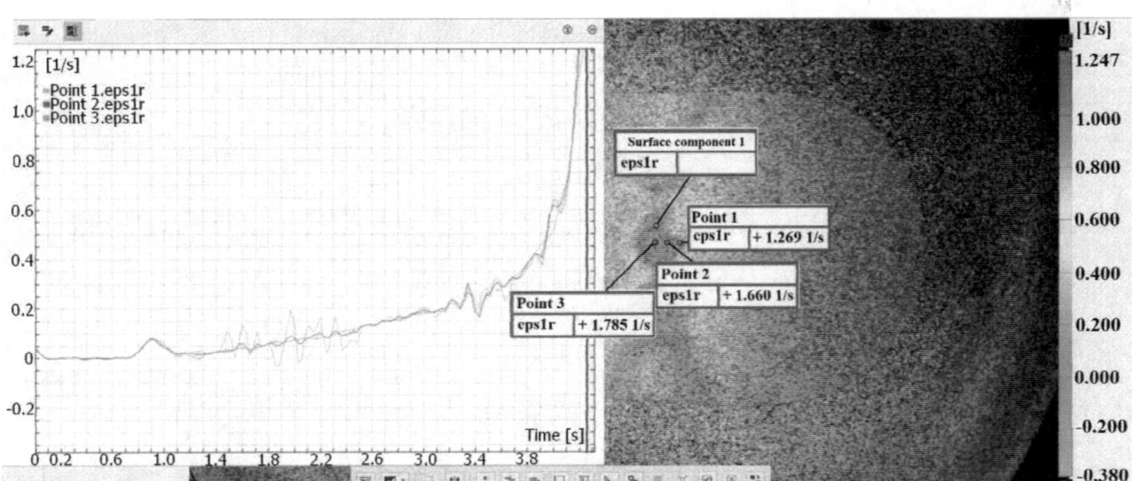


**Figure 3.** Evaluation of the influence of the strain rate in real time for the punch speed 10 mm/s and the DX57D Geometry 3

## 3 Influence of strain rate on the result of numerical simulation

The next step to verify that the strain rate is one of the parameters affecting the FLC position was to verify this assumption in a numerical simulation, specifically in the AutoForm R8 software. For individual pressed parts, their final height was measured (at the moment of crack initiation), and this height of the pressed part was entered into the numerical simulation. To perform numerical simulations, it was first necessary to create tools, ie punch, die and blank holder, see **Figure 4.A** As a representative for this simulation, a specimen with a remaining blank width of G3 = 105 mm. The simulation was created only for material DX57D. The next step was the choice of process parameters, ie initial conditions. The friction setting was constant in this part of the analysis, it was always the friction against the sheet, which was 0.5 for the die, 0.03 for the punch, and 0.5 for the blank holder. Defining the formed material was also a necessary step to perform the simulation. Different groups of materials are available in the software, as well as different material models for the same materials. [3 – 4]

To fully cover this study, performed simulations for multiple material models. able to observe not only their behavior when changing the speed of the test, but also the differences between the individual models. In **Table 4**. There are average values from the points subtracted in the area of the peak of the specimen. You can see an example of numerical simulations in **Figure 4.B**

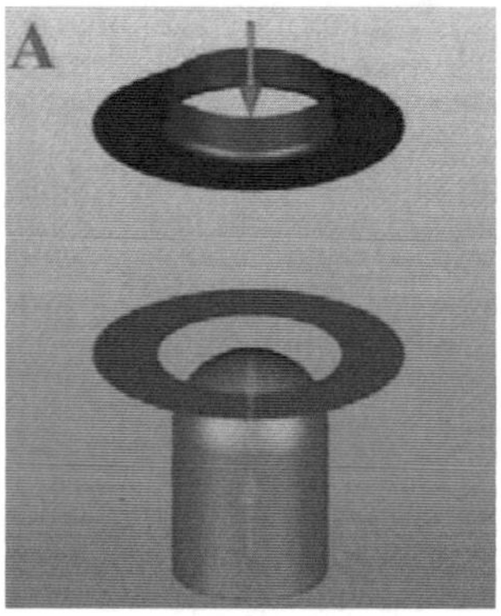


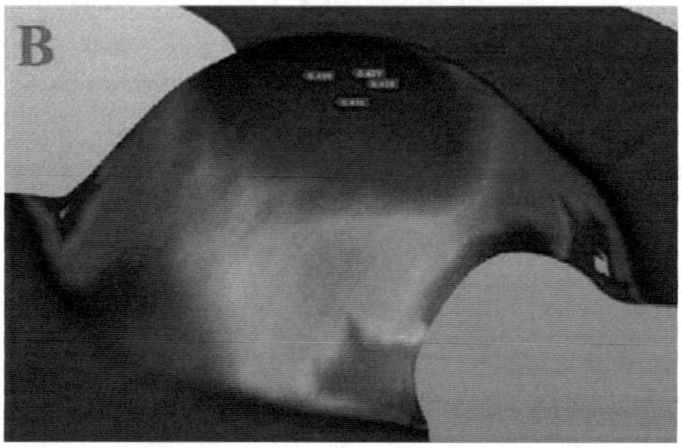


**Figure 4. B)** Numerical simulation in AutoForm R8 software for punch speed 200 mm/s for material model DX57 Tata Steel Strain rate Sensitive

**Table 4.** Influence of punch speed in numerical simulation for G3 and material DX57D Tata Steel Strain rate Sensitive

| Material card | Punch speed | | | |
|---|---|---|---|---|
| | 1 mm/s | 200 mm/s | 1 mm/s | 200 mm/s |
| | Major strain (-) | | Minor strain (-) | |
| **DX57 Tata Steel Strain rate Sensitive** | 0.467 | 0.430 | - 0.014 | - 0.016 |

According to the results of the numerical simulation, it seems that in most cases the change in the strain rate does not have a significant effect on the amount of strain. The only material model with a significant decrease in major deformation is the DX57 Tata Steel Strain rate sensitive. It is also the only model that has hardening curves for different strain rates listed in the material card. An analysis of the behavior of various material models of material quality DX57D, which are part of the material database, was performed. The initial results of the numerical simulation did not show differences in the magnitude of the achieved deformation of the pressed part when changing the punch speed speed, therefore a significantly larger speed difference was intentionally chosen than in the real experiment, punch speed was chosen 1 and 200 mm/s. Furthermore, it turned out that not all material models show sensitivity to the strain rate. In this respect, the numerical simulation was focused on the material DX57 Tata Steel Strain rate Sensitive.

## 4 Conclusion

This article deals with the issue of the influence of the strain rate on the formability of sheets expressed by the Forming Limit Curves. The introduction provides basic information about this problem. Furthermore, the principle of determining FLC using the Nakajima test and ARAMIS system is described, as stated in the standard ISO 12004-2:2008. The tested materials were DX57D steel and HCT490X dual-phase steel. The material HCT490X has a smaller supply of plasticity than DX57D. An answer was found to the question of the difference in the amount of deformation between numerical simulation and a real experiment. It has been shown that in the case of the strain rate, the factor must be approached as a value which also depends on the pressure of the holder, the surface roughness of the tools and the blank and the speed of mutual movement of the pressed part relative to the punch. Another problem was the absence of a failure criterion in numerical simulations, which can be, for example, the determined value of limit deformations (FLC). For this reason, it is not possible to use it without this criterion to determine the moment of crack formation. The results showed that with increasing punch speed, the FLC curve shifts to lower values of the major strain, which means a lower supply of plasticity at a higher strain rate. This trend was most visible in the area of negative minor strain (uniaxial deformation), and with the decrease of major strain, the difference between the individual velocities decreased. It was also possible to observe the heating of the specimens and larger changes in the lubricant system at a higher strain rate. Along with the fact that the coefficient of friction varies depending on speed and pressure, these factors may also have affected the result.

**Acknowledgment**

The research was financed by the project SGS19/163/OHK2/3T/12 "Research, optimization and innovation of production processes".

**References**

[1] Diehl A, Vierzigmann U and Engel U 2009 Characterisation of the mechanical behaviour and the forming limits of metal foils using a pneumatic bulge test. *International Journal of Material Forming* pp. 605-608 DOI: 10.1007/s12289-009-0608-4. ISSN 1960-6206.

[2] Kim S B, Huh H, Bok H H and Moon M B, Forming limit diagram of auto-body steel sheets for high-speed sheet metal forming. *Journal of Materials Processing Technology* pp. 851-862

[3] Tomáš M, Evin E, Kepič J and Hudák 2019 Physical modelling and numerical simulation of the deep drawing process of a box-shaped product focused on material limits determination. *Metals* Volume 9, Issue 10, Article number 1058

[4] Sobotka J, Solfronk P, Kolnerova M and Korecek D 2019 nput data acquisition possibilities for numerical simulation of drawing process by means of the contact-less optical system and thermo-camera *Manufacturing Technology* pp. 144-149

[5] EN ISO 12004-2: 2008 Metallic materials, Sheet and strip, Determination of forming-limit curves, 2008

[6] ČSN EN 10346:2016 Continuously hot-dip coated steel flat products for cold forming - Technical delivery conditions, 2016

[7] Valeš M, Novák V, Tatíček F, Šanovec J and Chrášťanský L 2019 Outer car body panels made of dual-phase steel. *Proceedings of the 28th Conference on Metallurgy and Materials* pp. 507-512

[8] Novák V, Valeš M, Tatíček F, Šanovec J and Chrášťanský L 2019 The effect of strain rate on position of forming limit curve. *Proceedings of the 28th Conference on Metallurgy and Materials* pp. 450-454

[9] Valeš M, Chrášťanský L, Tatíček F and Pačák T 2018 Comparison of numerical simulation and deep drawing test of DP500 steel. Proceedings of the 27th Conference on Metallurgy and Materials pp. 501-506

# Direct comparison between Co-28Cr-6Mo alloy prepared by Selective Laser Melting and traditional investment casting

**M Roudnicka[1,2], J Bigas[1], V Sreibr[3], D Palousek[3] and D Vojtech[1]**

[1]Department of Metals and Corrosion Engineering, University of Chemistry and technology Prague, Technicka 5, 166 28 Prague 6, Czech Republic

[2]FZU – Institute of Physics of the Czech Academy of Sciences, Na Slovance 1999/2, 182 21 Prague 8, Czech Republic

[3]Institute of Machine and Industrial Design, Brno University of Technology, Technicka 2896/2, 616 69 Brno, Czech Republic

E-mail: michaela.roudnicka@vscht.cz

**Abstract**. Thanks to high mechanical properties stable at elevated temperatures, high abrasion resistance, suitable corrosion resistance and biocompatibility, ternary cobalt alloy Co-28Cr-6Mo is a widely used material in the field of aviation industry and medical implants production. Due to the difficult formability and machinability of this alloy in combination with complex shapes of produced parts, investment casting is the mostly employed production process. However, additive manufacturing provides new processing opportunities. E.g. Selective Laser Melting (SLM) technology fusing metallic powders by precisely focused and controlled laser beam enables the production of highly complex parts with high precision. In this contribution, we bring a direct comparison between the microstructures and mechanical properties of Co-28Cr-6Mo alloy prepared by SLM and investment casting. Due to significant differences in microstructure fineness and phase composition, SLM is revealed to increase not only material strength and hardness but concurrently plasticity. As a result of rapid solidification during SLM, very fine cellular microstructure with Mo microsegregation at cell walls is formed without the occurrence of any secondary phases. The distinctive microstructure of the additively manufactured alloy can be expected to be the cause of many other variances in the material behaviour.

## 1 Introduction

Due to bad machinability, formability and weldability, cobalt-chrome alloys are mostly processed by casting. From the time and economic point of view, casting is the best solution, especially in case of investment casting which yields products of final shapes with a good surface condition. Only minor finishing operations are thus necessary. However, lower material performance is usually obtained by casting compared to different production techniques, such as forming or powder metallurgy, which limits some demanding applications [1].

New opportunities have been brought by the implementation of additive manufacturing (AM) into the processing of cobalt-chrome alloys. Powder-based techniques of AM, such as Selective Laser Melting (SLM) or Electron Beam Melting (EBM), enable direct production of complex-shaped parts with high precision thanks to the successive processing of small amounts of fine powders by a

precisely focused energy beam. In addition, they offer other benefits highly valued by high-performance applications in the fields as aviation industry or medical implants production.
E.g. the possibility of topology optimization and introduction of porous structures may significantly reduce the weight of aircraft parts and thus reduce fuel consumption and operating costs or bring better fixation and thus longevity of bone implants [2, 3].
The ternary Co-28Cr-6Mo alloy is one of cobalt-chrome alloys frequently applied for such applications, e.g. knee and hip arthroplasty or gas turbine parts [4]. Therefore, this paper is focused on its processing by AM, specifically by SLM, and comparison with parts produced conventionally by investment casting. Although several papers have been published on the SLM of this alloy, only a few show directly the comparison between completely different microstructures and the effect of such difference on mechanical performance [5]. Understanding the difference is a prerequisite for extensive testing of many other material manifestations that can crucially influence the confidence of potential part operation. More background data are thus desirable.

## 2 Materials and Methods

### *2.1 Studied materials*

The material studied in this work is a Co-28Cr-6Mo alloy conforming to the ASTM F75 standard. To provide a comparison between the microstructure and mechanical properties of this material prepared conventionally by investment casting or prepared by additive manufacturing (AM), commercially available materials have been used. The material produced by investment casting was provided by Prospon s.r.o. company; the exact production parameters are thus not available. In case of AM, a commercially available gas-atomized powder of Co-28Cr-6Mo alloy with a particle size of 10–45 μm was purchased from SLM Solutions and was further processed. The chemical composition given by the producers which conforms to the ASTM F75 standard is given in table 1 together with the exact composition determined by X-ray fluorescence (Axios FAST). The composition of both materials meets the standard. The cast material contains little more admixtures than the AM material.

**Table 1.** Chemical composition (wt.%) of the Co-28Cr-6Mo alloy.

| | Co | Cr | Mo | Si | Mn | Fe | Ni |
|---|---|---|---|---|---|---|---|
| **ASTM F75** | bal. | 27.0–30.0 | 5.0–7.0 | max. 1.0 | max. 1.0 | max. 0.75 | max. 0.5 |
| **cast** | bal. | 29.4±0.3 | 5.9±0.1 | 0.82±0.03 | 0.69±0.04 | 0.18±0.03 | 0.25±0.04 |
| **SLM** | bal. | 28.3±0.30 | 6.3±0.1 | 0.41±0.02 | 0.34±0.03 | 0.07±0.02 | 0.17±0.03 |

Quite surprising founding was provided by the combustion analysis (Bruker G4 Icarus HF) for the determination of C, S, N and O content. The content of these impurities is given in table 2 together with the ASTM F75 prescription. The standard allows relatively high content of C which serves as the main precipitation-strengthening element. However, compared to the cast material, the AM material is almost free of C and S as well. That indicates the impossibility of carbide formation and suggests a different material response to heat treatment. In case of gaseous elements, N and O, the rapidly solidified AM material is more prone to their capture.

**Table 2.** Impurity content (ppm) in the Co-28Cr-6Mo alloy.

| | C | S | N | O |
|---|---|---|---|---|
| **ASTM F75** | max. 3500 | max. 100 | max. 2500 | – |
| **cast** | 2300±130 | 38±1 | 245±5 | 62±3 |
| **SLM** | 70±40 | 7±2 | 639±4 | 345±57 |

Additive manufacturing via the SLM technology was carried out on an SLM Solutions 280 HL machine with the process parameters set-up given in table 3. Testing samples directly in the shape of round tensile test specimens (figure 1) were built in the vertical orientation using a stripe scanning strategy. The working chamber was supplied with a nitrogen atmosphere to prevent oxidation. The building plate was preheated to 100 °C.

**Table 3.** SLM process parameters.

| **Laser power (W)** | **Scanning speed (mm s$^{-1}$)** | **Hatching distance (mm)** | **Layer thickness (mm)** |
|---|---|---|---|
| 275 | 800 | 0.12 | 0.05 |

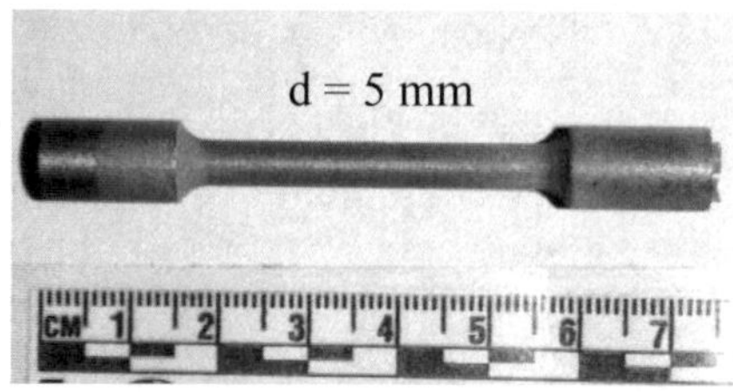


**Figure 1.** A testing sample produced by SLM.

### *2.2 Microstructure characterization*

As-polished metallography sections were captured by optical microscopy (OM, Zeiss Axio Lab.A1) and subjected to image analysis (ImageJ) for porosity determination. No porosity was detected in the cast material and the additively manufactured material showed a high relative density too; porosity was 0.03%. Microstructures were etched by aqua regia (31% HCl + 65% $HNO_3$, 3:1) and observed by OM as well as by a scanning electron microscope (SEM, JEOL JSM-6010PLUS/LA) supplemented by EDS analyser. Phase composition was determined by X-ray diffraction (XRD, PANalytical X'Pert PRO with Cu anode). Grain size was determined by the image analysis of etched longitudinal sections.

### *2.3 Mechanical testing*

On the prepared metallographic sections, Vickers hardness HV10 was measured using a Wilson VH3300 hardness tester. Uniaxial tensile and compressive tests were carried out on a universal loading machine Instron 5882. Constant deformation rate of 0.001 s$^{-1}$ was set for all measurements. Tensile and compressive testing was done only with the load acting in the building direction of the samples. For tensile testing, additively manufactured samples were used directly (figure 1), with polishing the surface to the final diameter of 4.8 mm. For compressive testing, cylinders with a height-to-diameter ratio of 1.5 were prepared by cutting the gauge of the as-built tensile samples.

## 3 Results and Discussion

### *3.1 Microstructure of cast Co-28Cr-6Mo alloy*

The microstructure of the cast Co-28Cr-6Mo results from a relatively slow melt solidification (tens of °C/min) from the casting temperature of ~ 1550 °C. It is thus formed by coarse (100–1000 μm) equiaxed grains (figure 2a) of dendritic nature. Despite relatively slow cooling, high-temperature FCC γ-phase is retained due to the sluggish nature of γ-FCC → ε-HCP transformation [6]. The proportion of ε-phase depends on the cooling rate; for casting, it is in the order of % units (e.g. 4% in ref. [7]). As this phase is much harder and significantly promotes tribological properties, enhancing its content is desirable for biomedical applications such as joint replacements [6]. That can be achieved by isothermal heat treatment below transformation temperature (~ 970 °C [7]), which was most probably applied on the material obtained for this work. The XRD analysis showed the Co matrix to be formed by the FCC γ-phase from 72% and HCP ε-phase from 28%. López and Saldivar-Garcia [7] reached such content of ε by annealing at 850 °C for 12 h. The nucleation of ε needles occurs at stacking faults of FCC grains, preferably in the areas where Cr and Mo content are not reduced by carbide formation. Carbides are formed in the interdendritic spaces due to Cr and Mo segregation (figure 2a). Interdendritic carbides are thus situated in FCC areas while needle-like ε-phase fills the areas in between (darker matrix areas).

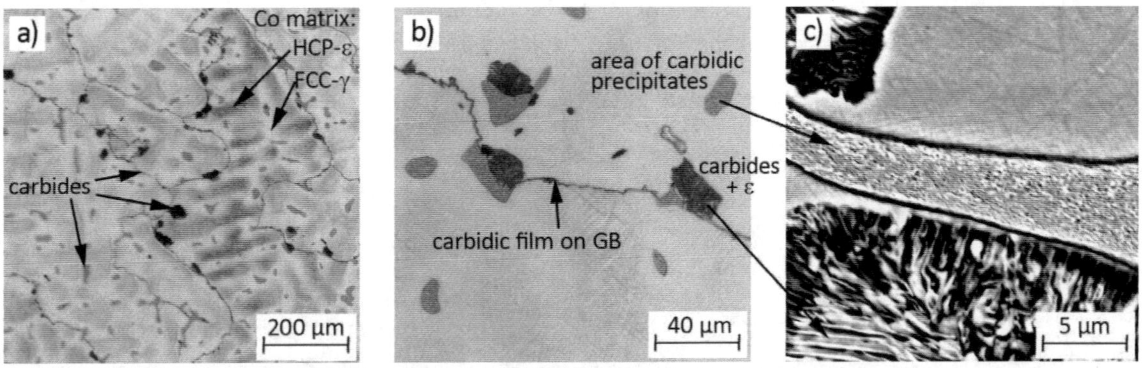


**Figure 2.** Microstructure of the Co-28Cr-6Mo casting.

XRD proved the presence of $M_{23}C_6$ and $M_6C$ carbides, which contents were determined to be 5.0 and 1.3%. $M_6C$ carbides form mainly a continuous film on grain boundaries (figure 2b). The relatively coarse (10-20 µm) interdendritic carbidic particles are of two different types. Grey carbidic areas (figure 2b) are areas of nanosized precipitates of $M_{23}C_6$ (figure 2c) formed probably by the decomposition of metastable σ-phase below 1150 °C; $\sigma + C \rightarrow M_{23}C_6$ [8]. The XRD analysis did not prove the presence of tetragonal σ-phase; the decomposition was complete. Lesser amount of black carbidic areas closely adjacent to the grey carbide particles (figure 2b) then represents lamellar mixture of carbides and ε-phase formed by the concurrent nucleation of ε-phase and carbide precipitation during isothermal annealing [6]. Table 3 shows the average chemical composition in these carbidic areas in comparison to the composition of the Co-matrix determined by multiple EDS point analyses. The values of C content are only comparative due to the insensibility of EDS to C analysis. The areas of carbide precipitates show the highest content of Mo and Cr.

**Table 3.** Chemical composition (wt.%) of different areas of the cast Co-28Cr-6Mo alloy.

| | **Co** | **Cr** | **Mo** | **C** | **Si** |
|---|---|---|---|---|---|
| **Areas of carbide precipitates** | bal. | 46.1±1.2 | 26.1±0.5 | 7.5±0.5 | 1.3±0.2 |
| **Lamellar mixture of carbides + ε** | bal. | 40.3±7.4 | 11.6±3.0 | 6.3±0.5 | 0.7±0.1 |
| **Co-matrix** | bal. | 28.4±0.9 | 6.7±1.4 | 3.8±1.1 | 0.8±0.1 |

### *3.2 Microstructure of powdered and subsequently additively manufactured Co-28Cr-6Mo alloy*

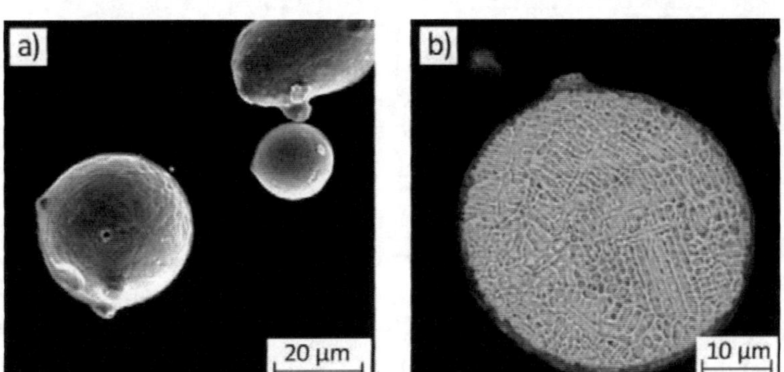


**Figure 3.** Powder of the Co-28Cr-6Mo alloy for SLM: (a) particle shape, (b) microstructure.

Figure 3 shows the appearance and microstructure of the as-purchased Co-28Cr-6Mo powder prepared by gas atomization. The dendritic structure is visible on the particle surface already (figure 3a). It results from rapid solidification (about $10^4$ °C/s [9]) during gas atomization. XRD analysis confirmed that all material is in the state of single FCC γ-phase although microsegregation at interdendritic spaces is obvious from figure 3b. The microsegregation of Mo was confirmed by multiple EDS analyses which showed 7.4±0.9 wt.% of Mo in the interdendritic spaces and 4.9±0.2 wt.% in the dendrites.

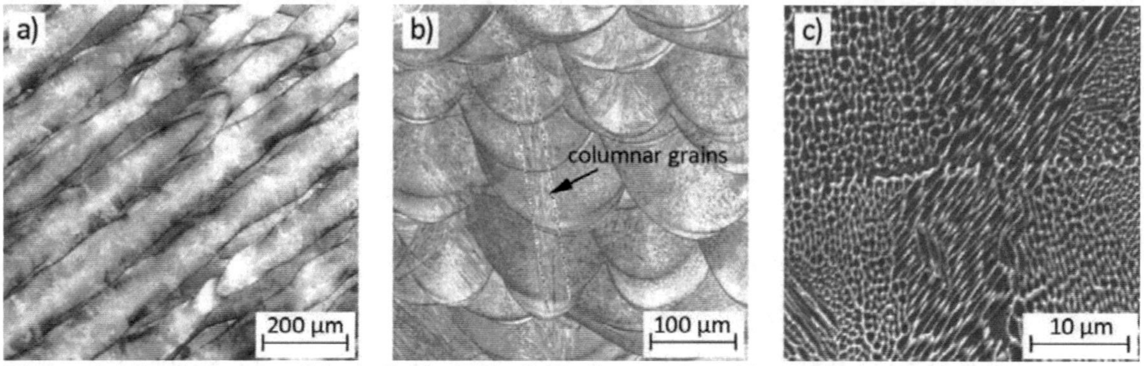


**Figure 4.** Microstructure of the bulk Co-28Cr-6Mo alloy produced by SLM: (a) cross section, (b) longitudinal section, (c) cellular microstructure.

The bulk material produced by SLM is documented by the images in figure 4. In the first magnification, the characteristic features of layer-by-layer production are visible; single tracks oriented in the scanning direction can be observed in the cross section (figure 4a) while a layered structure of overlapping melt pools is observed in the longitudinal section parallel to the building direction (figure 4b). The depth of the melt pools is approx. double the layer thickness, which ensures proper layers interconnection. Columnar grains sometimes growing across several melt pool depths can be distinguished (figure 4b). The grains of 18±2 µm in thickness can reach length of hundreds of micrometres. Columnar growth is given by the layer-wise production in which already formed grains in a previous layer represent nucleation sites for melt crystallization [10].

The SEM image (figure 4c) shows the cellular microstructure formed due to very high cooling rates during SLM. In each grain, cells show a different orientation. Usually, the crystallographic direction ⟨100⟩ which is the closest to the temperature gradient becomes the cell growth direction [11]. The EDS point analyses suggest the intercellular network to be formed by Mo segregation (7.6±0.5 wt.% compared to 6.2±0.6 wt.% in the cells). Other works [12, 13] report on segregation of Mo together with C, however, due to the negligible C content in the initial powder used for SLM (table 2), C segregation was not observed in this work.

The XRD analysis detected 70% of HCP ε-phase and only 30% of γ-phase. Such a high amount of ε-phase can be attributed to the repeated heat affection of the already solidified material by laser melting of newly deposited layers. In combination with the preheating of the building plate, temperature can fluctuate at a high enough level for sufficiently long time to induce $\gamma \rightarrow \varepsilon$ transformation similarly to isothermal annealing below the transformation temperature [14].

### *3.3 Mechanical properties*

Comparison between the mechanical properties of the cast and SLM Co-28Cr-6Mo alloy is shown in the bar chart in figure 5. Hardness showed to be isotropic for the whole SLM material no matter on the tested section as it is ruled by the fine cellular microstructure which does not show any directionality. The average hardness value for the SLM material was 357±3 HV10. Hardness higher by 50 units compared to the cast material can be attributed to the fine cellular microstructure with Mo segregation forming cell boundaries and to the higher content of harder HCP ε-phase. Similarly, it is with the yield and ultimate strength under both tensile and compressive loading. Yield strength showed an increase compared to the cast material by 12-19%, ultimate tensile and compressive strength increased by 50%. Material plasticity was favourably affected by the absence of brittle secondary phases such as coarse carbide particles and continuous carbidic film on grain boundaries in the cast material.

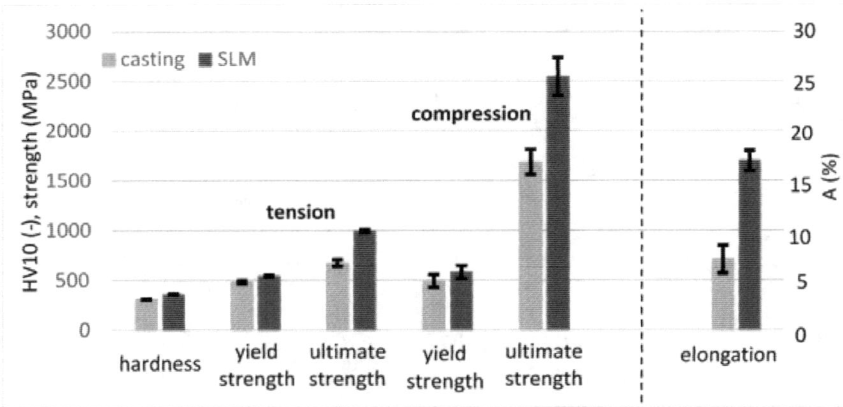


**Figure 5.** Mechanical properties of Co-28Cr-6Mo alloy prepared by SLM and by investment casting.

## 4 Conclusion

Microstructure refinement due to the rapid solidification during SLM together with the absence of secondary phases have brought increase not only in the strength but also in the plasticity of the Co-28Cr-6Mo alloy conventionally processed by investment casting. The distinctive microstructure of the additively manufactured alloy together with the detected carbon absence can be expected to be the cause of many other variances in the material behaviour, such as during heat treatment, corrosion exposition etc., which is the subject of our ongoing studies.

## Acknowledgements

This work was supported from the grant of Specific university research – grant No. A2_FCHT_2020_026.

## References

[1] Giacchi J V, Morando C N, Fornaro O and Palacio H A 2011 *Mater. Charact.* **62** 53-61
[2] Narushima T, Ueda K and Alfirano 2015 *Advances in Metallic Biomaterials: Tissues, Materials and Biological Reactions,* ed M Niinomi, T Narushima and M Nakai (Berlin: Springer) pp 157-178
[3] Pollock T M 2016 *Nature Mater.* **15** 809-815
[4] Arcam EBM System *ASTM F75 CoCr Alloy* http://www.arcam.com/wp-content/uploads/Arcam-ASTM-F75-Cobalt-Chrome.pdf
[5] Mergulhão M V, Podestsá C E and Neves M D M 2018 *Biomaterials in Regenerative Medicine* ed L A Dobrzański (Rijeka: InTech) chapter 6 pp 61-185
[6] Song C B, Park H B, Seong H G and López H F 2006 *Acta Biomater.* **2** 685-691
[7] López H F and Saldivar-Garcia A J 2008 *Metal. Mater. Trans.* A **39** 8-18
[8] Rosenthal R, Cardoso B R, Bott I S, Paranhos R P R and Carvalho E A 2010 *J. Mater. Sci.* **45** 4021-28
[9] Mullis A M, Farrell L, Cochrane R F and Adkins N J 2013 *Metall. Mater. Trans.* B **44** 992-999
[10] Zhang J, Song B, Wei Q, Bourell D and Shi Y 2019 *J.Mater. Sci, Technol.* **35** 270-284
[11] Chen Z W, Phan M A L and Darvish K 2017 *J. Mater. Sci.* **52** 7415-7427
[12] Song C, Zhang M, Yang Y, Wang D and Jia-kuo Y 2018 *Mater. Sci. Eng.* A **713** 206-213
[13] Li J, Ren H, Liu C and Shang S 2019 *Materials* **12** 1321
[14] Zhang M, Yang Y, Song C, Bai Y and Xiao Z 2018 *J. Alloys Compd.* **750** 878-886

IOP Conf. Series: Materials Science and Engineering 1178 (2021) 012047 doi:10.1088/1757-899X/1178/1/012047

# Determination of mechanical properties of steam turbine rotor materials

**M Rund[1], J Džugan[1], P Konopík[1], M Nesládek[2], J Kuželka[2], M Růžička[2], P Měšťánek[3]**

[1]COMTES FHT a.s., Průmyslová 995, 33441 Dobřany, Czech Republic
[2]Czech Tehnical University in Prague, Faculty of Mechanical Engineering, Technická 4, 16607 Praha 6, Czech Republic
[3]Doosan Škoda Power s.r.o., Tylova 1/57, 301 28 Plzeň, Czech Republic

E-mail: martin.rund@comtesfht.cz

**Abstract**. Increasing share of renewable energy sources in the electricity production results in a pressure on the thermal power plants to cover the power fluctuation very frequently. The increasing number of the start-ups and shutdowns as well as requirements for prompt ramping and for higher operation temperature has cardinal influence on the lifetime of the steam turbine. This all brings new challenges in to the designing process of the new steam turbines and lifetime analysis of currently operated ones. This paper aims to present experimental work, basic material parameters such as tensile test properties and FATT are determined here. Additionally, advanced material properties are probed by fracture toughness tests, fatigue crack growth tests and stress intensity factor threshold tests under various temperature ranges and cycle asymmetries R. The study is carried out for several steels used for the turbine shafts design.

## 1 Introduction

Increasing share of renewable energy sources in the electricity production results in a pressure on the thermal power plants to cover the power fluctuation [1]. The increasing number of turbine start-ups and shutdowns as well as requirements for prompt ramping and for higher operation temperature has cardinal influence on the lifetime of the steam turbine [2],[3]. This all brings new challenges in to the design process of the new steam turbines and lifetime analysis of currently operated ones. Experimental data are a key aspect. Their influence on the accuracy of the resulting analysis is crucial. In order to provide useful and realistic input data for the FEM based design, a close relationship between the testing laboratory and the simulation team is always necessary. Testing condition must be clearly specified, and in many cases must correspond with the operation conditions of given component, like temperature, load rates, testing frequency etc. [4][5]. Any deviation can then lead to over-conservative results, or even worse, to too optimistic ones.

In the case of this project there was requirement to describe transition properties of given rotor steels and to characterize fracture behaviour during operation conditions of the rotor materials. Critical for each mechanically loaded component is to avoid operation in temperatures at which the material exhibits brittle or ductile-to-brittle transition characteristics. Therefore, the so-called fracture appearance transition temperature (FATT) was determined by means of standard Charpy impact tests

IOP Conf. Series: Materials Science and Engineering 1178 (2021) 012047 doi:10.1088/1757-899X/1178/1/012047

as the first parameter on which testing temperatures for subsequent tests were based. Subsequently, tensile tests were performed for temperature range from 20 to 600°C for basic characterization providing also necessary input data for further tests and simulations.

The Fatigue crack growth tests (FCG) were subsequently performed at temperatures FATT-30°C, FATT, FATT+30°C and operating temperature for considered materials. These tests were subsequently supplemented by measurements of stress intensity factor (SIF) threshold range $\Delta K_{th}$ in the same temperature interval. It is assumed that during the start-stop cycles, the rotor is loaded by a superposed high-frequency load, therefore, in order to get relevant data, testing at various load asymmetries R was carried out. Another important material parameter for numerical simulations, which represents the relationship between the toughness, stress and flaw size is fracture toughness $K_{IC}$, which was measured and evaluated for above mentioned temperature intervals.

The experiments described above were carried out on four materials typically applied in steam turbine design.

## 2 Experiments - overview

Four different steam turbine (ST) materials were investigated in this project. The detailed parameters of individual above-mentioned tests are given in individual chapters below.

Description and justification of applied different temperature ranges is presented in Figure 1.

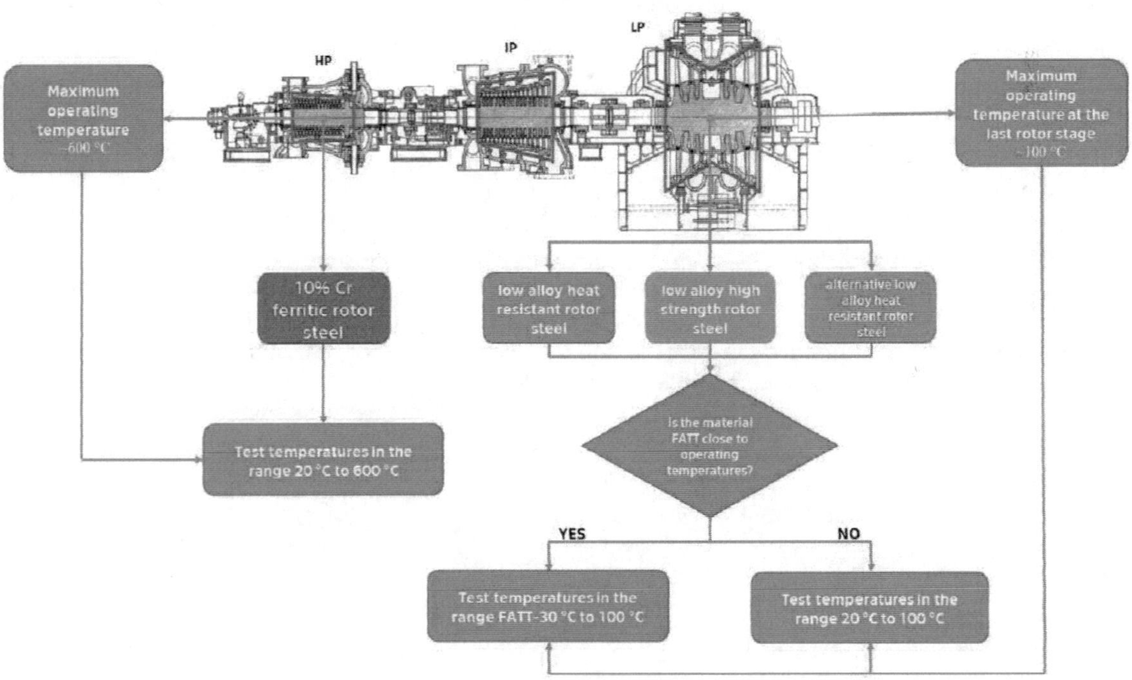


**Figure 1.** Justification of the temperature test ranges for testing of material fracture mechanics properties.

### *2.1 Tensile tests*

Tensile tests were carried out according to the standard CSN EN ISO 6892-1 and CSN EN ISO 6892-2 in temperature range 20, 200, 400 and 600 °C. Geometry of the specimen is presented in Figure 2.

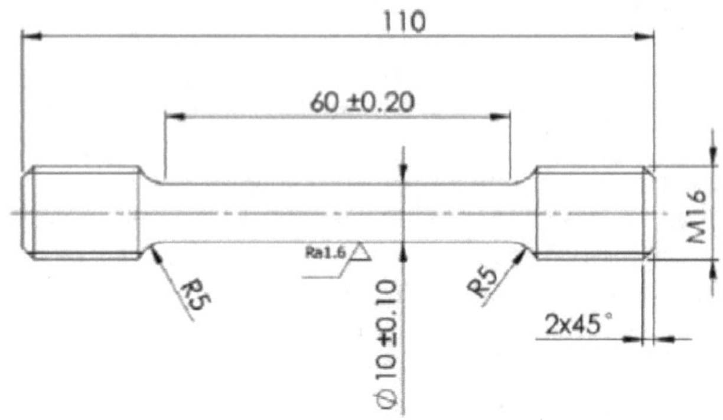


**Figure 2.** Geometry of the tensile test specimens

## *2.2 Determination of the FATT*

Tests were performed in accordance with the standard ČSN EN ISO 148-1 and ČSN EN ISO 14556 on impact pendulum with the energy capacity of 300 J. Standard Charpy geometry with V-notch and dimensions of 10x10x55 mm was used. The temperature interval of the tests corresponded to the needs of individual materials for the full determination of FATT.

## *2.3 Fatigue Crack Growth tests (FCG) and identification of ΔKth*

Test procedure for measurement of FCG rates (i.e. Paris' curves) and $\Delta K_{th}$ are given in the ASTM E645-15 standard. Tests were performed using magneto-resonance testing machine Rumul with adapted split furnace or cooling chamber. Crack growth was monitored indirectly by means of measurement of the specimen's compliance using the Crack Opening Displacement (COD) extensometer. Geometry of the specimens is shown in Figure 3a.

In all cases the specimens were precracked up to a/W = 0.3 with resulting SIF range 16 MPa√m. Tests were carried out at temperatures FATT-30°C, FATT, FATT+30°C, 20 and 600 °C and asymmetry ratios R applied here are 0.1 and 0.7. Temperature of individual tests and coefficient of the asymmetry are presented in Table 1 and Table 2.

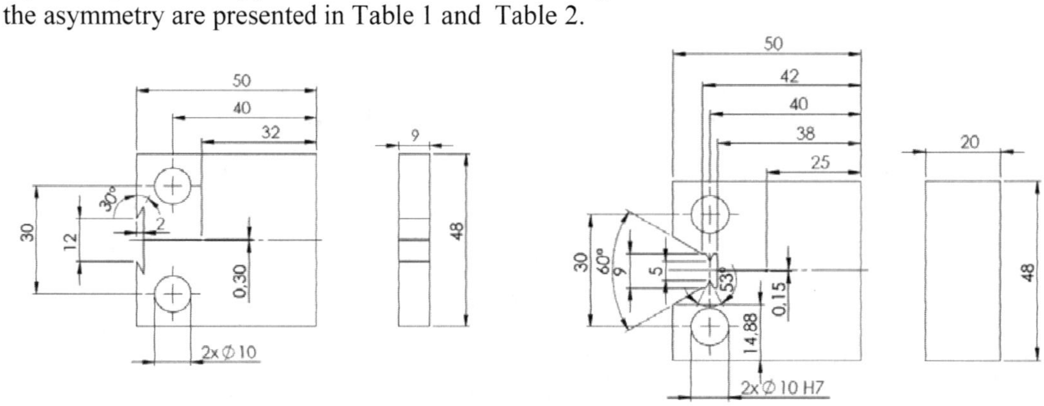


a) b)

**Figure 3.** Geometry of the specimen for tests: a) FCG and $\Delta K_{th}$ ; b) fracture toughness $K_{IC}$

## *2.4 Fracture toughness measurement - $K_{IC}$*

In the transition temperature range the fracture toughness was measured according to the standard for Master curve evaluation ASTM E1921-16. In the ductile region the J-R curve concept according to the

ASTM E1820-11 was followed. In all cases the CT specimen geometry was used (Figure 3b). Specimens were precracked up to a/W = 0.5 and the final SIF range was 16 MPa√m. Before testing, side grooves were machined with total depth of 2 mm from each side to ensure mode I crack growth throughout the entire specimen cross section. Tests were carried out in the temperature range FATT± 30°C. *10% Cr ferritic rotor steel* was tests at temperature 20 and 600 °C. Test conditions are summarized in Table 3.

**Table 1.** Test conditions for measurement of FCG rates

| Test parameters | low alloy heat resistant rotor steel | low alloy high strength rotor steel | alternative low alloy heat resistant rotor steel | 10% Cr ferritic rotor steel |
|---|---|---|---|---|
| T=FATT - 30°C, R=0,1 | X | | X | |
| T=20°C, R=0,1 | | X | | X |
| T=FATT-30, R=0,7 | X | | X | |
| T=20°C, R=0,7 | | X | | |
| T=100°C, R=0,1 | X | X | X | |
| T=600°C, R=0,1 | | | | X |

**Table 2.** Test conditions for measurement of $\Delta K_{th}$

| Test parameters | low alloy heat resistant rotor steel | low alloy high strength rotor steel | alternative low alloy heat resistant rotor steel | 10% Cr ferritic rotor steel |
|---|---|---|---|---|
| T=FATT – 30°C, R=0,1 | X | | X | |
| T=20°C, R=0,1 | | X | | X |
| T=FATT-30°C, R=0,7 | X | | X | |
| T=20°C, R=0,7 | | X | | |
| T=100°C, R=0,1 | X | X | X | |
| T=200°C | | | | |
| T=600°C, R=01 | | | | X |

**Table 3.** Test conditions for measurement of fracture toughness $K_{IC}$

| Test parameters | low alloy heat resistant rotor steel | low alloy high strength rotor steel | alternative low alloy heat resistant rotor steel | 10% Cr ferritic rotor steel |
|---|---|---|---|---|
| T=FATT | X | X | X | |
| T=FATT - 30°C | X | X | X | |
| T=FATT + 30°C | X | X | X | |
| T=20°C | | X | | X |
| T=600°C | | | | X |

## 3 Results and discussion

In Figure 4, the comparison of the static material properties in relation to the temperature is depicted. Highest values of $R_m$ and $R_{p0.2}$ exhibit material *Alternative Low alloy heat resistant steel*. As expected, material *Low alloy heat resistant steel* has the lowest $R_m$ and $R_{p0.2}$ in the temperature range 20 – 400 °C.

FATT results are summarized in Table 4. Except the *Low alloy high strength steel*, all materials exhibit ductile-to-brittle transition temperature at approx. room temperature. Great emphasis must be placed on handling the LP part of the rotors at this temperature, not only during the installation process, but even more at the operating temperatures where may approach the transition region and where the roots of the blades are stressed mechanically, due to centrifugal force and vibrations.

In the HP and IP part of the steam turbine the operating temperature is above the transition region. The domain loading is due to the thermal stress, which are caused be start-ups, shot-downs and load changes.

Results obtained from the fracture toughness measurement are presented in Table 4. The *Low alloy high strength steel* and *10% Cr ferritic rotor steel* only exhibit the value of fracture toughness higher than 200 MPa√m at FATT temperature.

SIF threshold value $\Delta K_{th}$ is ranging from 7.3 to 10.3 MPa√m at temperature FATT-30°C and R = 0.1 (Figure 5). It was observed that $\Delta K_{th}$ drops by approx. 40% at R = 0.7 and at the same temperature level. The similar effect of R was observed on FCG rates, i.e. with increasing R the crack exhibits a higher crack growth rate.

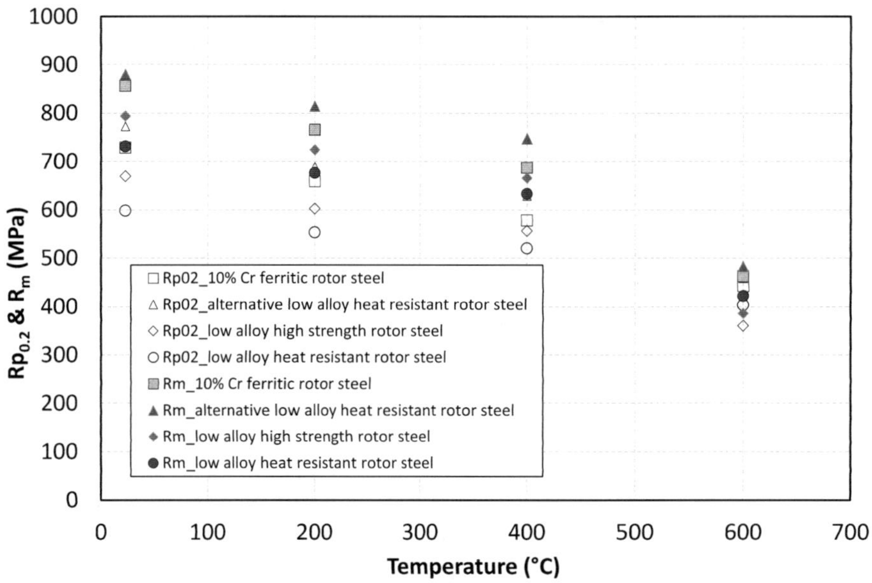


**Figure 4.** Comparison of the average tensile tests results for four investigated materials

**Table 4.** Summarization of transition temperature FATT and fracture toughness tests $K_{IC}$

| Designation of the steel | FATT temperature (°C) | | Fracture toughness (MPa√m) at given temperature | | | | |
|---|---|---|---|---|---|---|---|
| | Centre | Circumference | FATT-30 | FATT | FATT+30 | 20 | 600 (°C) |
| low alloy heat resistant | 27.0 | 32.0 | 94.2 | 129.2 | 262.0 | | |
| low alloy high strength rotor steel | -114.0 | -152.0 | 153.0 | 248.2 | 508.6 | 292.2 | |
| alternative low alloy heat resistant rotor steel | 26.0 | 2.0 | 103.3 | 174.6 | 178.6 | | |
| 10% Cr ferritic rotor steel | 3.6 | | | | | 246.0 | 211.8 |

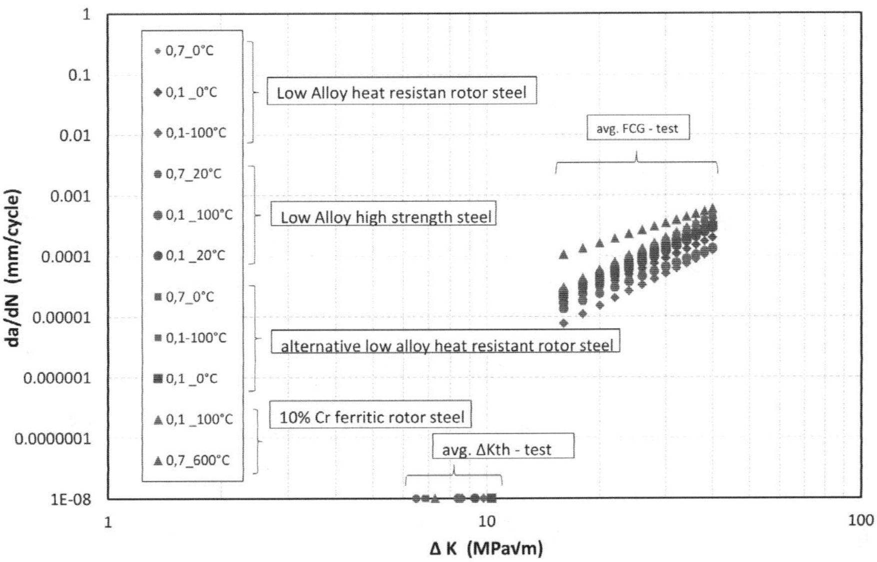


**Figure 5.** Summarization of the test results from FCG and $\Delta K_{th}$ measurements

## 4 Conclusion

This paper aim is to summarize the test campaign performed in order to provide material input data for numerical simulations of FCG rates in ST rotors. The tests prove that the test conditions must not be underestimated, especially the effect of other parameters such as R and temperature. Especially, increasing R has negative influence on the resulted FCG parameters as well as on $\Delta K_{th}$. At the same time it was proven, it is very useful to support tests with tensile and FATT tests, which are not time or money consuming and whose benefits are undeniable.

Tests results will be used as input parameters for FEA simulations of residual service life of in service steam turbine rotors.

## Acknowledgement

This project has received funding from the European Union's Horizon 2020 research and innovation programme under the Grant Agreement 764545 – TURBO-REFLEX.

## References

[1] Share of renewables in energy consumption in the EU reached 17% in 2016. http://ec.europa.eu/eurostat/documents/2995521/8612324/8-25012018-AP-EN.pdf/9d28caef-1961-4dd1-a901-af18f121fb2d. Accessed:2019-18-04.
[2] Nesládek, M., Jurenka, J., Bartošák, M., Růžička, M., Lutovinov, M., Papuga, J., Procházka, R., Džugan, J., Měšťánek, P. 2018. Thermo-Mechanical Fatigue Analysis of a Steam Turbine Shaft. Acta Polytechnica CTU Proceedings, 20, pp. 56-64.
[3] Kuželka, J., Nesládek, M., Lutovinov, et. all. 2019. Numerical Simulations of Fatigue Crack Growth in a Stream Turbine rotor blade groove
[4] Dzugan J., Prochazka R., Konopik P. 2017 Low cycle fatigue tests with the use of miniaturized test specimens, Proceedings of the ASME Pressure essels and piping conference, VOL 1A
[5] Viehrig, H. -W, Boehmert, J., Džugan, J. 2002, Some Issues by using the Master curve concept, Nuclear Engineering and Design 212(1):115-125

IOP Conf. Series: Materials Science and Engineering 1178 (2021) 012048 doi:10.1088/1757-899X/1178/1/012048

# Tensile properties evaluation of additively manufactured AISI 316L thin wall and bulk material using various miniaturized specimen geometries

**S Rzepa[1], D Melzer[1], M Koukolíková[1], P Konopík[1], M Brázda[1], J Vavřík[1]**

[1]COMTES FHT a.s., Průmyslová 995, Dobřany, 334 41, Czech Republic

E-mail: sylwia.rzepa@comtesfht.cz

**Abstract**. Additive Manufacturing is an innovative technology, which allows creating structures of complex and unique geometries layer by layer. The mechanical properties of deposited objects can vary depending on their length, thickness and deposition orientation due to different cooling rates and various thermal conductivity. Austenitic steel AISI 316L, deposited using direct energy deposition technology, was an experimental material. The technology employs a high-power laser beam as an energy source for the powder melting. During the process, a powder is blown through the nozzle by protective gas into the processing area and subsequently molten. The aim of the study was to compare basic mechanical properties of the specimens extracted from a thin wall and a bulk material. The specimens were tested in horizontal and vertical orientations in relation to the build orientation in machined and as-deposited states. The tensile characteristics of proportional and non-proportional specimens were investigated within the study. The miniaturized tensile test specimens were employed in order to examine the strength and elongation of the experimental material. The mechanical testing results were complemented by microstructural analysis using a light microscopy (LM). Based on the obtained results, it can be concluded that the specimen proportionality affects not only the specimen elongation, but also the strength values of investigated material.

## 1 Introduction

Additive manufacturing (AM) is a novel technology allowing creation of unique, three-dimensional objects of complicated geometries that found an application in automotive, medical and aerospace industries. In this technology, a process planning is not required since a model created by means of Computer Aided Design (CAD) can be manufactured directly, layer-by-layer.

The influence of the deposition parameters and scanning strategy on the material microstructure, mechanical properties, defects content and residual stresses for both, direct energy deposition (DED) and powder bed fusion (PBF) technologies is a topic that focuses an attention of many researchers [1],[2]. During the manufacturing process, a microstructure that varies over one deposited object can be created due to different heating and cooling conditions, which are influenced by an object dimensions, geometry and contact with supporting structures. This phenomenon significantly affects the mechanical properties of created part.

In the study, a direct energy deposition (DED) technology was employed for the deposition of experimental material, in which the powder is blown through a nozzle with inert gas, deposited in spot and melted using a laser beam as an energy source. The DED systems can be applied for

repairing of remanufacturing of already existing parts as well as creation of functionally graded materials that are advantages specific only for this technology [3],[4].
A single additively manufactured structure of complex geometry (such as turbine or propeller) is characterized by various wall thicknesses within the whole part. In many cases, it is impossible to manufacture a testing specimen with the dimensions defined by standard due to insufficient available material. In addition, the extraction of proportional specimens of minimal size results in modification of the surface conditions that is disadvantageous when the AM-ed part is intended to be used in as-deposited state and the in-service conditions have to be reproduced during mechanical testing. *Džugan et al.* [5] studied the effect of surface grinding on the miniaturized tensile test results. They concluded that the surface condition and defects play a great role in material behaviour. It is also worth noting that the mechanical properties of parts manufactured by different AM technologies were characterized by various sensitivity on the amount of removed by machining material. The specimen proportionality effect was also studied by *Moura et al.* [6]. They stated that the proportionality coefficient $k$ (according to ISO 6892-1) can significantly influence the tensile test results. According to their research, with increasing $k$ coefficient, the elongation and tensile toughness decrease. The trend can be extrapolated by a logarithmic function.
The aim of the study was to investigate the mechanical properties of AM-ed 316L thin wall and bulk material as well as the effect of a specimen size and proportionality on obtained tensile characteristics. The specimens were tested in machined and as-deposited states in vertical and horizontal orientations. The mechanical behaviour of the material was studied based on the results of miniaturized tensile test (MTT). The microstructure was examined by means of a light microscopy (LM).

## 2 Materials and methods

The experimental material was 316L stainless steel (SS 316L), which can be distinguished by relatively high strength and ductility as well as great corrosion resistance. The material is widely used in petrochemical industry, pharmaceutical and medical equipment or marine applications and others.

The experimental material was deposited by InssTek MX600 Powder Blown DED machine equipped with a 2 kW fibre-guided Ytterbium laser. The deposition was conducted under protective atmosphere of argon. The particle size of the powder used for the deposition ranged from 45 to 105 μm. The laser power was set up as 400 W and 800 W for the thin wall with dimensions of 40 × 30 × 2.9 mm and the bulk material with dimensions 35 × 35 × 35 mm, respectively.
The microstructural investigation was conducted my means of a light microscope NIKON ECLIPSE MA200 cooperating with NIS Elements 5.2 software. The metallographic samples were subjected to the standard metallographic preparation.
The mechanical properties of the AM-ed 316L experimental materials were examined based on the miniaturized tensile test (MTT) results. The method was widely used in other studies and documented as reliable way of material characterization [7]-[9]. The specimens were manufactured by a wire electric discharge machine (WEDM) according to the cutting schema presented in **Figure 1**. The experiment was carried out under quasi-static conditions at room temperature according to the methodology described in the standard ISO 6892-1 using a universal testing machine TiraTest with a load capacity of 10 kN. The deformation was recorded by a DIC system Mercury RT calibrated in 2D mode. Prior to testing, the specimens were polished, cleaned and measured in three different locations (thickness and width). Subsequently, the specimens were covered with a stochastic black and white pattern that allowed non-contact deformation measurement. After testing, the specimen dimensions were measured again using a stereomicroscope and the tensile parameters were evaluated (yield and ultimate tensile strength, uniform and total plastic elongation designated as YS, UTS, UE and El, respectively).
The machined specimen geometries were proportional and non-proportional according to the standard ISO 6892-1. The specimen designation system followed the document ASTM WK4922 [10].

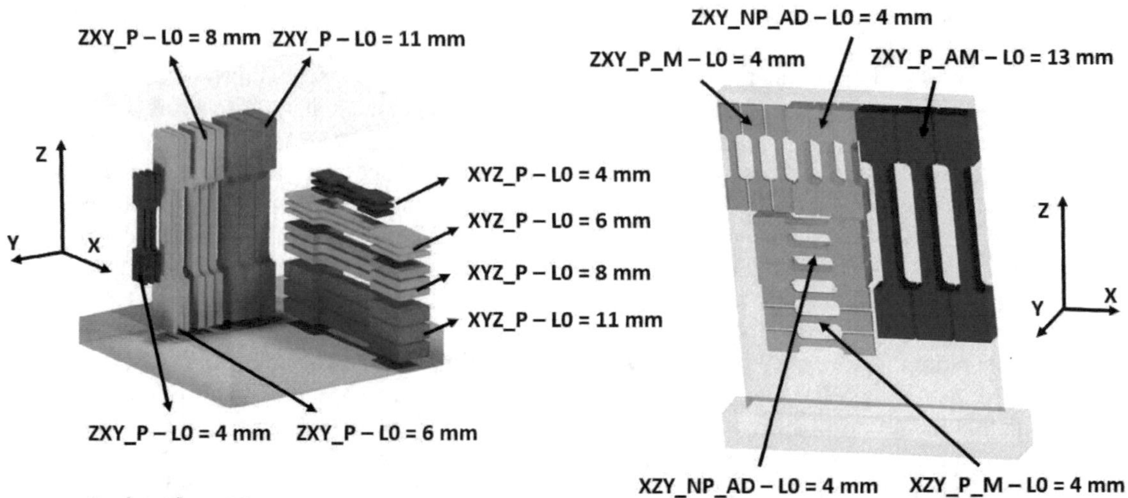


**Figure 1.** Additively manufactured bulk material and thin wall – cutting schema.

## 3 Results and discussion

The results of microstructural analysis are presented in **Figure 2** and **Figure 3**. The maximal depth of crated melt pools of 600 μm and 600-1000 μm were observed for the thin wall and the bulk material, respectively. The elongated grains growing epitaxial across many deposited layers were noticed for both microstructures. In addition, many fine grains were found in close proximity of the dendritic ones.

At high magnification, a cellular structure was identified for the thin wall. The porosity measurement revealed globular pores evenly distributed. The total porosity was evaluated as 0.015% and 0.183% with a maximal pore size of 40 μm and 85 μm for the thin wall and bulk material, respectively.

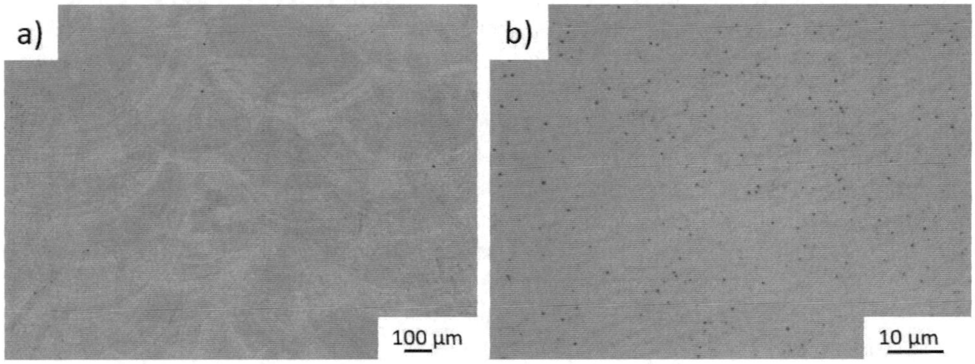


**Figure 2.** Results of microstructural analysis of the thin wall: a) melt pool detail, b) microstructure detail.

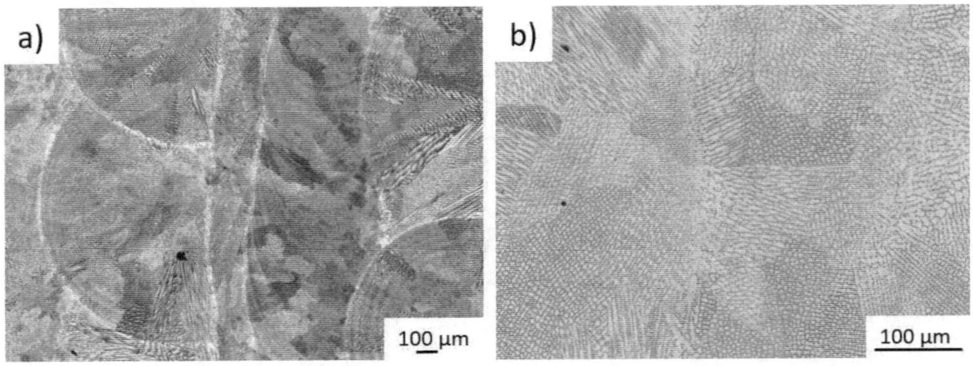


**Figure 3.** Results of microstructural analysis of the bulk material: a) melt pool detail, b) microstructure detail.

The results of MTT are presented in **Figure 4 - 6**. Three different specimen geometries were used for a thin wall mechanical properties investigation: machined (thickness of 0.5 mm, L0 = 4 mm), as-deposited non-proportional (original wall thickness of 2.9 mm, L0 = 4 mm) and as-deposited proportional (original wall thickness of 2.9 mm and L0 = 13 mm). An anisotropy of the tensile properties was observed as the XZY-oriented specimens yielded about 50 MPa higher values of both, yield and ultimate tensile strength, in comparison to ZXY-oriented specimens. Furthermore, the elongation of XZY-oriented specimens was about 4 and 10 % lower in comparison to ZXY-oriented specimens for both, as-deposited and machined states, respectively. In addition, the non-proportional (NP_AD) MTT specimens yielded systematically higher values of all evaluated tensile parameters compared to the machined proportional specimens. In the case of non-machined specimens, the reduction of the tensile parameters could be expected since very rough and defected surface of AM-ed objects can act as a stress concentrator and cause earlier material failure. Nevertheless, the results of as-deposited proportional (P_AM) specimens correspond to those for the machined proportional (P_M) specimens. This phenomenon demonstrates that the different values obtained for NP_AD and P_M batches are a result of various proportionality coefficient *k*. This observation is in a good agreement with the results published by *Fotovvati et al.* [11]. They presented the results for Ti-6Al-4V, which revealed that the YS and UTS decrease with the increasing proportional coefficient.

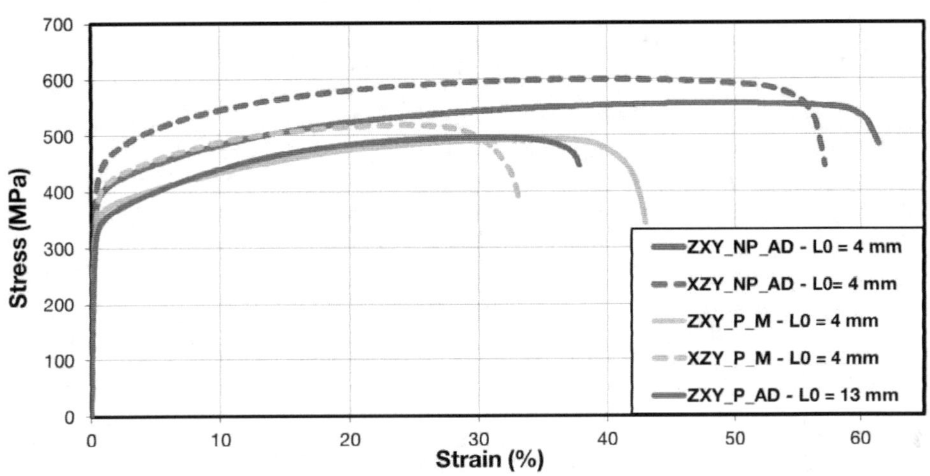


**Figure 4.** Engineering stress-strain curves for representative specimens extracted from additively manufactured 316L thin wall (P – proportional, NP – non-proporional, AD – as-deposited, M – machined).

Four different specimen geometries were tested in order to investigate the influence of the specimen thickness on the tensile test results for bulk material. The specimens with thicknesses of 2, 1, 0.5 mm (width of 2 mm) as well as the specimens 0.5 mm thick (width 1 mm) were machined. All specimen geometries were designed to be proportional according to $k = 5.65$. The evolution of YS, UTS and UE is presented in **Figure 5** for ZXY and XYZ orientation, respectively. The observed trends show continual increase of YS and UTS values with increasing gauge length for both ZXY- and XYZ-oriented specimens. For the greatest gauge length of 11 mm, the UTS and YS values significantly decrease. This tendency does not apply to UTS value of XYZ-oriented specimens. In the case of UE, the trend is the same for XYZ-oriented specimens and for ZXY-oriented is opposite.

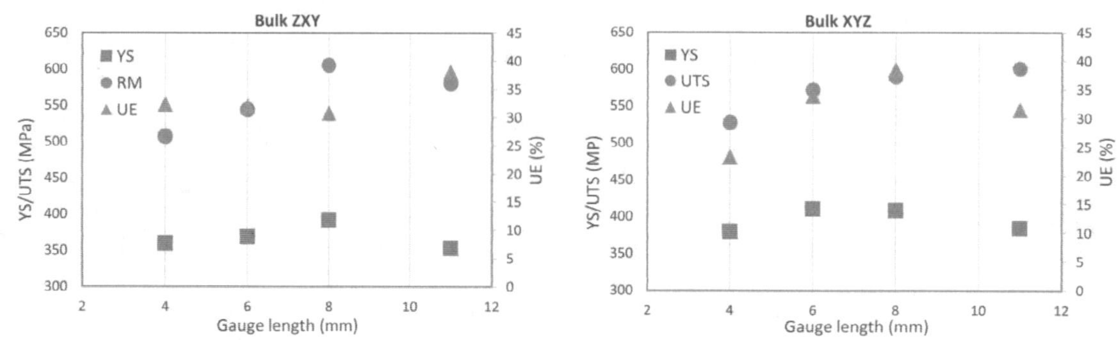


**Figure 5.** Summary of the strength values and uniform elongation in dependence on specimen cross-section (and resulting gauge length) for additively manufactured 316L bulk material in ZXY and XYZ orientations.

The results of MTT for proportional machined specimens were compared between thin wall and bulk material. The tensile stress–strain curves are presented in **Figure 6**. The specimens extracted from bulk material yielded slightly higher YS and UTS values but also slightly lower UE. Nevertheless, the results are in good agreement with small deviation regarding the fact that the thin wall and bulk material were deposited applying different deposition parameters.

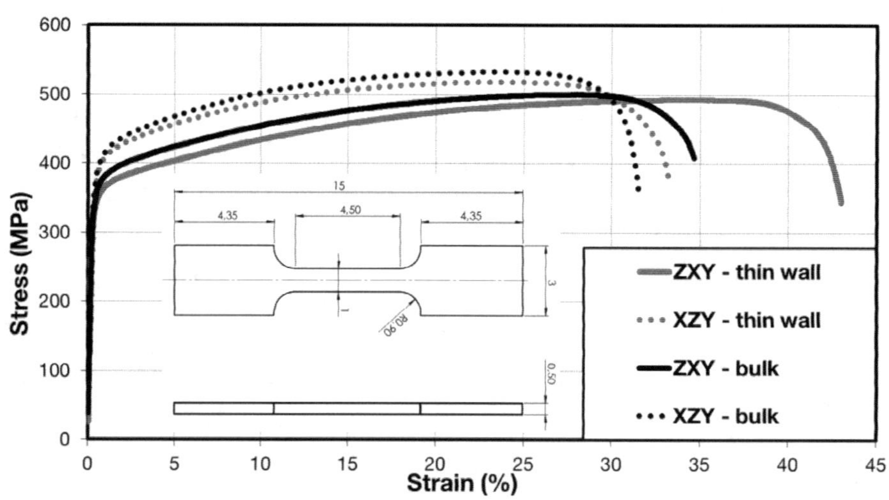


**Figure 6.** Engineering stress-strain curves for representative MTT (L0 = 4 mm) specimens – comparison of the tensile characteristics between bulk material and thin wall.

## 4 Conclusions

The aim of the study was to investigate the effect of the different tensile specimen geometries on the mechanical properties of the additively processed bulk and thin walled 316L material. The conclusion points are as follows:

- The results of the tensile test of the thin wall showed that the as deposited non-proportional tensile specimens yielded higher strength and elongation values compared to the machined-proportional ones. Furthermore, the as deposited proportional specimens yielded very similar values to the proportional machined ones. It demonstrates that the proportionality of the specimen has significant influence not only on the measured elongation but also on the strength parameters. Furthermore, the effect of the original as deposited surface is negligible for the measured tensile parameters.
- The results for proportional specimens extracted from the bulk material with different thicknesses showed a trend of increasing YS and UTS values with increasing gauge length (specimen cross-section) regardless of the orientation to the point of the gauge length of 11 mm where a sudden drop of the tensile parameters was observed. The same trend was observed for UE of the XYZ-oriented specimens. In the case of ZXY-oriented specimens UE systematically decreased with gauge length increase.
- The comparison of the MTT results between bulk and thin wall additively processed 316L showed that the results are in good agreement. Thus, the different additive manufacturing parameters applied for the bulk and thin wall manufacturing has negligible effect on the material performance.

**Acknowledgement**
This paper was developed thanks to the project of Pre-Application Research of Functionally Graduated Materials by Additive Technologies, No.: CZ.02.1.01/0.0/0.0/17_048/0007350, financed by the Ministry of Education of the Czech Republic.

**References**
[1] Saboori A, Piscopo G, Lai M, Salmi A, Biamino S 2020 *Materials Science and Engineering A* **780** 139179
[2] Shipley H, McDonnell D, Culleton M, Coull R, Lupoi R, O'Donnell G and Trimble D 2018 *International Journal of Machine Tools and Manufacture* **128** pp. 1-20
[3] Mathohoab I, Akinlabib E T, Arthura N and Tlotlengab M 2020 *CIRP Journal of Manufacturing Science and Technology* available online
[4] Dzugan J, Melzer D, Koukolikova M, Vavrik J and Seifi M, *Structural Integrity of Additive Manufactured Materials & Parts* **STP1631** pp. 247-256
[5] Dzugan J, Seifi M, Prochazka R, Rund M, Podany P, Konopik P and Lewandowski J J 2018 *Materials Characterization* **143** pp. 94-109
[6] Moura L, Vittoria G, Guimarães G A, Bertoni da Fonseca E, Pellizzer G L, Webster T, Najar L and Eder S 2020 *Journal of Materials Science* **55(22)** pp. 9578-9596
[7] Dzugan J, Sibr, M Konopík, P, Procházka R and Rund M 2017 Mechanical properties of AM components. IOP Conference Series: Materials Science and Engineering 179 012019
[8] Gussev M N, Busby J T, Field K G, Sokolov M A and Gray S E 2015 Small Specimen Test Techniques **6** pp. 31-49
[9] Gotterbarm M, Seifi M, Melzer D, Džugan J, Salem A, Liu Z and Körner C 2020 *Additive Manufacturing* **36** 101449
[10] ASTM WK49229. Standard Guide for Orientation and Location Dependence Mechanical Properties for Metal Additive Manufacturing – work in progress (2017)
[11] Fotovvati B, Alireza S and Asadi E E 2019 *Materials Science and Engineering: A* **760** pp. 431-447

# The use of induction heating in reaction apparatus small tonnage production

**A S Savina[1], V V Men'shikov[1,2], V S Boldyrev[3], A A Razvodova[3], N A Bogatov[1,3] and A P Zotkin[1]**

[1]D. Mendeleev University of Chemical Technology of Russia, Moscow, Russia
[2]Scientific Production Association «Lakokraspokrytiye», Khotkovo, Moscow region, Russia
[3]Bauman Moscow State Technical University, Moscow, Russia

assavina@muctr.ru

**Abstract**. Currently, in the planning and further implementation of technological processes that require heating of products to operating temperatures of about 1300 C, the issue of selecting energy-efficient environmentally friendly reactors using various heating methods is an acute issue. This scientific article discusses the effective use of electric and non-electric heating methods. Recommendations on the practical application of these methods are given, based on the final product and the specific features of the technological cycle of a particular enterprise. The question of the compliance of both methods with modern international requirements for the small-tonnage chemistry devices under consideration is considered. On the example of the Scientific Production Association «Lakokraspokrytiye» the main advantages in the application of induction heating in reactors used in technological lines for the production of paints and varnishes and coatings are shown. Examples are given of devices that are currently being developed and implemented in the Scientific Production Association «Lakokraspokrytiye», their basic characteristics are described. The described small-tonnage chemistry devices are widely used in the Russian Federation, CIS countries and Eastern Europe.

## Introduction

In modern conditions of industrial development and increasing prices for energy resources, one of the primary tasks of equipment design is the optimization of energy costs during the operation of industrial equipment. In the paint and varnish industry, the most energy-intensive equipment should be considered from the standpoint of optimizing energy consumption. The main methods for optimizing energy consumption for reactors with induction heating can be divided into two directions: structural (modernization of the ventilation system for purging the reactor shell, "stretching" the turns in the coil, i.e. winding each layer with a forced step) and automation (using modern measuring instruments, control and automation of the resin synthesis production process).

## 1 The problem of induction heating in reaction apparatus

IOP Conf. Series: Materials Science and Engineering 1178 (2021) 012049 doi:10.1088/1757-899X/1178/1/012049

Currently, in the planning and further implementation of technological processes that require heating of products to operating temperatures of about 1300 C, the issue of selecting energy-efficient environmentally friendly reactors using various heating methods is an acute issue. This scientific article discusses the effective use of electric and non-electric heating methods. Recommendations on the practical application of these methods are given, based on the final product and the specific features of the technological cycle of a particular enterprise. The question of the compliance of both methods with modern international requirements for the small-tonnage chemistry devices under consideration is considered. On the example of the Scientific Production Association «Lakokraspokrytiye» the main advantages in the application of induction heating in reactors used in technological lines for the production of paints and varnishes and coatings are shown. Examples are given of devices that are currently being developed and implemented in the Scientific Production Association «Lakokraspokrytiye», their basic characteristics are described. The described small-tonnage chemistry devices are widely used in the Russian Federation, CIS countries and Eastern Europe.

When carrying out technological processes associated with the need to heat chemical products to a temperature of 125-140 ° C, it is necessary to make a decision on the choice of a method for heating a product located in a reaction apparatus (reactor) using electric and non-electric heating methods.

With the non-electric heating method, it is necessary to distinguish between devices with direct heating with an open flame and using intermediate heat carriers, which include flue gases, steam, hot water and high-temperature organic heat carriers, heated by heat obtained by burning gas, coal or fuel oil.

The direct use of an open flame has some disadvantages:

- uneven heating;
- large energy losses with exhaust gases;
- high capital costs;
- the need for preliminary heating of the linings due to their large thermal inertia;
- difficulty in adjusting the temperature regime;
- compliance with sanitary and hygienic and environmental requirements.

The use of intermediate coolants is effective only in cases where their heating occurs due to a more complete use of available energy resources, which include exhaust flue gases, steam or hot water from the cooling system.

## 2 The electrical types of heating apparatus

According to the method of converting electricity to heat and taking into account the nature of the transfer of heat to a heated apparatus, the following electrical types of apparatus heating are distinguished: indirect and direct (electrical) heating based on ohmic resistance and induction heating.

For indirect heating of resistance, tubular electric heaters (TEH) are used, made in the form of a wire, rod or tape made of special alloys (usually nichromes), which are distinguished by increased electrical, fire and explosion safety and are widely used in chemical industries. Indirect heating allows you to heat both the apparatus and intermediate heat transfer media directly. The disadvantage of indirect heating is a TEH small resource of work.

Direct heating involves passing an electric current through a heated metal product (wire, tape, pipe) using a step-down transformer. This method is very economical, since the heat is released directly in the heated object, and the equipment used is simple to implement. The disadvantages of a direct electric heater are the presence of electric potential on the equipment and the possibility of heating only certain types of apparatus, since the electrical resistance of the heated object must be higher than the resistance of the current leads to achieve a high efficiency.

In the apparatus with induction heating, the output to the working heating occurs directly in the wall of the reactor. Thus, when heated, only one temperature transition is overcome: between the wall of the apparatus and the product. The magnitude of the temperature difference depends primarily on the magnitude of the transmitted power and the heat transfer coefficient from the wall of the apparatus to the product. Therefore, when using induction heating, a significantly higher specific heat load of the heating surface can be achieved with minimal heat loss. With a changing level of product loading, the induction heating power can change due to the sequential shutdown of the induction coils manually or in automatic mode.

When using induction heating in two or more stages of the technological process (for example, in the synthesis of alkyd resins), the process can easily be programmed with the necessary heating rates and temporary exposure times of the product at given points, which is quite difficult with other heating systems. Induction heating in comparison with other systems is inertia-free, which allows you to quickly and accurately adjust process parameters. Along with a sufficiently high efficiency, induction heating allows minimizing the need for production facilities and installing devices under a canopy in open areas. In addition, induction heating has high operational safety, a long service life, ease of maintenance, economy and environmental friendliness [1, 2].

## 3 The issue of cooling inductors

Due to the fact that the electrical resistivity and relative magnetic permeability of the material of the walls and bottom of the reactor (structural steel) depend on temperature (figure 1 and figure 2) [3], and electrical insulation is designed for a certain temperature, one of the main problems is the question cooling inductors.

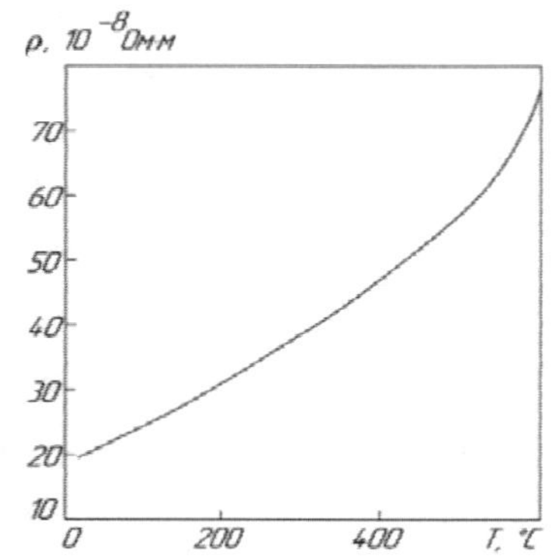


**Figure 1.** The dependence of the electrical resistivity $\rho$ of structural steel on temperature $T$.

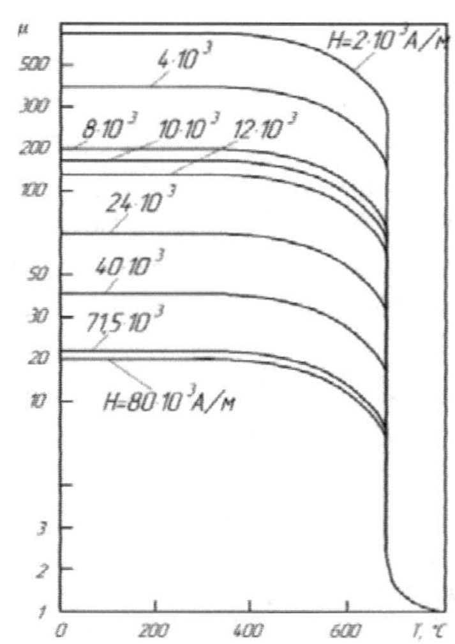


**Figure 2.** Dependence of the relative magnetic permeability $\mu$ of structural steel for various values of magnetic tension $H$ on temperature $T$.

In practice, induction heating is widely used water cooling of inductors, and for large specific heating powers, also use forced (figure 3) [4] or natural air cooling, which simplifies the design and operation of the apparatus.

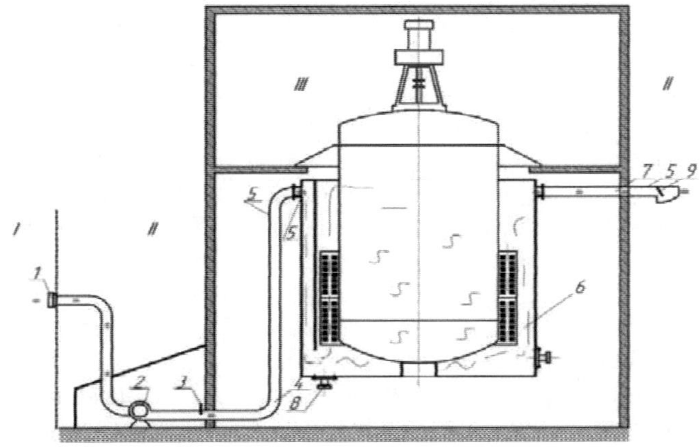

**Figure 3**. The scheme for providing the protective gas (air) of the inductor in the shell, blown under excess pressure, combined with the cooling circuit. (I - space in the non-explosive zone from which atmospheric air is taken for use as a protective gas; II - non-explosive zone; III - explosive zone. 1 - filter; 2 - fan supply of protective gas (air); 3 - shutter for regulating the amount of protective gas (air); 4 - inlet duct; 5 - fitting for connecting gauges for monitoring excess pressure in the shell of the inductor; 6 - inductor; 7 - exhaust duct; 8 - inductor input device).

Modernization of the ventilation system for purging the reactor shell (figure 4) [4, 5] using the latest heat and electrical insulation materials with high mechanical and thermal resistance in the structural elements of the inductor can reduce the power of the fans used. This leads to energy savings and reduced heat loss with the purged air.

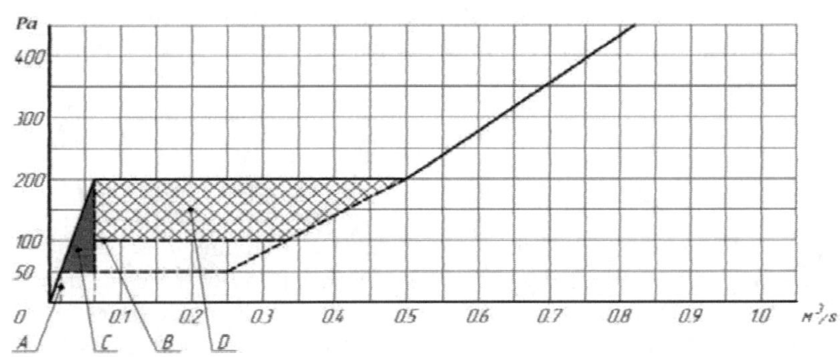


**Figure 4.** The dependence of the pressure of the protective gas (air) at the entrance to the shell in the shell of the inductor on its flow rate. (A - minimum permissible values of overpressure and flow rate of the protective gas (air) at which interlocks should operate; B - the minimum values of the overpressure and the flow rate of the protective gas (air) during normal operation of the inductor, taking into account the differential value of the overpressure control device; C - working area after modernization; D - working area before modernization).

In devices with induction heating, non-productive heat consumption is mainly associated with the flow rate of the shielded gas (air) blown through the casing to create explosion safety. These losses are estimated from 3.5 to 7.0% of the installed inductor power. For example, for a reactor with a capacity of 6.3 m$^3$, they will amount to 16 kW/h in heating mode.

At the three largest paint and varnish factories (Yaroslavl PA Lakokraska, Lida PA Lakokraska and JSC Empils), the workshops were executed according to a single project and equipped with devices with

IOP Conf. Series: Materials Science and Engineering 1178 (2021) 012049 doi:10.1088/1757-899X/1178/1/012049

induction heating produced by the GDR. Initially, reactors with a capacity of 12 $m^3$ had an installed power of 560 kW and a blowdown system based on a fan No. 6.3-7.5 kW, which was justified for the adopted design of the inductor. After replacing the impotent defective inductor with a domestic one with a closed magnetic system, the power decreased to 480 kW without deteriorating performance. For reactors of 6.3 $m^3$, this figure changed from 310 to 240 kW. At the same time, it became possible to reduce the volume of air blown through the casings of the modernized inductors by 2 times. For example, at OOO Kraski KVIL, a fan No. 3 with a 1.5 kW motor is used for a 16 $m^3$ reactor.

For induction-heated reactors, magnetic shielding of induction coils can provide significant economic benefits. Magnetic shielding reduces operating currents and Joule losses in each induction coil. The main purpose of the magnetic system is to localize electromagnetic energy in the boiler wall, eliminate the influence of adjacent phases on each other and, as a result, increase the electrical efficiency. and the natural power factor of the reactor as a whole.

The most important advantage of such magnetic shielding is the independence of the geometric dimensions and placement of the induction coils relative to the case, a sharp decrease in the air gap, the ability to comply with the "Electrical Installation Rules" and "Rules for the manufacture of explosion-proof and mine electrical equipment" in relation to the permissible minimum distances between the turns. "Stretching" the turns in the coil - winding each layer with a forced pitch ensures the following effects are achieved: allows you to make the coil removable, i.e. to reduce the repair time by more than 30 times, significantly reduce the cost of it, ensure the possibility of repairs without interrupting production, and also eliminate transportation costs; facilitates the working conditions of the insulation of each inductor and increases its operational reliability; improves heat dissipation during ventilation; reduces electrodynamic forces (shock loads when switching on and vibration during operation) by increasing the distance between the turns.

Automation of the production process also plays an important role in reducing energy costs. Depending on the requirements for a particular production, our enterprise has identified various economically justified directions in the automation of both the entire line and a single reactor. To classify these areas, we have introduced the concept of "Small automation" and "Integrated automation". Both automation options have a three-level structure: the lower level is the periphery (sensors, converters); middle level - hardware (for complex automation - these are controllers from SIEMENS of the S7-300 family, for small automation - this is a microprocessor-based measuring regulator from Aries); upper level - control and data input - SCADA (personal computer with installed SCADA visualization system GTI Procon).

The introduction of an automatic control system of each of the two types will ensure the achievement of the following goals: increasing the efficiency of the resin synthesis process control; archiving data and process history; setpoint control of the values of technological parameters and timely warning alarms, allowing to avoid emergency situations or to shutdown equipment as a result of triggering the emergency protection alarm; processing and analysis of information with subsequent optimization of operating modes; formation and issuance of reporting and archival documentation.

In addition, complex automation ensures the achievement of a number of additional goals: coordination of the operation of technological systems and coordination of the operation of all technological equipment; reducing the influence of the human factor in the preparation and decision-making.

## Conclusion

Based on the experience gained, we proposed to introduce into the automation system, which maintains the blowdown parameters, the ability to change and regulate the temperature and flow rate of air outgoing from the casing during the modernization of reactors with induction heating. Optimization consists in choosing the specified parameters, determining various options for their influence on each other, outputting these parameters to a personal computer for monitoring the process and the possibility of developing an individual program for controlling the process or its individual stages.

Correct optimization of energy losses guarantees their reduction by 50%. To determine the readiness of the product, graphs of the dependence of the degree of readiness on viscosity, acidity and electrical conductivity are used.

Chemical apparatuses with induction heating with a capacity from 1.0 to 12.5 $m^3$ that are being developed and manufactured by Scientific Production Association «Lakokraspokrytiye» fully meet modern requirements. These devices are made of stainless steel, equipped with mixers and coils for cooling or preheating the reaction mass. The use of domestic reactors with induction heating made it possible to completely abandon the purchase of similar equipment abroad, and the re-equipment of existing imported apparatus eliminated the cost of acquiring spare parts. Long-term operation in an industrial environment showed that the technical solutions made it possible to develop devices that were not inferior in their characteristics to Siemens (Germany), but even surpassed them in some respects. The use of a number of technical solutions protected by copyright certificates has made it possible to increase the efficiency, durability, safety of operation of the reactors, reduce the duration and improve the quality of repair work. Equally important are the use of high-quality materials, operational control, the use of specially designed equipment, acceptance tests at a special stand with certified devices, and highly qualified designers and manufacturers.

It should be noted that the explosion safety of the reactor is ensured by connecting induction coils to power cables through a cable entry placed in a casing blown with protective gas (air) under excessive pressure through an open ventilation cycle, as a result of which access to the casing of combustible gases and vapors and the formation of explosive mixtures there with by air. The inductor casing has a degree of protection IP54 according to GOST 14254-80.

The purge of the inductor casing is carried out using fans. Two fans are installed in the system, one of which is redundant, automatically turning on when the excess static pressure in the inductor shell decreases, as well as air ducts within the explosive zone below 50 Pa.

During the operation of devices with induction heating, there were no stops due to breakdowns of the inductor. These devices are used not only in Russia, but also in countries such as Ukraine, Belarus, Poland, Latvia. Since the beginning of the development of production, more than 500 devices of this type have been manufactured and implemented.

Currently, based on original developments, apparatuses with a wall temperature of up to 500 ° C are manufactured instead of the usual 300 (for example, evaporators made of cast iron with a wall thickness of 75 mm for the production of resorcinol). The accumulated experience allows us to design and manufacture inductors for devices from various materials and ferrous metal, clad and enameled containers with a volume of up to 32 $m^3$.

All designs of inductors according to individual customer requirements are developed and have a great advantage over foreign ones - the ability to mount and dismantle on site without the use of hot work. Devices with induction heating are environmentally friendly and do not have electromagnetic radiation outside the protective casing, even if it is partially destroyed.

**References**

[1] Bogomolov B B, Zubarev A M, Meshalkin V P, Men'shikov V V, Boldyrev V S 2019 Intelligent logical information algorithm for choosing energy- and resource-efficient chemical technology *Theoretical Foundations of Chemical Engineering* Vol 53 No 5 pp 709-718

[2] Boldyrev V S, Averina Yu M, Menshikov V V, Kuznetsov S V, Kolubanov K Y 2020 Technological and organizational engineering of paint processing *Theoretical Foundations of Chemical Engineering* Vol 54 No 3 pp 420-424

[3] Kuvaldin A B, Zhukovsky V E, Gorbatkova S A, Mineev V E 1985 Chemical apparatuses with induction heating *M.: Chemistry*

[4] Karuzsky V O, Nozhkin A Yu 2007 Technology of coatings *M.: LKM-Press* pp 56-63

[5] Averina Yu M, Kalyakina G E, Menshikov V V, Kapustin Yu I, Boldyrev V S 2019 Neutralization process design for electroplating industry wastewater containing chromium and cyanides *Herald of the Bauman Moscow State Technical University, Series Natural Sciences* № 3 pp 70-80

IOP Conf. Series: Materials Science and Engineering 1178 (2021) 012050 doi:10.1088/1757-899X/1178/1/012050

# Characterization of Microstructure Variations by Laser-Ultrasound during and after the Heat Treatment of Metals

**Scherleitner E[1], Kerschbaummayr C[1], Haderer W[1], Reitinger B[1], Mitter T[2], Gruensteidl C[1]**

[1]Research Center for Non Destructive Testing, Altenberger-Str. 69, 4040 Linz, Austria
[2]voestapine Stahl GmbH, voestalpine-Str. 3, 4020 Linz, Austria

edgar.scherleitner@recendt.at

**Abstract**. We demonstrate the potential of laser ultrasound for non-destructive testing applications in steel production and processing in two use cases: (1) Imaging of the grain structure transition between the thermally hardened surface and the unaffected core material with a comparison to sectional views by an optical microscope. We found very good correlation between the laser ultrasonic image and the microscopic view at the cross section. (2) In situ monitoring of Poisson's ratio of steel sheets during thermal processes by evaluation of resonant frequencies of Lamb waves. In our ongoing research the results will be correlated to metallurgical properties of interest.

## 1 Introduction

Extended knowledge about steel production and processing is the basis for meeting the increasing requirements in ever shorter development cycles and increasing energy efficiency, where in particular the steel industry has a high potential for savings. Laser-induced ultrasound is a non-contact technology that can provide relevant, supplementary insights into the complex microscopic processes of metals [1][2]. In our presented applications laser pulses of 5-10 ns and up to 100 mJ per pulse are fired at the surface from several centimetres distance and thus generate broadband ultrasonic waves by thermoelastic expansion and slight ablation. The surface displacement caused by the generated ultrasound waves is detected by a laser vibrometer. In the presented cases we used a robust, industrial interferometer developed to measure on rough surfaces [3][4]. Laser ultrasound (LUS) typically provides higher bandwidth signals than contact ultrasound and also allows for remote measurements on hot samples [5][6]. These properties are of great importance for the presented applications.

We show two use cases, where we use LUS to analyse the microstructure and the metallurgical phases in steel. These are (1) imaging of hardness penetration depth, (2) determination of the Poisson's ratio of sheet steel during heat treatment. The state-of-the-art technique to obtain information on the microstructure of metals are etched microsections which can be prepared to show grains, texture, phase, precipitations and more [7]. To receive this information the sample needs to be cut and specifically prepared for microscopic analysis [8]. During heat treatment this is not directly possible. Instead, samples are quenched at certain points of the thermal process before being analysed ex situ. This requires a model-based reconstruction of the high-temperature condition, which introduces uncertainties and provides results only at the selected points [9].

LUS has the potential to provide in situ information with a sampling rate up to hundreds of Hz. The challenge and focus of current investigations are the correlation of observable characteristics of the

IOP Conf. Series: Materials Science and Engineering 1178 (2021) 012050 doi:10.1088/1757-899X/1178/1/012050

ultrasonic wave propagation with metallurgical properties of interest. We want to show that LUS can complement existing methods for the determination of the hardness penetration depth [10][11][12][13] and provide cost saving alternatives for in situ methods having their individual advantages and disadvantages.

## 2 Use case 1: Imaging of hardness penetration depth

Thermal hardening is commonly used to increase the surface resistance of steel components in heavily stressed areas such as bearings. Carbon-containing steel is heated and then cooled abruptly, so that a martensitic structure, strained by the carbon, is formed from the austenitic phase. We have developed a LUS setup, **Figure 1**, and applied a method to measure and image the grain structure transition between the finer grained, hardened zone near the surface and the unaffected core. For excitation we used a pulsed Nd:YAG laser, the Quantel Q-smart, with 532 nm, 10 ns pulses and a maximum energy per pulse of 430 mJ. The LUS detection was the Tecnar Lus system with an 80 µs long-pulse laser at 1064 nm.

The short generation laser pulse produces a broadband acoustic wave package inside the sample. According to scattering theory the waves are scattered at the grain boundaries due to changing elastic properties. Scattering strongly depends on the average grain diameter $D$ and the acoustic wavelength $\lambda$, thus, for the evaluated bandwidth of 10 MHz to 40 MHz, it mainly occurs in the non-hardened core regions with coarser grain structure. For the grain diameter in the hardened layer holds $D \ll \lambda$ and therefore the acoustic waves are scattered very weakly compared to the core region. This behavior can be seen in **Figure 2** (right). It shows a B-Scan with the lateral position on the sample plot on the x-axis, the depth down into the sample on the y-axis and the signal amplitude in grey scale. The backscattered signal starts occurring at approximately 3 mm depth which is in good agreement with the microscopic view shown in **Figure 2** (left).

**Figure 1.** Laboratory setup for LUS measurements on a thermally hardened part of a crankshaft. The laser paths of excitation (green) and detection (red) are indicated and intersect each other at the current measurement position.

In some cases, the hardened zone is already optically visible on the cross section as it is the case with this induction hardened steel sample **Figure 2** (left). A scan line of 15 mm length is performed in the vicinity of this sectional cut consisting of 150 point measurements with a total acquisition time of 15 s. There is excellent agreement between the depth and shape of the transition zone of the LUS image compared to the microscopic view on the cut surface. It can be clearly seen that the coarser grain of the unaffected core causes stronger back-scattering than the hardened layer closer to the surface, whereby the first millimetre of the ultrasonic image can hardly be evaluated due to the intensive ultrasonic generation effect and it appears as horizontal white lines in the image.

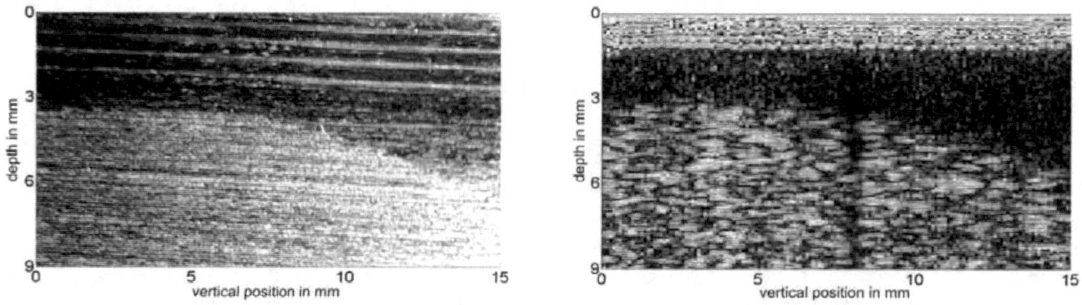


**Figure 2.** Left: Microscopic view on cross section of an induction hardened steel sample with visible fine grained (hardened) and coarser grained (unaffected) core. Right: Reconstruction of LUS data.

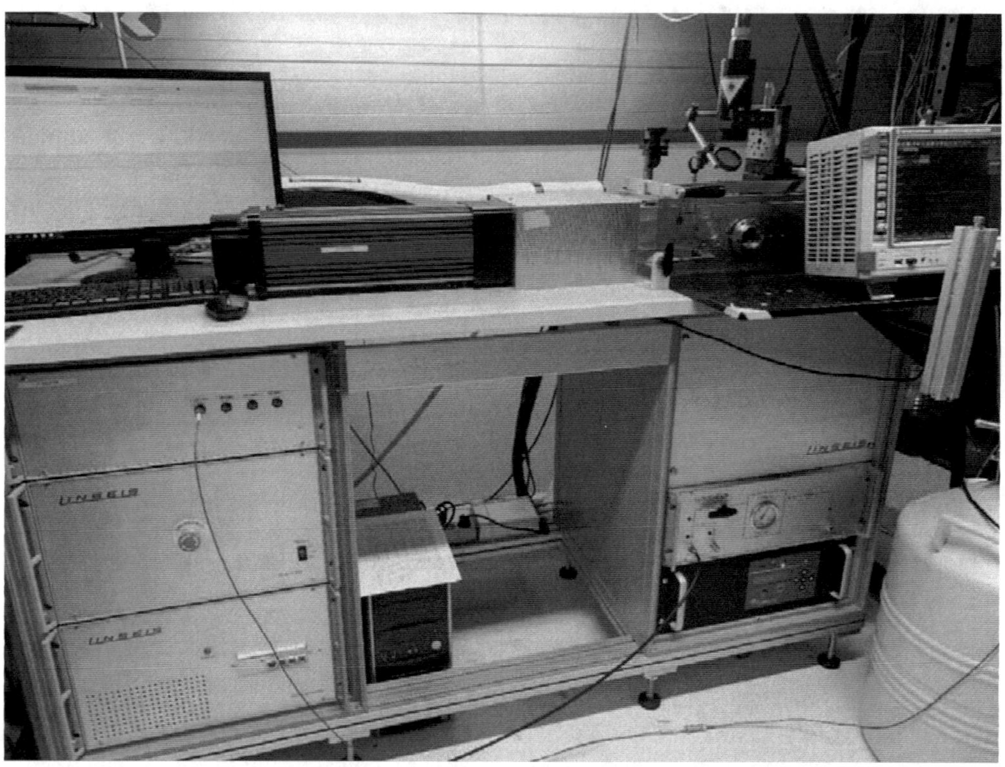

**Figure 3.** Deformation and quenching dilatometer based on Linseis L78 adapted for LUS application.

## 3 Use case 2: Determination of the Poisson's ratio of sheet steel during heat treatment

Sheets or plates are a common product of the metal industry. This geometry sustains guided waves e.g. Lamb waves, which can be used in many NDT applications. In our presented approach we measure resonant frequencies of Lamb waves [14] [15][16][17] instead of the time-of-flight of longitudinal pulses. Using the central frequency of two resonances, the Poisson's ratio of a sample can be determined with high accuracy [15][16]. For this analysis knowing the thickness of the sample (which is subject to thermal expansion during heat processes) is not required. The resonances occur in the range <10 MHz for typical thickness in the mm range. This allows for a strong limitation of the detection bandwidth and thus reduces electronic measurement noise compared to time-of-flight measurements.

A LINSEIS dilatometer type L78 has been adapted for in situ LUS [18][19] measurements, see **Figure 3**, adapted for inductive heating and cooling of sheets by a flat coil and both lasers coming confocally from the top window (2-3mm spot size each), see **Figure 4**.

The right side of **Figure 5** shows a typical measured response spectrum with peaks caused by resonant zero group velocity (ZGV) resonances (~3 MHz) and by longitudinal thickness resonances (~6 MHz and ~9 MHz). Their frequency positions correlate to the local minimum (ZGV) and cut-off frequencies of the propagating modes (thickness resonances) in the dispersion curves of the sheet (**Figure 5**, left). The peaks could be resolved with satisfying SNR, without coming applying 50 mJ for the excitation pulses.

**Figure 4.** Left: LUS setup on dilatometer accessing the sheet sample through the top window. Excitation- (green) and detection laser (red) are indicated and aim at the same spot. Right: zoom of the steel sheet above flat, inductive coil with holes for gas cooling.

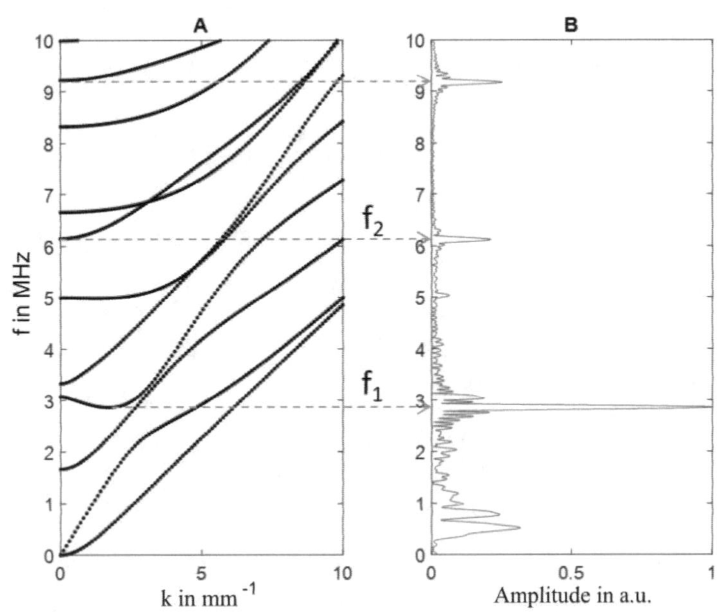


**Figure 5.** Left: dispersion curves for a 1 mm thick steel sheet with $c_L$=6147.5 m/s, $c_T$=3329.3 m/s and ρ=7870 kg/m³. Right: response spectrum of sample at room temperature.

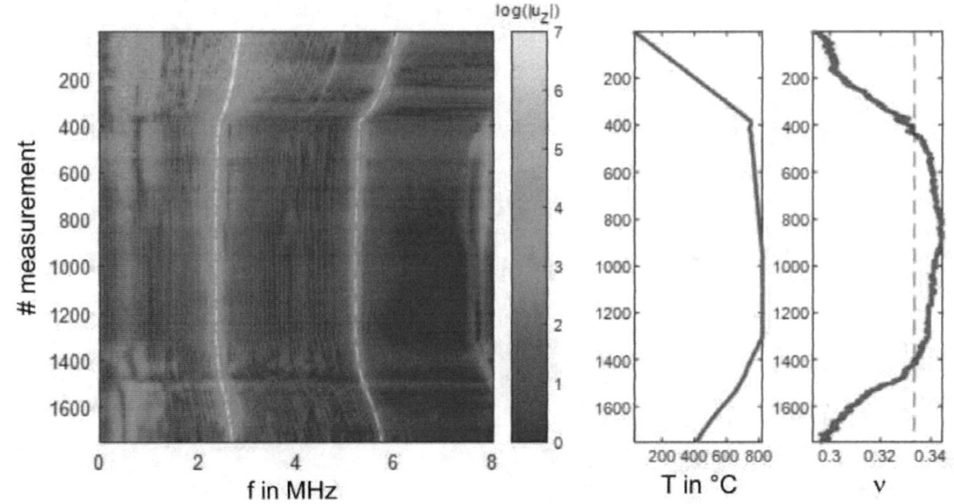


**Figure 6.** Left: Response spectra of a plate during a thermal cycle. Right: Temperature and calculated Poisson's ratio.

**Figure 6** shows the response spectra of a plate during a thermal cycle. From the maxima positions $f_1$ and $f_2$ of the peaks in **Figure 5** we obtained the Poisson's ratio $\nu$ of the sample during the process, which is shown in the right subfigure. The data also reveals additional resonant modes appearing in certain ranges values of $\nu$, which is in agreement with literature [14] and increases confidence in our results. The correlation of the Poisson's ratio with metallurgical reference measurements is currently in progress.

## 4 Conclusion

LUS measurements were carried out on three different use cases to demonstrate the potential of this technology for the characterisation of steel properties. The results of use case 1 show the ability of LUS to image the hardness penetration depth on thermally hardened steel samples. The imaged depth is in very good correlation with the microscopic view on the cross section. By use case 2 we presented an approach for plate geometries to determine the Poisson's ratio in situ which is independent of the plate thickness. This ongoing research aims to develop a robust method to correlate metallurgical parameters in sheet steel production.

## Acknowledgements

Financial support was provided by the Austrian Research Promotion Agency (FFG) under the scope of the COMET programme within the research project "Photonic Sensing for Smarter Processes (PSSP)" (contract number 871974). This programme is promoted by BMK, BMDW, the federal state of Upper Austria and the federal state of Styria, represented by SFG.

## References

[1] Kruger S E, Moreau A and Monchalin J P 2008 Laser-Ultrasonic Monitoring of Metallurgcal Transformations in Advanced Ultra-High Strengh Steels *1st International Symposium on Laser Ultrasonics, Science, Technology and Applications* (Montreal: Laser Ultrasonics)

[2] Hinterlechner I, Barriobero-Vila P, Reitinger B, Fromherz T, Requena G and Burgholzer P 2018 In-situ monitoring of phase transformation in Ti-6Al-6V-2Sn using laser ultrasonics *Nondestructive Testing and Evaluation* **33** 130–8

[3] Monchalin J P, Neron C, Bussiere J F, Bouchard P, Padioleau C, Heon R, Choquet M, Aussel J D, Durou G and Nilson J A 1998 Laser-ultrasonics: from the laboratory to the shop floor *Advanced Performance Materials* **5** 7–23
[4] Zamiri S, Reitinger B, Berer T, Bauer S and Burgholzer P 2010 Laser Ultrasonic Measurements of Metal Thickness Using a Photorefractive Crystal *Informationstagung Mikroelektronik 2010* 172–6
[5] Choquet M and Schramm H G 2007 Wall thickness measuring on hot tubes with laser-ultrasonic technology **57** 32–5
[6] Roither J, Mitter T, Reitinger B, Wiesinger A, Hofer C, Grün H and Burgholzer P 2016 Detection of Casting Defects in Aluminum Slabs by Laser Ultrasonic Measurements *19th World Conference on Non Destructive Testing* 1–7
[7] Schönmaier H, Pölzgutter M and Schnitzer R 2020 Etching methods for the microstructural characterization of a heat resistant 2.25Cr-1Mo-0.25V weld metal *Praktische Metallographie/Practical Metallography* **57** 5–30
[8] Ryde L 2006 Application of EBSD to analysis of microstructures in commercial steels *Materials Science and Technology* **22** 1297–306
[9] Sanz L, López B and Pereda B 2018 Characterization of Austenite Microstructure from Quenched Martensite Using Conventional Metallographic Techniques and a Crystallographic Reconstruction Procedure *Metals* **8** 294
[10] Salchak Y A, Sednev D A, Kroening M and Ardashkin I B 2015 Method of case hardening depth testing by using multifunctional ultrasonic testing instrument *IOP Conference Series: Materials Science and Engineering* **81**
[11] Schneider D, Hofmann R, Schwarz T, Grosser T and Hensel E 2012 Evaluating surface hardened steels by laser-acoustics *Surface and Coatings Technology* **206** 2079–88
[12] Moorthy V, Shaw B A and Brimble K 2004 Testing of case depth in case carburized gear steels using magnetic barkhausen emission technique *Materials Evaluation* **62** 523–7
[13] Bernard M, Reimche W and Bach F 2006 Non-Destructive In-Process Determination of Hardness Values by Harmonic Analysis of Eddy Current Signals *9th European Conference on NDT* 1–12
[14] Prada C, Clorennec D and Royer D 2008 Local vibration of an elastic plate and zero-group velocity Lamb modes *The Journal of the Acoustical Society of America* **124** 203–12
[15] Grünsteidl C, Murray T W, Berer T and Veres I A 2016 Inverse characterization of plates using zero group velocity Lamb modes *Ultrasonics* **65** 1–4
[16] Grünsteidl C, Berer T, Hettich M and Veres I 2018 Determination of thickness and bulk sound velocities of isotropic plates using zero-group-velocity Lamb waves *Applied Physics Letters* **112** 251905
[17] Laurent J Ô, Royer D and Prada C 2014 Temporal behavior of laser induced elastic plate resonances *Wave Motion* **51** 1011–20
[18] Scherleitner E, Reitinger B, Gruber M and Burgholzer P 2018 Implementation and Use of a Laser-Ultrasonic System in a Deformation- and Quenching Dilatometer *Materials Science Forum* **941** 2423–8
[19] Linseis C, Linseis F, Raithel M R B 1918 Vorrichtung und Verfahren zur Bestimmung elastischer Eigenschaften und/oder von Gefügezuständen von Proben *Europäisches Patent Amt*

# Corrosion Resistance of 18Ni-300 maraging steel manufactured by LPBF

**T Simson[1], J Koch[1], M Zetek[2], I Zetková[2]**

[1]OTH Amberg-Weiden, Kaiser-Wilhelm-Ring 23, Amberg 92224, Germany

[2]University of West Bohemia, Universitní 22, Pilsen 306 14, Czech Republic

E-mail: t.simson@oth-aw.de

**Abstract**. The study describes the corrosion behaviour of LPBF parts made of MAR 300 (EN 1.2709) in sulphuric acid. It includes the influence of the heat treatment on the part's properties as solution annealed and age hardened with reference to conventional samples. Laser remelted surfaces were examined as well. Due to the layered production of AM components, which is characteristic for this process, high surface roughness occurs. Therefore, for many applications and mostly on functional surfaces, complex post-processing of the surfaces is necessary. The reliable use of the manufactured components, demand knowledge of the specific influence of the selected post-processing methods on the electrochemical behaviour. The corrosion investigation takes place on different sample surfaces, such as milled, turned, sandblasted, untreated, grinded and laser-machined. The results show that there is no correlation between surface treatment and corrosion resistance. However, an increase in corrosion rate due to age hardening of the material can be observed for both LPBF and conventionally produced samples. By laser surface remelting of these age hardened samples, the corrosion resistance of these parts could be increased again, to the same level as in the solution annealed state.

## 1 Introduction

The laser powder bed fusion (LPBF) process is a part of additive manufacturing (AM). The advantages of AM processes are i.a. geometrical freedom, material flexibility, creating opportunities to fabricate innovative parts with enhanced functionality. [1] MAR 300 is a well-known maraging steel, which belongs to a special class of ultra-high strength steels, which differs from conventional steels by hardening via intermetallic precipitations without carbon. The quality of the used powder feedstock as well as the achievable product properties, corresponds to the powder morphology, including particle size distribution and shape. Therefore, for many applications and mostly on functional surfaces, complex post-processing of the surfaces is necessary. This post-processing is often carried out by conventional manufacturing and mechanical surface treatment techniques (e.g. grinding, milling, drilling or blasting). But also, laser treatment as a thermal post-processing method can smooth the surface and influence material properties. [2] For the reliable use of the manufactured components and due to the increasing relevance of AM processes in the industry it is necessary to know the specific influence of the selected post-processing methods on the electrochemical corrosion behaviour.

R. Schaller et al. [3] report on microstructural inhomogeneities such as pores or oxides on the sample surface effecting the corrosion behaviour of materials. Due to the manufacturing process by LPBF, samples are susceptible to pores, oxides and interconnectivities in the matrix which can affect the corrosion behaviour. According to the tested MAR 300 some publications e.g. A. Ettefagh et al. [4] analyse that samples from stainless steel AISI 316L show decreased corrosion rates in comparison to conventionally manufactured ones. J. Suryawanshi et al. [5] on the other side mentioned that AM parts of MAR 300 show lower corrosion resistance caused by the higher porosity level compared with conventionally manufactured samples. In order to get to a substantiated statement of the comparability between the corrosion behaviour of LPBF and conventionally manufactured samples, corrosion measurements on the MAR 300 were carried out.

## 2 Experimental procedure

### *2.1 Material and Samples*

The material used for this investigation is martensitic steel MAR 300 (EN 1.2709) with the chemical composition shown in Table 1. The powder was bought by Electro Optical Systems (EOS) Finland Oy and produced using gas atomization resulting in spherical particles. According to the manufacturer the employed particle size (D50) was 30 µm. All test samples used to determine the corrosion resistance were produced by an EOS M290 direct metal laser sintering system in nitrogen atmosphere.

**Table 1.** Chemical composition of new MAR 300 powder in wt%

| Fe | Co | Ni | Mo | Ti | Al | Cu | Cr | Mn | Si | C | P | S |
|---|---|---|---|---|---|---|---|---|---|---|---|---|
| Balance | 8.76 | 18.25 | 4.80 | 0.64 | 0.06 | 0.03 | 0.12 | 0.06 | 0.02 | 0.01 | <0.01 | 0.01 |

The laser scanning strategy used zigzag pattern with 33.3° rotation between the superimposed layers. The process parameters used were laser power (*P*), scan speed (*v*), hatch spacing (*h*) and layer thickness (*d*), the values are shown in Table 2. Before starting the LPBF process, the base plate was preheated to 40°C by a heater placed inside the building platform, and the oxygen content in the process chamber was maintained below 0.04 % by pumping in nitrogen continuously.
Further investigations by computer tomography determined for all samples a porosity in average of 0.0073 % which corresponds to a structure density, higher than 99.99 %.

**Table 2.** Process parameters for different pattern

| Pattern | Laser power (W) | Scan speed (mm/s) | Hatch spacing (mm) | Layer thickness (mm) |
|---|---|---|---|---|
| In Fill | 285 | 960 | 0.11 | 0.04 |
| UpDown | 153 | 600 | 0.11 | 0.12 |
| DownSkin | 145 | 2400 | 0.11 | 0.16 |

After the manufacturing process (as-built state) the samples were first solution annealed at 820 °C for 1 h in Argon atmosphere by using a gas-tight chamber furnace of the PKR 35/11 type from LAC followed by air-cooling down to room temperature. According to the testing conditions some samples were age hardened at 490 °C for 6 h. The aging was carried out using a chamber furnace from Linn High Therm GmbH with the type designation HK-70. The volume is 70 litres and the heating power 18 kW.
The post-processing of to sample surface was prepared with the same cutting edge radius by turning, milling and the cutting conditions were optimized to obtain the same surface input properties. The surface modifications were performed using an ANCA MX7 grinding machine with the cutting conditions, cutting speed ($v_c$) 50 m/s, feed rate ($v_f$) 100 mm/min and depth of cut ($a_e$) 0.01 mm. Milling was done with a DMU 40 and turning with a multitask turning centre CTX BETA 1250TC. The cutting conditions were chosen to be the same for both machining with $v_c = 110$ m/s, $v_f = 140$ mm/min and depth of cut ($a_p$) 0.5 mm. A manually controlled machine with a pressure of 4 bar and a Garnet 100 blasting abrasive was used for sandblasting. The laser treatment (LT) was carried out by a diode-pumped fibre laser from IPG Photonics of the type YLR-150/1500-QCW-MM-AC. The processing parameters selected were $P = 93$ W, $v = 125$ mm/s, $h = 0.09$ mm, the surfaces were remelted twice, each with an angular offset of 90°. Before laser processing, the samples were grinded with silicon carbide abrasive paper with a grain size of P180 (75 µm).

IOP Conf. Series: Materials Science and Engineering 1178 (2021) 012051 doi:10.1088/1757-899X/1178/1/012051

The surface roughness of the samples was determined with the Perthometer H1 from Garant. This is integrated into the system (Perthometer Concept) of the company Mahr (Mahr M1 Perthometer). The Rz values were evaluated and logged using the MarSurf 20 XR 20 V1.40-3 (SP1) software and determined by an average value from three measured points. The surface morphology was analysed by a scanning electron microscope (SEM) Leo Stereo Scan 440 (Figure 1).

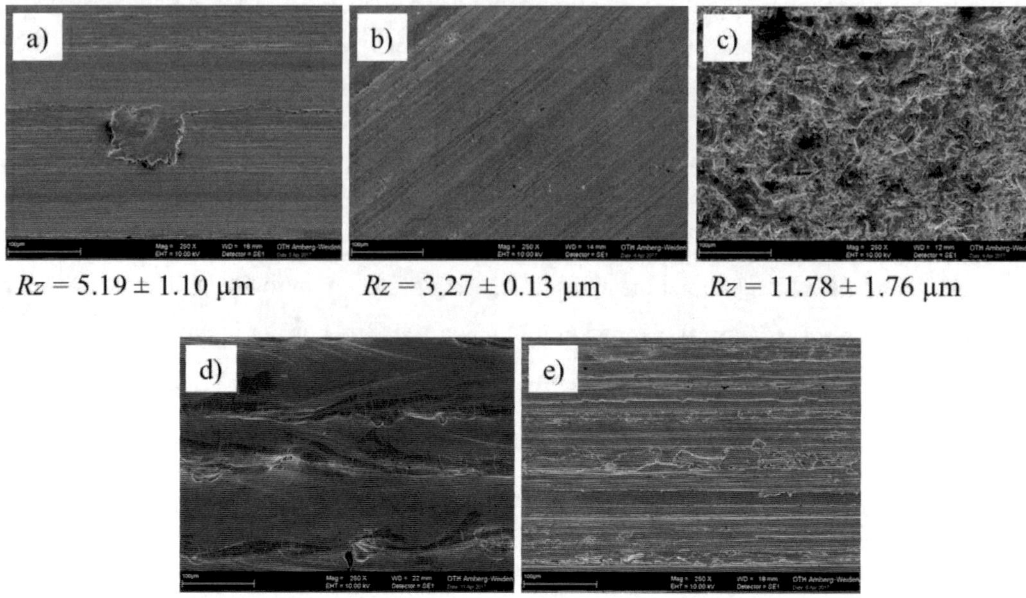


$Rz = 9.99 \pm 1.66$ µm $Rz = 7.65 \pm 1.14$ µm

**Figure 1.** SEM micrographs 200-fold magnification with values of surface roughness (a) milled, (b) turned, (c) sandblasted, (d) as-build, (e) grinded

The laser treated surfaces were analysed by microscope images displayed in Figure 2. The evaluation of the remelting depth was carried out on the cross-section (perpendicular to the building direction) using the BX60M reflected-light microscope from Olympus and the DMC2900 camera from Leica. For image processing the software IMS Client V17Q4 from Imagic was used. Figure 2 a) shows a partial area of the laser treated surfaces with a roughness of 2.64 ± 0.62 µm. To determine the melting depth of the laser treatment metallographic examinations were carried out. Figure 2 b) shows a melting depth *d* of 75 µm, the heat-affected zone is 46 µm.

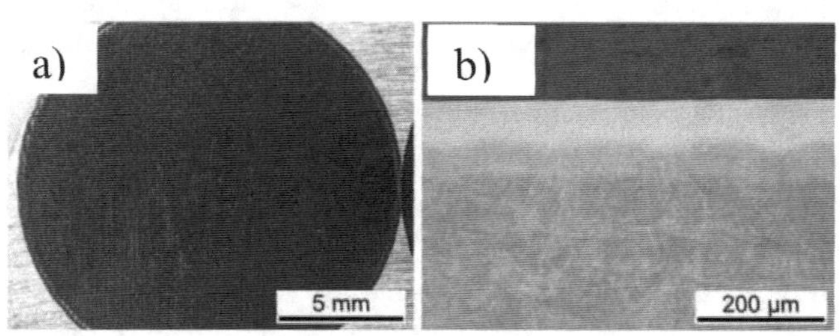


$Rz = 2.64 \pm 0.62$ µm $d = 75$ µm

**Figure 2.** Micrographs with values of surface roughness (a) laser treated, (b) microsection of laser treatment

*2.2 Corrosion measurements*

The corrosion behaviour was investigated by using a potentiostatic measurement system, type Avesta Cell. The performed research work examined corrosion potentials for different surface structures of

IOP Conf. Series: Materials Science and Engineering 1178 (2021) 012051 doi:10.1088/1757-899X/1178/1/012051

solution annealed LPBF parts in comparison to conventionally produced material. The parts for the corrosion investigations (diameter $D$ 35 mm, thickness $t$ 16 mm) were prepared to have different surface qualities, such as milled, turned, sandblasted, as-build, grinded and laser treated ones. The test set-up includes the potentiostat PGS 95 and the software CPC934s1 V 5.1 from Bank Elektronik. The reference electrode used was Ag/AgCl in a $1nH_2SO_4$ electrolyte. Experiments were carried out at 21°C. In addition, a Haber-Luggin capillary was included in the structure. The samples were cathodically polarized at the beginning of the measurement at a potential $\varepsilon_R$ (starting in the anodic area) and driven with $d\varepsilon/dt$ from 0.2 mV/s to $\varepsilon_R \pm 200$ mV. The total measuring area of the individual samples is 0.28 cm². The evaluation of the current density-potential curves was done with Microsoft Excel. In this case, current per area $I/A$ (mA/cm²) is plotted logarithmically against the normal hydrogen electrode (*NHE*). The potentiodynamic polarisation parameters, corrosion potential ($E_{corr}$) and current density ($I_{corr}$), are evaluated values for assessing the corrosion behaviour determined by Tafel plots according to Figure 3. The intersection of the cathodic (yellow line) and the anodic (green line) tangents determine the corrosion behaviour for all measurements. [6] All sample were measured three times to calculate an arithmetic mean value of the corrosion behaviour.

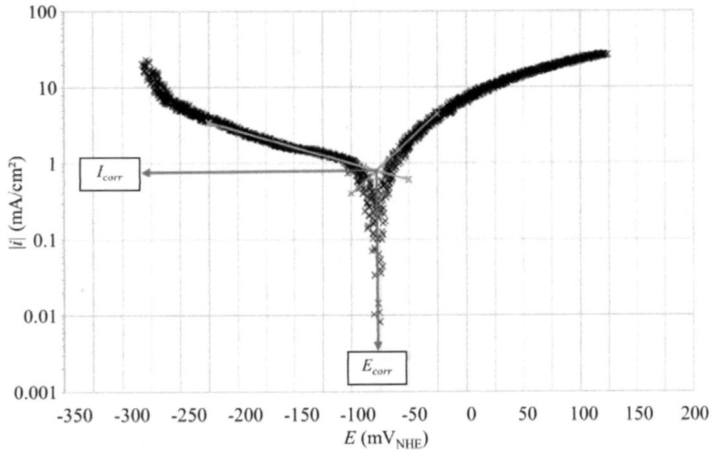


**Figure 3.** Corrosion evaluation by Tafel plots on an additive sample (grinded surface and solution annealed heat treatment)

## 3 Results and discussion

According to the post-processes such as milled, turned, sandblasted, as-build, grinded and laser treated (compare Figure 1 and 2), different surface roughness was realized. The arithmetic mean values of the electrochemical measurements on the different post-processes for annealed samples were summarised in Table 3. The results of the corrosion measurements show that the corrosion potential $E_{corr}$ – varying in a range of -78 ~ -74 mV – is comparable for all post-processes. Also, the current density as a value for the corrosion rate $I_{corr}$ – varying between 0.38 ~ 0.70 mA/cm² – does not show any influence by the surface roughness. The results of the samples and the wrought material show that there is no correlation between the surface roughness and $E_{corr}$ or $I_{corr}$.

**Table 3.** Average $\boldsymbol{E_{corr}}$ and $\boldsymbol{I_{corr}}$ of annealed LPBF samples with different surface qualities

| Surface treatment | Sample | Heat treatment | $E_{corr}$ (mV$_{NHE}$) | $I_{corr}$ (mA/cm²) |
|---|---|---|---|---|
| Milled | Additive | Solution annealed | -75 +/- 2.0 | 0.48 +/- 0.03 |
| Turned | Additive | Solution annealed | -75 +/- 1.5 | 0.55 +/- 0.15 |
| Sandblasted | Additive | Solution annealed | -78 +/- 0.5 | 0.70 +/- 0.10 |
| As-build | Additive | Solution annealed | -74 +/- 1.0 | 0.38 +/- 0.03 |
| Grinded | Additive | Solution annealed | -78 +/- 0.5 | 0.65 +/- 0.05 |
| Grinded + LT | Additive | Solution annealed | -77 +/- 1.5 | 0.50 +/- 0.10 |

The measured corrosion behaviour of the conventional and LPBF samples in the heat treatment stages solution annealed and age hardened is summarized in Table 4 including the mechanical (grinding) and thermal (laser treated) post-processes. It can be observed that, contrary to the statement of Jyoti Suryawanshi [5], the LPBF samples do not deteriorate the corrosion resistance compared with conventional manufactured samples. The investigations revealed that the manufacturing methods have no measurable effect and show similar corrosion behaviour.

**Table 4.** Average $E_{corr}$ and $I_{corr}$ of LPBF and conventional manufactured samples in different heat treatment stages and mechanical and thermal post-processed surfaces

| Surface treatment | Sample | Heat treatment | $E_{corr}$ ($mV_{NHE}$) | $I_{corr}$ ($mA/cm^2$) |
|---|---|---|---|---|
| Grinded | Conventional | Solution annealed | -74 +/- 1.0 | 0.53 +/- 0.03 |
| Grinded + LT | Additive | Solution annealed | -77 +/- 1.5 | 0.50 +/- 0.10 |
| Grinded + LT | Conventional | Solution annealed | -81 +/- 2.0 | 0.70 +/- 0.05 |
| As-Build | Additive | Age hardened | -79 +/- 3.5 | 1.25 +/- 0.05 |
| Grinded | Additive | Age hardened | -73 +/- 2.5 | 1.20 +/- 0.15 |
| Grinded | Conventional | Age hardened | -71 +/- 1.5 | 1.55 +/- 0.15 |
| Grinded + LT | Additive | Age hardened | -78 +/- 0.5 | 0.40 +/- 0.05 |
| Grinded + LT | Conventional | Age hardened | -79 +/- 0.5 | 0.50 +/- 0.05 |

However, it can be observed that the corrosion of the solution annealed samples is lower than that of the age hardened samples. The corrosion potentials $E_{corr}$ for the samples according to Figure 3 are in the range of -81 ~ -71 mV (values referring to the potential of the normal hydrogen electrode *NHE*). The corrosion current densities of the solution annealed samples are in a range of $I_{corr}$ = 0.50 ~ 0.7 mA/cm². The corrosion current densities of the age hardened samples (without LT) vary between $I_{corr}$ = 1.2 ~ 1.55 mA/cm², which corresponds to a factor of about two. The corrosion current density is directly proportional to the corrosion rate, which doubles in this case. This may be due to the fact that these samples are solution annealed and that this heat treatment results in a homogeneous microstructure. A single phase or homogeneous metal resisted the corrosion attack better than a multiphase or inhomogeneous metal. [7]

In this condition, the samples contain a martensitic matrix consisting predominantly of iron and nickel. As a result, the corrosion behaviour is comparable to that of iron and nickel. In the case of MAR 300, the precipitations formed during aging have been identified as complex of nickel-cobalt-molybdenum-titanium ($Ni_3Mo$ and $Ni_3Ti$). [8]–[9] These intermetallics have different compositions therefore their electrochemical behaviour differs from the matrix. [10]–[13] Due to the different crystal structures and lattice parameters, stress fields formed around the precipitations. These composition differences lead to a lower corrosion resistance of age hardened samples due to the galvanic effect than the solution annealed ones. [14]–[17] The corrosion current densities of the age hardened and subsequently laser treated samples show a similar corrosion behaviour as the solution annealed states ($I_{corr}$ = 0.4 ~ 0.5 mA/cm²). As a result of the laser treatment, the precipitations in the melted area of the microstructure dissolve again and thus increase the corrosion resistance of the material again. In principle, all corrosion attacks are uniform surface corrosion.

## 4 Conclusion

The results demonstrate that samples from MAR 300 processed by LPBF with high structure density and with a low number of pores deliver similar results as conventionally manufactured ones.

- It can be observed that different surface qualities such as milled, turned, sandblasted, as-build and grinded have no significant influence on the corrosion resistance.
- However, the corrosion resistance of age hardened samples is lower than of solution annealed samples. This leads to a lower corrosion resistance of the age hardened specimen due to a galvanic effect. For the solution annealed samples, the corrosion current densities are in a range of $I_{corr}$ = 0.38 ~ 0.7 mA/cm². The corrosion current densities, approximately twice as high, were measured for them, $I_{corr}$ = 1.2 ~ 1.55 mA/cm².

- By laser treatment, the precipitations could be dissolved in a thin surface layer of 75 µm and thus the corrosion rate decreases to the values of the solution annealed condition. The influence of the laser treated surface on the mechanical properties must be examined further.

The manufacturing method of the additive or conventional material samples has no measurable effect on the corrosion resistance. The investigations show that LPBF components show a similar corrosion behaviour as the conventionally produced samples.

**Acknowledgements**
This study was supported by the BTHA, "New materials in the additive manufacturing" with grateful acknowledgement.

**References**
[1] Rüßmann M, Lorenz M, Gerbert P, Waldner M, Justus J, Engel P and Harnisch M 2015 *Industry 4.0: The Future of Productivity and Growth in Manufacturing Industries* Boston Consulting Group: Boston MA USA **Volume 9**
[2] Frenk A, Kurz W 1992 *Laser Surface Treatment: Microstructural Aspects,* Springer Science and Business Media Dordrecht
[3] Schaller R F, Taylor J M, Rodelas J and Schindelholz J 2016 *Corrosion Properties of Powder Bed Fusion Additively Manufactured 17-4 PH Stainless Steel* Sandia National Laboratories Albuquerque
[4] Ettefagh A H and Guo S 2016 *Electrochemical behavior of AISI316L stainless steel parts produced by laser-bed powder bed fusion process and the effect of post annealing process* ScienceDirect Additive manufacturing: Louisiana State University Baton USA
[5] Suryawanshi J, Baskaran T, Prakash O, Arya S B and Ramamurty U 2018 *On the corrosion resistance of some selective laser melted alloys* ScienceDirect Materialia: Indian Institude of Science Bangalore India
[6] Buchanan R A and Stansbury E E 2012 *Handbook of Environmental Degradation of Materials* Elsevier **2nd ed.(4)** pp 113-117
[7] Wadhwa V C 1965 *Corrosion behavior of heat treated 18 per cent nickel maraging steel*
[8] Floreen S 1968 *The physical metallurgy of maraging steels* Metallurgical Reviews **13:1** pp 115-128
[9] Tewari R, Mazumder S, Batra I S, Dey G K and Banerjee S 2000 *Precipitation in 18 wt% Ni maraging steel of grade 350* Acta Mater **48(5)** pp 1187–1200
[10] Shoesmith D W 1992 *Metals Handbook* ASM International 9th ed. **volume 13**
[11] Sedriks A J 1996 *Corrosion of Stainless Steels* Electrochemical Society **2nd ed.** New York: Wiley
[12] Chan K and Tjong S 2014 *Effect of secondary phase precipitation on the corrosion behavior of duplex stainless steels* Materials **7(7)** pp 5268-5304
[13] Poornima T, Nayak J and Shetty A N 2010 *Corrosion of aged and annealed 18 Ni 250 grade maraging steel in phosphoric acid medium* Int. J. Electrochem. Sci **5** pp 56-71
[14] Kumar P and Shetty A N 2013 *Corrosion behaviour of 18% Ni M250 grade maraging steel under welded condition in hydrochloric acid medium* Portugaliae Electrochimica Acta **31(1)** pp 21-32
[15] Ansell T Y, Ricks J P, Park C, Tipper C S and Luhrs C C 2020 *Mechanical Properties of 3D-Printed Maraging Steel Induced by Environmental Exposure* Metals **10(2)** 218.
[16] Tan C, Zhou K, Kuang M, Ma W and Kuang T 2018 *Microstructural characterization and properties of selective laser melted maraging steel with different build directions* Science and Technology of Advanced Materials **19(1)** pp 746-758
[17] Avelino A F, Araújo W S, Dias D F, dos Santos L P M, Correia A N and de Lima-Neto P 2018 *Corrosion investigation of the 18Ni 300 grade maraging steel in aqueous chloride medium containing H2S and CO2* Electrochimica Acta **286** pp 339-349

IOP Conf. Series: Materials Science and Engineering 1178 (2021) 012052 doi:10.1088/1757-899X/1178/1/012052

# The influence of the temperature on the mechanical properties of Sanicro 25 steel

**M Sroka**[1]

Silesian University of Technology, Akademicka 2A, 44-100 Gliwice, Poland

Corresponding author's e-mail address: marek.sroka@polsl.pl

**Abstract**. Austenitic steel 22Cr25NiWCoCu (Sanicro 25) is one of the newest and most promising steels for use in supercritical and ultra-supercritical power units. High resistance of Sanicro 25 steel to corrosion and oxidation in the steam atmosphere at up to 700 °C is ensured by chromium content at 21.5-23.5 wt%. Also, higher chromium content provides less mass decrement during high-temperature corrosion at 700 °C. Currently, the Sanicro 25 steel is characterised by the highest creep strength among the commercially used creep-resistant stainless steels. In this work, the study of precipitation processes and microstructure stability in the delivery state and after the ageing process in time up to 20,000 h at 750 °C was performed. Identification of secondary phases was performed using electron diffraction on thin films using transmission electron microscopy and X-ray phase composition analysis. Long-term ageing depending on time and temperature showed that there are four secondary phases in Sanicro 25: $M_{23}C_6$, MX, NbCrN, σ-phase and Cu_rich. The precipitation and increase in the size of secondary phase particles during ageing results in a decrease in matrix saturation with alloying elements, which leads to a reduction in hardening by solid solution mechanism. These processes and effects result in changes in mechanical properties. The alloy overageing effect observed for the ageing temperature of 750 °C results in obtaining strength properties similar to those of the steel in the as-received condition.

## 1 Introduction

In the analysis of the operation of the national power system within the next several years, the following factors should be taken into account: European Union law, world and national climate policies, Poland's energy security, ability to improve technical and economic indicators, development and availability of materials and operating technology. These factors and significant depreciation of the national power generation infrastructure must be taken into account in the development of the method for achieving the technical and economic objectives and meeting the increasingly stringent EU requirements focused on the implementation of the so-called Climate and Energy package concerning, but not limited to, the maximum greenhouse gas emission limits [1-4].

Despite the increase in harmful gas emission prices, which were to curb the development of the conventional power generation, there are no signs that the proportion of electricity generation sources could significantly change in the coming years. The electricity production in Poland is still based on coal and lignite. Due to having the national coal and lignite reserves, they are the primary source of energy. This results in adverse emission of significant amounts of pollutants into the atmosphere, such as carbon dioxide ($CO_2$), sulphur dioxide ($SO_2$), nitrogen oxide ($NO_x$) and different types of dusts. These pollutants can be reduced, for example, by increasing steam parameters which affect the increase in boiler operation efficiency and reducing coal consumption.

The other important aspects are the optimisation of processes and system modules, strict control of the operation, an increase in thermal flexibility and introduction of the carbon capture technology [5-9].

The economic and environmental considerations and continuous improvement in the thermal efficiency of power units force the introduction of newer and newer materials for use in modern conventional power industry which includes, but are not limited to, the newly developed austenitic steels [10-12].

The 22Cr25NiWCoCu (Sanicro 25) steel was developed by AB Sandvik Material Technology in Sweden. This steel was produced under the European Therme AD700 program aimed at the development of a new type of power unit, and thereby the structural materials with a stable microstructure, high mechanical properties and significant resistance to corrosion and steam oxidation during the operation at 700 °C [13].

## 2 Materials and methods

The subject of the research was ϕ38x8.8mm samples taken from the superheater coils made of creep-resistant austenitic steel 22Cr25NiWCoCu (Sanicro 25). The chemical composition of the investigated steam superheater sample is presented in Table 1. The scope of the research included the investigations of both the microstructure and mechanical properties of Sanicro 25 after long-term ageing at 750 °C. The microstructure of the Sanicro 25 steel was observed with Inspect F scanning electron microscope (SEM) on conventionally prepared electrolytically, etching in the Mi19Fe reagent, metallographic microsections.

The identification of precipitates was carried out using the selected area electron diffraction through a transmission electron microscope Titan 80-300 (TEM) with thin foils, which were electrolytically thinned to perforation using a solution of 20% perchloric acid in ethanol at a temperature of approx. -30 °C and voltage of approx. 20 V. Long-term ageing is one of the most appropriate methods to simulate the operating conditions of materials used in the power sector. The investigations were carried out for material in the as-received state and aged for up to 20,000 h. Ageing at high-temperature levels for up to 20,000 h was to simulate changes in the microstructure and mechanical properties at a temperature similar to the long-term service temperature. The application of ageing temperature higher than the maximum expected service temperature was to accelerate the degradation processes in the test steel without changing the nature of the phenomenon.

**Table 1.** Chemical composition of investigated material, wt.%

| Chemical composition, wt.% | | | | | | | | | | | | |
|---|---|---|---|---|---|---|---|---|---|---|---|---|
| **C** | **Si** | **Mn** | **P** | **S** | **Cu** | **Cr** | **Ni** | **W** | **Co** | **Nb** | **B** | **N** |
| 0.06 | 0.25 | 0.50 | 0.01 | <0.01 | 2.9 | 23.0 | 24.1 | 3.2 | 1.4 | 0.4 | 0.005 | 0.17 |

## 3 Results and discussion

In the as-received condition (figure 1), the Sanicro 25 steel had an austenitic matrix with a grain size of 7 according to the scale of ASTM E112-13 standards (which corresponds to the average diameter of 31.2 μm). Fine-grain structure of austenitic steels contributes to an increase in their strength and plastic properties and impact energy, but also have a positive impact on an increase in oxidation resistance compared to the coarse-grained steels. The improved oxidation resistance of fine-grained austenitic steel results from numerous and easy ways of diffusion, such as grain boundaries. However, during ageing/operation, grain boundaries as defects which allow faster diffusion compared to the grain interior affect the reduction in σ-phase precipitation incubation time. In the microstructure of the test steel in the solution-treated condition, primary NbX and NbCrN precipitates (Z phase) are observed.

Longer ageing times result in the occurrence of the particle with different morphology, i.e. σ-phase, at the grain boundary. The amount and size of precipitates of this phase increase with the extension of the ageing time, which results from high susceptibility of the σ-phase to coagulation. At a higher ageing temperature, 750 °C, the relative amount and size of the σ-phase precipitates (figure 2) at the grain boundaries are greater than those in the material aged at 700 °C. Like $M_{23}C_6$ carbides, the σ-phase is precipitated at the grain boundaries, and the privileged site is the interface of three-grain boundaries.

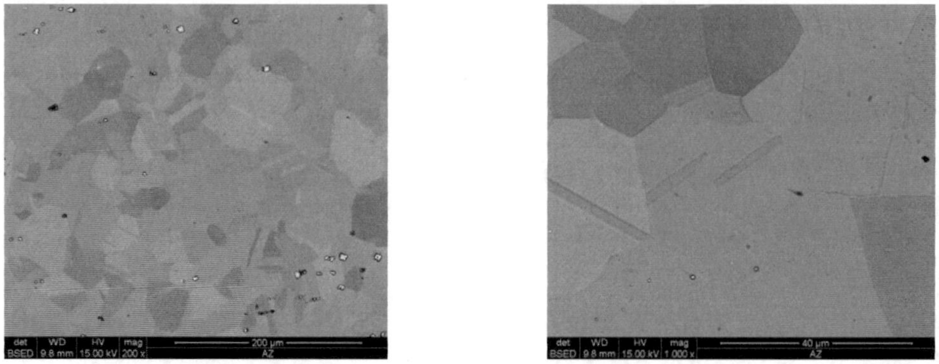


**Figure 1.** The microstructure of Sanicro 25 in the as-received condition, SEM

In addition to ε_Cu precipitates, the presence of finely-dispersed secondary Z-phase precipitates at 750 °C was revealed too. The finely-dispersed, or worm-shaped, the form of these precipitates and their relatively high stability make both single particles and clusters of these precipitates effectively inhibit and limit the dislocation movement, and thus contribute to precipitation hardening of the alloy. The interaction between a precipitate and dislocation may take place by the Orowan mechanism or cutting through a particle. The precipitation hardening is assumed to be mainly related to interaction by the Orowan mechanism because secondary precipitates in austenitic steels are too hard to be cut by dislocations [14,15].

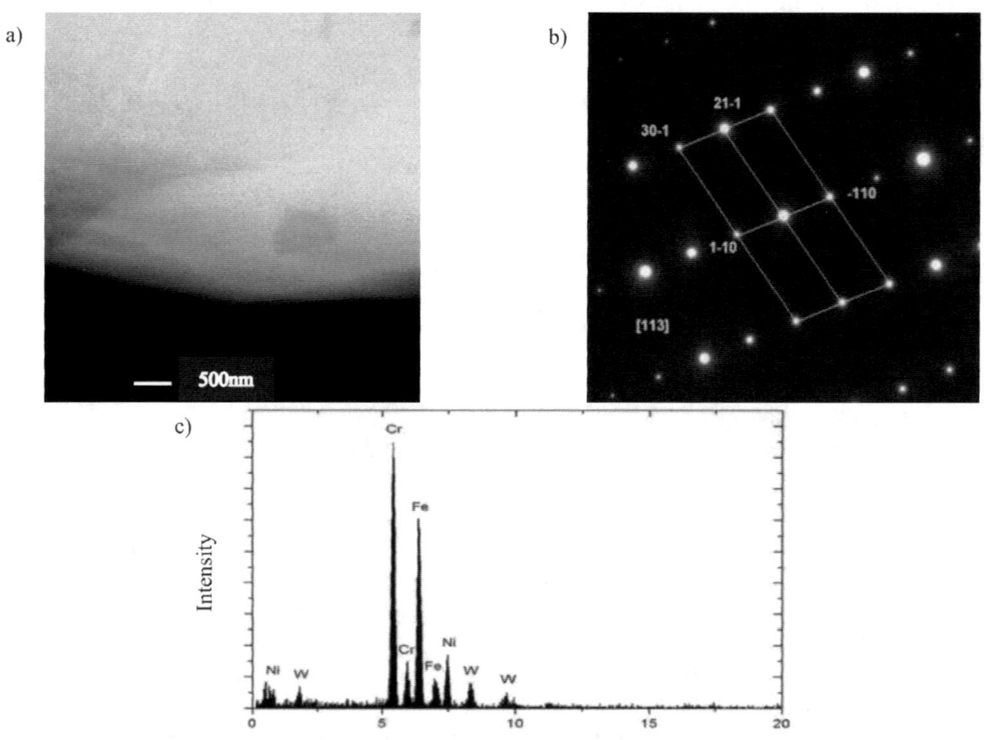


**Figure 2.** σ-phase precipitate in Sanicro 25 steel after 20,000 h ageing at 750 °C: a) bright field, b) resolved diffraction pattern from the area in figure a, c) EDS spectrum

In addition to the ε_Cu phase, Laves phase and secondary Z-phase particle precipitates, the existence of composite complexes of precipitates: primary Z-phase precipitates – Laves phase – $M_{23}C_6$ carbide precipitates, where the $M_{23}C_6$ particles and the Laves phase nucleate heterogeneously on the primary Z-phase precipitate, were revealed within the grains in the test alloy after 20,000 h ageing at 750 °C (figure 3).

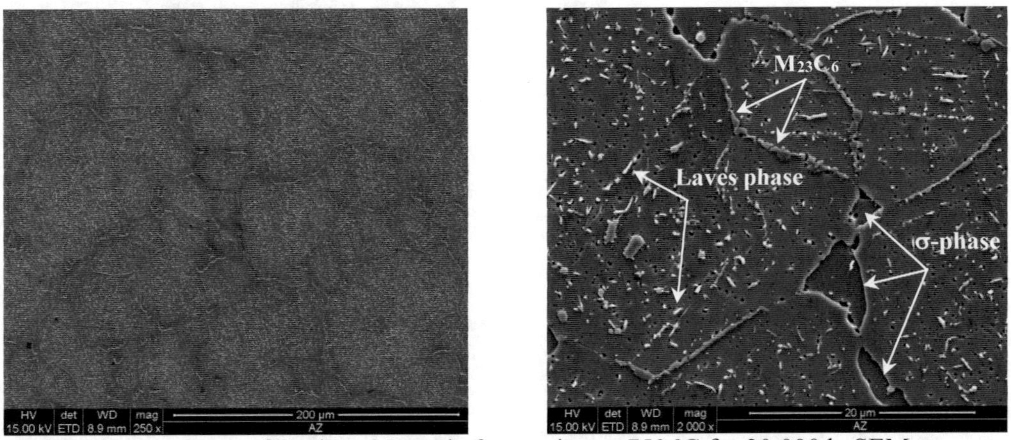


**Figure 3.** The microstructure of Sanicro 25 steel after ageing at 750 °C for 20,000 h, SEM

Changes in microstructure of the steel affected the reduction in its mechanical properties. In the initial period of ageing at 750 °C, the strength properties are increased, and plastic properties are decreased.

The effect of ageing time at 750 ° C on the tensile strength, yield strength and elongation determined at room temperature is shown in figures 4-6. The increase in tensile strength after 5,000 h ageing can be seen, which was about 9% of the properties as-received. After this time, a slight decrease in tensile strength was observed, which is about 5% after ageing of 20,000 h. A similar tendency was observed for the yield strength, which initially increases by approx. 11% for 5,000 h of ageing, and then slightly decreases (by about 6%) for the material aged 20,000 h. This trend is very similar to the determined hardness values and takes place as a result of the appearance of a large number of precipitates of different morphology. The elongation, on the other hand, decreased by about 60% after the 20,000 h ageing compared to the elongation of the material as-received. This decrease is due to the coagulation and spheroidization of the precipitates and the processes of twins decay. After 20,000 h of ageing, the tensile strength, yield strength values are still above the minimum [16], while the plasticity of the tested steel has dropped by half.

The precipitation and increase in the size of secondary phase particles during ageing results in a decrease in matrix saturation with alloying elements, which points to a reduction in hardening by solid solution mechanism. Also, changes in the dislocation substructure, i.e. dynamic recovery and recrystallisation processes, loss of twins and formation and increase in width of the precipitation-free zone, were visible in the test steel. These processes and effects result in changes in mechanical properties. The rate of the changes in the microstructure of the test steel depends on the ageing temperature. At a higher temperature, the steel microstructure degradation was more advanced, which resulted in a gradual reduction in proof stress and tensile strength determined at room temperature.

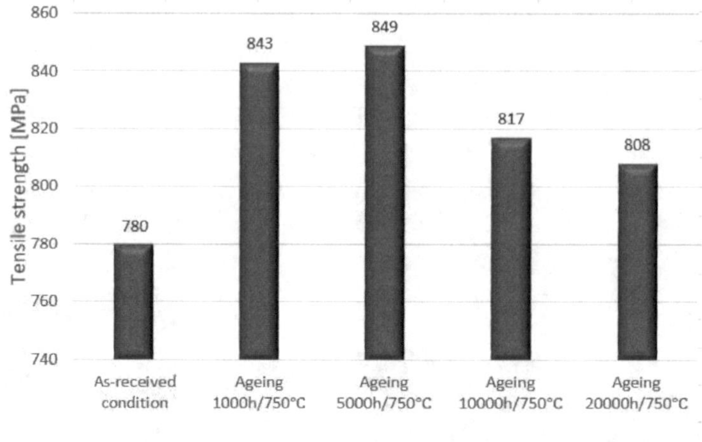


**Figure 4.** Change in tensile strength of Sanicro 25 steel after long-term ageing at 750 °C

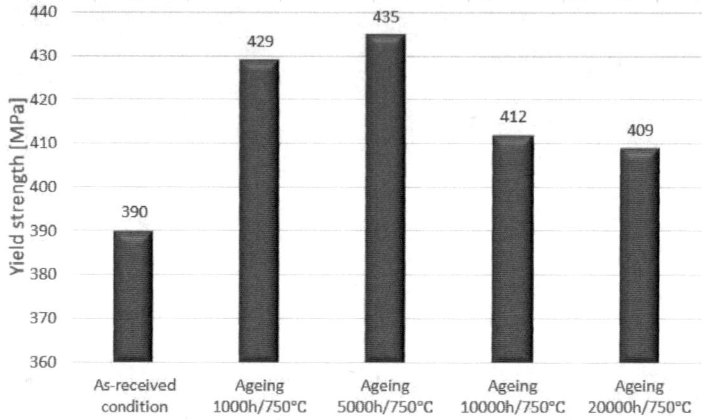


**Figure 5.** Change in yield strength of Sanicro 25 steel after long-term ageing at 750 °C

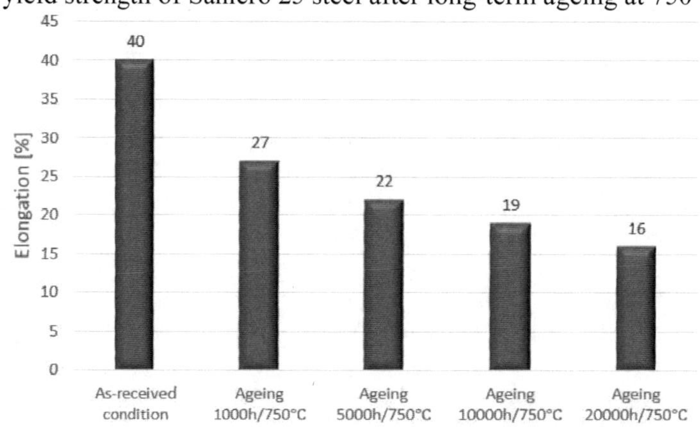


**Figure 6.** Change in elongation of Sanicro 25 steel after long-term ageing at 750 °C

## 4 Conclusion

Due to its good mechanical properties, including high creep strength with simultaneously high resistance to high-temperature corrosion and economically well-balanced chemical composition, the Sanicro 25 steel is an alternative to creep- and heat-resistant nickel alloys which are currently used in the power industry for components of the pressure equipment designed to work at elevated temperature.

The alloy overageing effect observed for the ageing temperature of 750 °C results in obtaining strength properties similar to those of the steel in the as-received condition.
The metallographic examinations carried out for Sanicro 25 steel in the as-received condition and after ageing have shown that ageing of up to 20,000 h at 750 °C leads to the precipitation of copper-rich (ε_Cu) particles, Laves phase, σ-phase and secondary Z phase precipitates in the matrix, while at the grain boundaries the occurrence of $M_{23}C_6$ and Laves phase.

**Acknowledgements**
The results in this publication were obtained as part of the research co-financed within the framework of the statutory financial grant supported by the Faculty of Mechanical Engineering of the Silesian University of Technology 2020.

*References*

[1] Sroka M, Zieliński A and Mikuła J 2016 *Arch. Metall. Mater.* **61** 969
[2] Król M, Tański T and Sitek W 2015 IOP Conference Series-Materials Science and Engineering (MODTECH 2015) Vol. 95 012006
[3] Śliwa A, Sroka M, Bloch K, Sandu I G, Abdullah M M AL B and Sandu A V 2018 *Rev. Chim-Bucharest.* **69** 324
[4] Snopiński P, Król M, Tański T and Krupińska B 2018 *J. Therm. Anal. Calorim.* **133** 379
[5] Dobrzański LA, Borek W and Mazurkiewicz J 2016 *Arch. Metall. Mater.* **61** 725
[6] Sroka M, Zieliński A and Golański G 2019 *Acta Phys. Pol. A* **135** 207
[7] Sozańska-Jędrasik L, Mazurkiewicz J, Borek W and Matus K 2018 *Arch. Metall. Mater.* **63** 265
[8] Sourmail T *Mater. Sc. Techn.* 2001 **14** 1
[9] Chi Ch–Y, Yu H–Y, Dong J–X, Xie X–S, Chen X–F and Lin F 2011 *Acta Metall. Sin.* **24** 141
[10] Plaut R L, Herrera C, Escriba D M, Rios P R and Padilha A F 2007 *Mater. Research* **10** 453
[11] Zieliński A, Miczka M and Sroka M 2016 *Mater. Sci. Tech-Lond.* **32** 1899
[12] Golański G, Zielińska-Lipiec A, Zieliński A and Sroka M 2017 *J. Mater. Eng. Perform.* **26** 1101
[13] Chai G and Forsberg U 2017 Sanicro 25: An advanced high-strength, heat-resistant austenitic stainless steel ed. A. Di Gianfrancesco (Cambridge: Woodhead Publishing) chapter 12 pp 391-421
[14] Wang J-Z, Liu Z-D, Bao H-S, Cheng S-C and Wang B *J.* 2013 *Iron Steel Res. Int.* **20** 54
[15] Bai J W, Liu P P, Zhu Y M, Chi C Y, Yu X Y, Xie X S and Zhan Q 2013 *Mat. Sci. Eng. A-Struct.* **584** 57
[16] Materiały Sandvik Materials Technology, Sanicro TM 25 (UNS S31035) – A new austenitic tube grade for use in coal-foired boilers at material temperatures up to about 700°C.

IOP Conf. Series: Materials Science and Engineering 1178 (2021) 012053 doi:10.1088/1757-899X/1178/1/012053

# Problems with Surface Protection of Welded Joints by Vitreous Enamel Coating

**Sternadelová K, Mohyla P, Kozák J, H Krupová and L Krejčí**

VSB - Technical University of Ostrava, Faculty of Mechanical Engineering, 17. listopadu 2172/15, Ostrava - Poruba, 708 00

E-mail: kristyna.sternadelova@vsb.cz

**Abstract**. This work deals with the application of a vitreous enamel coating on welded joints, whereas vitreous enamel being intentionally applied in high thicknesses. As a base material, we used low-carbon steel Kosmalt E 300 T. Two different processes were chosen for pre-treatment - chemical and mechanical. The resulting enamel coating consisted of two successively applied layers of base and cover enamel coating. The thickness and microhardness of the coating were measured. The metallographic analysis was also performed on selected samples. The results show that the high thickness of the enamel coating significantly impairs the useful properties of the product.

## 1 Introduction

The basic raw material for enamel production is an enamel frit (vitreous glass granules). It is produced by melting a mixture of various materials, such as silica, borax, potassium chloride, alumina, titanium dioxide, and other metal oxides. Enamel frit is a key ingredient for making enamel slip which is a mixture that is suitable for applying to a metal surface. Enamel slip is created by milling enamel frit with water, clays, pigments, and electrolytes into wet suspension in a ball mill that contains porcelain-grinding materials [1].

The main reason for the wide use of enamel coatings in the industry is their corrosion resistance. The enamel coatings also have great resistance to mechanical abrasion and chemical resistance. Due to their glassy nature coatings exhibit excellent impermeability, washability, high-temperature resistance, and fire resistance. Another advantage of enamel coating is its ability to withstand UV radiation with the characteristics of the surface unchanged in time [2]. This feature together with wear and chemical resistance makes enamel coatings suitable for the pharmaceutical, food, or chemical industry.

In these industries, there are cases where it is necessary to form an enamel coating on a product that contains welded joints, such as pressure vessels for various chemicals or water heaters. The welded joint will create chemical and geometric inhomogeneity on the surface of the product. One problem is the strong increase in coating thickness around the weld bead. The recommended total maximum thickness of the enamel coating for consumer goods is 450 microns according to [3]. For use in aggressive environments, it is permitted to create a higher thickness, but in the vicinity of the weld bead, the thickness of the coating may increase several times. Another problem is chemical inhomogeneity in the area of the welded joint.

The presented work focuses on the application of a vitreous enamel coating of high thicknesses. In the experimental part of this work, very thick coatings were intentionally created, corresponding to the geometric shape of the weld bead. In the extreme case, the thickness of the enamel coating reached almost 1500 microns.

## Experimental

### 1.1 *Base material*

As the base material for welding was used low carbon steel Kosmalt E 300 T. Table 1 shows the chemical composition of the base material as well as the weld metal. Glow-discharge optical emission spectrometry (GDOES) analysis was performed using a Spectrum Analytic optical emission spectrometer according to [4, 5].

**Table 1.** Composition of the base metal and weld metal in Wt.%.

| Element | C | Mn | Si | P | S | Al | Cu | Ti |
|---|---|---|---|---|---|---|---|---|
| **Base Metal** | 0.039 | 0.195 | 0.027 | 0.013 | 0.011 | 0.048 | 0.021 | 0.060 |
| **Weld Metal** | 0.064 | 0.942 | 0.605 | 0.011 | 0.014 | 0.009 | 0.026 | 0.007 |

### *1.2 Preparation of samples*

Samples for welding were prepared by cutting from a whole piece of Kosmalt E 300T steel sheet with a thickness of 5 mm. The individual samples had dimensions of approximately 198x90 mm. Before welding, the edges of the weld surface were chamfered to 30 ° and stitched. The ABB IRB 1660 ID-4 welding robot was used for welding. Method 135 (MIG / MAG) was used for all samples. A solid wire from ESAB with the designation OK AristoRod 12.50 (C = 0.10 %, Mn = 1.50 %, Si = 0.90 %) was used as additional material. The size of the weld gap ranged from 1.0 to 1.6 mm.

The welding current ranged from 210 to 250 A and the welding voltage from 18 to 24 V. Shielding gas M21 (82 % Ar + 18 % $CO_2$) with a flow rate of 12 l/min was used to protect the weld according to the standard EN ISO 14 175. The welding speed was 5 mm/s.

### 1.3 *Pretreatment*

For the first set of samples (K-01, R-01A, R-02A, R-03A, and R-04A), the pretreatment consisted of the following steps. Degreasing by an aqueous solution of Simple Green in ratio 1:15 at room temperature (approximately 20 °C) for 20 minutes. Then samples were pickled in 10 % HCl solution at 27 °C for 18 minutes. Following neutralization was carried out in 10 % NaOH solution at 40 °C for 4 minutes. Finally, samples were rinsed with demineralized water and dried in a drying oven at 100 °C.

The pretreatment of the second set of samples (K-02, R-01B, R-02B, R-03B, and R-04B) consisted of blasting the surface after welding. Silica sand was used as the blasting material. Before enameling, all samples were degreased in an aqueous solution of Simple Green in a ratio of 1:15 for 20 minutes. Then rinsed with demineralized water and dried in a drying oven at 100 °C.

### 1.4 *Enameling*

The enamel suspension was applied by the wet method. Both the base and cover enamel layers were applied by hand spraying with a pressure gun.

After application and drying the blue Ferro base enamel, the samples were placed into a furnace and fired at 840 °C for 12 minutes. The cover green Mefrit enamel was applied after samples cooled down to room temperature and then dried and fired at 830 °C in the furnace for 9 minutes burning times and temperatures were determined based on previous experience. Samples R-02A and R-02B were selected for further analysis.

## 2 *Results*

### 2.1 *Thickness measurement*

Elcometer 456 digital thickness gauge was used to measure the thickness of the enamel coating layer. The measurement was performed according to the ISO 2178 standard. The samples were divided into 3 measuring areas - above the weld (Area 1), in the weld area (Area 2) and below the weld (Area 3), see Figure 1. In each area, fifty measurements were performed.

The measured thickness of the enamel coating in Area 1 ranged from 775 μm to 1498 μm. In Area 2 it ranged from 718 μm to 1495 μm. The coating thicknesses ranging from 556 μm to 1478 μm were measured in Area 3. Summarized measured coating thicknesses are listed in Table 2.

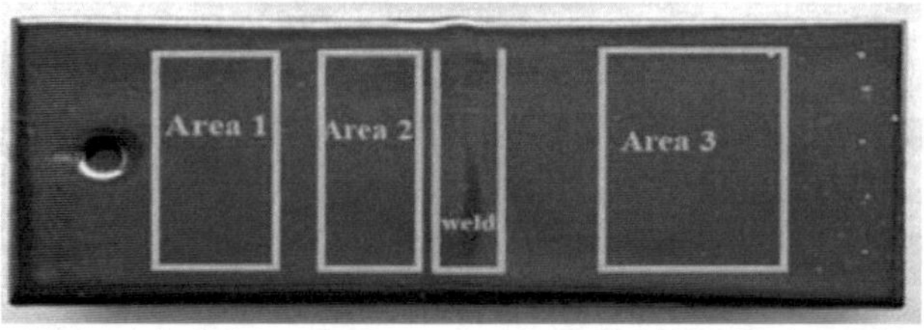


**Figure 1:** Measuring areas – Area 1, Area 2, Area 3.

**Table 2.** Measured mean values of coating thickness.

| **Sample** | **Area 1** [μm] | | **Area 2**[μm] | | **Area 3**[μm] | |
|---|---|---|---|---|---|---|
| | **Mean** | **Std. Dev.** | **Mean** | **Std. Dev.** | **Mean** | **Std. Dev.** |
| **K-01** | 1391 | 76 | 1362 | 93 | 1271 | 107 |
| **R-02 A** | 893 | 97 | 830 | 69 | 610 | 31 |
| **K-02** | 1210 | 66 | 1395 | 62 | 1015 | 64 |
| **R-02 B** | 1404 | 44 | 1359 | 63 | 861 | 83 |

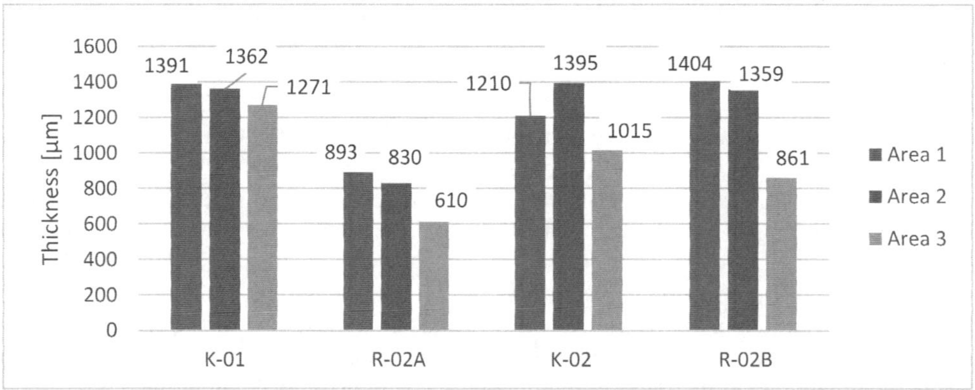


**Figure 2.** Comparison of mean values of measured thicknesses in three measured areas.

Figure 2 shows a comparison of the measured thickness of the enamel coating on selected prepared samples. The coating thickness on the R-02A sample reached the lowest values in all measured areas. Unlike the other samples, the enamel coating of sample R-02A did not peel off in preparation for further analysis. When comparing the thicknesses, it is clear the peeling occurred due to higher values of the thickness of the enamel coating.

## 2.2 *Microhardness measurement*

The microhardness measurement according to Vickers was performed following the EN ISO 4516 standard. The microhardness was measured on a LECO microhardness tester LM 247 AT with a load of 100 g. Eight indentations were made in one line, out of the weld area. Table 3 shows the comparison of microhardness measured values of R-02A and R-02B samples.
According to the paper [6], the microhardness of the enamel is, inter alia, dependent on the temperature and the firing time. The measured values correspond to the microhardness given in the [6] work and, after conversion to megapascals, also partially correspond to the results obtained in the [7] work.

**Table 3.** Microhardness measurement, $HV_{0.1}$

| | 1 | 2 | 3 | 4 | 5 | 6 | 7 | 8 | Mean | Std. Dev. |
|---|---|---|---|---|---|---|---|---|---|---|
| | | | | | **R-02A** | | | | | |
| **Base layer** | 573 | 594 | 585 | 578 | 551 | 595 | 546 | 597 | 577 | 20 |
| **Cover layer** | 632 | 621 | 608 | 593 | 610 | 626 | 579 | 640 | 614 | 20 |
| | | | | | **R-02B** | | | | | |
| **Base layer** | 568 | 607 | 573 | 568 | 612 | 627 | 646 | 648 | 606 | 33 |
| **Cover layer** | 646 | 676 | 738 | 570 | 618 | 570 | 698 | 521 | 630 | 74 |

The sample R-02A (pretreated by pickling) was measured without more frequent deviations. Compared to that the coating on sample R-02B pretreated by blasting contained a larger number of pores and was, therefore, more difficult to measure.
The indentor often penetrated directly into the pore. The increased number of pores in the coating of sample R-02B could be caused by different pretreatment.

## 1.3 *Metallographic analysis*

For better visualization of the enamel coating on the welded joints were samples subjected to metallographic analysis. The samples were cut and encapsulated. Prepared capsules were subsequently mechanically treated by grinding and polishing. To highlight the resulting structure, the samples in the capsules were etched with Nital etchant. Figure 6 shows a split sample R-02A.

**Figure 6.** View of the split sample R-02A.

Figures 7 and 9 show the macrostructure of the cross-section of samples R-02A and R-02B, respectively. There is a clear difference in the thickness of the coating on both samples. The cross-section of sample R-02A (Figure 7) clearly shows the individual layers of the enamel coating. The figure shows the thicknesses of the base and cover enamels, the base enamel ranged from 237 to 588 microns, the cover enamel ranged from 205 to 269 microns. The basic layer of enamel is blue and the covering layer of enamel is green.

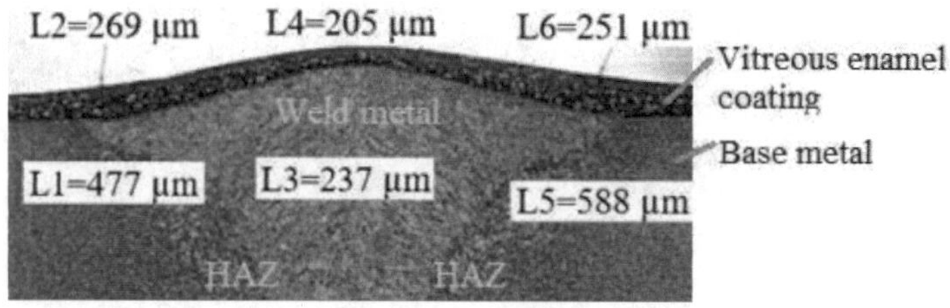


**Figure 7.** Macrostructure of the sample R-02A.

Figure 8 shows divided sample R-02B on which the enamel peeled off during cutting. It is clear the thickness of the layer significantly affects the adhesion of the coating.

**Figure 8**. Divided sample R-02B.

The cross-section of sample R-02B (Figure 9) clearly shows the individual layers of the enamel coating. The figure shows the thicknesses of the base and cover enamels, the base enamel ranged from 488 to 1175 microns, the cover enamel ranged from 376 to 471 microns. The basic layer of enamel is blue and the covering layer of enamel is green.

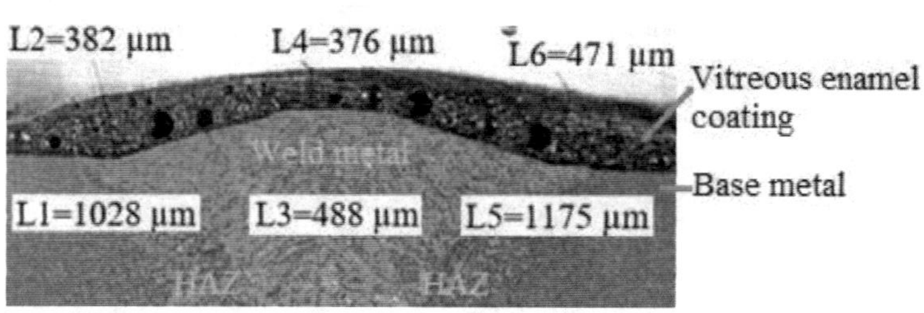


**Figure 9.** Macrostructure of the sample R-02B.

## 4 Conclusion

This study deals with the application of high thickness vitreous enamel coating on welded joints. Very thick coatings were created within the experiment, corresponding to the geometric shape of the weld bead.

The lowest measured thickness of the enamel coating was 610 ± 31 μm on the sample R02-A. The highest measured thickness of the enamel coating was 1404 ± 44 μm on the sample R02-B. The cutting process of the samples confirmed that a higher thickness of the enamel leads to greater destruction of enamel coating. For samples with higher thickness of the coating the enamel peeled off during cutting.

Samples R-02A and R-02B which differed used pretreatment before application of enamel coating were used for further research. Both samples included pores of different amounts and sizes. The presence of pores was higher in the sample R-02B.

Furthermore, the microhardness of the base and cover layer of the enamel was measured by the Vickers method with a load of 100 g. The mean microhardness of the base layer of the sample R-02A was 577 $HV_{0.1}$ and the mean microhardness of the cover layer of the same sample was 614 $HV_{0.1}$. The mean microhardness of the base layer of the sample R-02B was 606 $HV_{0.1}$ and the mean microhardness of the cover layer of the same sample was 630 $HV_{0.1}$.
The study shows that the coating thickness should be uniform and without major fluctuations. This could be achieved either by high-quality spraying (requires good skills of the person) or, at best, by applying powder enamel in a high-voltage electrostatic field [8]. In addition, this technology is not so sensitive to the quality of the surface before enameling (pretreatment) [8]. The practical results of powder enameling in a high-voltage electrostatic field have been confirmed on production lines in the production of steel bathtubs [9], or in the production of welded boilers for hot water preparation [10].
It is clear the higher thickness of the coating significantly worsens the adhesion of the coating. From the point of view of welding technology, the use of low-carbon welding wire can be recommended, as the increased carbon in the weld metal can increase the formation of pores in the enamel coating [2]. For the same reason, it would be appropriate to use an inert gas (pure argon) instead of a mixed gas ($Ar\text{-}CO_2$). Of course, it would be beneficial to mill the weld to the plane of the base metal, this variant will be discussed in the next study.

**Acknowledgements**
This work was supported by SP2020/106 Research and Development of Modern Technologies of Welding, Forming and Surface Treatment Including Production Process Control.

**References**

[1] Zucchelli A, Dignatici M, Montorsi M, Carlotti R and Siligardi C. Characterization of vitreous enamel-steel interface by using hot stage ESEM and nanoindentation techniques. Journal of the European Ceramic Society. Volume 32, Issue 30, August 2012, p 2243-2251.

[2] Podjuklová, J., Hrabovská, K., Barčová, K., Štepánek, I. Nové obzory v poznání vlastností sklokeramických smaltových povlaků. Ostrava: VŠB - Technická univerzita Ostrava, 2010. ISBN 978-80-248-2339-3.

[3] Bouše, V., Krajina, A., Moravčík, A. Smalty a jejich použití v protikorozní ochraně. Praha: SNTL – Nakladatelství technické literatury, 1986.

[4] Vontorová, J., Mohyla, P. Use of GDOES method for evaluation of the quality and thickness of hot dip galvanized coating. Transactions of the Institute of Metal Finishing. 2018, vol. 96, no. 6, pp. 313-318. DOI: 10.1080/00202967.2018.1520531, WOS:000447137500007.

[5] Vontorová, J., Váňová, P. Determination of carburized layer thickness by GDOES method, AIMS Materials Science. 2018, vol. 5, no. 1, pp. 34-43. DOI: 10.3934/matersci.2018.1.34. WOS:000428531900002.

[6] Wang, D. Effect of Crystalization on the Property of Hard Enamel Coating on Steel Substrate. Applied Surface Science 255 (2009), pp. 4640-4645.

[7] Podjuklová, Jitka et al. An Effect of Size Clay Particles on Functional Properties of Vitreous Enamel Coats. COMAT 2012.

[8] Mohyla, M. Příspěvek ke studiu nanášení práškových sklovitých smaltů v elektrickém poli vysokého napětí. Silikáty 29, 1985. str. 41_49

[9] Mohyla, M., Moravčík, A. Anwendung der Pulverelektrostatik bei der Gussemaillierung. Mitteilungen, No. 45. 1997. pp. 1 – 4.

[10] Pirulli V. Automatische Pulveremeillieranlage für Warmwasserbereiter. Email – Mitteilungen, No 67. 2019. pp. 90-94

IOP Conf. Series: Materials Science and Engineering 1178 (2021) 012054 doi:10.1088/1757-899X/1178/1/012054

# The mechanical properties of coatings sprayed by electric arc for use in coal power plants

**P Šulcová[1], Š Houdková[1], J Duliškovič[1]**

Research and Testing Institute Pilsen Ltd., Pilsen, Czech Republic

sulcova@vzuplzen.cz

**Abstract**. The components in coal power plants such as fire sides in boilers, preheaters, or heat exchangers are significantly mechanically and chemically stressed. The main stress originates from high temperature corrosion caused by the aggressive atmosphere inside the boiler, but other influences such as erosion and abrasion by coal and flying ash, heat shocks, etc. also have significant negative effect. One of the options for increasing the service life and improving the utility properties of these components is a thermal spray coating. Because the disassembly of these components is impossible, the mobile TWAS deposition technology was used. In this paper, different types of TWAS- sprayed coatings suitable for application in such environment are compared and evaluated. Two types of coating materials were evaluated, FeCr-based ones and NiCr-based ones, in terms of microstructure and mechanical properties such as hardness, adhesion, abrasion and erosion resistance. The SEM microscopy evaluation was used to evaluate the wear mechanism. It was found that the coatings with lower hardness but higher toughness resist the solid particle erosion better than hard but more brittle coatings. The experiences with the application of these coatings in selected areas on the fire side inside the boiler in the coal power plant are briefly reported.

## 1 Introduction

In a coal-fired power plant, there are a lot of components whose service life and performance can be increased by the appropriate use of modern technologies of surface protection. One of the options of this protection is to apply coatings with specific properties by thermal spraying. This paper is focused on proper choice of coatings materials that should serve as a protection of surfaces of components in coal-fired power plant which are significantly chemically and mechanically stressed. These components include, for example the fire side of water wall in boilers, components of superheaters, reheaters, or evaporators. Most of these components are impossible to disassemble, or the disassembling is very difficult. Because of this, the method Twin Wire Arc Spraying for coating of these components has to be used. The spraying device only needs a supply of electricity and compressed air to operate, and thus it can be used for spraying outside a specialized workplace.

In this paper, the commercially available materials for this application are described and their mechanical properties are evaluated and compared. These materials can be divided into two groups: FeCr-based and NiCr-based. The number of mechanical properties such as adhesion of coating, microhardness, abrasion and erosion wear resistance was tested and the results are described in this paper. The advanced SEM microscopy analysis was done for evaluating wear tracks.

## 2 Materials of coatings

Fe28Cr (+SiMnB) is a composite wire which produces a highly alloyed chrome steel alloy coating. This coating is a hard, abrasion and corrosion resistant, useable up to 925°C and it has also exhibited the unique quality of increasing hardness in service. This material finds its application for example for coating of exhaust fans, pump components, coal-fired boilers, superheaters, economizer waterwalls, or boiler tubes. [1]

Fe18Cr (+NiCAlVSi) make a thermally stable austenitic steel alloy coating, which is finding its application for power generation boiler tubes and waste-to-energy boiler tubes. The coating is erosion and corrosion resistant even at high temperatures (600°C). The significant advantage of this coating is a readability of the thickness of the coating by standard electromagnetic thickness gauge on as-sprayed coating and also on coating exposed up to 800 °C in one calibration of the thickness gauge.[2]

NiCrFeSi is a composite wire which can be used to protecting the components from high-temperature corrosion and erosion wear. Compared to similar coatings, the coating made by this material is harder, and more wear resistant. [2]

The coating made by NiCr wire is high temperature corrosion and erosion resistant. Its typical application is such as bond coat under ceramic coatings in high temperature or chemically aggressive environments. The amount of chromium is about half (20%) compared to next, very similar, evaluated material (NiCrTi) [3].

Next material of the coating is nickel alloy with a composition similar to Inconel 625. Coating made by this material is appropriate for oxidation and hot gas corrosion at elevated temperatures up to 870 °C and it has also a very good erosion resistance. This coating is finding its application for example for restoring components made by nickel-based superalloys, or bond coat for ceramic coating. [3]

NiCrTi is a wire developed specifically for boiler applications. This coating is highly corrosion resistant to sulfur/vanadium atmospheres up to 980 °C. The material is pre-alloyed with titanium to produce superior bond strengths. The amount of chrome (more than 40%) is sufficient to form enough chromium oxide to prevent the breakdown of nickel into nickel sulfide which is not a protective layer. [4]

## 3 Experimental

The microstructures for each material were evaluated with using an optical microscope on sample's cross sections, which were ground and polished by standard metallographic procedure. For testing mechanical properties of these coatings, four tests were done. The adhesion was tested according to ASTM C633 – 13 [5]. Hardness was tested according to ČSN EN ISO 6508-1 (420360) [6]. Wear resistance was investigated by Erosion resistance test [7] and by Dry Sand/Rubber Wheel abrasion test [8]. For these two tests, the abrasive/erosive medium was white corundum with grain size F70. The tracks after the wear test were analyzed by scanning electron microscope. The abrasion resistance was tested according to the ASTM G-65. The evaluating factor is cumulative volume loss per abrasive distance. Three samples were tested for each material and the resulting value is the average of these three measurements. During the Erosion resistance test the erosive media impacted the samples at different angles (90°, 60°, 45°, 30° and 15°). For each material of coating, two samples were tested. Three cycles were done for each sample. The duration of one cycle was 2 minutes. The average volume loss per one cycle was evaluated.

## 4 Results and discussion

The microstructures of each coating are shown in Figure 1. The structures of all coatings are without any cracks, however, typical to a coating made by TWAS: laminar, with well-distinguishable particular splats and with a content of pores and oxides. Pure NiCr and NiCrTi have the most significant content of oxides at splat boundaries because of a chemical reaction of molten particles during spraying. On the other hand, the Fe-based coatings are mostly without oxides.

For the correct evaluation of the results of the Adhesion/Cohesion Strength Test, it is necessary to know the location of initialization of the crack, and due to this the measured value is related to the adhesive or cohesive strength. For all evaluated coatings, the locations of initialization of the crack were inside the coating, so the values shown in Figure 2 – left are related to cohesive strength. The NiCrFeSi and NiCr coatings had excellent cohesive strength (more than 50 MPa). The cohesive strengths of other coatings are adequate. [9]

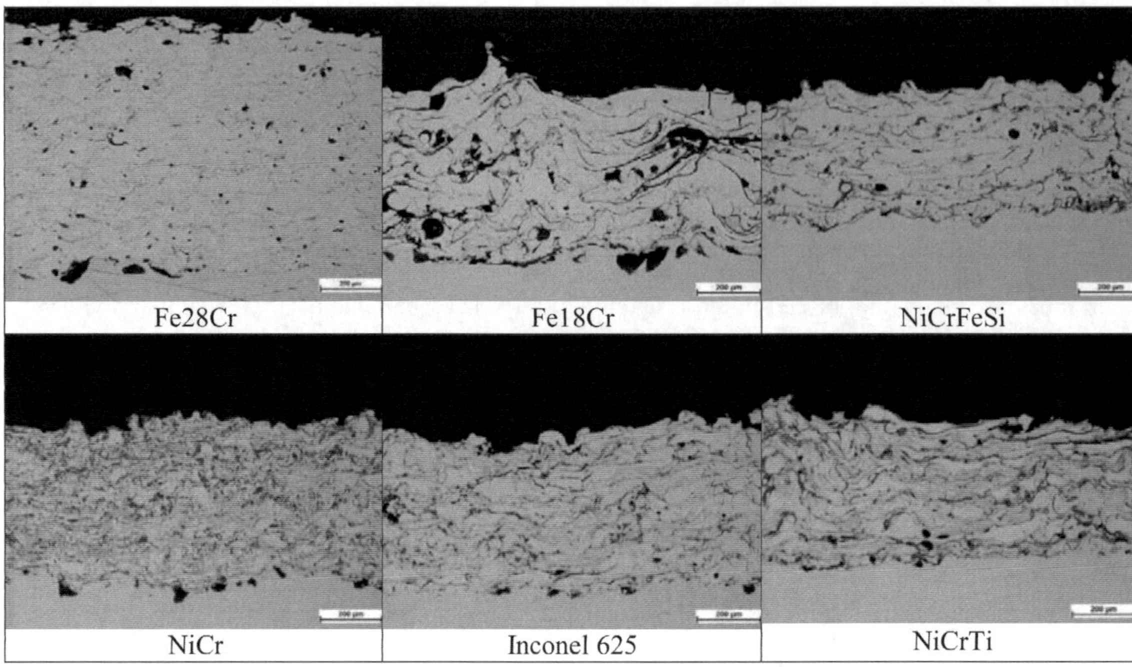


**Figure 1**. The microstructures of coatings

The results of microhardness are shown in Figure 2 – right. The highest microhardness had coatings made by composite wires: Fe-based coating with a 28% Cr and NiCrFeSi because of highly-alloyed coating. The lowest microhardness have ductile NiCr and NiCrTi coatings.

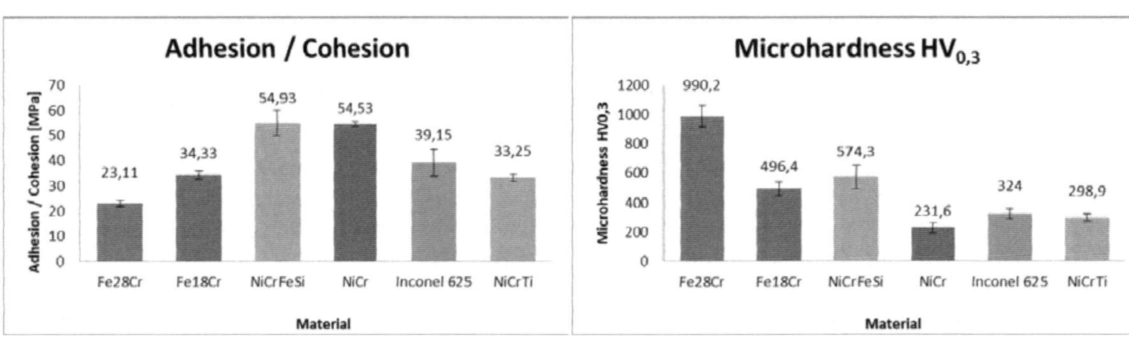


**Figure 2.** The result of Adhesion/cohesion Strength Test (left) and Microhardness HV0,3 (right)

The result of this test is shown in Figure 3. The best abrasion resistance had the NiCrFeSi coating and the Fe28Cr. These results are in correlation with the results of microhardness and their good abrasion resistance is caused by a high hardness of the coating. Tough NiCr based coatings had lower abrasion resistance. The abrasive wear tracks were evaluated by SEM microscopy (Figure 4). The wear tracks of coatings of hard coating (Fe28Cr and NiCrFeSi) are smoother with a minimum visible plastic deformation compared to tough materials (NiCr, Inconel 625 and NiCrTi).

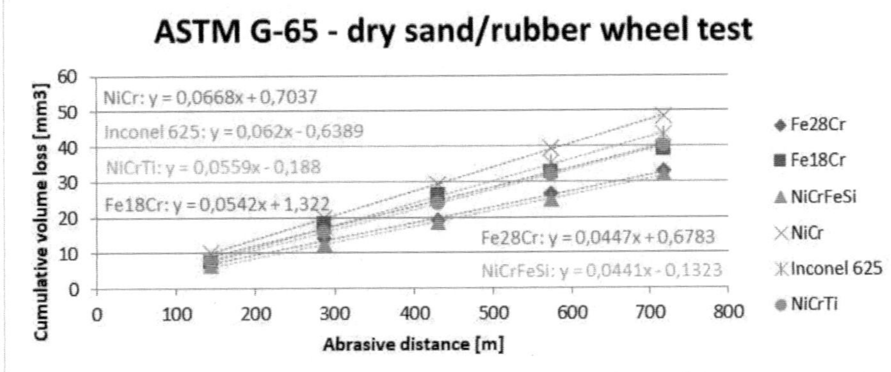


**Figure 3**. The result of Abrasive Resistance Test

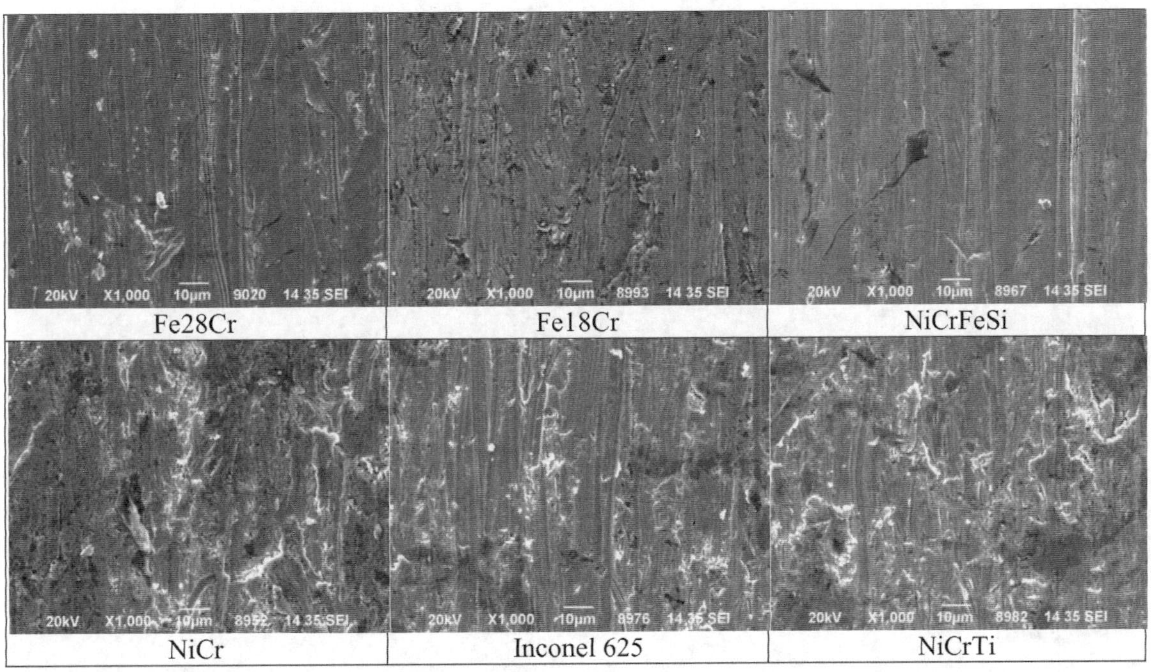


**Figure 4.** SEM analysis surface after Abrasive Wear Resistance Test

The results of the Erosion Wear Resistance test are shown in Figure 5. All coating proved a really good erosion resistance. The higher volume loss was lower than 3.5 $mm^3$ per 2 minutes of intensive actuation of the erosive medium. The lowest erosion resistance had the Fe28Cr coating because of lower cohesion strength (as also shown the result of the Adhesion/Cohesion Strength Test). The dependence of volume loss per angle of impaction is different for different materials. For the Fe28Cr coating the volume loss is decreasing with decreasing angle of impaction. The same relation, although not as significant, had the NiCrFeSi and Fe18Cr coatings. This dependence is typical for brittle material because the material is removed due to the formation and intersection of cracks that extend from the point of impact of the particle. Therefore, these coatings had lower resistance when the angle of impaction of erosive media is perpendicular. This phenomenon also confirms the SEM analysis (Figure 6) of the surface after the impaction of erosive media at 90° angle, where the delamination of whole splats for these materials is visible. The dependence of volume loss per angle for tough materials is converse. In this case, the wear mechanism is characterized by plastic deformation (Figure 6), in which the material of coating is removed or the material is cut out by particles of the erosive medium, therefore, the volume loss is increasing with decreasing angle of impact. This dependence is well noticeable for NiCr, NiCrTi and Inconel 625 coatings. The best erosion resistance had NiCr coating.

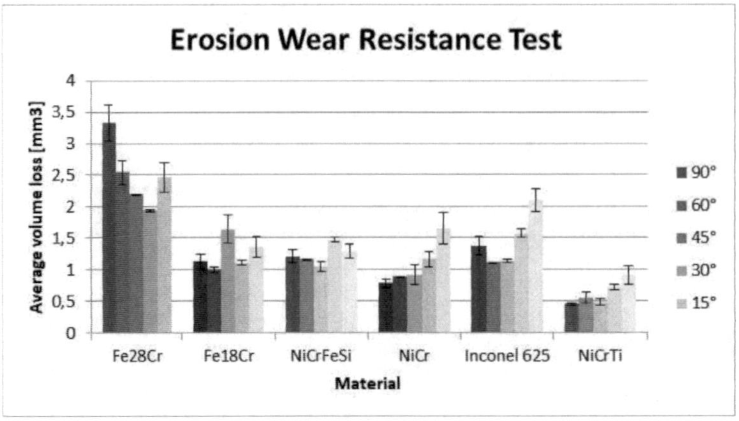


**Figure 5**. Result of Erosion Wear Resistance Test

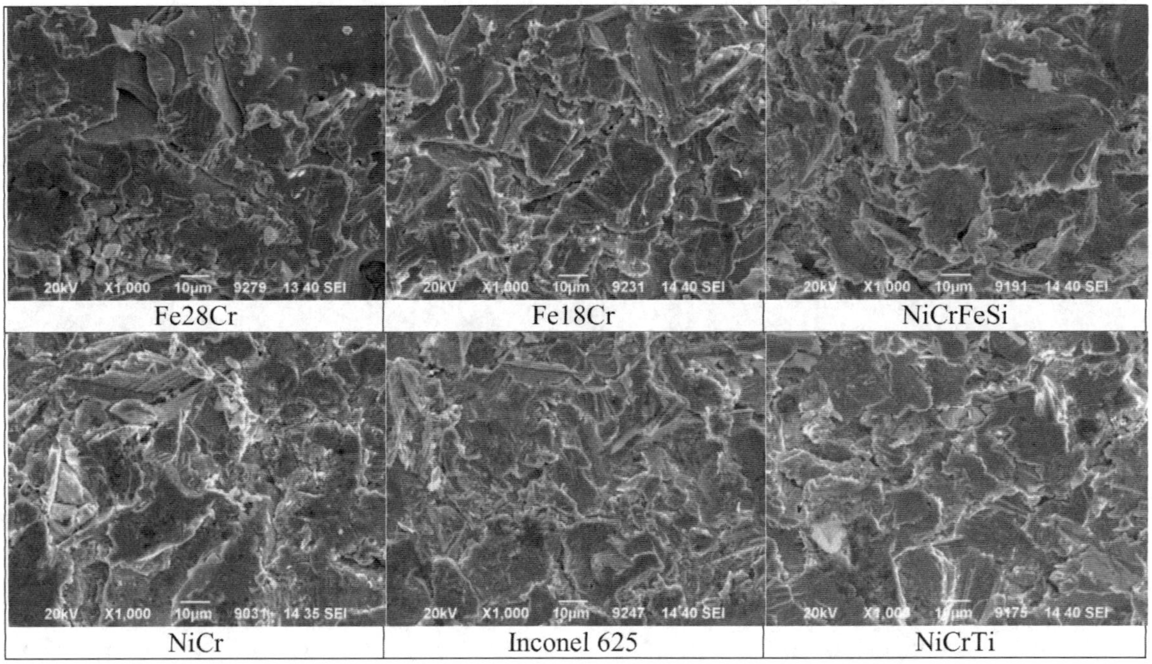


**Figure 6.** SEM analysis of surface after Erosive Wear Resistance Test

## 5 Real application

As a part of the experimental program, these coatings were applied on the fire side of water wall inside the coal-fired boiler to prove the coatings applicability outside the specialized workshop. The coatings were applied to selected areas where the water wall is significantly stressed by mechanical and chemical strain. To further increase the service life, the protective ceramic coating was applied on top of all thermal sprayed coatings.

The revision of these coatings was done after one year of operation of the boiler. The coatings were evaluated visually and their thicknesses were measured. The results of this revision showed, that different areas inside the boiler are exposed to different levels of stress. Therefore, it is not possible to determine precisely the degree of degradation of each coating, however all these coatings showed great resistance in these difficult conditions.

## 6 Conclusion

The results of selected tests of mechanical properties of TWAS-sprayed coatings for application in coal power plant boilers are described in this paper. All of the tested coatings showed prominent erosive resistance, very good abrasive resistance, and adequate microhardness and cohesive strength. These coatings can be divided into two groups: hard, brittle coatings and tough coatings. The Fe28Cr coating had the highest microhardness and really good abrasive resistance, but the lowest erosive resistance caused by lower cohesive strength and delamination of splats. Similarly, the NiCrFeSi coating showed high abrasive wear resistance due to the high microhardness. Moreover, this coating was also resistant against erosion wear thanks to the higher cohesive strength. On the other hand, ductile NiCr, Inconel 625 and NiCrTi have a lower hardness and abrasive resistance. The dependence of volume loss per impact angle during the Erosion Wear Resistance Test was typical for this type of materials: increasing with decreasing impact angle. The NiCrTi coating has higher hardness and better abrasive and erosive resistance compared to the NiCr coating. It is probably caused by the double content of chromium in the material composition. The applications of these coatings inside the coal-fired boiler showed a great resistance in these difficult conditions. However, is not possible to exactly evaluate the comparison of tested coatings because the chemical and mechanical load varies in different areas inside the boilers.

**Acknowledgements**

The paper has originated in the framework of the project of Technology Agency of the Czech Republic no. TN01000007 - National Center for Energy.

## References

[1] Castolin Eutectic, Eutronic Arc 595, Castolin Corporation, USA, 2008
[2] Oerlikon Metco, DSM-0214.0 – Austenitic Thickness Readable Alloy, Oerlikon Metco, 2018
[3] Oerlikon Metco, DSMTS-0052.7 – Nickel Chromium Thermal Spray Wires, USA, 2016
[4] Praxair Surface Technologies, Praxair and TAFA TAFALOY® 45CT® Wire, Praxair Technology, USA 2000
[5] ASTM C633-13(2017), Standard Test Method for Adhesion or Cohesion Strength of Thermal Spray Coatings, *ASTM International*, West Conshohocken, PA, 2017, www.astm.org
[6] ČSN EN ISO 6508-1,Kovové materiály - Zkouška tvrdosti podle Rockwella - Část 1: Zkušební metoda, *Úřad pro technickou normalizaci, metrologii a státní zkušebnictví*, 2017
[7] Legoux G., Arsenault B., Hawthorne H., Immarigeon J-P., Erosion behavior of WC-10CO-4Cr HVOF Coatings, Thermal Spray Technology, ASM International, USA, 2003
[8] ASTM G65-16e1, Standard Test Method for Measuring Abrasion Using the Dry Sand/Rubber Wheel Apparatus, *ASTM International*, West Conshohocken, PA, 2016, Available from www.astm.org
[9] Murakami R, Koinkar P M, Fujii T, Kim T G, Abdullah H 2019 NAC 2019: Proceedings of the 2nd International Conference on Nanomaterials and Advanced Composites (Singapore: Springer Proceedings in Physics) Chapter 11 ISBN 978-981-15-2294-9

# Comparison of the optical and electrical energy band gap of a thin layer of polyelectrolyte based on MEH-PPV and KI

M M Szindler[1] and M Szindler[2]

[1] Department of Engineering Materials and Biomaterials, Silesian University of Technology, 18a Konarskiego Str., 44-100 Gliwice, Poland
[2] Scientific and Didactic Laboratory of Nanotechnology and Material Technologies, Faculty of Mechanical Engineering, Silesian University of Technology, 7 Towarowa Str., 44-100 Gliwice, Poland

**Abstract**. A thin layer of polyelectrolyte based on Poly[2-methoxy-5-(2-ethylhexyloxy)-1,4-phenylenevinylene] (MEH-PPV) and potassium iodide (KI) was prepared using the sol-gel method. Comprehensive studies of the optical and electrical energy gap of the produced material were carried out. Using the Thermo Scientific Evolution 220 spectrophotometer equipped with a xenon lamp in the wavelength range from 190 to 1110nm, the absorbance A was measured. The optical energy gap ($Eg_{opt}$) was calculated. With the change in the concentration of potassium iodide, the change in the energy band gap of polyelectrolyte thin film was registered. In this paper, the cyclic voltammetry (CV) for testing electrochemical properties was used. The location of energy levels HOMO, LUMO and the value of the energy band gap were calculated, which is very important for polymeric materials that conduct electricity and the possibility of their use in solar cells.

## 1. Introduction

Polymeric materials thanks to their properties find application in various fields of modern technology, being an important material for an engineer constructor. The development of science about polymeric materials and their mass production can undoubtedly be considered as one of the greatest technological achievements of the 20th century. At the turn of the 20th and 21st centuries, their production amounted to about 180 million tons, and in 2006 it already reached the level of 245 million tons. In recent years, this production has been increasing every year by 5-5.5%, and the consumption of polymeric materials per capita in highly industrialized countries exceeded 150 kg. Of the general-purpose polymeric materials, many products of great practical importance are manufactured, among others packaging, tires, conveyor belts, drive belts, seals, thermal insulation (foams) and electrical (cable coatings), textiles, furniture, linings, casings, window frames, toys, etc. An important element in the development of production of more specialized polymers and resins is the mutual struggle for a place on the market of large recipients of polymeric materials, such as the aerospace, automotive, construction, electronic, electric and photovoltaic industries. Until recently, electronics was based on materials of inorganic origin (primarily on metals, as well as physical phenomena occurring in semiconductors and dielectrics). However, the characteristic properties of metals, i.e. opacity, lack of flexibility, the inability to obtain a strong conductivity anisotropy, the lack of possibility of processing in solutions, high density or susceptibility to corrosion, often constitute a barrier to their use [1-9]. Technological progress, continuous improvement and minimization of devices, forces searching for new materials and improving existing ones. The rapid development of electronics and materials engineering, especially regarding semiconductor technology and chemistry of polymer materials is related to the introduction of modern engineering materials, including polymeric materials that conduct electricity. Professor Hideki Shirakawa is considered to be the discoverer of conductive polymers. In 1975 he made acetylene polymerization, resulting in the reaction of polyacetylene (an organic conductive polymer). In 1976, to collaborate on conducting polymers joined by Professor Alan G. MacDiarmid and Professor Alan J. Heeger, which resulted in the doping polyacetylene with iodine and bromine, significantly increasing its electrical conductivity. In 2000, "for the discovery and development of conductive polymers" professors Hideki Shirakawa, Alan G. MacDiarmid and Alan J. Heeger received the Nobel Prize. This led to increased interest in this group of materials [7-16]. Polymeric materials with semiconductor properties show some common feature: they are made of macromolecules in which, in spite of all complexity, more or less regular repeating sequences of specific atomic groups can be distinguished, e.g. chains of conjugated bonds. The greater its extent, the more favorable are the electron flow conditions through the material, and hence its higher conductivity. Conductive polymers in the undoped state are multiparticulates semiconductors. They are part of a wider group of organic semiconductors. In the group of electronically conducting polymers, Π-electron aliphatic systems are often distinguished. By doping a polymer material, it is possible to obtain an electrical conductivity similar in size to the conductivity of metals. An example is poly (acetylene). This polymer in the undoped state is a semiconductor with a specific conductivity of $10^{-7}$ S/cm. The introduction of an acceptor-like admixture (eg $AsF_5$) or a donor (Li, K) to poly (acetylene) increases its conductivity by $10^{-12}$ orders of magnitude [11-20].

The main problem in the development of solar cells with polymer material is to choose the appropriate energy levels of each material to ensure the oriented movement of electrons. In these kinds of solar cells the HOMO and LUMO energy levels of each component must be selected that guarantees the oriented transport of electrons. A novelty in the article is the preparation of a thin layer of MEH-PPV polymer doped with potassium iodide which can be used in polymeric and dye sensitized solar cells. Then, the optoelectric properties of this layer were characterized. Optical and electrical energy band gap and HOMO and LUMO levels were determined.

## 2. Materials and methodology

Poly[2-methoxy-5-(2-ethylhexyloxy)-1,4-phenylenevinylene], anhydrous potassium iodide (KI) and other chemicals were supplied by Sigma Aldrich. N-Methylpyrrolidine (NMP) was used to dissolve the polymer and salt, and the liquid solutions were spin-coated to form polyelectrolyte thin films. Using the Thermo Scientific Evolution 220 spectrophotometer equipped with a xenon lamp in the wavelength range from 190 to 1110nm, the absorbance A was measured. This is the logarithm of the ratio of the incident radiation ($I_0$) to the intensity of the outgoing radiation ($I_t$), which is defined as the decimal logarithm of the transmittance:

$$A = log\frac{1}{T} \tag{1}$$

The optical energy gap ($Eg_{opt}$) was calculated by plotting on the normalized absorption spectrum a straight line indicating the wavelength ($\lambda_{onset}$) for the recorded absorption edge and using the formula:

$$Eg_{opt} = \frac{h \cdot c}{\lambda onset} \tag{2}$$

The location of energy levels HOMO, LUMO and the value of the energy band gap are very important for polymeric materials that conduct electricity and the possibility of their use in solar cells. One indirect methods of determining energy levels are electrochemical measurements. In this paper, the cyclic voltammetry (CV) for testing electrochemical properties was used. In the voltammetric measurements, the dependence of the intensity of the electric current on the applied voltage or potential of the electrode is examined. The change in potential causes that the molecules chemically take or get rid of the electron, which means that they can undergo oxidation and reduction processes. As a result, an electric current flows, which depends on the rate at which the electrons move at the electrode-solution interface. The tests were performed in a three-electrode system with a platinum auxiliary electrode and a silver-chloride reference electrode. The working electrode was an ITO tile covered with a thin layer of polymer. In order to calculate the position of energy levels, first the

IOP Conf. Series: Materials Science and Engineering 1178 (2021) 012055 doi:10.1088/1757-899X/1178/1/012055

oxidation process of $E_{ox}$ was determined, which is connected with the removal of the electron from the HOMO level, and the beginning of the first reduction process $E_{red}$, which corresponds to the introduction of the electron to the LUMO level.

## 3. Results

The recorded electron absorption spectra are presented as a function of the dependence of the absorbance on the wavelength (Figure 1). The thin layer of MEH-PPV polymer material, dissolved in NMP, has the highest absorption band at a wavelength of 416 nm. The highest absorption band of the MEH-PPV layer dissolved in the same solvent but with 5% KI was recorded at 428 nm. The highest absorption band for the MEH-PPV layer with 10% KI was recorded at the wavelength of 431 nm, and for the layer with 15% KI at 435 nm. Along with the increase in the proportion of potassium iodide in the polymer material in the recorded spectra, bathochromic shifts of bands towards longer wavelengths were found. In the analysed spectra, along with the increase in the percentage of KI, a small hyperchromic effect of the absorption bands was registered. A slight widening of the absorption band was observed along with the increase in KI content.

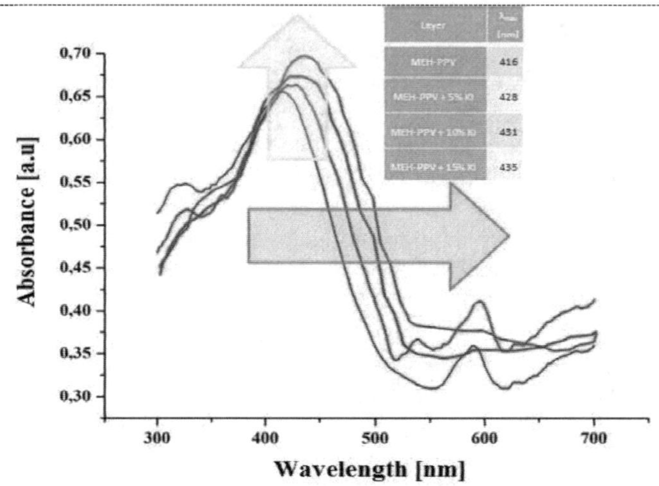


**Figure 1.** The dependence of the absorbance on the wavelength for a thin layer of MEH-PPV deposited on a glass substrate together with MEH-PPV graphs with an increasing contents of potassium iodide

The optical energy gap ($Eg_{opt}$) was calculated by plotting on the normalized absorption spectrum a straight line indicating the wavelength ($\lambda_{onset}$) for the recorded absorption edge, and the results obtained are summarized in Table 1. For the thin layer of MEH-PPV dissolved in NMP, the wavelength of the absorption edge was 510 nm, and calculated on this basis the optical band gap is 2.43 eV. The MEH-PPV layer with 5% KI is characterized by $\lambda_{onset}$ = 520 nm, and the optical band gap at this wavelength is equal 2.39eV. For a 10% share of KI in MEH-PPV, the graphically determined wavelength for the absorption edge was 535 nm, and the optical band gap is equal 2.32 eV. The wavelength of the absorption edge determined from absorption of MEH-PPV with 15% was 540 nm, and the $Eg_{opt}$ calculated on this basis is 2.3 eV. Along with the change in the share of KI in the polymeric material, its optical energy break has changed (Table 1). These changes were small, by 0.13 eV. However, it has an impact on its electrical conductivity. With the change in the concentration of KI in the polymeric material, changing its optical energy band gap (Figure 2). The measurements of electrochemical properties of polymeric materials deposited by the sol-gel method on a glass substrate with an ITO layer were made by means of cyclic voltammetry. The results of measurements are presented in the form of current-voltage diagrams (Figure 3).
The potentials of reduction and oxidation of polymer layers and the energy levels HOMO and LUMO calculated on their basis, as well as the value of the energy band gap are summarized in table 2.

IOP Conf. Series: Materials Science and Engineering 1178 (2021) 012055 doi:10.1088/1757-899X/1178/1/012055

The oxidation and reduction potentials for the thin layer of non-doped MEH-PPV polymer material are 0.69 and -1.47 respectively. The calculated HOMO energy level is 5.09 eV, and the LUMO energy level is 2.93 eV. The contribution of 5% potassium iodide influenced on the energy levels of the MEH-PPV by increasing the HOMO level to 0.08 eV, while the LUMO level by increasing to 0.29 eV. The energy band gap has consequently decreased from 2.16 eV to 1.95 eV. Further doping up to 10% KI had a slight impact on the HOMO and LUMO energy levels of the MEH-PPV polymer material. For a thin layer of polymer material MEH-PPV with 10% potassium iodide the energy gap was equal 1.88 eV.

**Table 1.** Values determined on the basis of absorbance curves, wavelengths and optical band gap of thin layers calculated on this basis

| Layer | $\lambda_{max}$ [nm] | $\lambda_{onset}$ [nm] | $Eg_{opt}$ [eV] |
|---|---|---|---|
| **MEH-PPV** | 416 | 510 | 2.43 |
| **MEH-PPV + 5% KI** | 428 | 520 | 2.39 |
| **MEH-PPV + 10% KI** | 431 | 535 | 2.32 |
| **MEH-PPV + 15% KI** | 435 | 540 | 2.30 |

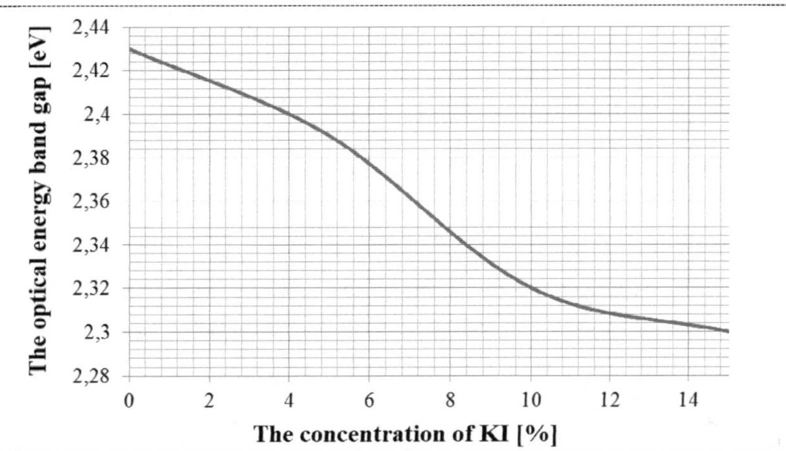


**Figure 2.** The influence of KI doping on the optical band gap value of the polymer thin film MEH-PPV

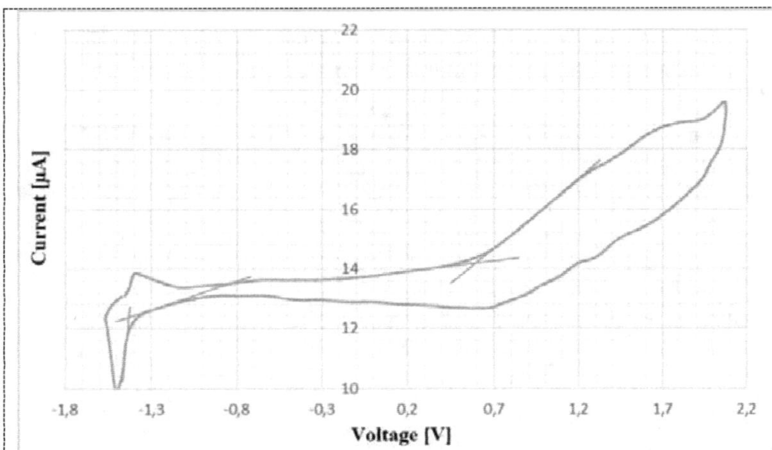


**Figure 3.** A cyclic voltamperogram of a thin layer of MEH-PPV deposited on a glass substrate with ITO

**Table 2.** Comparison of the values of reduction potentials and oxidation of thin polymer layers and the energy levels of HOMO, LUMO and energy gap values calculated on their basis

| Layer | $E_{ox}$ [V] | $E_{red}$ [V] | $E_{Homo}$ [eV] | $E_{Lumo}$ [eV] | $E_g$ [eV] |
|---|---|---|---|---|---|
| **MEH-PPV** | 0,69 | -1,47 | 5,09 | 2,93 | 2,16 |
| **MEH-PPV + 5% KI** | 0,77 | -1,18 | 5,17 | 3,22 | 1,95 |
| **MEH-PPV + 10% KI** | 0,78 | -1,1 | 5,18 | 3,3 | 1,88 |

## 4. Conclusion

The suitability of the tested materials for the construction of dye sensitized solar cells was confirmed using a UV-VIS spectrophotometer, determining their optical properties. In the analyzed UV-VIS spectra, together with the increase in the concentration of potassium iodide in the polymer material, batochromic shifts of bands towards longer wavelengths and the hyperchromic effect of absorption bands were registered. In polymeric materials, the shift of absorption bands is related to the change of the chain length, the shifting of the bands towards the longer wavelengths indicates the increase of the chain length in the oligomers. The optical energy band gap for the material of MEH-PPV is 2.43 eV, which is confirmed in the literature. With the change in the concentration of KI in the polymeric material, changing its optical energy band gap. Significance for polymeric conductive materials and the possibility of their use in solar cells not only has an energy band gap, but also the location of their energy levels HOMO and LUMO. By adding potassium iodide to the meh-ppv polymer it is possible to reduce its energy band gap from 2.16 eV to 1.88 eV and slightly influence on the HOMO and LUMO energy levels.

**Acknowledgements:** Thin publication was supported under the own scholarship fund of the Silesian University of Technology in the field of scientific research and development works, Year 2020.

## References

[1] Kumar V V, Balaganesan G, Kong Yoong Lee J, Esmaeely Neisiany R, Surendran S, Ramakrishna S 2019, A Review of Recent Advances in Nanoengineered Polymer Composites, *Polymers* 11 644.

[2] Zenkiewicz M, Zuk T, Pietraszek J, Rytlewski P, Moraczewski K, Stepczynska M 2016 Electrostatic separation of binary mixtures of some biodegradable polymers and poly(vinyl chloride) or poly(ethylene terephthalate), *Polymers* 61/11-12 835-843.

[3] Moraczewski K, Mroz W, Budner B, Malinowski R, Zenkiewicz M 2016 Laser modification of polylactide surface layer prior autocatalytic metallization, *Surface & Coatings Technology* 304 68-75.

[4] Rytlewski P, Moraczewski K, Zenkiewicz M 2014 Effects of Coffee on the Stability of Accelerated Aged Poly(acrylonitryle-butadiene-styrene), *Journal Of Applied Polymer Science* 131/4.

[5] Paula D.R., Robeson L.M. 2008, Polymer nanotechnology: Nanocomposites, *Polymers* 49/15 3187-3204.

[6] Akinci A 2009 Mechanical and structural properties of polypropylene composites filled with graphite flakes, *Archives of Materials Science and Engineering* 35/2 91-94.

[7] Hak D J, Mauffrey C, Seligson D, Lindeque B 2014 Use of carbon-fiber-reinforced composite implants in orthopedic surgery. *Orthopedics* 37 825–830.

[8] Weszka J, Szindler M, Śliwa A, Hajduk B, Jurusik J 2011 Reconstruction of thin films polyazomethine based on microscopic images, *Archives of Materials Science and Engineering* 48/1 40-48.

[9] Shirakawa H 1995 Synthesis and characterization of highly conducting polyacetylene, Synthetic Metals 69 3-8.

[10] Kroekchai Inpor, Vissanu Meeyoo, Chanchana Thanachayanont 2011 Enhancement of photovoltaic performance using hybrid CdS nanorods and MEH-PPV active layer in ITO/$TiO_2$/MEH-PPV:CdS/Au devices, *Current Applied Physics* 11 11/1 171-174.

[11] Sirimanne P.M., Premalal E.V.A., Pitigala P.K.D.D.P., Tennakone K. 2006, Utilization of MEH-PPV as a sensitizer in titana-based photovoltaic cells, *Solar Energy Materials and Solar Cells* 90/11 1673-1679.

[12] Weszka J, Szindler M, Chwastek-Ogierman M, Bruma M, Jarka P 2012 CLSM and UV-VIS researches on polyoxadiazoles thin films, *Archives of Materials Science and Engineering* 55 53-61.

[13] Wang M, Greatzel C, Zakeeruddin S.M., Greatzel M., 2012 Recent developments in redox electrolytes for dye-sensitized solar cells, *Energ. Environ. Sci.*, 5, 9394-9404.

[14] Su'ait M.S., Rahman M.Y.A., Ahmad A., 2015 Review on polymer electrolyte in dye-sensitized solar cells (DSSCs), *Sol. Energ.* 115, 452-470.

[15] Fischer A., Pettersson H., Hagfeldt A., Boschloo G., Kloo L., Gorlov M., 2007 Crystal formation involving 1-methylbenzimidazole in iodide/triiodide electrolytes for dye-sensitized solar cells, *Sol. Energ. Mat. Sol. C*, 91/12, 1062-1065.

[16] Gratzel M., Dye-sensitized solar cells, *J. Photoch. Photobio. C*, 2003, 4/2, 145-153.

[17] Bella F., Ozzello E.D., Sacco A., Bianco S., Bongiovanni R., 2014 Polymer electrolytes for dye-sensitized solar cells prepared by photopolymerization of PEG-based oligomers, *Int. J. Hydrogen Energ.*, 39/6, 3036-3045.

[18] Bella F, Mobarak NN, Jumaah FN, Ahmad A, 2015 From seaweeds to biopolymeric electrolytes for third generation solar cells: an intriguing approach, *Electrochim Acta*, 151, 306-31.

[19] Darvishzadeh P, Redzwan G, Ahmadi R, Gorji NE 2017 Modeling the degradation/recovery of short-circuit current density in perovskite and thin film photovoltaics, *Org. Electron*, 43, 247-252.

[20] Holt A L, Leger J M, Carter S A 2005 Electrochemical and optical characterization of p and n-doped poly[2-methoxy-5-(2-ethylhexyloxy)-1,4-phenylenevinylene], *J. Chem Phys* 123 11-17.

IOP Conf. Series: Materials Science and Engineering 1178 (2021) 012056 doi:10.1088/1757-899X/1178/1/012056

# Influence of ceramic counter electrode on the properties of dye sensitized solar cell

**M Szindler[1], M M Szindler[2], A Drygała[2], P Boryło[2] and M Domański[3]**

[1] Scientific and Didactic Laboratory of Nanotechnology and Material Technologies, Faculty of Mechanical Engineering, Silesian University of Technology, 7 Towarowa Str., 44-100 Gliwice, Poland
[2] Department of Engineering Materials and Biomaterials, Silesian University of Technology, 18a Konarskiego Str., 44-100 Gliwice, Poland
[3] Centre of Polymer and Carbon Materials, Polish Academy of Sciences, 34 Marie Curie-Skłodowska str., 41-819 Zabrze, Poland

**Abstract.** Dye sensitized solar cells consisting of a sandwich structure have been a topic outspread in the science for above twenty years and they start a new development trend in obtaining energy from the sun. One the examined aspects of their application is building-integrated photovoltaics. Despite the fact that the higher efficiency of energy conversion was obtained with this type of substrate, glass, due to the cost of production and energy consumption, is disadvantageous in the case of production on an industrial scale. The article presents the possibility of replacing a classic glass substrate with an counter electrode based on a building ceramic tile material. A dye sensitized solar cell was built layer by layer on the tile. A method of atomization was used to deposit fluorine doped tin oxide. After that, a thin layer of platinum was printed. It does not require the use of high temperatures, complicated devices and high costs. Scanning electron microscope, atomic force microscope and Raman spectroscopy were used to evaluate the influence of the transparent conductive oxide layer on the photovoltaic properties of counter electrode tile. Electrical parameters of as prepared DSSCs with and without ceramic tile were characterized by measurements of current-voltage characteristics under standard AM 1.5 radiation. A dye sensitized solar cell integrated with building ceramic tile material was prepared successfully, which efficiency was over 4%.

## 1. Introduction

An alternative to the solar cells of the 1st and 2nd generation are dye sensitized solar cells classified as 3rd generation and are perceived as an interesting direction in the development of obtaining electricity from the sun. The principle of their operation is often compared to the phenomena observed in the plant environment (the difference is that the donor-acceptor structures used in them convert the energy of light into electric power, and not into biochemical energy as in photosynthesis) [1-4]. One of the most interesting applications is their possible use in the architecture of building construction [5]. Their construction based on glass electrodes allows them to be used in glazed buildings as functional elements, and thanks to the use of various types of dyes, they can also have an aesthetic function. In the Building Integrated Photovoltaics (BIPV) the developed solutions consist in mounting solar cells with a steel structure on the finished roof or partially replacing the tile and are based on silicon. [5-7]. Trying to eliminate the liquid electrolyte or replacing the glass electrode with other materials, attempts are made to extend the use of dye sensitized solar cells in the construction industry. Replacing glass with other materials such as: polymer foil, metals, steel, paper which can be additionally lighter or flexible may cause cost decrease because glass substrate represents 15-20% of the price of a ready

DSSC [8-14]. This article presents the findings on the modification of a experimental dye sensitized solar cells based on a cheap building ceramic tile material. The preparation of a counter electrode based on a ceramic tile as a substrate requires, in the first place, a layer of transparent conductive oxide (TCO) [15-17]. These types of layers are mainly used in electronics and optoelectronics as elements of any displays, monitors, etc. Another area of applications are photovoltaic cells. Here it is not only the high conductivity of the layer but also high transmission glass it's very important. Novelty is the use of TCO in the field of architectural glass as laminated glass structure with built-in light-emitting diodes. The use of RGB LEDs allows to obtain any colour. These windows are used to display information and commercials, as well as an element of architectural decoration. Commonly used transparent conductive layer is indium tin oxide (ITO). These thin films are distinguished by a very high transmittance of visible light (> 85%) and low resistivity ($10^{-4}$ Ω cm) [13-14, 16]. Considering the difficult availability and prices of Indium, other solutions are sought. Encouraging appear to be a thin film prepare from fluorine doped tin oxide. This material has optical and electrical characteristics approximate to the well known ITO. The transparency of the FTO layer is equal about 82%, and the resistivity in the range from 4.0 $10^{-4}$ Ω cm to 6.0 × $10^{-4}$ Ω cm. There are various methods for producing transparent conductive oxide layers. There are known methods such as sol-gel, electron beam evaporation, ion deposition, magnetron sputtering, epitaxial growth, and many others [15-17]. Such a variety of techniques used due to the fact that each of them has its advantages and disadvantages, which significantly influence on properties of the obtained coatings. In the framework of this articles, it is proposed to use the atomization method due to its universality, repeatability and stability of work [18].

## 2. Material descriptions and research methodology

Classic solar cells are constructed with the use of commercially available glass plates with an FTO layer which are the basis of the photoanode and counter electrodes. In this article, fluorine doped tin oxide coated glass slide (FTO) with surface resistivity 7 Ω/sq and thickness 2.3 mm were used (purchased from Sigma Aldrich company). In the counter electrode, glass was replaced with ceramic tiles (Figure 1) and it's surface was covered with FTO by atomization method (Figure 2). This method involves embedding the sol on the surface of the element using a container with an atomizer. The implementation of this method makes it possible the control of temperature of the element in a broader range than with typical spraying deposition methods. Besides, the deposition by this method not require an additional gas lines with a transporting medium.

The building integrated dye sensitized solar cells (BIDSSC) according to the invention consists of three elements: the first element is a photoanode, the second element is a liquid electrolyte and the third is a counter electrode. A transparent, electrically conductive thin film was deposited on the ceramic tile in order to prepare the counter electrode. The fluorine doped tin oxide (FTO) was embedding by using a container with an atomizer. This required the preparation of two solutions which were then mixed together. The first solution is Dibutyltin Diacetate (DBTDA) with ethanol and the second is NH4F with water. After joining them together they were agitated ultrasonically for 45 min. This solution was sprayed onto anot heated and heated substrate (surface temperature = 100°C) by the container with an atomizer. Then platinum paste with nanoparticles (Merck) was used as a catalytic thin film. Platinum paste was deposited on substrate with FTO by screen-printing method. In this work Screen-printing station MS300FRO was used. After printing, the thin layer was dried in 125°C for 10 minutes and two more films were deposited. Summarizing, three layers of platinum paste were sintered in an oven. The oven was heated from room temperature to 500°C with heating rate of 15°C/min and holded at 500°C for 30 minutes. After that, in order to prepare a photoelectrode, on the commercially available FTO glass plate, a titanium oxide thin film was deposited by ALD method. Next a layer of porous nanocrystalline titanium oxide was deposited by screen printing method (18 NR-T, Greatcell Solar). The dye (N719, Merck) was then applied by dipping the substrate with embedded layers in an ethyl alcohol dye solution to get the final stage of photoelectrode. At the end, in order to get experimental building integrated dye sensitized solar cells, the electrodes were joined together with layers directed inwards, and liquid electrolyte (EL-HSE, Merck) containing a $I^-/I^{3-}$ redox couple was placed between them by spotting. The construction of as prepared DSSCs was shown in

Figure 3. Combination of a scanning electron microscope (SEM) equipped with an in lens detector for the examination of thin films and the energy-dispersive spectrometer (EDS) enabled the analysis of the surface morphology and the identification of the chemical composition. The structure of the deposited layers was examined. For this purpose, Raman spectrometer equipped with an green laser with a 514.5 nm length was used. The results were analyzed with WiRE3.1 program. Electrical parameters of prepared experimental DSSCs were characterized by measurements of current-voltage (I-V) characteristics using PV Test Solutions Tadeusz Zdanowicz Solar Cell I-V Tracer System and Keithley 2400 source meter under standard AM 1.5 radiation.

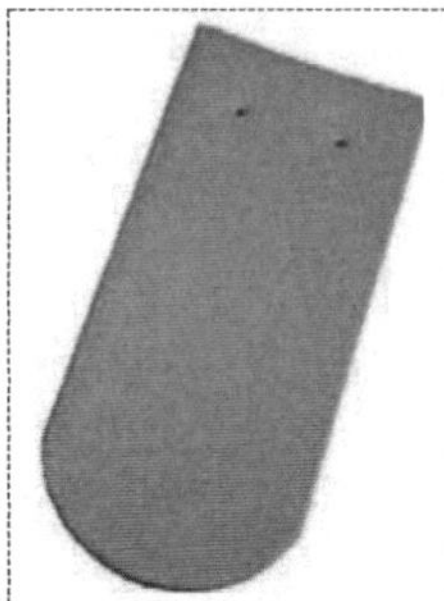

**Figure 1.** View of ceramic tiles.

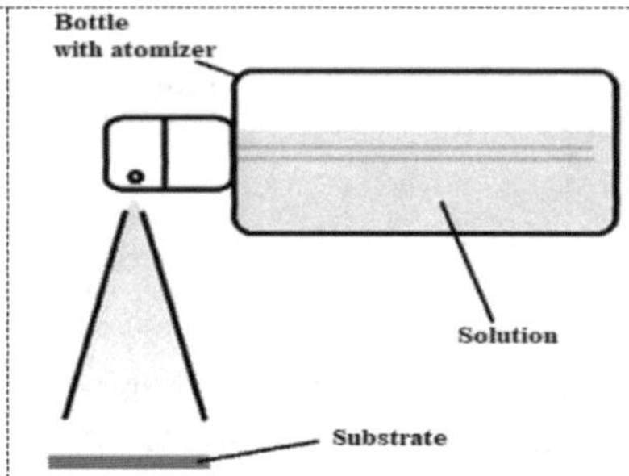


**Figure 2.** Scheme of FTO layer deposition by atomization method.

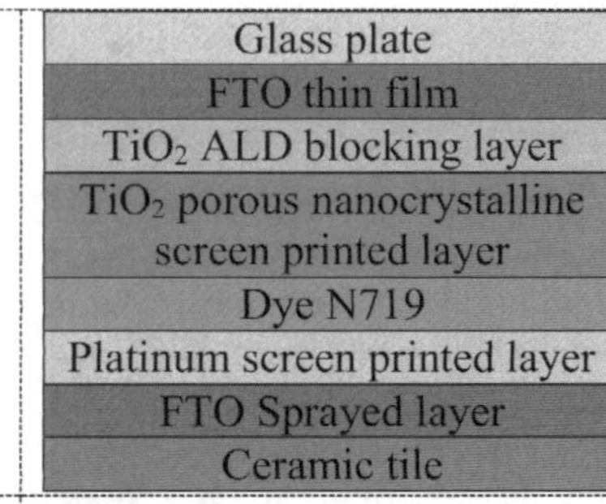


**Figure 3.** The construction of as prepared building integrated dye sensitized solar cells.

## 3. Results and discussion

All the deposited layers were characterized by continuity and repeatability of the structure. No defects were observed. (Figure 4). On an unheated substrate, the structure of the FTO layer looks similar to partially amorphous (Figure 4a). In turn, the increase of the substrate temperature started the crystallization of the TCO layer. Grain size not exceeding 100 nm was registered (Figure 4b).

a)

b)

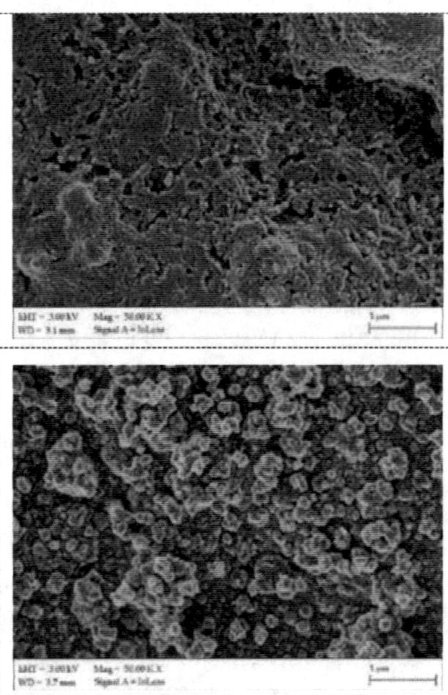

**Figure 4.** SEM surface topography images of the FTO sprayed layer on the substrate without (a) and with heated substrate (b)

a)

b)

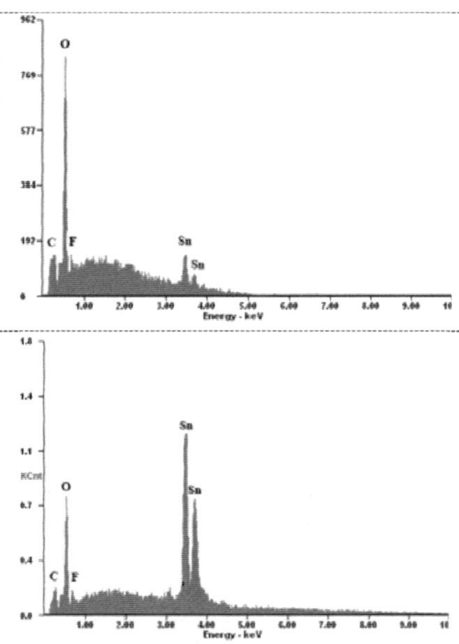


**Figure 5.** EDS spectrum of the FTO sprayed thin film on the substrate without (a) and with heated substrate (b)

An EDS analysis was performed, recording reflections characteristic of tin, fluorine and oxygen from layer (Figure 5). For tin at 3.444 eV (spectrum line $L\alpha_1$) and at 3.663 eV (spectrum line $L\beta_1$) were registered. Oxygen classically at 0.525 eV (spectrum line $K\alpha_1$). At 0.677 eV (spectrum line $K\alpha 1$) a fluoride reflex was identified.

Using the Raman spectrometer, the spectrum of ceramic tile and the spectrum of ceramic tile with deposited ITO layer were recorded (Figure 6). To produce a ceramic tile, building materials such as clay, minerals (e.g. quartz) and various types of coloring substances are used. Clay is mainly silicon oxides ($SiO_2$ approx. 50%), rare earth oxides ($R_2O_3$ approx. 25%), aluminum oxides ($Al_2O_3$ approx. 15%), and other (e.g. iron oxide) and sometimes elements like calcium and potassium (approx. 10%). Therefore, ceramic tile is a product that consists of many different molecules, as shown in the Raman spectrum. The bands derived from the vibrations of the $SiO_2$ molecule were recorded at 321, 335, 354, 393 and 454$cm^{-1}$ wavelengths. Rare earth molecules, depending on their composition, give bands in the range of 95-660 $cm^{-1}$. Bands originating from vibrations of ferrum oxide molecules were identified at 220, 300, 410, 500 and 620 $cm^{-1}$. At 150, 400, 520 and 645 $cm^{-1}$, bands from vibrations of titanium dioxide molecules were also recorded. The vibrations of the alumina molecules give bands at 360, 400, 460, 610, 620 $cm^{-1}$. The vibrations of the magnesium oxide molecule give bands at 420, 476, 594$cm^{-1}$. Small bands from the vibrations of the calcium oxide molecule at 464, 522 and 553 $cm^{-1}$ were also identified. Vibration bands from the SnO, $SnO_2$ and $In_2O_3$ molecules were recorded in the spectrum of the ITO layer. Vibration bands from the SnO molecule were recorded at 115 and 220$cm^{-1}$, and vibration bands from the $SnO_2$ molecule were recorded at 450, 518, 560, 680$cm^{-1}$. The vibrations of the $In_2O_3$ molecule give a wide band between 200 and 300 $cm^{-1}$ and between 600 and 700 $cm^{-1}$. At 109, 133.231, 307, 366, 433, and 584 $cm^{-1}$, small bands were also recorded from indium oxide vibrations.

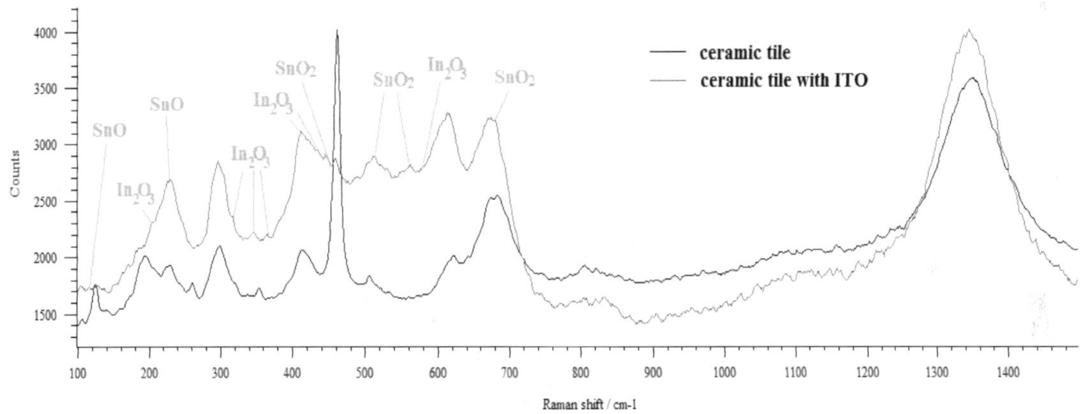


**Figure 6.** Raman spectra of ceramic tile and ceramic tile with ITO

Platinum has a tendency to increase as the so-called "islands". Figures 7a and 7b are demonstrated surface morphology of the FTO with deposited islands of platinum. Despite this, the higher densely packed was identified on FTO embedded on the heated ceramic tile substrate. To document that the observed layers are platinum were performed quality chemical composition study using Energy Dispersive Spectrometer – EDS. In Figure 8 spectre with the reflections characteristic for platinum at 2.127 keV (spectrum line $M\beta_1$) were documented. The use of ceramic plate as the cathode affect on the shape of the current-voltage characteristic. It is less rectangular than the dssc with a classic glass cathode (Figure 9). The temperature of the substrate in atomization method has a slightly influence on short circuit current an open circuit voltage. It all has an impact on the efficiency of dye sensitized solar cells. The increased temperature of the substrate during the deposition of the FTO layer increased the efficiency of the solar cell by 0.15%. Basic electrical parameters of as prepared experimental solar cells are documented in Table 1. Analysing the measured electrical parameter can be noticed that the highest efficiency has a solar cell with an FTO layer deposited with heated substrate from among the as prepared dye sensitized solar cells with a ceramic counter electrode.

a)
b)

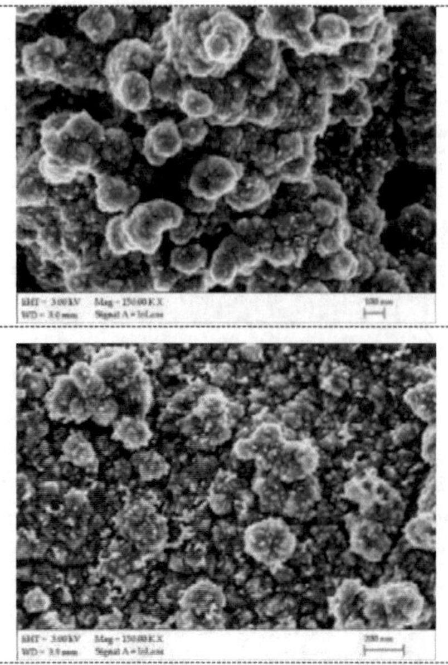

Figure 7. SEM surface topography images of the Pt on FTO/ceramic tile without (a) and with heated substrate (b)

a)
b)

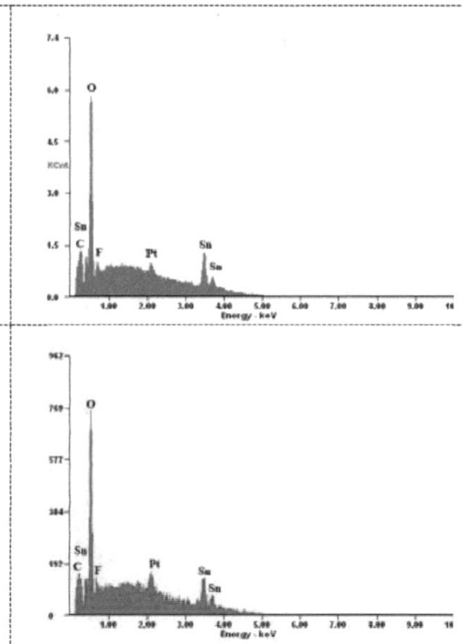


**Figure 8.** EDS spectrum of the Pt on FTO/ceramic tile without (a) and with heated substrate (b)

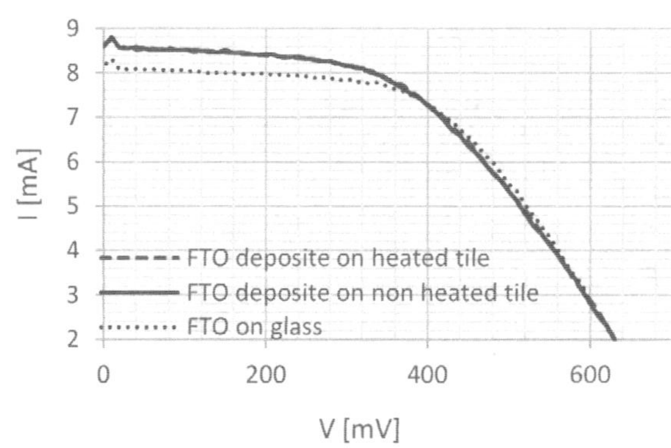


**Figure 9.** Current-voltage characteristics of experimental DSSCs

**Table 1.** Electrical properties of experimental DSSCs

| | Isc [mA] | Voc [mV] | Imax [mA] | Vmax [mV] | Pmax [mW] | FF | Eff [%] |
|---|---|---|---|---|---|---|---|
| Non heated tile | 8.695 | 628.998 | 6.989 | 417.956 | 2.817 | 0.54 | 4.07 |
| Heated tile | 8.728 | 629.988 | 7.018 | 417.248 | 2.838 | 0.54 | 4.22 |
| Glass | 8.737 | 631.821 | 6.906 | 414.501 | 2.969 | 0.53 | 4.89 |

## 4. Conclusion

The experimental integrated dssc were prepared based on ceramic building material. The strong point of this configuration is high level of integration with building object with the same time generation of electricity. In the counter electrode, glass was replaced with ceramic tiles and it's surface was covered with FTO by atomization method. The crystalline structure of deposited FTO was documented. Then platinum paste with nanoparticles was used as a catalytic thin film. Platinum paste was deposited on substrate with FTO by screen-printing method. The SEM research have documented that a thin film of FTO was embedded without major impurities and defects. Grains smaller than 100 nm in diameter have been documented. The electrical properties of as prepared experimental DSSCs with ceramic tile as the cathode are slightly lower than the efficiency of solar cell using typical glass as the cathode. Such solution give a prospects of the future preparation of stable, cheap and efficient building integrated dye sensitized solar cells.

**Acknowledgements:** Thin publication was supported under the own scholarship fund of the Silesian University of Technology in the field of scientific research and development works, Year 2020.

**References**

[1] O'Regan B, Graätzel M, A low-cost, high-efficiency solar cell based on dye-sensitized colloidal TiO2 films. Nat. **353** (1991) 737–740
[2] Desilvestro H., Hebting Y., Khan M., Milliken D., Understanding and successfully applying materials for dye-sensitized solar cells. Mat. Mat. **9/1** (2014) 14-18
[3] Al-Kahlout A, ZnO nanoparticles and porous coatings for dye-sensitized solar cell application: Photoelectrochemical characterization. Thin Solid Films **520/6** (2012) 1814-1820
[4] Wang M, Huang C, Cao Y et al., Dye-sensitized solar cells based on nanoparticle-decorated ZnO/TiO2 core/shell nanorod arrays, Journal of Physics D: Applied Physics **42/15** (2009) 155104
[5] Jelle BP, Breivik C, Røkenes HD, Building integrated photovoltaic products: A state-of-the-art review and future research opportunities, Solar Energy Mater. Solar Cells **100** (2012) 69–96
[6] Yen-Chieh H and et al., Development of building integrated photovoltaic (BIPV) system with PV ceramic tile and its application for building façade, Energy Procedia **61** (2014) 1874-1878.
[7] Debbarma M. and et al., Comparison of BIPV and BIPVT: A review. Resource-Efficient Technologies **3/3** (2017) 263-271
[8] Toivola M, Ahlskog F, Lund P, Industrial sheet metals for nanocrystalline dye-sensitized solar cell structures. Solar Energy Materials & Solar Cells **90** (2006) 2881-2893
[9] Yamaguchi T, Nobuyuki T, Matsumoto D, Takuma and et al., Highly efficient plastic-substrate dye-sensitized solar cell with validated conversion efficiency of 7,6%, Solar Energy Materials & Solar Cells **94** (2010) 812-816
[10] Hasitha CW, Fuzhi H, Chenga YB, Fabrication of flexible dye sensitized solar cells on plastic substrates, Nano Energy **2/2** (2013) 174-189
[11] Kanta AF, Decroly A, Stainless steel electrode characterizations by electrochemical impedance spectroscopy for dye sensitized solar cells, Electrochimica Acta **56** (2011) 10276-10282
[12] Drygała A, Szindler M, Szindler MM and Jonda E, Atomic layer deposition of TiO2 blocking layers for dye-sensitized solar cells, Microelectronics International **37/2** (2020) 87-93
[13] Drygała A, Dobrzański LA, Szindler M, Prokopiuk Vel Prokopowicz M, Pawlyta M and Lukaszkowicz K, Carbon nanotubes counter electrode for dye-sensitized solar cells application, Archives of Metallurgy and Materials **61/2** (2016) 803-806
[14] Drygała A, Dobrzański LA, Prokopiuk Vel Prokopowicz M, Szindler M, Lukaszkowicz K and Domański M, A carbon-nanotubes counter electrode for flexible dye-sensitized solar cells, Materiali in Tehnologije **51/4** (2017) 623-629
[15] Ali HM, Mohamed HA, Mohamed SH, Enhancement of the optical and electrical properties of ITO thin films deposited by electron beam evaporation technique, Eur. Phys. J. Appl. Phys. **31/2** (2005) 87-93
[16] Wei L, Shuying C, Photoelectric properties of ITO thin films deposited by DC magnetron sputtering, J. Semicond **32/1** (2011) 013002
[17] Oshima M, Yoshino K, Characteristic of low resistivity fluorine-doped SnO2 thin films grown by spray pyrolysis, Jpn. J. Appl. Phys. **50/5S2** (2011) 05FB15
[18] Machalska G, Noworolnik M, Szindler M, Sitek W, Babilas R, Titanium dioxide nanoparticles and thin films deposited by an atomization method, Archives of Materials Science and Engineering **100/1-2** (2019) 34-41.

IOP Conf. Series: Materials Science and Engineering 1178 (2021) 012057 doi:10.1088/1757-899X/1178/1/012057

# Investigation of the influence of calcination temperature on morphology and structure of electrospun 1D $SnO_2$ nanostructures

**T Tański[1], W Smok[1], W Matysiak[1]**

Silesian University of Technology, Akademicka 2A, 44-100 Gliwice, Poland

Corresponding author's e-mail address: weronika.smok@polsl.pl

**Abstract**. The aim of the study was to manufacture the $SnO_2$ one-dimensional nanostructures via the sol-gel technique and electrospinning methods from a solution of polyvinylpyrrolidone (PVP), tin chloride pentahydrate ($SnCl_4 \cdot 5H_2O$), *N, N*-dimethylformamide (DMF) and ethanol (EtOH). The as obtained fibrous mats consisting of polymer and precursors particles were subjected to the heat treatment in air in the following temperatures 500, 600 °C to obtain pristine tin oxide nanowires. The scanning electron microscope (SEM) was used to determine the effect of calcination temperature on the morphology and structure of produced 1D $SnO_2$. The energy dispersive spectrometry (EDX) and Fourier-Transform Infrared spectroscopy (FTIR) were performed to investigate the chemical composition and structure of manufactured nanomaterials, respectively.

## 1 Introduction

Recently, increased interest in one-dimensional nanomaterials has been observed, which is primarily due to their extraordinary properties that distinguish them from other types of nanomaterials. The limited electron movement to two quantum directions and the high density of energy states are often responsible for optical, electrical, magnetic, and electrochemical properties of these nanostructures, which are much better than zero, two and three-dimensional nanomaterials. Ceramic semiconductor nanomaterials based on such compounds as titanium oxide ($TiO_2$), zinc oxide (ZnO), silicon oxide ($SiO_2$), bismuth oxide ($Bi_2O_3$) as well as tin oxide ($SnO_2$) are particularly popular among one-dimensional nanostructures [1-7].

As of today, based on literature study, it should be noted that $SnO_2$, despite the equally interesting as other oxides, optical and electrical properties (energy band gap, refractive index and dielectric constant) is much less studied, and analysis mechanisms responsible for their structure and properties are still a challenge for scientists. This is why this material should be given special attention, as it can be an ideal alternative to popular $TiO_2$ or ZnO in terms of optical properties. Tin oxide nano-materials are n-type semiconductor characterised by large surface-to-volume ratio, high optical transmission in the visible range, energy band gap in the range of 3.0–3.9 eV and good mechanical, chemical and thermal stability [7, 8].

For this reason, nanostructures have a wide application potential, which includes the construction of modern electronic and optical devices such as sensors or photovoltaic cells, as well as photocatalytically assisted water and air purification processes [7-10].

Therefore, in order to obtain $SnO_2$ nanowires with the desired morphology and physical properties, the influence of the parameters of the manufacturing process should be carefully studied. The particular attention should be paid to the calcination temperature, which significantly affects the diameter and optical properties of $SnO_2$ nanowires, which was presented in this work.

## 2 Materials and methods

To prepare spinning solutions, in order to produce one-dimensional tin oxide nanostructures, the following reagents were used: poly(vinylpyrrolidone) (PVP, purity of 99%, Mw = 1,300,000 g/mole), *N,N*-dimethylformamide (DMF, purity of 99.8%), ethanol (EtOH, purity of 99.8%) and tin chloride pentahydrate ($SnCl_4 \cdot 5H_2O$), purity of 98%). All substrates were purchased from Sigma-Aldrich.

In the first step, the mixture of 6g of tin chloride pentahydrate and dimethylformamide (10 ml) was made using magnetic stirrer for 48 h; mixture of ethanol (10 ml) and 2 g of PVP was prepared and then added to the above mixture and such homogenous solution was subjected to magnetic stirring for 48 h. The composite nanofibers polymer/precursor were obtained using the FLOW – Nanotechnology Solutions Electrospinner 2.2.0-500 device, with the following parameters: the distance and voltage between the electrodes of 15 cm and 24 kV, respectively and the flow rate of 0.45 ml/h. Directly after the fibrous mat production process, drying at room temperature was performed. After this, calcination in a high-temperature furnace EasyTube 2000 provided by first nano at temperatures of 500 and 600 °C (heating rate of 2 °C/min) for 6 h in air was conducted. The manufactured samples were left in a furnace to cool down completely.

In order to image the topography nanostructures and analyse the chemical composition of the prepared composite and ceramic one-dimensional nanostructures, the (SEM) Zeiss Supra 35 scanning electron microscopes with the Trident XM4 series (EDX) X-ray spectrometer were used. Based on the SEM images, a measurement of one hundred randomly selected diameter values of the obtained nanomaterials using DigitalMicrograph was provided. In addition, to analyse the oscillatory transitions of atoms vibrating between the oscillatory levels in the molecules of the produced nanomaterials, and to determine the functional groups existing therein, Fourier-Transform Infrared spectroscopy (FTIR) (Nicolet™ iS™ 50 FTIR Spectrometer, Thermo Fisher Scientific) was used.

## 3 Results and discussion

### *3.1 Morphology, structure and chemical composition*

An analysis of the morphology and structure of nanofibers prior to the calcination process demonstrated that PVP/$SnO_2$ nanofibers are devoid of structural defects and have a constant diameter along the entire length, which results from the preparation of a solution of appropriate viscosity and the correct selection of electrospinning parameters (figure 1a). Using both calcination temperatures during the calcination process, of 500 and 600 °C, allowed for obtaining homogenous one-dimensional $SnO_2$ nanowires without visible structural defects in the form of agglomerates or welded nanowires (figure 2, 3).

The analysis of the morphology of the obtained PVP/$SnCl_4$ nanofibers made on the basis of a hundredfold measurement of the diameters of randomly selected one-dimensional composite nanostructures with a polymer matrix, showed that the nanofibers were characterised by diameters in the range of 40–320 nm, an average diameter was 180 nm. In addition, the largest group was nanofibers with diameters ranging from 160 to 240 nm (figure 1, histogram).

Analysis of the morphology of $SnO_2$ nanowires calcined in 500 °C, performed on the basis of 100-times measurements of their diameters, has shown that wires diameters ranged between 50 and 250, with wires ranging between 125 and 175 nm making up the most numerous group of 36% of nanowires. The average wire diameter was 138 nm (figure 2, histogram).

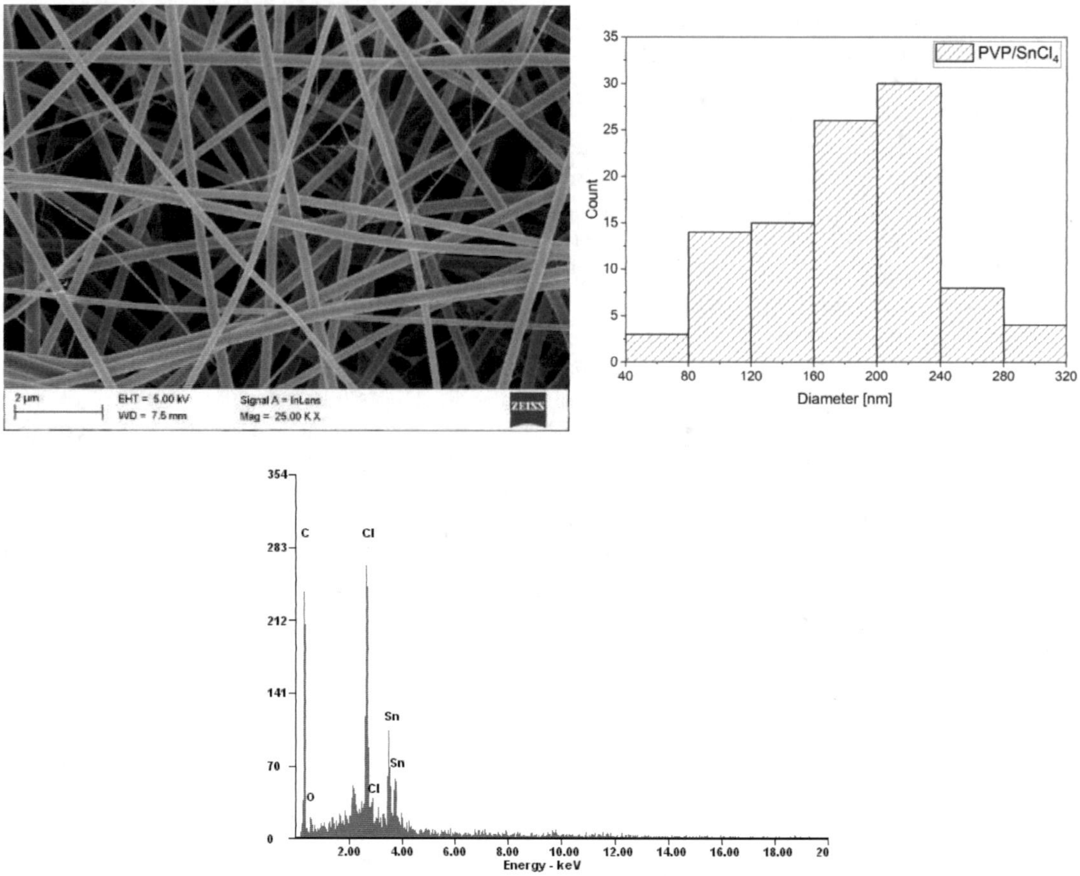


**Figure 1.** SEM surface morphology images, histograms showing the distribution of hundredfold measurement of the diameter and chemical composition of the produced PVP/$SnCl_4$ composite nanofibers

A hundredfold diameter measurement of random $SnO_2$ wires obtained from calcination in 600 °C showed that the measured diameters ranged from 15 to 75 nm, with the most common values for the diameter (60% of all measured values of diameters) contained in the range from 25 to 45 nm and an average diameter of 40 nm (figure 3, histogram). Therefore, a strong shrinkage of nanowires is noticeable as a result of the calcination process.

The observed decrease in the diameter of nanowires after calcination compared to the diameter of the fibers just after electrospinning results from the removal of the polymer matrix and obtaining pristine $SnO_2$ nanowires. In addition, the higher calcination temperature made it possible to obtain nanowires with a much smaller diameter than for wires calcined at 500 °C [11, 12].

The EDX spectra confirmed the formation of non-contaminated, pristine $SnO_2$ nanowires from z PVP/$SnCl_4$ nanofibers, because chlorine from the precursor nanoparticles does not appear in the spectra recorded for nanowires (figure 1-3).

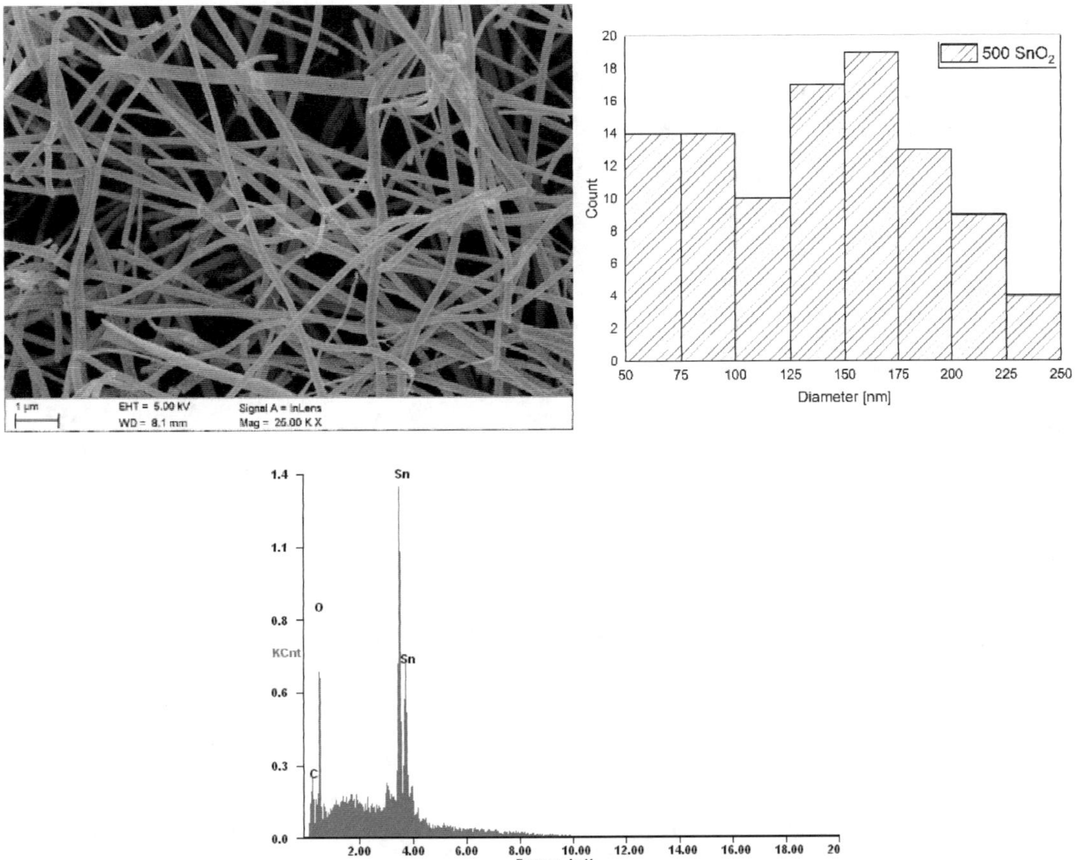


**Figure 2.** SEM surface morphology images, histograms showing the distribution of hundredfold measurement of the diameter and chemical composition of the produced $SnO_2$ nanowires calcined in 500 °C

The absorbance spectra in the function of wavenumber in the range of 400–3500 $cm^{-1}$ graphs were plotted for composite PVP/$SnCl_4$ nanofibers and $SnO_2$ nanowires with some characteristic peaks for individual vibration molecules or functional groups (figure 4). In the spectrum obtained for composite nanofibers we observe characteristic peaks at 558 $cm^{-1}$ that may be assigned respectively to N-C=O bending. The peaks at 1298 and 1632 $cm^{-1}$ proved the existence of C-N and stretching of C=O, respectively. All characteristic peaks for $-CH_2$ и$-CH$ groups of PVP at 1456, and 2957 $cm^{-1}$ were observed, respectively. The FTIR spectrum of PVP/$SnCl_4$ had a peak at 3169 $cm^{-1}$ which indicates O-H stretching. At 670 $cm^{-1}$ Sn-O bond from precursor occurrence in nanofibers can be observed [13, 14].

In the case of FTIR spectra recorded for both $SnO_2$ samples, the occurrence of broad peaks at 639 and 667 $cm^{-1}$ confirms generation of Sn-O bond. In addition, no characteristic peaks from PVP are observed, which confirms the degradation of the polymer during the calcination in both temperatures.

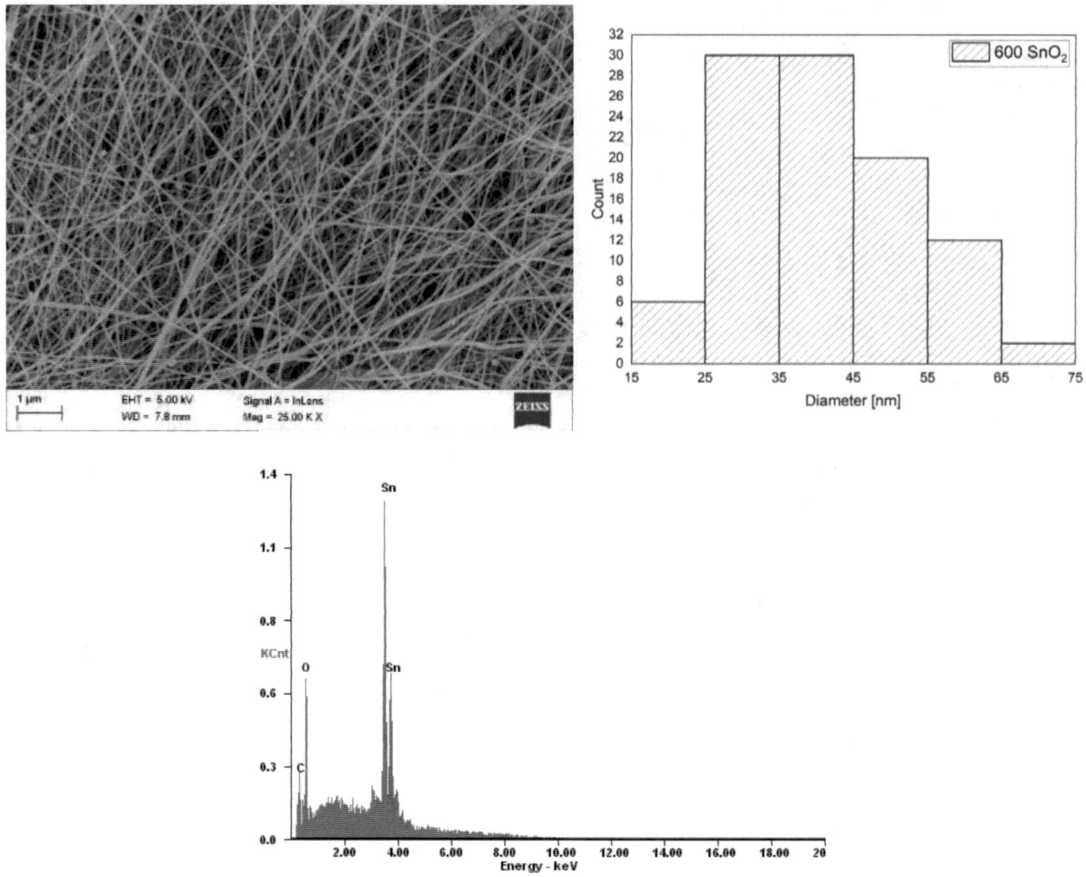


**Figure 3.** SEM surface morphology images, histograms showing the distribution of hundredfold measurement of the diameter and chemical composition of the produced $SnO_2$ nanowires calcined in 600 °C

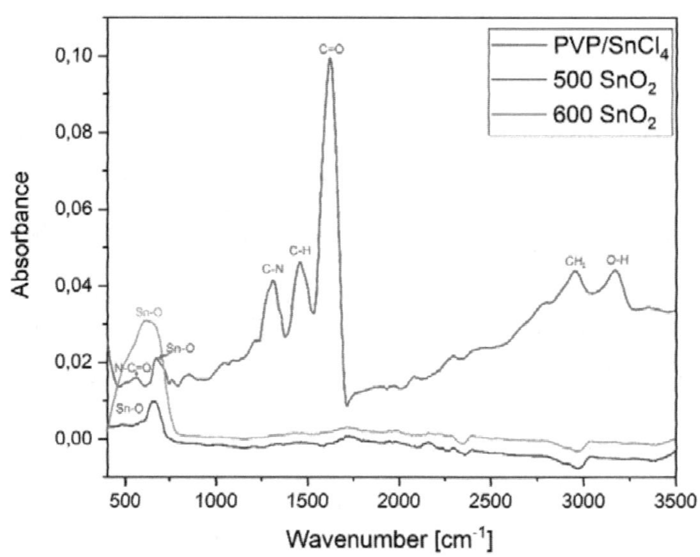


**Figure 4.** FTIR spectrum for $PVP/SnCl_4$ composite nanofibers and $SnO_2$ nanowires calcined in 500 and 600 °C

## 4 Conclusion

The aim of this work was to prepare, using combination method of sol-gel technique and electrospinning from solutions PVP/$SnCl_4$/DMF/EtOH, and examine the morphology and structure of $SnO_2$ nanowires. The performed morphology and structure analysis of the obtained after calcination in 500 and 600°C one-dimensional tin oxide nanostructures using scanning (SEM) revealed that $SnO_2$ nanowires were characterised by a lack of structural defects and constant diameter values along the entire lengths. Moreover, the increase in the annealing temperature is associated with a decrease in the nanowire diameter values, respectively 138 nm for 500°C and 40 nm in the case of 600°C.

Both, EDS and FTIR analysis confirmed that pristine, undoped $SnO_2$ nanowires were obtained, which confirms that the presented method allows control the morphology and chemical composition of the produced one-dimensional nanostructures. This work is a good basis for further research on the electrical and optical properties of $SnO_2$ nanowires depending on the parameters of the electrospinning and calcination process.

*References*

[1] Someswararao M V, Dubey R S, Subbarao P S V, Singh S 2018 *Results Phys.* **11** 223-231
[2] Secundino-Sánchez O, Diaz-Reyes J, Aguila-López J, Sánchez-Ramírez J F. 2019 *J. Mol. Struct.* **1194** 163-170.
[3] Matysiak W, Tański T, Zaborowska M 2018 *Appl. Surf. Sci.* **442** 382-389
[4] Matysiak W, Tański T 2019 *Mater. Today* **7** 382-388
[5] Matysiak W, Tański T 2019 *Appl. Surf. Sci.* **489** 34-43
[6] Lim G D, Yoo J H, Ji M, Lee Y I 2019 *J.Alloys Compd.* **806** 1060-1067
[7] Matysiak W, Tański T, Smok W 2020 *Appl. Phys. A.* **126** 1-10
[8] Cheng L, Ma S Y, Wang T T, Li X B, Luo J, Li W Q, Gz D J 2014 *Mater. Lett.* **131** 23-26
[9] Wang X, Fan H, Ren P 2013 *Catal. Commun.* **31** 37-41
[10] Phuoc P H, Hung C M, Van Toan N, Van Duy N, Hoa N D, Van Hieu N 2020 One-step fabrication of SnO2 porous nanofiber gas sensors for sub-ppm H2S detection *Sens. Actuator A Phys.* **303** 111722
[11] Du Z, Guo L, Zheng T, Cai Q, Yang X 2019 *Ceram. Int.* **45** 23975-23983
[12] Thangavel K, Roshini T, Balaprakash V, Gowrisankar P, Sudha S, Mohan M 2020 Structural, morphological and antibacterial properties of ZnO nanofibers fabricated by electrospinning technique *Mater. Today*
[13] Koczkur K M, Mourdikoudis S, Polavarapu L, Skrabalak S E 2015 *Dalton Trans.* **44** 17883-17905
[14] Sagadevan S, Podder J 2016 *Mater. Res.* **19** 420-425

IOP Conf. Series: Materials Science and Engineering 1178 (2021) 012058 doi:10.1088/1757-899X/1178/1/012058

# Material characterization of Au/Ni nanocatalyst for low-temperature carbon dioxide methanation

**Błażej Tomiczek[1*], Marek Szindler [1], Mirosława Pawlyta[1], Paulina Boryło[1]**
[1]Department of Engineering Materials and Biomaterials, Silesian University of Technology, Konarskiego 18a Str. 44-100, Gliwice, Poland
E-mail: blazej.tomiczek@polsl.pl

**Keywords:** Nanoparticles, Nanotechnology, Renewable energy, Methanation, Nanocatalyst

**Abstract**. The nanocatalyst for the carbon oxide methanation process is used, among others, in environmental protection, chemical industry, and renewable energy sources. The use of a suitable catalyst allows a chemical reaction to be carried out between unreacted gaseous substrates. The most frequently studied monometallic catalysts are: Ni, Ru, Rh, Pt, Au, Cu, Fe. Bimetallic nanocatalysts are equally popular. Their catalytic properties differ from those of pure component metals. Numerous studies indicate a positive effect in catalysts containing particles: Au-Ag, Au-Pt, Au-Pd, Pd-Ni, Pd-Cu. A review of the literature indicates that examples of the use of metal nanoparticles of spherical shape deposited on a nickel substrate in the methanation process are known, but so far no attempt has been shown in publications to produce a catalyst based on gold nanoparticles with developed surface in the form of spiky (nanourchins, nanostars) on a nickel base as presented in the article. Gold nanourchins are deposited on a nickel substrate in the form of a nickel molecular mesh. The prepared nanocatalyst has been subjected to structural analysis using a transmission electron microscope (TEM). Scanning Electron Microscopic (SEM) images were taken with a Zeiss Supra 35. Qualitative studies of chemical composition were also performed using the Energy Dispersive Spectrometer (EDS). Based on the TEM results, the appearance of the X-ray diffraction pattern was computer modelled. A nanocatalyst was obtained with a high coverage of the nickel molecular mesh surface with gold nanoparticles not exceeding 50 nm in diameter.

## 1 Introduction

Nanoparticles are defined as structures up to 100 nm in diameter [1-3]. They often exhibit unique physical, chemical, and biological properties that differ significantly from their macroscale forms. Nanoparticles are characterized by a specific geometric structure characterized by a high surface area to volume ratio, which is inversely proportional to the particle diameter. This causes a significant increase in their activity and affects the absorption properties and reactivity [1-5]. The reduced size changes the total energy of the system responsible for thermodynamic stability, which is the result of disturbances in the wave function of the electrons. The area of nanoparticles applications is still growing. On the priority list of nanomaterials predicted, the largest commercial use in the near future, published as part of the work carried out in the OECD (Inter-Organization Program for the Sound Management of Chemicals - IOMC), the following metals and their oxides have been placed: silver, iron, aluminum, cerium, titanium, zinc or nickel [4-7]. Nanoparticles are used, among others in the cosmetics, pharmaceutical, automotive and chemical industries (composites, paints, varnishes), construction, electronics, biomedicine and bioengineering, environmental protection and many others [1-3]. Metal nanoparticles in particular are of great interest to researchers due to their unique physicochemical properties determined by their size. They can significantly increase the speed, selectivity, and efficiency of chemical processes while minimizing the amount of by-products when they act as a catalyst. The most intensively studied monometallic catalysts are: Ni, Ru, Rh, Pt, Au, Cu, Fe [8-12]. Bimetallic nanocatalysts are equally popular. Their catalytic properties differ from those of pure component metals. The use of bimetallic nanocatalysts with an appropriate carrier is designed to eliminate the problem of the possibility of blowing out nanoparticles with catalysis products. Numerous studies indicate a positive effect of catalysts containing particles: Au-Ag, Au-Pt, Au-Pd, Pd-Ni, Pd-Cu [9-15]. A review of the literature indicates that examples of the use of metal nanoparticles of spherical shape deposited on a nickel substrate in the methanation process are known, but so far no attempt has been shown in publications to produce a catalyst based on gold nanoparticles with developed surfaces in the form of spikes (nanourchins, nanostars) on a nickel base as presented in the article. Among the various geometries of gold nanoparticles, gold nanourchins, alternatively referred to as nanoflowers, nanostars, or multibranched gold nanoparticles, have received much attention in recent years [16-17]. This is because of their catalytic activity, molecular detection, and biological applications in immunoassays, dark field imaging of cells and in plasmonic biosensors.

## 2 Material descriptions and research methodology

Two groups of methods can be distinguished in the synthesis of nanourchins: the seeded-growth and non-seeded-growth methods. In this article, gold nanourchins synthesized by the method described by Weifeng Lv [17] were used to produce the Au /Ni nanocatalyst. It's non-seeded method in which as a source of gold, Tetrachloroauric (III) acid trihydrate ($HAuCl_4 \cdot 3H_2O$, 99.5%) was used. Ascorbic acid was used as a reducing agent. Then the ready nanourchins was suspended in phosphate buffered saline (PBS). The synthesized nanoparticles were deposited directly on a nickel substrate in the form of a molecular mesh (Figure 1).

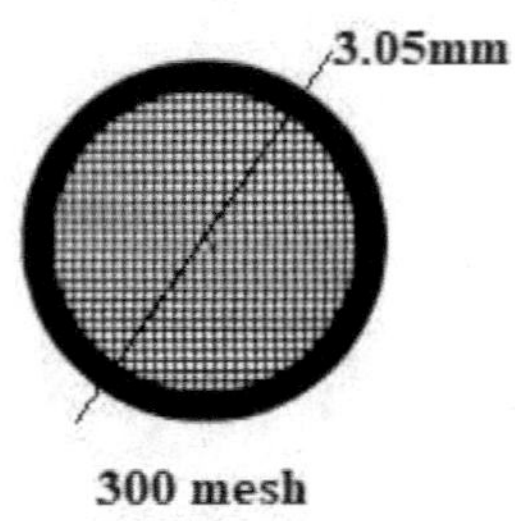


**Figure 1.** Nickel molecular mesh

Scanning electron microscopic (SEM) images were taken with a Zeiss Supra 35. The accelerating voltage was 5 kV. To obtain images of the surface topography, the detection of secondary electrons (by the detector In Lens) was used. Qualitative studies of chemical composition were also performed using the energy-dispersive spectrometer (EDS). The prepared nanocatalyst has been subjected to structural analysis using a transmission electron microscope (TEM). High-resolution transmission electron microscopy (HR-TEM) analysis was carried out using a transmission electron microscopy/scanning transmission electron microscopy (TEM/STEM) system (Titan 80-300, FEI Company, Hillsboro, Oregon, USA) with a super twin-lens operated at 300 kV and equipped with an annular dark field detector. The crystal structures and diffraction patterns of nanoparticles were simulated using VESTA software.

## 3 Results and Discussion

Clusters of palladium nanoparticles on most of the nickel surface were recorded. As can be seen from the image, there is a size distribution of particles below 50nm. However, there is a high preference for very small clusters (Figure 2).

a) b)

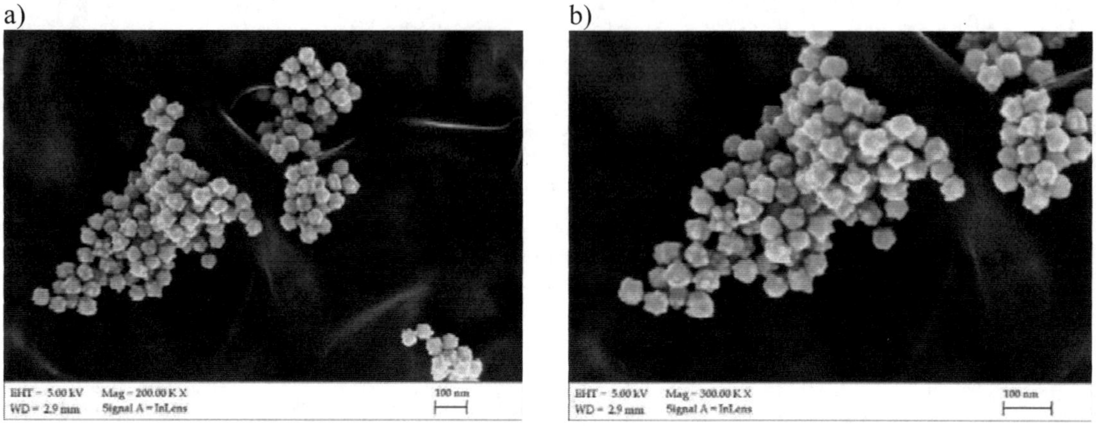


**Figure 2.** SEM surface topography images of the Au-Ni nanocatalyst with gold nanourchins

An EDS analysis was performed, recording the reflection characteristics of gold and nickel (Figure 3). For gold, 2.203 eV (spectrum line $M_{\beta 1}$) were registered. For nickel, 0.866 eV (spectrum line $L_{\beta 1}$), and 7.480 eV (spectrum line $K_{\alpha 1}$) were registered.

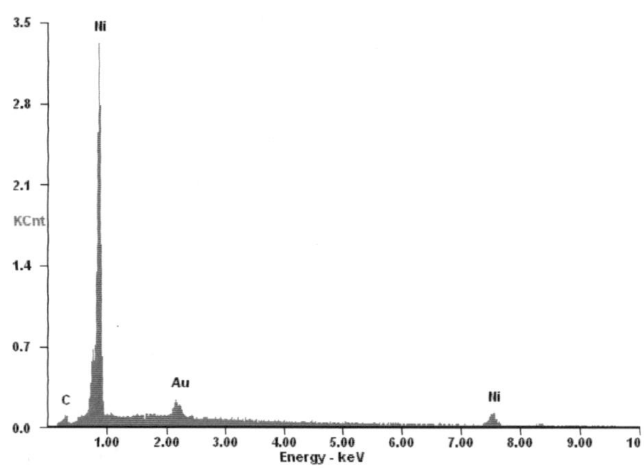


**Figure 3.** EDS spectrum of the Au-Ni nanocatalyst with gold nanourchins

IOP Conf. Series: Materials Science and Engineering 1178 (2021) 012058 doi:10.1088/1757-899X/1178/1/012058

Detailed results of surface morphology and structural studies were performed using the S/TEM. For research purposes, the STEM imaging mode was used. The Titan 80-300 microscope is equipped with three coaxial detectors dedicated to STEM mode: Bright Field–BF, Annular Dark-Field–ADF and High-Angle Annular Dark-Field–HAADF (Figure 4). The SAED pattern shows a clear appearance of diffraction rings from the Au. To solve the SAED pattern, JCPDS files were used, according to which were assigned the appropriate Miller indices (Figure 5b). The crystal structure of the gold nanoparticle was identified as face-centered cubic (Figure 6a). Based on the TEM results, the appearance of the X-ray diffraction pattern was computer modelled. Specialist VESTA software was used for this purpose. It is a 3D visualization program for structural models, volumetric data such as electron/nuclear density, and crystal morphology. VESTA allows to simulate X-ray and neutron powder diffraction patterns from lattice and structure parameters. The simulated diffraction pattern of gold nanoparticles is shown in Figure 6b. Lattice parameters which was used for simulation of diffraction pattern is collected in Table 1.

Table 1. Lattice parameters used for simulation of gold nanoparticle diffraction pattern [18]

| Crystal system: | Cubic | No. | h | k | l | d [A] | 2Theta[deg] | I [%] |
|---|---|---|---|---|---|---|---|---|
| Space group: | F m -3 m | 1 | 1 | 1 | 1 | 2,35039 | 44,739 | 100,0 |
| a (A): | 4,071 | 2 | 0 | 0 | 2 | 2,03550 | 52,138 | 48,4 |
| b (A): | 4,071 | 3 | 0 | 2 | 2 | 1,43932 | 76,848 | 29,8 |
| c (A): | 4,071 | 4 | 1 | 1 | 3 | 1,22745 | 93,563 | 34,0 |
| Alpha (°): | 90,0 | 5 | 2 | 2 | 2 | 1,17520 | 99,132 | 9,8 |

a) b)

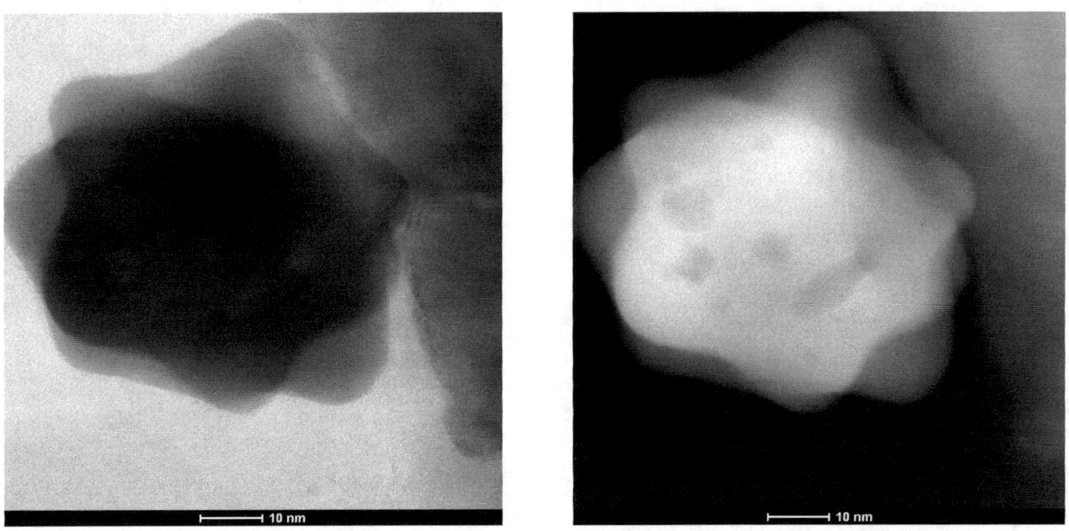


**Figure 4.** The BF_DF (a) and HAADF (b) images of Au-Ni nanocatalyst with gold nanourchins registered in STEM mode

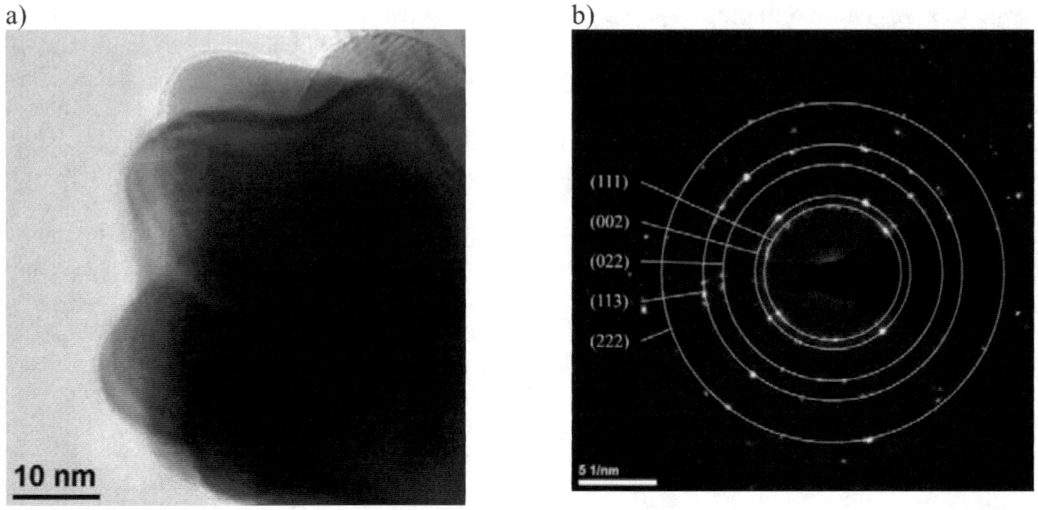


**Figure 5.** The HRTEM image (a) and SAED pattern (b) of Au-Ni nanocatalyst

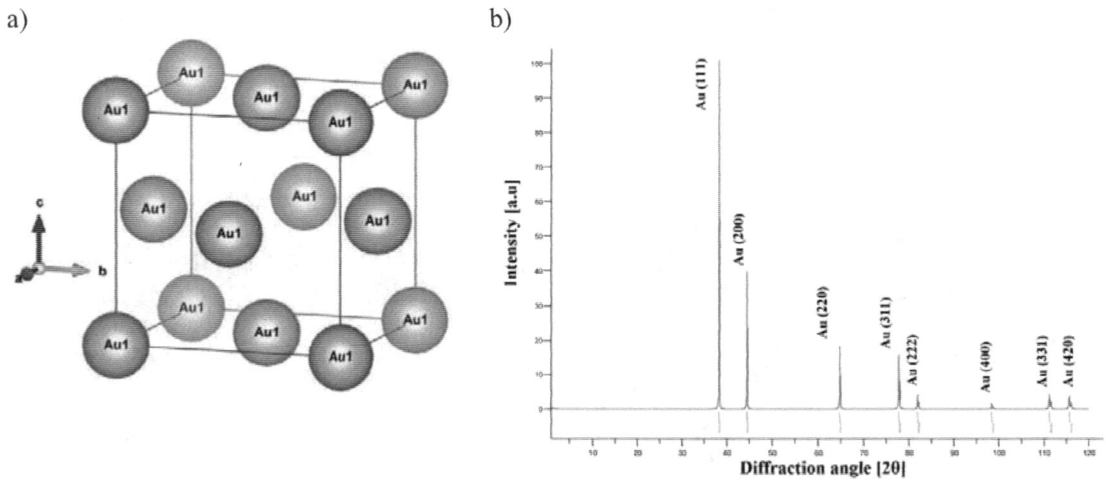


**Figure 6.** The face-centered cubic crystal structure (a) and diffraction pattern (b) of gold nanourchins simulated using VESTA software

## 4 Conclusions

Review of the literature indicates that there are known examples of the use of spherical-shaped metal nanoparticles deposited on a nickel substrate in the methanation process, but so far no attempts have been made in the publications to produce a catalyst based on gold nanoparticles with a developed surface in the form of spiked nanostars (nanourchins) on a nickel substrate. In summary, we prepared nano Au supported by Ni molecular mesh. By using electron transmission microscopy and XRD simulation, the phases in the nanocatalyst were identified. A nanocatalyst was obtained with a high coverage of the nickel molecular mesh surface with gold spiked from nanoparticles not exceeding 50 nm in diameter. The results obtained by the authors of this paper provide a promising perspective for research and indicate that this is an attractive direction in the search for new solutions in environmental protection, chemical industry, and renewable energy sources.

**Acknowledgements**

This work has been implemented within the framework of the project entitled "New unalloyed nano-metal conjugates on three-dimensional supports for catalytic processes of methanation and reduction of nitrogen oxide", funded by the Polish National Science Centre under grant no. UMO-2018/29/B/ST8/02303 in the framework of the "OPUS" competitions.

## References

[1] Mohan YM, Raju KM, Sambasivudu K, Singh S, Sreedhar B, Preparation of acacia-stabilized silver nanoparticles: A green approach, Journal of Applied Polymer Science **106** (2007) 3375-3381
[2] Shaalan NM et al., Synthesis of metal and metal oxide nanostructures and their application for gas sensing, Materials Chemistry and Physics **127** (2011) 143–150
[3] Huang J, Wan Q, Gas sensors based on semiconducting metal oxide one-dimensional nanostructures. Sensors **9** (2009) 9903-9924
[4] Tavakoli A, Sohrabi M, Kargari A, A review of methods for synthesis of nanostructured metals with emphasis on ironmcompounds. Chemical Papers, **61/3** (2007) 151-170.
[5] Szindler M, Szindler MM, Dobrzański LA, Jung T, NiO nanoparticles prepared by the sol-gel method for a dye sensitized solar cell applications, Archives of Materials Science and Engineering **92/1** (2018) 15-21
[6] Ali S et al., Ni-based nano-catalysts for the dry reforming of methane, Catalysis Today **343** (2020) 26-37
[7] Więcław-Solny L, Wilk A and et al., Catalytic carbon dioxide hydrogenation as a prospective method for energy storage and utilization of captured CO2, Journal of Power Technologies **96/4** (2016) 213–218
[8] Korzeniowska A, Grzybek J and et al., The structure-catalytic activity relationship for the transient layered zeolite MCM-56 with MWW topology, Catalyst **345** (2020) 116-124
[9] Kim A, et al., Mesoporous TiO2 Support Materials for Ru-Based CO2 Methanation Catalysts, ACS Appl. Nano Mater **2/5** (2019) 3220-3230
[10] Swalus C, et al., CO2 methanation on Rh/γ-Al2O3 catalyst at low temperature: "In situ" supply of hydrogen by Ni/activated carbon catalyst. Applied Catalysis B Environmental **125** (2012) 41–50
[11] Wang A., et al., Au-Ag alloy nanoparticle as catalyst for CO oxidation: Effect of Si/Al ratio of mesoporous support. Journal of Catalysis **237/1** (2006) 197-206
[12] Yang HL, Recent progress in synergistic catalysis over heterometallic nanoparticles. J. Mater. Chem. **21** (2011) 13705-13725
[13] Bernareggi M, Cu and Pt clusters deposition on TiO2 powders by DC magnetron sputtering for photocatalytic hydrogen production. Catalysis Today **326** (2019) 15-21
[14] Wei Z, Bimetallic catalysts for hydrogen generation, Chemical Society Reviews 41 (2012) 7994-8008
[15] Wang ZQ, et al., High-Performance and Long-Lived Cu/SiO2 Nanocatalyst for CO2 Hydrogenation. ACS Catal. **5/7** (2015) 4255-4259
[16] Phiri MM, Mulder DW, Vorster BC, Seedless gold nanostars with seed-like advantages for biosensing applications, R. Soc. open sci. **6** (2019) 1-9
[17] Weifeng Lv et al., One-Pot Synthesis of Multi-Branch Gold Nanoparticles and Investigation of Their SERS Performance, Biosensors **8** (2018) 1-10.
[18] Salamakha L.P., Bauer E., Mudryi S.I., Gonçalves A.P., Almeida M., Noël H., Isothermal section of the Ce–Au–Sb system at 870 K, Journal of Alloys Compd., **479** (2009) 184 - 188

IOP Conf. Series: Materials Science and Engineering 1178 (2021) 012059 doi:10.1088/1757-899X/1178/1/012059

# Influence of micro- and nanoparticles on mechanical properties of magnesium and magnesium alloys

**Z Trojanová[1,2*], P Lukáč[1,2] and J Džugan[2]**

[1]Charles University, Faculty of Mathematics and Physics, Ke Karlovu 3, 121 16 Praha 2, Czech Republic,

[2]COMTES FHT, Průmyslová 995, 223 41 Dobřany, Czech Republic

*ztrojan@met.mff.cuni.cz

**Abstract**. Magnesium and magnesium alloys were reinforced with various micro- and nanoparticles (SiC, Inconel718, $Al_2O_3$, $ZrO_2$, BN, graphite) Composites were prepared by squeeze casting and using powder metallurgy followed by hot extrusion. Tensile and compression tests were performed at elevated temperatures up to 300 °C. The influence of micro- and nanoparticles on mechanical characteristics is discussed on the base of physical models. Bonding at the interface between particles and magnesium matrix plays an important role for the resulting composite properties.

## 1 Introduction

Metal matrix composites (MMCs), metals, and alloys reinforced with particles, fibres, platelets, or whiskers with different physical and chemical properties are intensively studied in the last two decades. If at least one size of the reinforcing phase is in the micrometer range, we deal with the metal matrix microcomposite (mMMS) or if it exhibits a size up to 100 nm, this is a metal matrix nanocomposite (nMMC). Magnesium and magnesium alloys reinforced with various micro- and nanoparticles (mp and np) are very attractive materials with improved mechanical properties. The specific stiffness and strength are better compared to the monolithic materials, whereas the low density is preserved. Magnesium based MMCs usually have limited ductility. The strengthening of MMCs is a complex problem. Various mechanisms responsible for the strengthening of MMCs are described in the literature [1-4]. The problem how to combine the individual strengthening terms was not up to now satisfactorily solved. In the present study, the mechanical characteristics of magnesium-based mMMCs and nMMCs, prepared with different methods, are presented. Various strengthening mechanisms are discussed with the aim to find the main effects decided by the composites` resultant properties.

## 2 Manufacturing methods

The main problem in the manufacturing of MMCs is the uniform distribution of reinforcing phase particles or fibres in the metallic matrix. For composites preparation methods in situ where particles are formed in the melt solidification via a chemical reaction, or ex situ where the ready particles are inserted into the molten matrix. Usually, simple mixing of both phases is not an effective method due to the low wettability of particles or their tendency to form clusters. One possibility, how to avoid this problem, is using a prefabricated preform – some structure containing particles and a binder. The metallic melt is pressed into the preform in the squeeze casting process. In many cases, powder metallurgical methods are utilized for the manufacturing of mMMCs or nMMCs. In this case, the metallic powder is mixed with the particles, then compacted and hot extruded. The milling of powders with particles is often used for a better distribution of particles in the matrix. The original technique for the preparation of mMMCs and nMMCs was developed by Gupta and coworkers [1]. In the disintegration melt deposition (DMD) method, the melt containing the particles is disintegrated by the Ar current and then compacted and hot extruded. While MMCs prepared using squeeze casting exhibit no preferential orientation, the MMCs manufactured by the powder metallurgical (PM) methods exhibit some texture. This texture is the reason for the often observed tension compression anisotropy of magnesium extruded materials [5]. In textured Mg materials, the basal planes are parallel to the extrusion direction (ED) (the <c> axis is perpendicular to the ED), while the majority of the prismatic planes rotate around the <c> axis.

## 3 Mechanical properties

### *3.1 Microcomposites manufactured using a preform.*

A commercially available AZ91 (Mg – 9%Al; 1%Zn; 0.3%Mn) was reinforced with 10 vol.% of SiC particles and an addition of 3wt% of Si. Composites were prepared by the squeeze casting method using a preform. The microstructure consists of SiC particles, δ-Mg grains (solid solution of Al in Mg), primarily segregated γ-phase ($Mg_{17}Al_{12}$) and discontinuous precipitates (lamellae of γ+δ phase). The addition of 3%Si gives rise to the $Mg_2Si$ phase. Details of the microstructure are visible in the scanning electron micrograph introduced in Figures 1a,b,c, and d. Similar preform containing SiC particles and an addition of C was used for the mMMC based on ZE41 alloy (Mg – 4%Zn – 1%RE ). The tensile yield stress (TYS) and ultimate tensile strength (UTS) estimated for both composites at various temperatures are reported in Table 1. For comparison, the values of characteristic stresses obtained for the monolithic alloys are shown. The reinforcing effect of particles decreases with increasing temperature.

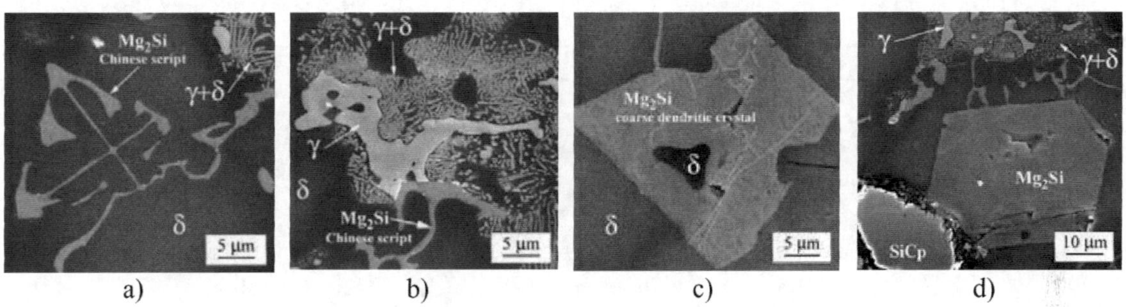


**Figure 1**. Details of microstructure visible in electron micrographs: a) $Mg_2Si$ eutectics in the form of Chinese script; b) $Mg_{17}Al_{12}$ continuous and discontinued precipitates; c) and d) $Mg_2Si$ and SiC particles.

**Table 1**. Mechanical characteristics of mMMCs based on Mg alloys investigated in tension.

| | AZ91+10SiC+3Si | | ZE41+4SiC+7C | | AZ91 cast | | ZE41 cast | |
|---|---|---|---|---|---|---|---|---|
| T (°C) | **TYS** (MPa) | **UTS** (MPa) | **TYS** (MPa) | **UTS** (MPa) | **TYS** (MPa) | **UTS** (MPa) | **TYS** (MPa) | **UTS** (MPa) |
| 23 | 175 | 359 | 183 | 394 | 141 | 237 | 122 | 264 |
| 50 | 182 | 328 | | | | | 122 | 272 |
| 61 | | | | | 138 | 237 | | |
| 100 | 146 | 337 | 164 | 328 | 132 | 208 | 89 | 213 |
| 150 | 132 | 237 | 136 | 222 | 108 | 166 | 114 | 230 |
| 200 | 105 | 141 | 89 | 139 | 102 | 122 | 104 | 188 |
| 250 | 82 | 98 | | | 78 | 91 | 97 | 161 |
| 300 | 53 | 57 | | | 57 | 66 | 67 | 92 |

### *3.2 Microcomposites prepared by the powder metallurgical technology*

Two mMMCs were prepared by mixing, compaction, and hot extrusion of the powders. Figure 2a shows the light micrograph of the as-prepared WE54+13vol%SiC mMMC. The SiC particles having a size of several μm are not uniformly distributed in the matrix; they form in many cases clusters. A significant amount of cuboidal fine particles (20-40 nm) was estimated distributed inside of the grains. On the basis of electron diffraction patterns, the particles were attributed to the β′ $Mg_{12}NdY$ phase. Another feature in the microstructure is a significant increase of the dislocation density visible in the left bottom corner in Figure 2c.

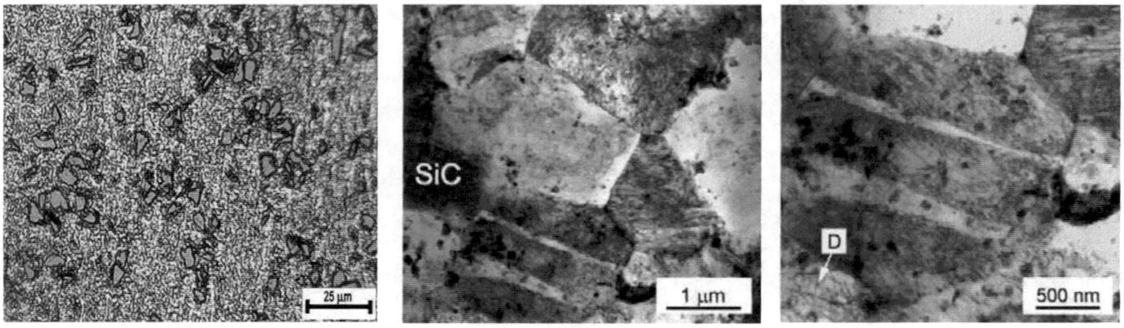

a) b) c)

**Figure 2**. (a) Light micrograph; (b), (c) electron micrographs of WE54/SiC mMMC (Significant dislocation density is depicted as D in figure 2c.)

**Table 2**. Mechanical characteristics of two mMMCs prepared with PM technology.

| T (°C) | WE54+13%SiC (PM) | | Mg-8Li+7%SiC (PM) | | WE54 | | Mg-8Li | |
|---|---|---|---|---|---|---|---|---|
| | **CYS (MPa)** | **CPS (MPa)** | **CYS (MPa)** | **CP S (MPa)** | **CYS (MPa)** | **CPS (MPa)** | **CYS (MPa)** | **CPS (MPa)** |
| 23 | 246 | 370 | 198 | 405 | 141 | 274 | 85 | 171 |
| 50 | 241 | 361 | 187 | 390 | 146 | 289 | 90 | 144 |
| 100 | 226 | 348 | 166 | 284 | 148 | 271 | 52 | 81 |
| 150 | 220 | 327 | 127 | 159 | 136 | 262 | | |
| 200 | 217 | 305 | 74 | 83 | 123 | 250 | 41 | 60 |
| 250 | 168 | 227 | 35 | 40 | 114 | 261 | | |
| 300 | 65 | 82 | 15 | 22 | | | | |

Similarly, the SiC particles were nonuniform distributed in the Mg-8Li alloy. This alloy is consisted of two phases: HCP α phase and BCC β phase. The compression yield strength (CYS) and compression peak stress (CPS) measured for two magnesium alloys and the unreinforced alloys are reported in Table 2. The reinforcing influence of particles is really high. Note that the values introduced in Table 2 for unreinforced alloys were obtained for cast alloys.

*3.3 Microcomposites prepared by disintegrated melt deposition.*

Pure magnesium reinforced with 0.7, 1.4, and 2.4 vol% of Inconel 718 particles was prepared using DMD technique. Magnesium melt together with the Inconel powder (0.2-5 μm) was disintegrated by the Ar gas before being deposited onto a metallic substrate [4]. Processed ingots were hot homogenised and hot extruded. Samples were deformed in tension and compression at elevated temperatures. Characteristic stresses are reported in Table 3 together with the stresses measured for pure Mg, prepared with the same method. Pronounced tension-compression asymmetry is appreciable. Such asymmetry is done by the basal texture which was formed in the extrusion process. The high reinforcing effect of In718 microparticles is obvious from Table 3 even by the small volume fraction of particles.

**Table 3**. Mechanical characteristics found for Mg-In718 mMMCs deformed in tension and compression.

| | TYS (MPa) | | | | UTS (MPa) | | | |
|---|---|---|---|---|---|---|---|---|
| T °C | 0 | 0.7 vol% | 1.4 vol% | 2.4 vol% | 0 | 0.7 vol% | 1.4 vol% | 2.4 vol% |
| 23 | 161 | 207 | 249 | 285 | 260 | 255 | 301 | 339 |
| 100 | 98 | 156 | 163 | 157 | 173 | 250 | 247 | 255 |
| 200 | 42 | 113 | 95 | 113 | 78 | 211 | 215 | 211 |
| 300 | 21 | 47 | 51 | 48 | 35 | 61 | 61 | 61 |
| | CYS (MPa) | | | | CPS (MPa) | | | |
| 23 | 88 | 131 | 152 | 185 | 383 | 295 | 326 | 362 |
| 100 | 91 | 122 | 154 | 196 | 241 | 239 | 280 | 313 |
| 200 | 69 | 109 | 112 | 139 | 103 | 153 | 168 | 183 |
| 300 | 34 | 41 | 108 | 70 | 40 | 48 | 137 | 78 |

*3.4 Nanocomposites prepared by milling and powder metallurgy.*

Magnesium was reinforced with ceramic np ($Al_2O_3$, $ZrO_2$, BN) and graphite via PM technology. Powders were mixed and milled together, compacted and hot extruded. TEM showed particles of $Al_2O_3$and $ZrO_2$ in the grain boundaries and some particles were found inside of grains. In the case of nMMCs with hexagonal BN (hBN) and Gr, the nanoparticles remained in the grain boundaries (see figure 3). Particles of hBN have a similar layered structure as Gr, within each layer, boron and nitrogen atoms are bound by strong covalent bonds, whereas the layers are held together by the weak van der Waals forces. The fine BN and Gr nanoparticles deformed during the milling and created intergrain material layers.

**Table 4**. Mechanical characteristics of $Mg/Al_2O_3$ and $Mg/ZrO_2$ obtained in tension and compression.

| | Tension | | | | | | Compression | | | |
|---|---|---|---|---|---|---|---|---|---|---|
| | μMg | | $Mg{+}3Al_2O_3$ | | $Mg{+}3ZrO_2$ | | $Mg{+}3Al_2O_3$ | | $Mg{+}3ZrO_2$ | |
| T (°C) | **TYS (MPa)** | **UTS (MPa)** | **TYS (MPa)** | **UTS (MPa)** | **TYS (MPa)** | **UTS (MPa)** | **CYS (MPa)** | **CPS (MPa)** | **CYS (MPa)** | **CPS (MPa)** |
| 22 | 229 | 257 | 269 | 306 | 176 | 208 | 189 | 415 | 162 | 236 |
| 77 | 171 | 186 | | | | | | | | |
| 100 | | | 173 | 206 | 115 | 126 | 141 | 211 | 125 | 136 |
| 126 | 116 | 137 | | | | | | | | |
| 200 | 78 | 85 | 86 | 91 | 68 | 73 | 96 | 111 | 80 | 85 |
| 300 | 49 | 54 | | 50 | 46 | 51 | | | 52 | 57 |

**Table 5**. Mechanical characteristics of μMg and μMg+3vol%BN.

| | μMg tension | | μMg compression | | Mg+3BN tension | | Mg+3BN compression | |
|---|---|---|---|---|---|---|---|---|
| T (°C) | **TYS (MPa)** | **UTS (MPa)** | **CYS (MPa)** | **CPS) (MPa)** | **TYS (MPa)** | **UTS (MPa)** | **CYS (MPa)** | **CPS (MPa)** |
| 22 | 254 | 284 | 229 | 317 | >450 | | 339 | 406 |
| 100 | 162 | 175 | 158 | 170 | 447 | 485 | 310 | 325 |
| 200 | 67 | 82 | 90 | 93 | 157 | 191 | 125 | 131 |
| 300 | 34 | 40 | 47 | 62 | 77 | 79 | 76 | 78 |

Samples with ceramic np were deformed in tension and compression at temperatures from RT up to 300 °C. The characteristic stresses obtained at various temperatures are reported in Table 4 together with the values for pure Mg, prepared with the same technique. Tension compression asymmetry is obvious at RT and 100 °C. The characteristic stresses found in tension and compression for Mg with 3 vol% of BN np are shown in Table 5. Note that the powder of pure Mg was only compacted and extruded (not milled) in contrast to the Mg+3vol%BN nanocomposite. The same preparation technology as in the case Mg/BN nMMC of Mg with Gr nanoparticles. Pure Mg was in this case milled and therefore the grain size of Mg samples was about 150 nm. Samples with Gr np were deformed in compression at RT and temperatures higher up to 300 °C. Comparing the YS measured for magnesium nMMCs with various np, we may state that Gr and hBN are materials with the excellent reinforcing effect (see Table 6).

**Table 6**. Mechanical characteristics of UFG-Mg and UFGMg+3vol.%Gr.

| | UFG-Mg compression | | UFG-Mg+3Gr compression | |
|---|---|---|---|---|
| T (°C) | **YS** (MPa) | **CPS** (MPa) | **YS** (MPa) | **CPS** (MPa) |
| 22 | 222 | 336 | 259 | 293 |
| 50 | 209 | 300 | | |
| 77 | | | 196 | 212 |
| 100 | 170 | 208 | | |
| 150 | 124 | 135 | | |
| 200 | 90 | 96 | 149 | 158 |
| 250 | 60 | 76 | | |
| 300 | 60 | 68 | 111 | 129 |

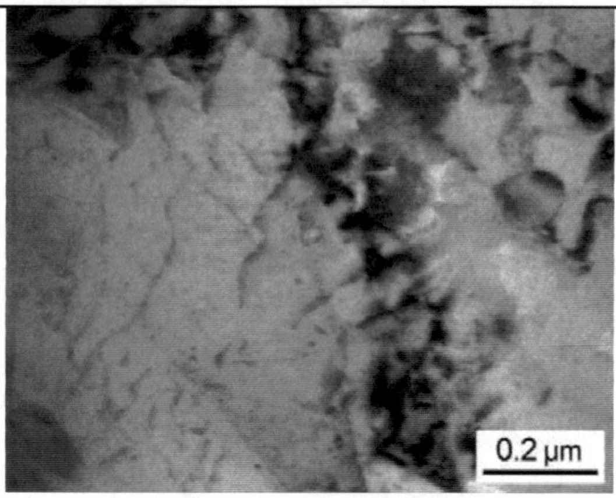

Figure 3. TEM micrograph of the sample deformed at room temperature.

## 4 Strengthening mechanisms

Speaking about the strengthening mechanisms in MMCs, many mechanisms must be considered: load transfer from the matrix to reinforcing phase particles (or fibres), increased density of thermal and geometrically necessary dislocations, Hall-Petch strengthening, and the influence of residual thermal stresses [4]. The special problem is how to combine the individual strengthening mechanisms. The particular strengthening terms expressed on the base of the physical models are summarised in Table 7.

**Table 7**. Contributions to strengthening following from the presence of the reinforcing phase.

| Mechanism | Relationship | Symbols |
|---|---|---|
| Load transfer | $\Delta\sigma_{LT} = 0.5\sigma_A f$ | $\sigma_A$ alloy stress, $f$ vol% of particles |
| Density of thermal dislocations | $\rho_T = 12 f \Delta\alpha\Delta T / b(1-f)\, s$ | $\Delta\alpha\Delta T$ thermal strain, b Burgers vector, $s$ size of particles |
| Enhanced dislocation density | $\Delta\sigma_D = \alpha_1 \psi G b(\rho_T + \rho_G)^{1/2}$ | |
| Density of geometrical dislocations | $\rho_G = f\, 8\varepsilon_p / b\, s$ | $\varepsilon_p$ plastic strain |
| Orowan strengthening | $\Delta\sigma_{OR} = \frac{Gb}{\Lambda} + \frac{5}{2\pi} Gf\varepsilon_p$ | G shear modulus, $\Lambda$ distance between particles |
| Residual stresses | $\sigma_{TS} = \frac{2}{3}\sigma_y ln\left(\frac{1}{f}\right)\frac{f}{1-f}$ | $\sigma_y$ yield stress in the matrix |
| Hall-Petch strengthening | $\Delta\sigma_{GS} = K_y\left(d_2^{-1/2} - d_1^{-1/2}\right)$ | $K_y$ Hall-Petch constant, d grain size |

The contribution to the strengthening from various mechanisms may be calculated according to the equations introduced in Table 7. The problem is how to combine individual strengthening terms. Possible solutions will be demonstrated in two examples. In the first, the strengthening effect of microparticles in Mg+2.4%In718 mMMC is calculated [6]. The simplest possibility is the plain summation of all contributions. The strengthening effect of mp, $\Delta\sigma_C$ calculated for the composite deformed in tension $\Delta\sigma_C(sum)= \sigma_C - \sigma_A = \Delta\sigma_D+\Delta\sigma_{HP}+\Delta\sigma_{LT}+\Delta\sigma_{OR}+\Delta\sigma_{TS}$ =33.2 + 84.2 + 2 + 2 +10 = 131.4 (in MPa). The experimental value $\Delta\sigma_C(exp)$= 285 – 161=124 MPa. The obtained result may be thought as a reasonable estimation, knowing that the particles have no unified size and are not evenly distributed in the matrix. The second possibility how to calculate the reinforcing contribution of particles, also mentioned in the literature, is the square root of quadrats of the strengthening terms. $\Delta\sigma_C(SR) = (\Delta\sigma_D^2 + \Delta_{HP}^2 + \Delta\sigma_{LT}^2 + \Delta\sigma_{OR}^2 + \Delta\sigma_{TS}^2)^{1/2}$ . The calculation gives for $\Delta\sigma_C(SR)$ a smaller value of 91.1 MPa. In the second example, the reinforcing effect of 13 vol% SiC mp is estimated: $\Delta\sigma_C(sum)$=28+52+11+4+28=123 and $\Delta\sigma_C(SR)$= 66.4 (in MPa). The experimental value exhibits $\Delta\sigma_C(exp)$=105 MPa. Comparing both results, the simple sum of all contributions seems to be the better solution. The strengthening effect of nanoparticles is much lower. In the case of Mg+3vol%$Al_2O_3$ and Mg+3vol.%Gr, it exhibits only about 40 MPa. It is remarkable that the TYS of Mg+3vol% of $ZrO_2$ is lower compared with pure Mg. The difference between TYS(Mg+3A12O3) – TYS(Mg+3$ZrO_2$)=93 MPa and CYS(Mg+3$Al_2O_3$) – CYS(Mg+3$ZrO_2$)=27 MPa. For such differences, there is no explanation on the base of the model presented above. This problem may be solved using internal friction (IF) measurements. While the IF, characterised by the logarithmic decrement, exhibits for Mg+3$Al_2O_3$ IF=0.00126 (at the strain amplitude ~$10^{-5}$), the IF value estimated for Mg+3$ZrO_2$ is approximately ten times higher IF=0.0128. The high values of IF are measured due to boundary sliding in the Mg/$ZrO_2$ interface. The bonding is weak in this case. On the other hand, nearly perfect bonding between Mg and $Al_2O_3$ may be considered. This strong bonding is caused by the chemical reaction: 3Mg+$Al_2O_3$ →3MgO+2Al. Similar reaction can be written for SiC particles: 2Mg+SiC→$Mg_2Si$ +C. An evidence of this reaction is visible in Figure 1d, where the interface between the SiC particles and the matrix is not smooth; it is decorated with a new compound which is a result of the chemical reaction of C coming from SiC particles with Al present in the δ-solid solution.

## 5 Conclusions

Magnesium and magnesium alloys based micro- and nanocomposites, prepared with different technologies were deformed in tension and compression in the temperature range from room temperature up to 300 °C. The reinforcing effect of particles strongly depends on the manufacturing process. The incorporated particles refined the grain size and contributed to an improvement of mechanical properties due to the Hall-Petch strengthening. An increase in the dislocation density plays also an important role in the composites strengthening. The grain boundaries and bonding between particles and the metallic matrix are deciding for the resultant composite properties.

**Acknowledgements.** The result was achieved with the use of support from the Ministry of Industry and Trade of the Czech Republic in the form of institutional funding.

## References

[1] Casati R and Vedani M 2014, *Metal* **4** 65
[2] Dieringa H 2011 *J. Mater. Sci.* **46** 289
[3] Gupta M and Nai SML 2011 Magnesium, magnesium alloys, and magnesium composite (John Wiley & Sons, Inc.: Hoboken, New Jersey)
[4] Trojanová Z, Száraz Z, Lukáč P, Drozd Z, and Džugan J 2020 *Crystals* **10** 466
[5] Máthis K, Čapek J, Zdražilová Z and Trojanová Z 2011 *Mater. Sci. Eng. A* **528** 5904
[6] Trojanová Z, Drozd Z, Lukáč P, Minárik P, Németh G, Seetharaman S, Džugan J and Gupta M 2020 *Materials* **13** 798

IOP Conf. Series: Materials Science and Engineering 1178 (2021) 012060 doi:10.1088/1757-899X/1178/1/012060

# Determination of Fracture Limit Line in Principal Strain Space by Shear Tests

**M. Urbánek, M. Fedorko and M. Rund**

COMTES FHT a.s., Prumyslova 995, Dobrany 334 41, Czech Republic

miroslav.urbanek@comtesfht.cz

**Abstract**. The main goal of this paper was to define a new limit line, which expands Forming Limit Curve (FLC) more precisely Fracture Forming Limit Curve (FFLC). The new limit line was defined by damage models transformed from triaxiality-fracture strain space to the major/minor strain space and we focused on the shear area. Because FLC and FFLC have some limitations, for example, some Advanced High Strength Steels (AHSS) fractures before thinning. Another limitation can be seen on the left side of the diagrams in the shear region, where the part can break in the process without crossing the FLC or FFLC. The different shapes of samples were used for finding a new line of the FFLC and Deep Drawing Test (DDT) was used for verification. The fracture points from measurement were used for calibration of simple uncoupled damage models. The simple model means model, which is calibrated based on one or two measurements (tensile test or tensile test + shear test). This paper was used an Advanced High Strength Steel (AHSS) DP1000 sheet with a thickness of 0.8 mm for the experiments.

## 1 Introduction

Deep drawing is one of the most widely-used sheet metal working processes in the automotive industry. Special material has even been developed for this process which exhibits high elongation and reduced strength. The demands on them have been increasing over time, leading to the introduction of AHSS, including CP1000, DP1000, DP1200, and other grades. Materials such as DP1400 appear certain to be adopted in the coming years. These high-strength steels are mostly used for structural parts and not as appearance grades. Owing to the specifics of their manufacturing, they are not governed by the same principles as the deep-drawing materials of the past. Failure without appreciable necking is the primary problem. The reason is that necking, or local thinning, is normally an important indication of material behavior and represents a threshold that is plotted in Forming Limit Diagrams (FLD). In FLDs, limit values are plotted across a range of states from uniaxial tensile loading, through plane-strain states, to biaxial tensile loading with linear loading paths. Shear loading is not captured in these Forming Limit Curves (FLC). Also, a distinction must be made between in-plane shear and out-of-plane shear (Figure 1), [1]. The testing of standard specimens usually involves in-plane shear.

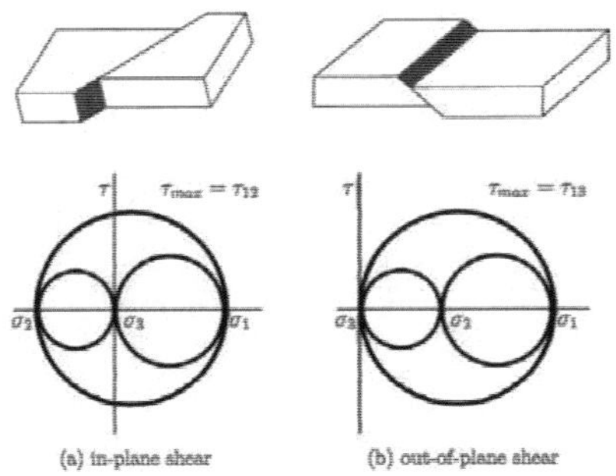


**Figure 1.** Definitions of shear mechanisms [1]

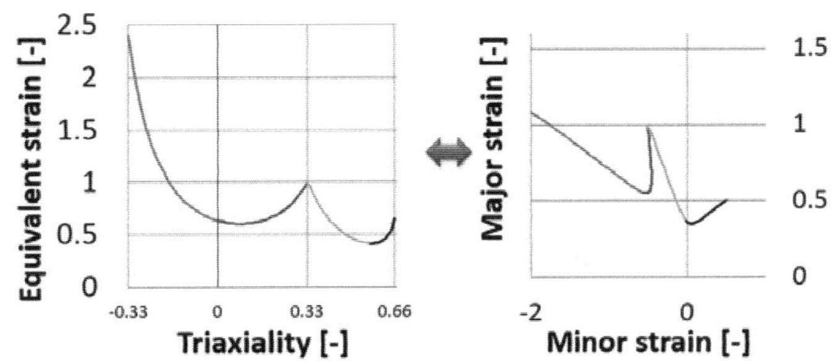


**Figure 2.** Translation between spaces

For practical purposes, the range should be expanded to include out-of-plane shear conditions, for instance by using a Deep Drawing Test (DDT). One of those who dealt with the out-of-plane shear region was Gorji [2]. He constructed FFLCs from data from a deep drawing test with small radii on the tooling. Hora [3] referred to the work of Wierzbicki and his collaborators Bai and Bao and their expansion of the Johnson-Cook damage model by incorporating Lode angle $\theta$ or Lode parameter $\xi$. This expansion brings the improved description of material behavior in different stress states. Wierzbicki's approach was previously studied and applied at COMTES FHT as part of a research project. The authors of this paper focused on the Fracture Forming Limit Curve (FFLC) which is defined in terms of fracture rather than mere thinning of the sheet. This paper deals with the shear deformation region, using simple damage models based on tensile test or tensile test + shear test for plotting the fracture limit line in this region. These models are defined in triaxiality-fracture strain space, therefore the limit curve was trans-formed to major-minor strains space, where was created the new line limit. These models will be compared with the result of DDT.

## 2 Fracture Forming Limit Curve

FFLC is determined using the same tests as those for FLC. However, the difference is that the test the evaluation focuses on fracture and not just thinning of the sheet specimen. As a result, principal strain space can be populated in those regions where conventional Nakajima or Marciniak tests do not apply. Thanks to the mapping relationships published by Stoughton [2] and Lee [3] and elaborated in [4], one can translate fracture points to major/minor strain space (Figure 2). One can therefore acquire additional fracture points from conventional mechanical test data and extend the fracture limit line. The expression 'conventional tests' refers generally to standardized tests on both standard and non-standard test specimens.

## 3 Translation of FLC data into Triaxiality-Lode angle space

Using the following relationships, FLCs can be translated into other spaces [2], [3], [4]. Plastic strain increments are defined by equation (1). Where $\alpha$ is a ratio of principal strains (2), more precisely principal strain increments, assuming that the material is incompressible and isotropic. The material can then be characterized in terms of von Mises equivalent stress $\bar{\sigma}$ as in equation. Where $\sigma_{ij}$ are stress components, $\sigma_1$, $\sigma_2$, and $\sigma_3$ denote principal stresses, $S_{ij}$ is stress deviator components, d$\Lambda$ denotes the plastic multiplier, and d$\varepsilon$ij stands for plastic strain increments (3). In equation (2), the range of values of $\alpha$ is expected to include both negative and positive values. Relevant stress ratios (1) have been defined for further work. Triaxiality $\lambda$ can be formulated using stress ratio $\beta$ and the following equations, where $\sigma_m$ is the mean stress and $\bar{\sigma}$ denotes equivalent stress. The equivalent strain increment is defined in terms of increments of principal strain components. Substituting equation (1) produces an expression for the equivalent strain (5). Under plane stress conditions, the strain ratio $\alpha$ is constant. Equivalent strain $\bar{\varepsilon}$ can then be expressed in terms of principal strain $\varepsilon$1 and parameter $\alpha$ (6). FLD can be translated into the space of ($\varepsilon$, $\lambda$) using equations (6) and (4). The procedure involves the use of equation (4) and elimination of the $\beta$ parameter among the three equations. Besides the above approach, one can also explore material behavior in other spaces. In the present case, the single most important alternative is the translation from triaxiality-strain space to principal strain space using a rearranged version of (4). Once the stress ratio $\beta$ has been determined using triaxiality, ratio $\alpha$ can be expressed. Based on the equivalent strain, the major strain was found using (6), and the then minor strain was determined from (1) as described in the literature [2], [3], [5].

$$\alpha = \frac{d\varepsilon_2}{d\varepsilon_1} \quad \frac{\sigma_2}{\sigma_1} = \beta \tag{1}$$

$$\bar{\sigma} = \sqrt{\frac{1}{2}\left[(\sigma_1 - \sigma_2)^2 + (\sigma_2 - \sigma_3)^2 + (\sigma_3 - \sigma_1)^2\right]} \tag{2}$$

$$d\varepsilon_{ij} = d\Lambda \frac{\partial \bar{\sigma}}{\partial \sigma_{ij}} = \frac{3}{2}\frac{d\Lambda}{\bar{\sigma}} s_{ij} \tag{3}$$

$$\lambda = \frac{\sigma_m}{\bar{\sigma}} = \frac{\beta + 1}{3\sqrt{\beta^2 - \beta + 1}} \tag{4}$$

$$d\bar{\varepsilon} = \sqrt{\frac{4}{3}}\left(1 + \alpha + \alpha^2\right) d\varepsilon_1 \tag{5}$$

$$\bar{\varepsilon} = \frac{2\varepsilon_1}{\sqrt{3}}\sqrt{1 + \alpha + \alpha^2} \tag{6}$$

## 4 Material

Fracture limit lines in the shear region were determined for the AHSS DP1000 sheet 0.8 mm in thickness. Material characteristics of the steels were determined using tensile testing (Figure 3a) on specimens oriented in three basic directions: Rolling Direction (RD), Diagonal Direction (DD), and Transverse Direction (TD). They are reported in the following Table 1, there is basic material characterization, as yield strength $\sigma_y$, ultimate tensile strength UTS, elongation A and reduction area RA. The purpose was to obtain fracture points for fitting material damage models, which were transformed into the major/minor space.

**Table 1.** Material characteristics

| **Material** | | $\sigma_y$ [MPa] | UTS [MPa] | A [%] | RA [%] |
|---|---|---|---|---|---|
| | RD | 730 | 1049 | 14 | 11 |
| **DP1000** | DD | 706 | 1018 | 15 | 16 |
| | TD | 740 | 1048 | 9 | 12 |

## 5 Shear tests

Data for plotting the fracture limit line in the shear region were gathered by shear testing. These tests were performed on the various shape of specimens with in-plane shear being the dominant mechanism. One of them conformed to ASTM B831–05 (Figure 3b). The second specimen (Figure 3c) was specially modified for DP1000 steel because the crack initialized in the narrow part of the sample body instead of in the side bridges. Tests were carried out under quasi-static conditions (approximately $0.0001 s^{-1}$) and at room temperature. The effects of the different deformation rates in the shear tests and the Nakajima test were not taken into account. Two shear specimens from DP1000 steel were tested. Their fracture strain values were similar (Figure 4) The goal of the measurement was to obtain load paths for individual specimens and compare each other. Fracture points were another result of the measurement. They were used for fitting curves from various fracture models, ensuring that the resulting curve is continuous and smooth in its separate parts.

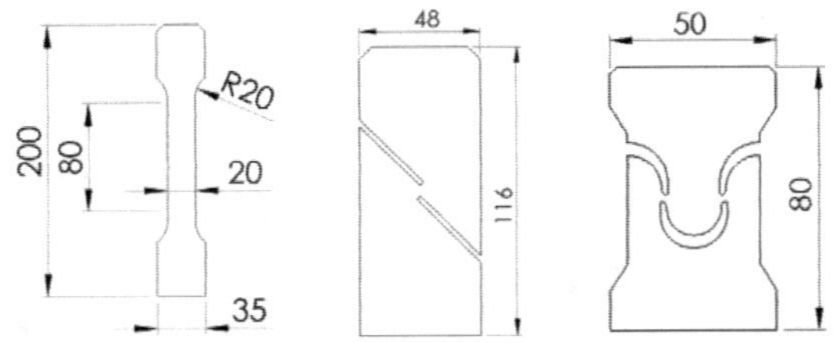


a) Tensile sample b) ASTM sample c) Optimized smile sample

**Figure 3.** Tested samples

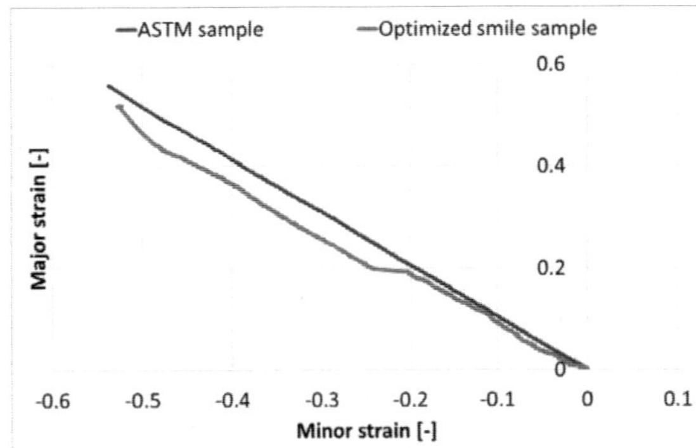


**Figure 4**. Load paths for shear tests

## 6 Damage models

The following models were used for plotting the fracture limit line: Bao-Wierzbicki [6] and MAX shear [7]. They were only applied to the triaxiality range less than 1/3 to provide a continuous description from pure tension through shear to compression.

### 6.1 *Simplified Bao-Wierzbicki fracture model in the shear region*

The Bao-Wierzbicki fracture model was calibrated against tensile tests only. Regions II and III were used, i.e. the triaxiality range of $-1/3 < \eta < 1/3$ was covered. The fracture initiation locus for the Bao-Wierzbicki model is defined in terms of equations (7). Constant *a* is the plastic strain at fracture in pure shear ($\eta = 0$). Constant *b* is the plastic strain at fracture under uniaxial tension ($\eta = 1/3$). Constant *b* can be derived from tensile test data. It is the ratio of the cross-sectional area before the test $A_0$ and at the fracture location after the test $A_f$. Width and thickness after the test were measured by microscope. Constant *a* is calculated using the following formula. Details and the method by which the formula was derived are described by Lee [8].

$$\bar{\varepsilon}_{fBWII} = b + (a-b)\left(\frac{\eta}{\eta_0}\right)^2 \quad \bar{\varepsilon}_{fBWIII} = \frac{a}{1+3\eta} \qquad (7)$$

$$\bar{\varepsilon}_{f\,MSII} = C\left\{\frac{\sqrt{1+\alpha+\alpha^2}}{1-\alpha}\right\}^{1/n} \qquad (8)$$

### 6.2 *Simplified Max shear fracture model in the shear region*

The MAX shear fracture model was calibrated against the shear test only (8). The model is also referred to as the Tresca criterion. It postulates that fracture occurs when the highest shear stress reaches a certain threshold. The model was applied over the triaxiality range $-1/3 < \eta < 1/3$ where C is a material constant, which was calibrated by measurement.

## 7 Evaluation and transformation

Curves for both models were fitted to the fracture strain points found by mechanical testing (Figure 5). These points were obtained from the tensile test on the flat specimen and two shear tests (ASTM and optimized smile samples). Calibrated damage models as Bao-Wierzbicki and MAX shear were transformed into the principal strain space and it was produced a new segment of the fracture limit line in this space. With these models, the line appears smooth and continuous.

IOP Conf. Series: Materials Science and Engineering 1178 (2021) 012060 doi:10.1088/1757-899X/1178/1/012060

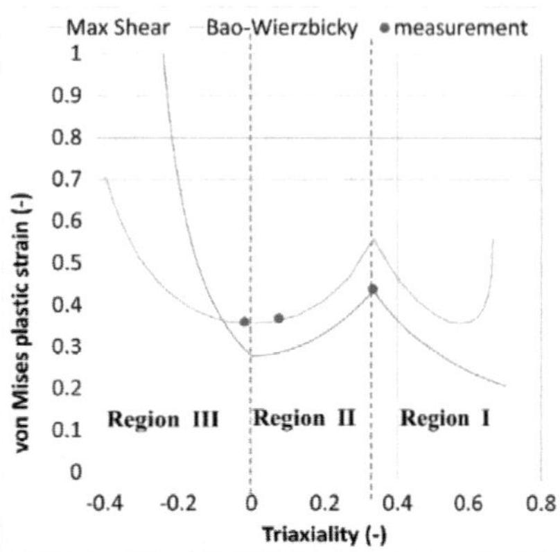


**Figure 5.** Fracture model curves fitted to points for DP1000 steel

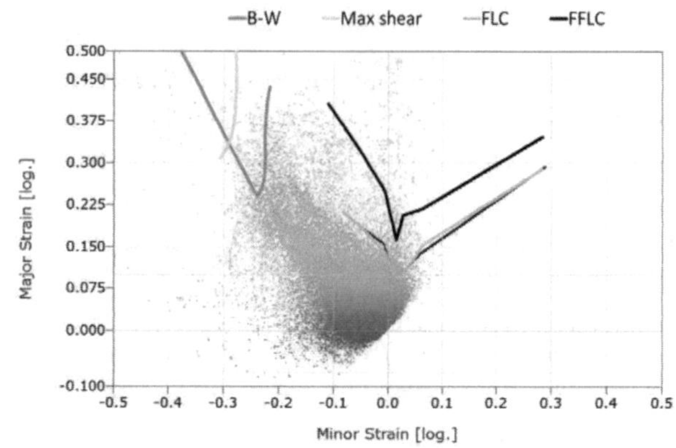


**Figure 6.** Comparison of new fracture limit lines and strain distribution after the test draw –DP1000

## 8 Deep drawing tests

Accuracy of the above-characterized models was validated against a test in a Deep Drawing Test Device (DDTD). The part drawn in the test was a triangle cup, as defined by Hora [3]. The purpose of the test was to assess fracture strain in the drawn part. Its shape produces various types of strain in the part, which enables the researcher to assess in-plane shear and out-of-plane shear states. A schematic illustration of the DDTD is shown in Figure 7.

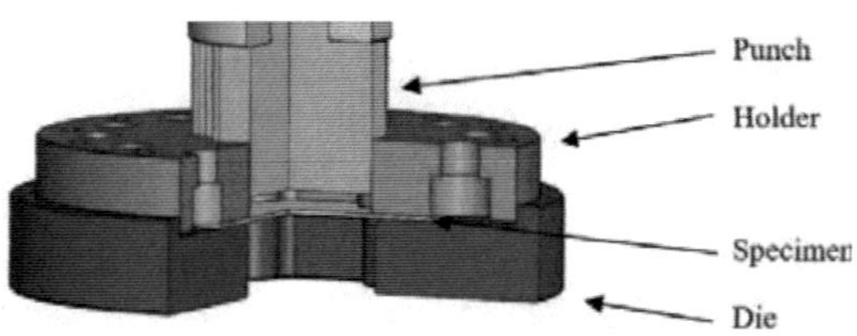


**Figure 7.**CAD drawing of the Deep Drawing Test Device

**Figure 8.**Triangle cup with R10 radii

The resolution of strain on the shape of the drawn cup is presented in Figure 8. The tooling in this experiment had a larger radius in the punch and the die than that published by Hora [5]. The R10 radius facilitates the drawing process and is acceptable for high-strength materials, but we are working on the dies with a smaller radius like Hora. The strain on the cup was evaluated using ARGUS and quality was dependent on the size of etched points and distance between them on the undeformed sheet. The used size was 1mm and the distance was 2mm.

## 9 Results and discussion

The main goal of the paper was defining a new fracture limit line, which expands Fracture Forming Limit Curve (FFLC). The new limit line was defined by damage models transformed from triaxiality-fracture strain space to the major/minor strain space and we focused on the shear area. These simple models could predict cracks in the shear area and their verification was performed by DDT. As noted, DDT produces various types of strain in the part and predominantly shear in the out-of-plane region, which cannot be measured by specimens, which were shown in Figure 3. A comparison of new fracture limit lines and strain distribution after the test draw was shown in Figure 6. The new fracture limit line from the Bao-Wierzbicki model provides an opportunity to used known damage models in principal strain space The prediction of fracture was successful for steel DP1000. Bao-Wierzbicki model has a good prediction, but its applicability will have to be verified for various conditions, such as holding force, strain rate, temperature, punch shape, etc.

## 10 Conclusions

A method for finding a fracture limit line for the shear strain region using fracture models was presented. It was derived from the definitions of the individual models. The models were calibrated against data from mechanical tests. These tests included tensile tests on flat specimens and specimens with a hole, shear tests, and plane strain tests. The measured fracture strains were plotted against triaxiality values. Fracture limit lines derived from the models were then translated into principal strain space. The line obtained with the Bao-Wierzbicki model was validated against a deep drawing test with a triangle cup. Thanks to the special shape of the cup, fracture occurred at the desired points. The procedure presented above was applied to the AHSS DP1000 sheet 0.8 mm in thickness.

## 11 Acknowledgement

The paper was developed with funding from The Technology Agency of the Czech Republic, the project name is "Development of a method for advanced characterization AHSS sheets with a focus on pressing of seat skeletons", No. FW01010643

## References

[1] Chung K, Kim H, Lee C 2014 Forming limit criterion for ductile anisotropic sheets as a material property and its deformation path insensitivity. Part I: Deformation path insensitive formula based on theoretical models. *International Journal of Plasticity 58*, p 3–34

[2] Gorji M, Berisha B, Hora P, Barlat F 2016 Modeling of localization and fracture phenomena in strain and stress space for sheet metal forming, *International Journal of Material Forming 9*, p 573–584

[3] Hora P, Gorji M, Berisha B 2016 Modelling of fracture effects in the sheet metal forming based on an extended FLC evaluation method in combination with fracture criterions. in IOP Conference Series: *Materials Science and Engineering 159*, (Institute of Physics Publishing)

[4] Španiel M, Prantl A, Džugan J, Růžička J, Moravec M and Kuželka J 2014 Calibration of fracture locus in scope of uncoupled elastic–plastic-ductile fracture material models, *Adv. Eng. Softw* 72 pp. 95-108

[5] Džugan, J. et al. 2018 Identification of ductile damage parameters for pressure vessel steel, *Nuclear Engineering and Design 328*, p 372–380

[6] Bao Y, Wierzbicki T 2004 Evaluation and calibration of seven fracture models, *J mech Phys*

[7] *Solds* Wierzbicki T, Bao Y, Lee Y-W, Bai Y 2005 Calibration and evaluation of seven fracture models. *International Journal of Mechanical Sciences 47*, p 719–743.

[8] Lee Y 2005 Fracture Prediction in Metal Sheets, *Massachusetts Institute of Technology 1–411*

IOP Conf. Series: Materials Science and Engineering 1178 (2021) 012061 doi:10.1088/1757-899X/1178/1/012061

# Production and Assembly of Car Body Fifth Door with Use of Dual-Phase Steel HCT500X

**M Valeš[1], J Havelka[1], L Chrášťanský[1], F Tatíček[1] and V Novák[1]**

[1]Czech Technical University in Prague, Faculty of Mechanical Engineering, Department of Manufacturing Technology, Technická 4, 160 00 Prague, Czech Republic

E-mail: michal.vales@fs.cvut.cz

**Abstract**. Due to environmental concerns and safety standards in the automotive industry, the development of durable and lightweight cars has become an important topic. Lowering the weight of the car body has become a crucial topic for all manufacturers, regardless of using combustion, electric or hydrogen engines. To fulfil this need a material with a higher strength to weight ratio must be used. Dual-phase steels seem to provide the right combination of mechanical properties for the production of particular car body panels. The goal of this research is to develop robust production of the fifth door part from HCT500X steel (DP500). The goal of this article is also to depict the forming of the selected part and the fifth door assembly for $4^{th}$ generation of ŠKODA Octavia. Numerical simulation is accomplished within the AutoForm Forming R8 and AutoForm Assembly R9 solutions. Different aspects influencing numerical simulation accuracy and topics relevant to the production chain are discussed in this article. This research is carried out in cooperation with ŠKODA AUTO, a.s.

## 1 Introduction

Digitalisation is an essence of ongoing Industry 4.0. This matter of fact impels OEMs and their suppliers towards modelling of manufacturing process of every car component. Within achieving this goal, digitalisation must cover all production processes that precede and follow-up the final product. Facts, as mentioned earlier, require software solution which covers the production of the whole Body-in-White (BiW). This backbone solutions must include forming, joining and painting and should be connected to ERP systems. Thus, modification of part geometry or part of the process should be right away projected into connected processes.

Another driving factor is that modifications of already manufactured forming tools are costly. Additional engineering and try-out (including milling, spotting, finishing, surface coating deposition) considerably lowers the profit and possibly delays subsequent processes. Digital modifications are considerably faster, cheaper and provide sought agility. Achieving a reliable digital design, designers must thoroughly set-up the process and to keep up with state-of-the-art tools.

The goal of this article is to depict the forming process design of outer car body part made of high-strength steel HXT500X and subsequent assembly process of fifth door outer for current generation of ŠKODA OCTAVIA.

## 2 Aspects of numerical simulation and its set-up

The techniques of modelling metal forming processes generally have three primary objectives. First and foremost, the analysis is focused on optimisation and the elimination of possible errors in the draft of the manufacturing process or tool. Secondly, this analysis allows the user to modify a body part so, that makes for simple and effective manufacturing. Thirdly, the simulation allows the user to resolve sudden issues, which can occur in an already established manufacturing process. Thus, the application of numerical simulations gives a better insight into the process at hand and also provides the means to reduce tooling costs.

Drawing operation is a very significant step in the stamping process. During this operation, the blank is formed into the primary shape of the final part, except specific surfaces and flanges that are formed during subsequent operations. This first operation is distinctive from the subsequent ones due to the considerably higher rate of reshaping. Due to this fact, the majority of part defects emerges during the drawing operation.

An essential aspect of the sheet metal forming process is the springback effect. This effect takes place due to the release of the stress and the ensuing equalisation of elastic strain, which results in shape changes. Precise prediction of the springback effect is necessary because the modern-day trend focuses on using advanced high-strength steels, see **Figure 1**. However, these advanced steel are considerably more prone to the effects of springback due to their higher mechanical properties, namely their yield strength. This tendency results in a higher shape warping due to the release of elastic strain. It can potentially jeopardise the part compatibility and thus directly affect the process of mechanical joining and hemming. Accurate springback analysis is also a crucial point in proper detection of visual defect and audit problems: surface imperfections detected with stoning (wrinkling, surface lows), projected light stripes evaluation in light tunnel (zebra lines).

Structural parts greatly influence the overall springback of the assembly. It is important to note that accurate springback results for assemblies can be achieved when all of the assembled parts contain deformation history, thus stamping process for each of these parts was simulated. [1, 2]

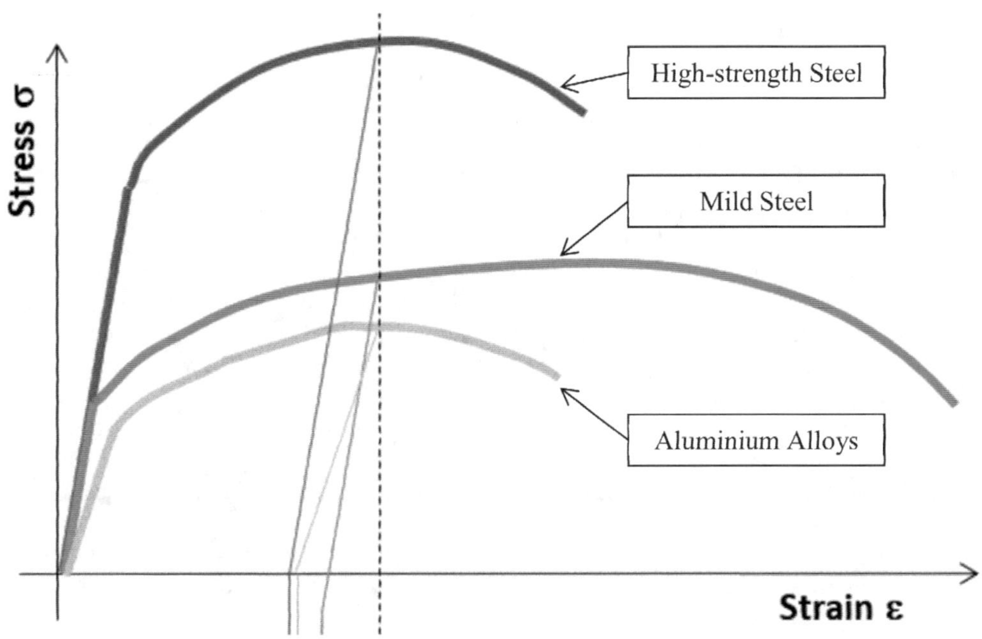


**Figure 1**. Springback effect, High-strength steel, Mild Steel, Aluminium Alloys

In numerical simulation, material definition significantly influences sheet behaviour. Material cards usually consists of hardening curve, yield locus and fracture limit. It is recommended to define a material card based on data measured for a particular material to obtain more accurate results. Available solutions allows press-shops to measure mechanical properties for each blank with eddy current (based on tensile test reference). This measurement can be amended with thickness, surface roughness and coating layer measurement. This method provides a relevant range of data which enter the stamping process and use it in numerical simulation set-up. Tensile test at elevated strain rate, tensile test with higher strain rate or tension/compression test which provides kinematic hardening curve can also significantly improve prediction of splits and springback. [3, 4]

Tribology is a complex item in numerical simulation. Single value of friction coefficient, $\mu$, does not correctly represent contact behaviour. $\mu$ is a function pressure, strain, temperature, relative velocity of tool and the sheet, amount of lubricant, sheet and tool roughness and should be calculated for each element. Effect of pressure and strain is shown in **Figure 2**.

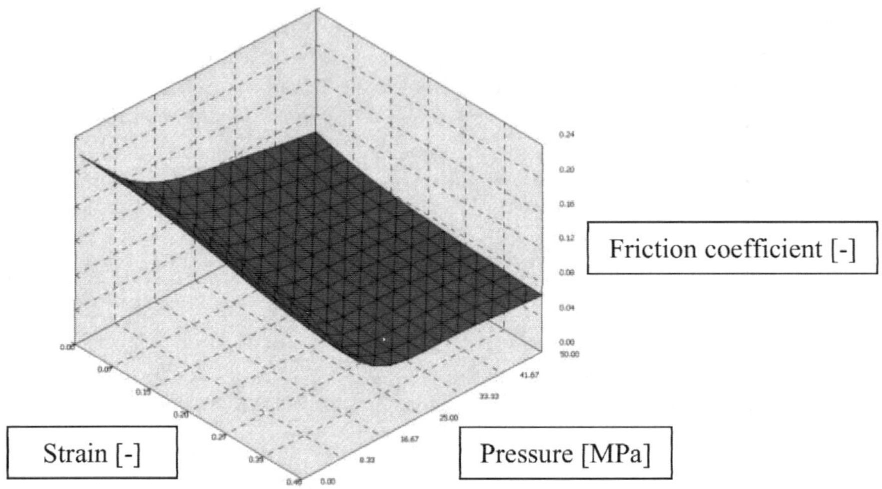


**Figure 2**. $\mu$ value, effect of pressure and strain (HCT500X)

Mesh settings (FEM element type and size) greatly influence computing time and should always respect the maturity of the process. Tool displacement step should be smaller at the end of the stroke, since a large portion of strain occurs at this moment.

As the computing capacity improves throughout the years, it allows calculation of robustness of the process. Since input in reality varies, the numerical simulation should also provide a cloud of results, e.g. concerning material parameters, blank shift and fluctuating blank holder force. Provided that the cloud of numerical results fits the cloud of real stamping results, simulation can be considered precise. The robustness is also an essential tool in respect to shrinkage of feasibility window as more demanding design features appear on the car.

Proper tool definition should always correspond with reality. Drawbead definition is a deciding factor for simulation accuracy and together with spotted, relieved and offset areas must always correlate with a reality. Thanks to modern methods of optical measurement, tool-shops can easily compare CAD data and the geometry of the tools. [5]

Results evaluation (post-processing) is a crucial step and must correspond to reality. The positioning of the measured part and the order of closing clamps play a significant role for thin outer car body parts.

## 3 HCT500X, fifth door outer lower (Heckklappe außen unten)

The driving factor for this research is an ongoing competitiveness in the automotive market and increasing restrictions for carbon footprint. European Union is charging car manufacturers for exceeding the set limit of $CO_2$ emission. The fleet-wide average emission target is 95 g $CO_2$/km in 2020. Thus, one of the OEM's R&D goals is to reduce the mass of Body-in-White. A partial solution is to use materials, which provide more favourable strength to weight ratio.

Part selected for this research is a fifth door outer lower and is highlighted in **Figure 3**.

**Figure 3**. 4th generation ŠKODA Octavia, fifth door outer lower

The production material is a dual-phase steel HCT500X (commonly referred as DP500 or CR290Y490T-DP according to standard VDA 239-100). HCT500X steel is a part of dual-phase steel family; the structure consists of ductile fine-grained ferritic matrix and hard martensitic phase dispersed in the form of islands. In comparison to mild steels (e.g. DC05), HCT500X provides higher strength properties; therefore, a thinner sheet is used (0.60 mm), and the blank cost is comparable to mild steel. Based on the ŠKODA methods and recommendations stated in chapter 2, the numerical simulation was set-up. Material card was verified within prior research task [7]. Due to reduced ductility of HCT500X, the part geometry had to be modified to fit inside of feasibility window. Surface shape, trim angles, flange length and radii had to be adjusted as seen in **Figure 4**. [6, 7]

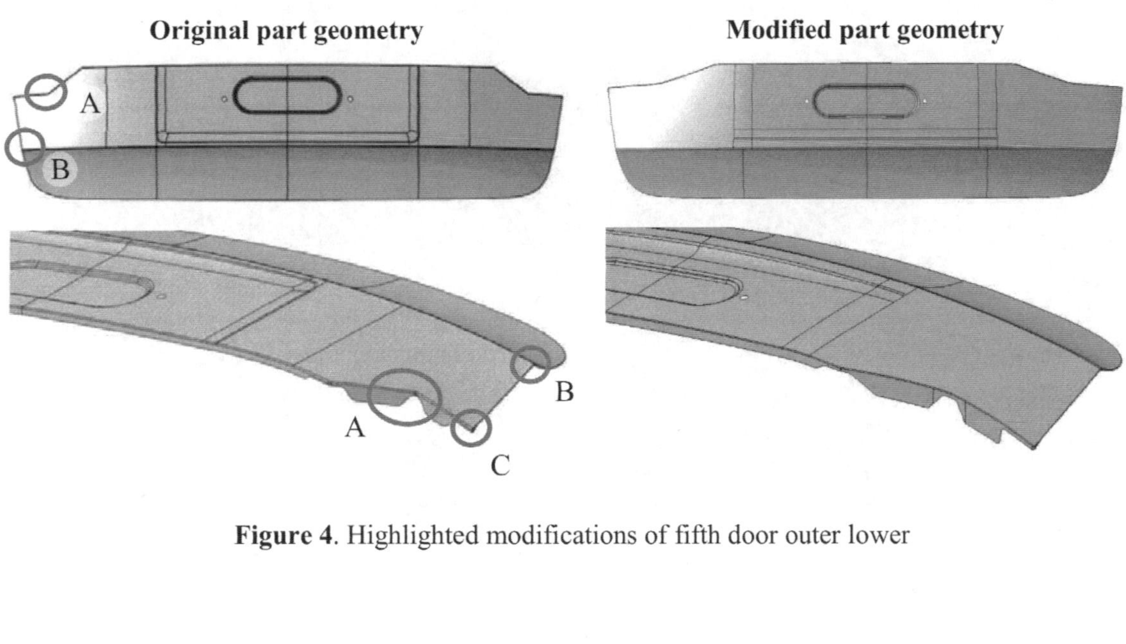


**Figure 4**. Highlighted modifications of fifth door outer lower

Forming simulation (AF R8, default Final Validation settings) does not exhibit risk of splits or wrinkles. A robustness analysis was performed to evaluate springback stability, since springback is very sensitive towards input noise. **Figure 5** shows springback results based on ±10% variation of mechanical properties, sheet thickness, blank holder force and blank shift. Input scatter is based on production of same type parts in ŠKODA AUTO. Springback specification limits (± 0.80 mm) were not exceeded, and this result is considered stable for production. When exceeding tolerance, either part geometry must be modified, or input variation must be restricted.

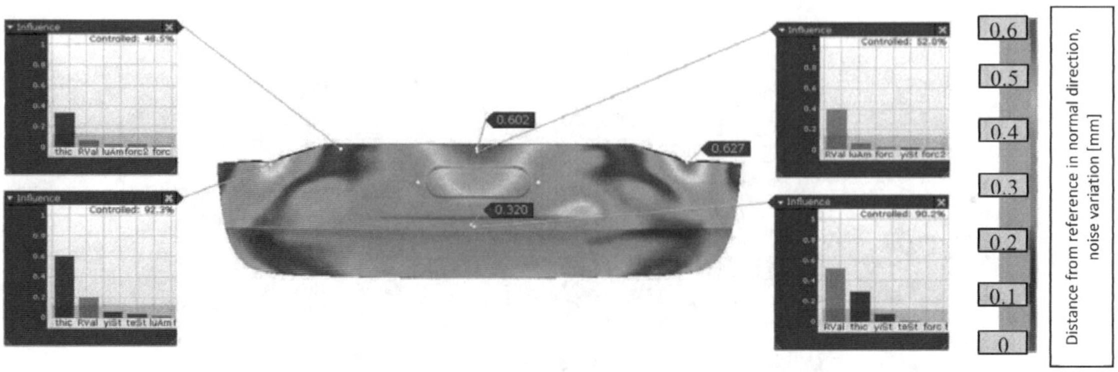


**Figure 5**. Springback variation; Distance from ref in the normal direction (AF Forming R8)

The Springback of fully assembled fifth door, including all the structural components, is represented in **Figure 6**. However, structural components from external contractors are defined as rigid since the simulation outputs were unavailable during the time of this research. Therefore, the behaviour of the entire fifth door assembly from the simulation cannot be fully compared to a result of the actual manufacturing process, since the behaviour of the individual structural part is omitted.

**Figure 6** shows a considerable amount of deviation in the highlighted areas. The normal distance from the reference reaches a maximum value -2.5 mm. The edge at the base of the roof also shows a significant deviation of nearly -1.4 mm.

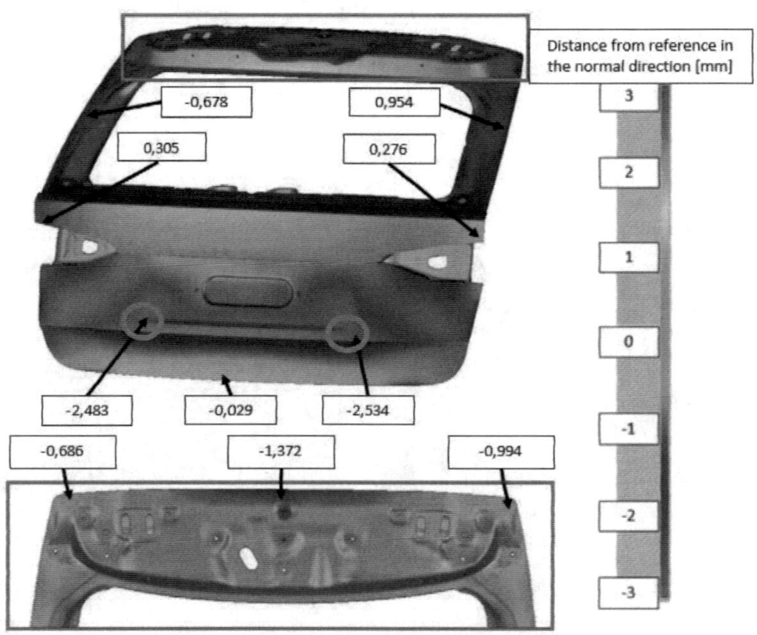


**Figure 6**. Fifth door assembly springback results (AF Assembly R9)

## 4 Conclusion

Obtained results demonstrate benefits and the requirement for digitalisation in manufacturing processes for all assembled parts in sheet metal forming and subsequent processes. Fifth door outer lower part made of HSS steel HCT500X meets springback specification limits. Since the springback results of assembled fifth door exceed specification limit for springback, springback must be compensated. It is efficient to compensate only the most influencing parts in the assembly in order to obtain a satisfying result (springback compensation of an assembly in not covered int this article). This approach can significantly reduce costs connected to tool modification and help to achieve a genuinely robust result of the simulation.

The field of numerical simulation is quickly developing and gradually tries to meet customer requests. In future, further attention must be paid towards, e.g. realistic stiffness of the tools, which is highly influenced by its construction, and towards temperature effects. Temperature plays a key role in friction during forming. Uneven temperature gradient during paint baking (simultaneously glue curing) can result in dimensional accuracy of BiW and visual defects in outer parts.

Presented results demonstrate the need for digitalisation of all parts, and its manufacturing process, that enters the assembly. Fulfilling this request will allow OEMs to reach an unequivocal and robust result. Therefore, a fully-fledged application of these simulations will require close cooperation with respective structural component suppliers.

**Acknowledgement**
The research was financed by SGS19/163/OHK2/3T/12. Research, optimisation and innovation of production processes.

**References**

[1] CHONGTHAIRUNGRUANG B et al. 2013 V Springback prediction in sheet metal forming of high strength steels. *Materials & Design*. 50, 253-266 10.1016/j.matdes.2013.02.060
[2] BANU M et al. 2006 Simulation of springback and wrinkling in stamping of a dual phase steel rail-shaped part. *Journal of Materials Processing Technology*. 173(2), 178-184. 10.1016/j.jmatprotec.2005.11.023
[3] ABSPOEL M et al. 2017 Accurate anisotropic material modelling using only tensile tests for hot and cold forming. *Journal of Physics: Conference Series*. 896 10.1088/1742-6596/896/1/012049
[4] EVIN E et al. 2012 Emil Prediction of Local Limit Deformations of Steel Sheets depending on Deformation Scheme *Chemicke listy*. 106(3) 401-404
[5] DANESHJO N et al. 2019 Application of Data Glove for Assembly Analysis of Selected Component. *MM Science Journal* 2019(03), 2946-2951doi:10.17973/MMSJ.2019_10_2019039
[6] KUZIAK R et al. 2008 Advanced high strength steels for automotive industry. *Archives of Civil and Mechanical Engineering* 8(2), 103-117. 10.1016/S1644-9665(12)60197-6
[7] VALES M 2019 Outer Car Body Panels Made of Dual-Phase Steel. *METAL 2019 Conference Proceedings*. 507-512 WOS 000434346900085, SCOPUS 2-s2.0-85079382094

# Ion-plasma modification of surface of ultrafine-grained titanium alloys: effect of substrate on coating properties

*R R Valiev [1,2], Iu M Modina [1], Yu M Dyblenko [1], I P Semenova [1]

[1]Institute of Physics of Advanced Materials, Ufa State Aviation Technical University, Ufa, 450008, Russia
[2]Saint Petersburg State University, St. Petersburg State University, 7/9 Universitetskaya nab., St. Petersburg, 199034 Russia

E-mail: rovaliev@gmail.com

**Abstract**. This article is focused on research of properties of vacuum-plasma coatings Ti+TiVN and V+TiVN applied to the surface of titanium alloys Ti-6Al-4V and VT8M-1(rus.) (Ti-5.7Al-3.8Mo-1.2Zr-1.3Sn) in ultrafine-grained (UFG) states obtained by two different severe plastic deformation (SPD) techniques - equal-channel angular pressing (ECAP) and rotary swaging (RS). The influence of substrate UFG structure on coatings service properties improvement has been established. It was demonstrated that complex hardening of titanium alloy including formation of bulk UFG structure and applying vacuum-plasma protective coating to the surface significantly increases resistance to erosive wear of the latter in comparison with the substrate having the initial coarse-grained structure.

## 1 Introduction

Two-phase titanium alloys have a vast field of application as constructional materials in aviation as well as engine manufacturing due to the most favorable ratio of strength, ductility, crack resistance, corrosion resistance and other characteristics that provide high weight efficiency, service life and reliability of gas turbine engine (GTE) products. Titanium alloy VT8M-1 (Ti–5.7Al–3.8 Mo–1.2Zr–1.3Sn) exhibits better properties in terms of heat resistance and thermal stability than popular Ti-6Al-4V alloy [1,2].
However, products and components of gas turbine engines may carry high cyclic, dynamic loads, or experience erosion and corrosion during operation and thus, the requirements to their structural strength tend to become more rigorous.
Processing by severe plastic deformation (SPD) has proved itself as one of the most efficient ways to considerably increase the mechanical and fatigue characteristics of metals and alloys by forming in such materials ultrafine-grained (UFG) structure with the grain size in submicron and nanocrystalline range with mostly high-angle grain boundaries (HAGB) [3-5]. At the same time a proper surface treatment can provide different properties, such as resistance to wear, erosion and corrosion and another way to protect materials in this paper deals with ion surface improvement by the deposition of vacuum-plasma

coating [6-8]. Therefore, combining the formation of UFG structure in the material and vacuum-plasma coatings on the surface can serve the purpose of complex strengthening of GTE parts.

## 2 Material and experimental procedure

The studies were conducted on titanium alloys Ti-6Al-4V and VT8M-1 (Ti-5.7Al-3.8 Mo-1.2Zr-1.3Sn) that are classified as hard-to-deform materials. Therefore, the rods of these alloys were quenched in water at 940 °C and annealed for 4 hours at 700 °C with cooling in air as a pretreatment for the formation of a duplex structure consisting of primary α-phase not exceeding 25% and a thin plate α+β structure, thus providing the conditions of plastic deformation [9].

Ultrafine-grained state was achieved by two techniques. The 30mm diameter Ti-6Al-4V alloy rods were subjected to equal channel angular pressing (ECAP) with 4 passes, route Bc at 750 °C (e=2.7 with a strain rate of 4 mm $s^{-1}$), channel intersection angle $\psi$ = 120°. The VT8M-1 alloy rods with 70mm diameter and 1000 mm in length were processed by rotary swaging with intermediate annealing at 750°C till the diameter was reduced to 60, 50, 40, 30 mm (e~1.7, the strain rate exceeding 300 mm $s^{-1}$). These procedures are described in more detail in [10,11].

For high surface quality, Ti alloys with coarse-grained (CG) initial and UFG structures were subjected to mechanical grinding and further electrolyte-plasma polishing (EPP) [12,13] before coating deposition. EPP of the samples was conducted in a 5% electrolyte solution (KF and NH4F) at 70...90°C by pulsed electric discharges at U=320-350V and I=0.2...0.5 A/$dm^2$ for 5 minutes [13], which interacted with the tops of microroughnesses and removed them. As result, the samples acquired a characteristic mirror-like metallic luster. Further, the CG and UFG samples of Ti alloys were sprayed with TiVN coating with two versions of sublayer on the Ferry-WATT machine, where in the first case the substrate was treated with Ti, and in the second one – with V, the deposition time of sublayers and functional coating layers was 8 and 30 minutes respectively. Please, refer to [14,15] for more details of the coating deposition technique. The coating was certified by measuring the thickness of the formed sub-layers/coating layers on the Calotest machine and the microhardness on the Struers Duramin machine at the applied load of 300g for 15s. Adhesion strength of the coating was tested by scratching at a continuously increasing load of 0.03 to 10 N/min at an indentation rate of 2 mm/min on a CSM Micro-scratch testing machine. Erosive wear testing was performed on the basis of the comparative test method according to ASTM G76-18 "Standard Test Method for Conducting Erosion Tests by Solid Particle Impingement Using Gas Jets". This standard cover metal materials and coatings and establishes a method for their testing for abrasive wear in a solid particle stream.

## 3 Results and discussion

### *3.1 Microstructure of SPD-processed alloys Ti-6Al-4V and VT8M-1*

Two-phase titanium alloys after heat treatment had duplex structure, the average size of primary α-phase was ~2.7 μm and its volume fraction equaled to 25%, and included the interlayers of the plate-like (α+β) structure (Figure 1a).

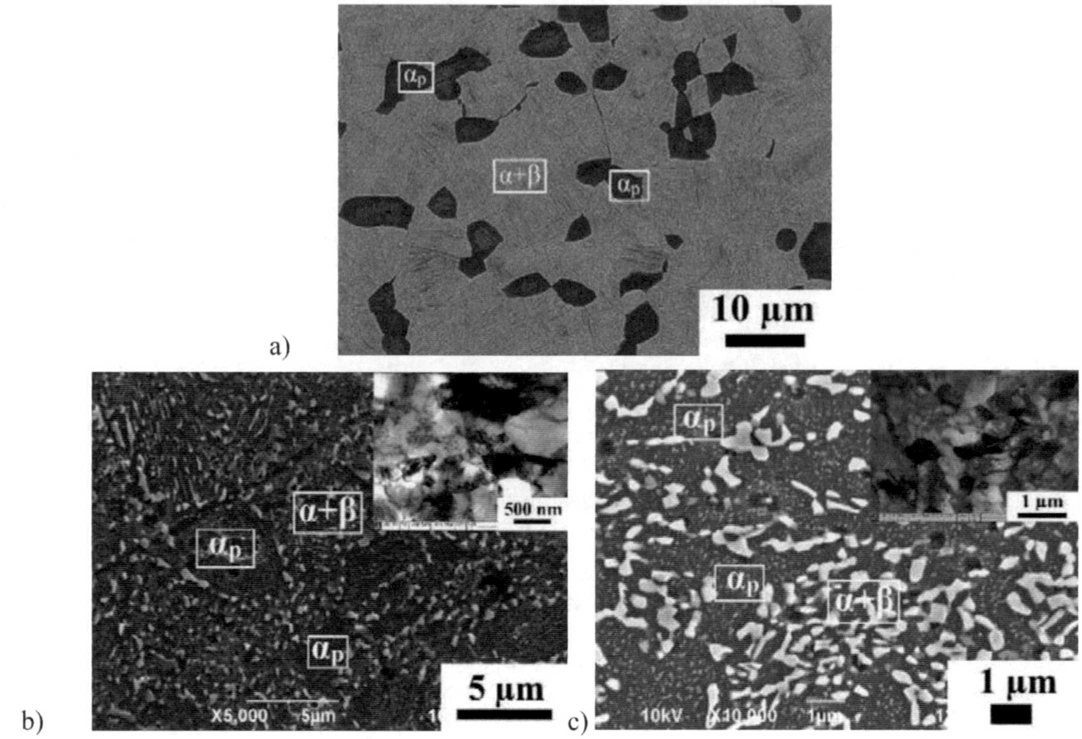


**Figure 1.** Microstructure of Ti-alloy (a) after heat treatment; and after subsequent SPD-processing: (b) Ti-6Al-4V by ECAP; (c) VT8M-1 by RS.

SEM and TEM analysis (Figure 1b, c) demonstrated the complete transforming of plate-like structure into a globular one after processing by ECAP and RS, with the grain and subgrain size ($\alpha+\beta$) of 0.48 and 0.25 μm respectively. The size of the primary $\alpha$-phase equaled to ~3 μm and was close to equiaxial, but some grains after RS processing were elongated in the direction of straining with the length up to 7 μm. The features of formation of these UFG structures are discussed in more detail in [11].

### 3.2 *Architecture and properties of vacuum-plasma protective coatings Ti+TiVN and V+TiVN*

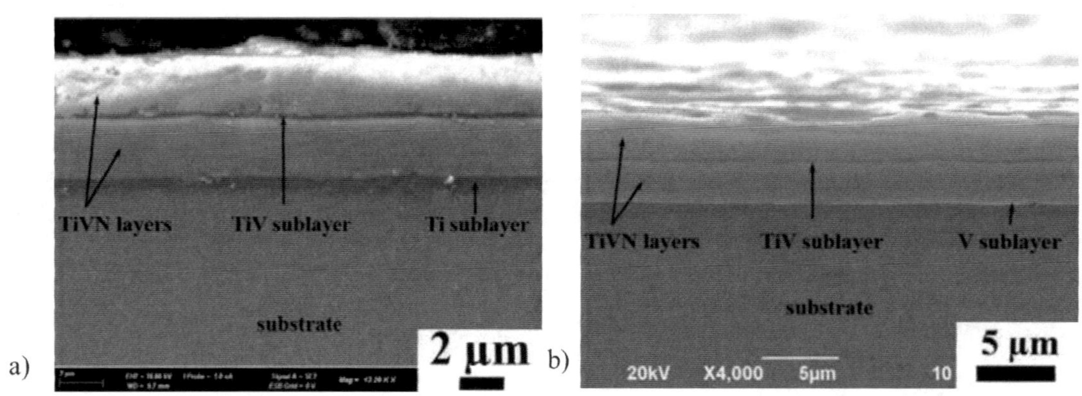


**Figure 2.** SEM image of UFG substrate + deposited coating: (a) Ti-6Al-4V + Ti+TiVN; (b) VT8M-1 + V+TiVN.

Peculiarities of the formation of these coatings were earlier investigated in our works [14,15]. The same coating structure was deposited on the samples of these alloys in coarse grained states.
Table 1 presents the average values of the microhardness for CG and UFG Ti alloys and coatings deposited on these substrates. As is seen, the coating hardness on the UFG alloy is higher than on the CG one, which speaks to the fact of the substrate hardness producing influence on that of the coating [16].

**Table 1.** Properties of vacuum-plasma coatings Ti+TiVN and V+TiVN on CG and UFG substrates of titanium alloys

| | **CG Ti-6Al-4V + Ti+TiVN** | **UFG Ti-6Al-4V + Ti+TiVN** | **CG VT8M-1 + V+TiVN** | **UFG VT8M-1 + V+TiVN** |
|---|---|---|---|---|
| **Coating thickness, h, µm** | 5.4 | 5.5 | 6.0 | 6.2 |
| **Coating hardness, MPa** | 10000±500 | 12000±500 | 16500±1000 | 26300±1000 |
| **Substrate hardness, MPa** | 3200±100 | 4050±100 | 2860±150 | 3350±160 |
| **Critical load during cohesive fracture of the coating, Lc1, N** | 3.9±0.12 | 4.5±0.10 | 5.0±0.15 | 5.1±0.10 |
| **Critical load during adhesive fracture of the coating, Lc2, N** | 6.9±0.20 | 14.2±0.30 | 7.6±0.2 | 18.1±0.15 |

As is seen in this work (Table 1), the adhesive strength and physico-mechanical properties of the applied coatings are considerably influenced by the substrate grain refinement. The experimental data obtained during the adhesive strength tests by scratching showed that the formation of UFG structure by grain refinement leads to an increase of the critical load to fracture (Lc2) by 2 and 2.5 times for coatings with sublayer of Ti and V respectively as well as to the enhancement of the critical load during adhesive fracture of the coating (Lc1). The effect of UFG structure on the hardness and adhesion strength of the coating is clearly of a general nature. This effect may be related to an increase in the crystallization rate of the coating material and its vacuum-plasma deposition on the UFG substrate, which leads to a decrease in the grain/crystal size in the coating [15].
The use of a V sublayer while vacuum-plasma protective coating TiVN deposition leads to increased erosion resistance due to the enhanced viscosity of vanadium [17]. Figure 3 demonstrates the results within 11 cycles of tests for erosive wear of vacuum-plasma protective coating V+TiVN on titanium alloy VT8M-1 in CG and UFG states as well as uncoated samples. The obtained data prove a sample from ultrafine-grained titanium alloy VT8M-1 with vacuum-plasma coating V+TiVN deposited on its surface to be the most resistant to abrasive wear.

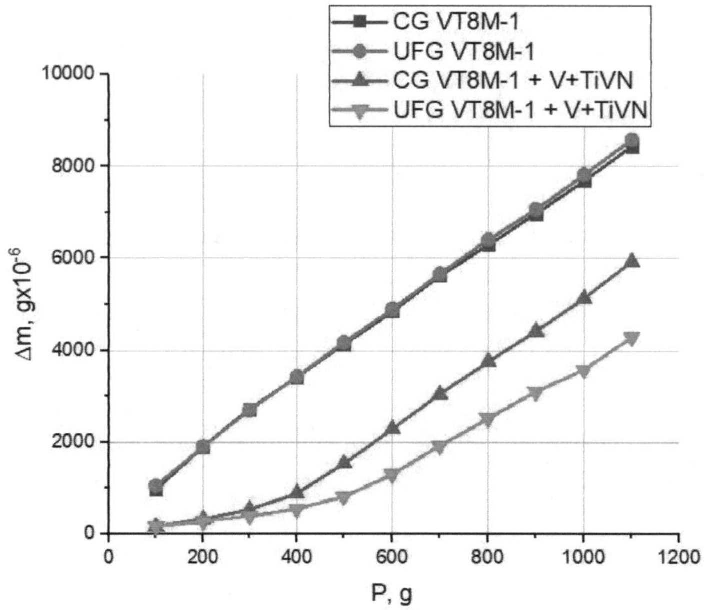


**Figure 3.** Dependence of mass loss of titanium alloy VT8M-1 samples in CG and UFG state with V+TiVN protective coating on erosive wear test cycle.

The results of erosion tests testify to the deposition of vacuum-plasma protective coating V+TiVN on titanium alloy VT8M-1 in CG and UFG states significantly reduces mass loss of the sample (Table 2) and twice increases erosion resistance of the material as compared to uncoated CG alloy. A substrate with a UFG structure is even more efficient for achieving this phenomenon, which is evident by the increased adhesion strength of the coating (Table 1).

**Table 2.** Erosive wear testing results of titanium alloy VT8M-1 in CG and UFG states with vacuum-plasma coating V+TiVN

| | **CG VT8M-1** | **UFG VT8M-1** | **CG VT8M-1 + V+TiVN** | **UFG VT8M-1 + V+TiVN** |
|---|---|---|---|---|
| **Total average mass loss, $\Sigma\Delta m_{average}$, g×10$^{-6}$** | 8585 | 8444 | 5931 | 4286 |
| **Comparative erosion resistance, K** | 1.0 | 1.02 | 1.45 | 2.0 |

Thus, the data on high hardness and enhanced adhesive and abrasive properties of the coating on titanium alloys in UFG state by means of severe plastic deformation processing techniques provide evidence that these coatings with sublayer V and Ti are quite effective and promising for application as constructional materials in aircraft engine building that involve extreme environment and in particular must perform under the harsh conditions of abrasive treatments.

**Acknowledgements**
The authors gratefully acknowledge the financial support from the Russian Science Foundation in the framework of the project no. 19-79-10108.

**References**

[1] Veiga C, Davim J P and Loureiro A J R 2012 Rev. Adv. Mater. Sci. **32** 133
[2] Antashev V G, Nochovnaya N A, Pavlova T V and Ivanov I 2007 (Entsiklopedicheskii Spravochnik, VIAM) http://viam.ru/public/files/2006/2006-204686.pdf (in Russian).
[3] Meyers M A, Mishra A and Benson D J 2006 Prog Mater Sci. **51** 427
[4] Valiev R Z, Estrin Y, Horita Z, Langdon T G, Zehetbauer M J and Zhu Y T 2015 Mater Res. Lett. **4** 1
[5] Valiev R Z, Estrin Y, Horita Z, Langdon T G, Zehetbauer M J and Zhu Y T 2006 JOM **58(4)** 33
[6] Styazhkin V A, Kopylov A A and Paleeva S Y 2000 Protect of Metals **36(3)** 292
[7] Bram M, Ahmad-Khanlou A, Buchkremer H and Stover D 2002 Mater Lett **57(3)** 647
[8] Ahmad Z, Khan A U, Farooq R, Saif T and Mastoi N R 2016 Mechanism of corrosion and erosion resistance of plasma-sprayed nanostructured coatings High temperature corrosion ed Z Ahmad (London: InTechOpen) chapter 7 pp 123-146
[9] Semenova I P, Saitova L R, Raab G I, Korshunov A I, Zhu Y T, Lowe T C and Valiev R Z 2006 Mater. Sci. Forum **503–504** 757
[10] Semenova I P, Polyakov A V, Polyakova V V, Huang Y, Valiev R Z and Langdon T G 2016 Adv. Eng. Mater. **18** 2057
[11] Modina I M, Polyakov A V, Dyakonov G S, Yakovleva T V, Raab A G and Semenova I P 2019 IOP Conf. Ser.: Mater Sci and Eng **461** 012056
[12] Nestler K, Böttger-Hiller F, Adamitzki W, Glowa G, Zeidler H and Schubert A 2016 Procedia CIRP **42(1)** 503
[13] Smyslova M K, Tamindarov D R, Plotnikov N V, Modina I M and Semenova I P 2018 In IOP Conf. Ser.: Mater Sci and Eng **461(1)** 012079
[14] Semenova I P, Valiev R R, Selivanov K S, Modina Y M, Polyakov A V, Smyslova M K, and Valiev R. Z 2017 Rev on Adv Mater Sci **48(1)** 62
[15] Valiev R R, Selivanov K S, Modina I M, Dyblenko Y M, Semenova I P and Valiev R Z 2020 Adv Eng Mater 2000121
[16] Wang C T, Gao N, Gee M G, Wood R J K, Langdon T G and Mech J 2013 Behav. Biomed. Mater. **17** 166
[17] Latellaa B A, Ganb B K, Daviesb K E, McKenzieb D R and McCullochc D G 2006 Surf and Coatings Technol **200(11)** 3605

# Mechanical properties of additively manufactured builds from different 316L steel powders

**Jaroslav Vavřík, Michal Brázda, Josef Hodek, Miroslav Urbánek, Antonín Prantl, Abhilash Kiran**

COMTES FHT - Dobřany, Czech Republic, EU

jvavrik@comteshft.cz

**Abstract**. In the present study, additively-manufactured (AM) specimens from 316L steel were investigated. The AM method used was Direct Energy Deposition (DED) where powder is fed continuously to a melt pool and is completely melted by high power laser. The melt pool rapidly solidifies on and beneath the substrate surface. To protect it from oxidation during deposition, argon is used as a shielding gas and is blown continuously to the melt pool. The DED process generates structures with minimum pores. 316L austenitic stainless steel was chosen for the experiments in order to avoid dealing with complex phase transformations during deposition. Four 316L powders supplied by four different companies were used and differences in the mechanical properties of the builds were examined. The process parameters were the same for each powder. There a number of studies on Selective Laser Melting (SLM) which explore the effects of powder properties on final mechanical properties of the builds. The underlying idea is to select a powder from a particular supplier to meet the customer required properties.

## 1 Introduction

In today's industry, additive manufacturing (AM) is becoming more popular day by day. It can provide solutions to various problems and can produce shapes unachievable by conventional methods. Its applications are found in different sectors such as electronics, motor vehicles, medical, aerospace, industrial machines, military and others [1, 2]. There are numerous AM techniques, such as DED (Direct Energy Deposition), SLM (Selective Laser Melting), Binder Jetting and Sheet Lamination. Each of them has its strengths as well as weaknesses [1, 2, 3]. Localized heating sources used in them lead to residual stresses, structural instability and crack formation. Residual stresses in the built structure are important to consider in order to ensure that they remain within the safe limits for the structure [4]. Preheating is one of promising methods to reduce residual stress [5]. DED offers unique applications, such as repair of damaged engineering parts and deposition of protective coating with direct control by digital data [3]. With DED, different powders can be mixed during deposition to produce gradient materials with significantly improved mechanical and wear properties. DED technology can even handle extraordinary material combinations, such as Ni with Ti [6], Ti6Al4V with TiC, Invar with TiC [3, 7], duplex steels [8] and many others as per the application requirement. The whole deposition process takes place under a protective atmosphere consisting of high-purity argon gas. The resulting 3D structure is built layer by layer.

DED uses powder which is melted by a high-energy source in the process. The powder properties govern the mechanical properties of the build. The presence of oxides, oxidized particle surfaces, satellites, or even undesirable particle shapes can lead to unsatisfactory mechanical properties and failure during deposition. The outcome of deposition also depends on the size of powder particles, their shape, chemical composition, their method of production, flowability, size distribution and other aspects. There are many producers of metal powders across the globe. Research into SLM has shown that differences in powder properties (size, shape and other aspects) could lead to very significant changes in mechanical properties and microstructure of builds [8, 9]. The present study compares four 316L powders from different producers to assess the impact of their differences on the deposited material.

## 2 Experiment

Builds of 316L steel powder were prepared using the DED method in the Insstek MX-600 3D printer. The size of the builds was 35×35×35 mm. Each of them was built separately on a new platform. The dimensions of each platform were 95×95×20 mm. The platforms were heat-treated at 400 °C for 4 hours to remove stress. Their chemical composition is listed in Table 1. It was measured using Bruker Q4 Tasman optical spectrometer. The 316L stainless steel powders were procured from four companies to compare the resulting mechanical properties: Sandvik Osprey LTD, Höganäs, H. C. Starck, and Prax Air. The powder particle size was in the range of 53-150 µm for all of them except Prax Air, where the range was 45-106 µm. The chemical composition of the powders is listed in Tab. 1.

**Table 1.** Chemical composition of powders and build platforms (wt. %)

| | Fe | Cr | Ni | Mo | Mn | Si | C |
|---|---|---|---|---|---|---|---|
| Sandvik | Bal. | 17.2 | 10.4 | 2.3 | 1.3 | 0.8 | |
| Höganäs | Bal. | 16.9 | 12.8 | 2.6 | 1.6 | 0.7 | 0.014 |
| H. C. Starck | Bal. | 17.1 | 11.8 | 2.3 | 1.5 | 0.7 | |
| Prax Air | Bal. | 17.0 | 12.1 | 2.5 | 1.4 | 0.6 | |
| Platform | Bal. | 16.24 | 10.49 | 2.14 | 1.12 | 0.44 | |

Computer models of the builds were developed using the Magic Materialize printing software. The deposition strategy was CF ZZ, where C stands for 'Contour', F means 'Fill' and ZZ denotes 'Zig Zag' instructions. The direction of the deposition path was rotated 90° between layers. Each track was 0.25 mm high and 0.8 mm wide. The hatch distance was 0.5 mm. The average laser power was 400 W.

For the purpose of mechanical testing, the builds were sectioned by wire electrical discharge machining (EDM) to obtain specimens with dimensions as shown in Figure 1. Five samples in each of the three directions were made. The directions were designated ZYX (in the z direction), YXZ (in the y direction) and XZY (in the x direction). Tensile tests were performed in 5 kN LabControl electrical testing machine. Mechanical properties of the builds were thus measured along all three axes with five tests for each direction.

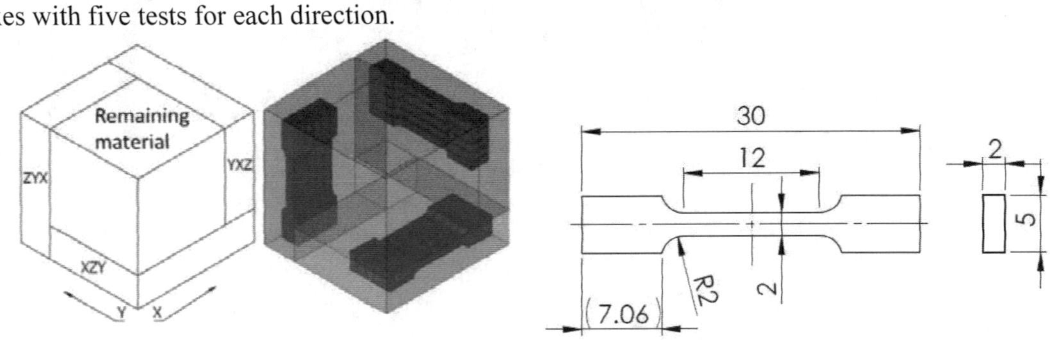


**Figure 1.** Test specimen orientations and test specimen drawing

## 3 Results

Mechanical properties measured in tension are summarized in Table 2.

**Table 2.** Mechanical properties of builds from powders by different producers

| Powder producer | Deposition strategy | Direction | $R_{p0.2}$ MPa | $R_m$ MPa | $A_g$ % | $A_5$ % |
|---|---|---|---|---|---|---|
| Sandvik | ZZ | ZYX | 460±8 | 640±2 | 45.0±0.8 | 66.8±0.7 |
| | | YXZ | 483±3 | 661±2 | 36.3±0.7 | 57.6±1.2 |
| | | XZY | 509±7 | 678±3 | 36.0±1.2 | 55.6±1.3 |
| Hoganas | ZZ | ZYX | 433±12 | 559±3 | 52.3±2.8 | 63.0±4.3 |
| | | YXZ | 454±12 | 603±11 | 24.7±3.4 | 34.3±5.4 |
| | | XZY | 476±12 | 606±3 | 22.5±3 | 33.0±5.9 |
| H. C. Starck | ZZ | ZYX | 395±3 | 493±3 | 53.3±4.4 | 65.6±6.1 |
| | | YXZ | 453±7 | 591±7 | 22.3±3 | 29.2±4.2 |
| | | XZY | 464±4 | 590±4 | 23.4±3.1 | 31.1±4.2 |
| Prax Air | ZZ | ZYX | 383±9 | 570±3 | 53.7±0.6 | 72.0±1.7 |
| | | YXZ | 406±3 | 589±3 | 45.3±0.3 | 61.9±3.6 |
| | | XZY | 423±5 | 595±3 | 43.8±0.9 | 60.9±1.5 |

Figure 2 illustrates a significant and recurring trend. The UST (ultimate tensile strength) increases between the directions ZYX→YXZ→XZY for all powders. This trend was invariable. By contrast, elongation values decreases between the same directions (ZYX→YXZ→XZY), as evidenced by Figure 3. Generally, specimens oriented in the ZYX direction had the highest elongation but some of the lowest UTS. It should be noted that the differences between the values for YXZ and XZY directions were minimal.

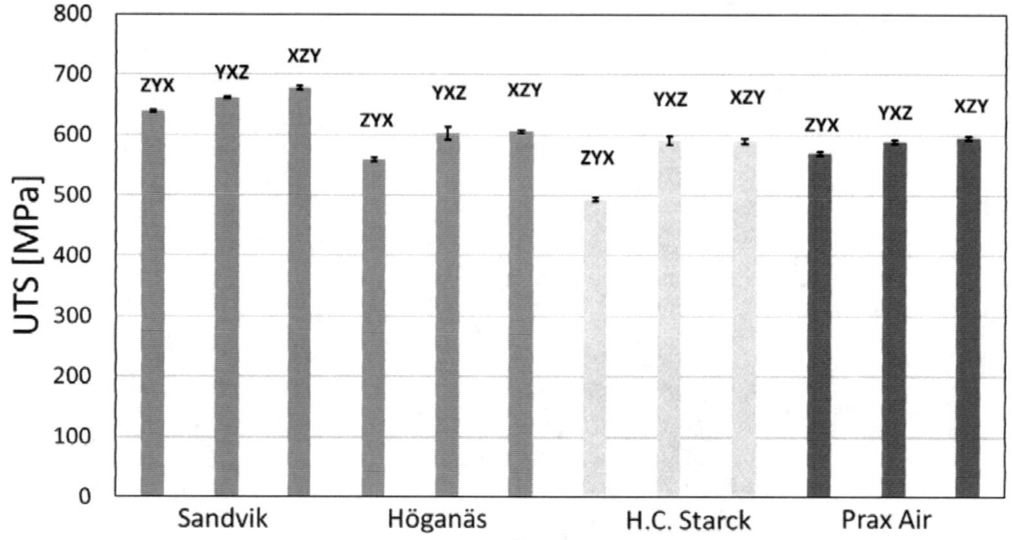


**Figure 2.** Ultimate tensile strengths of builds from powders by different producers

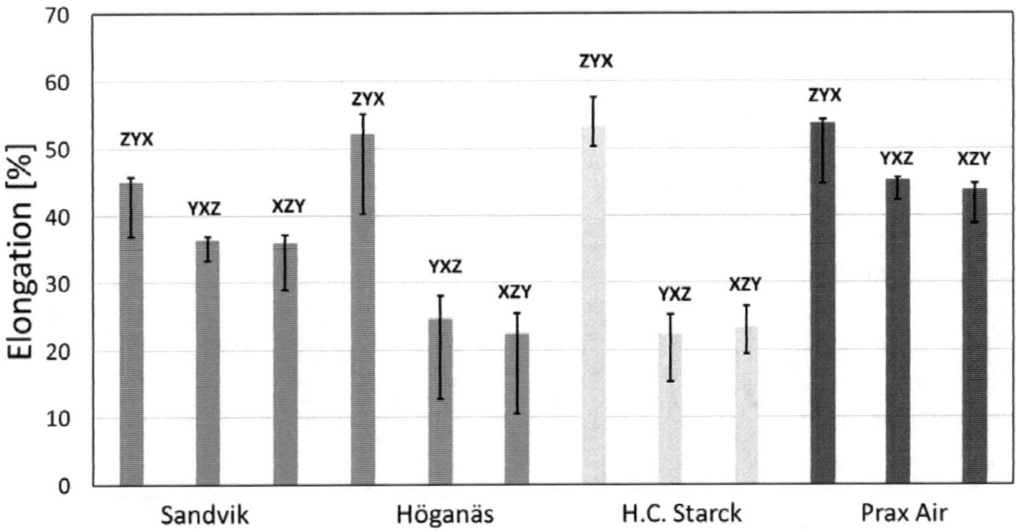


**Figure 3.** Elongations in builds from powders by different producers

Figures 4 and 5 show a comparison between the properties of builds in three directions. The build from Sandvik powder had higher UTS in all directions than those from the other powders. The one from Prax Air powder had the highest elongation by a large margin for all directions. The reason for this is presumably its smaller size of particles. The builds from Höganäs and H. C. Starck powders showed strengths similar to the other builds. In general, the differences in strengths are relatively insignificant but elongations differ among all the powders. The best performance appears to be that of the Sandvik powder. It has the highest UTS values and an elongation which is higher than in the other builds.

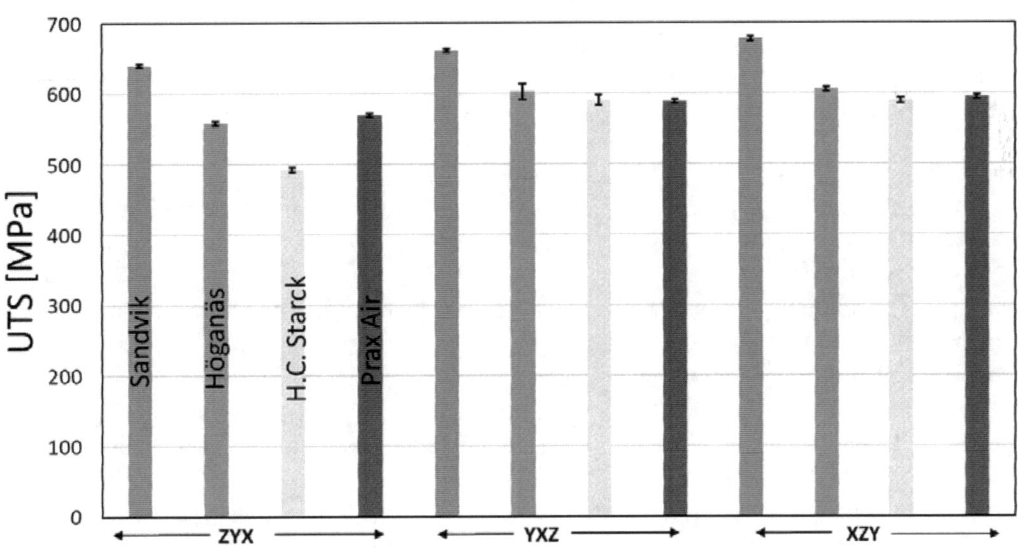


**Figure 4.** Comparison of UTS levels for builds from different powders and selected directions

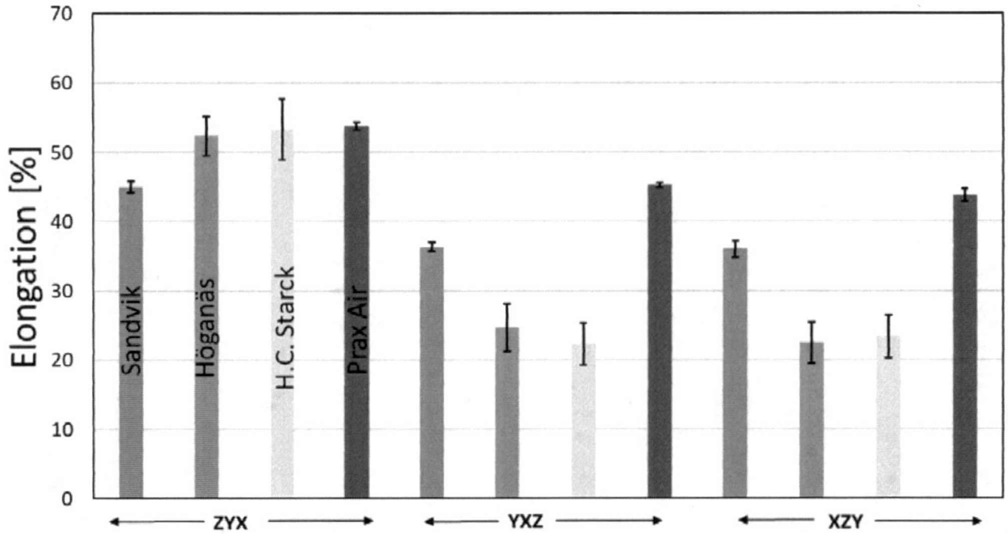


**Figure 5.** Comparison of elongation levels for builds from different powders and selected directions

## 4 Conclusions

Four cubes from 316L steel powders from different producers were built using the DED process for the purpose of comparing their mechanical properties. All the samples showed approximately equal UTS. However, their elongation behavior was significantly different. The build from Prax Air powder had the highest elongation, presumably due to its smaller powder particles. The build from Sandvik powder had the highest UTS and very high elongation. There was a trend of growing strength in the sequence of directions ZYX→YXZ→XZY and a corresponding opposite trend for elongation. However, it should be noted that the differences between the values in the YXZ and XZY directions were almost negligible.

**Acknowledgements**

This paper was developed thanks to the project entitled Pre-Application Research of Functionally Graduated Materials by Additive Technologies, No.: CZ.02.1.01/0.0/0.0/17_048/0007350, financed by the Ministry of Education of the Czech Republic.

## References

[1] M. Jiménez, L. Romero, I. A. Domínguez, M. d. M. Espinosa and M. Domínguez, "Additive Manufacturing Technologies: An Overview about 3D Printing Methods and Future Prospects," *Complexity,* p. 30, 2019.

[2] T. D. Ngo, A. Kashani, G. Imbalzano, K. T. Nguyen and D. Hui, "Additive manufacturing (3D printing): A review of materials, methods, applications and challenges," *Composites Part B: Engineering,* pp. 172-196, 2018.

[3] C. Zhang, F. Chen, Z. Huang, M. Jia, G. Chen, Y. Ye, Y. Lin, W. Liu, B. Chen, Q. Shen, L. Zhang and E. J. Lavernia, "Additive manufacturing of functionally graded materials: A review," *Materials Science and Engineering: A,* vol. 764, 2019.

[4] A. Kiran, J. Hodek, J. Vavřík and M. Urbánek, "Numerical Simulation Development and Computational Optimization for Directed Energy Deposition Additive Manufacturing Process," *Materials,* vol. 13, 2020.

[5] J. Vavřík and J. Hodek, "Residual stress investigation in additively manufactured samples," *IOP Conference Series: Materials Science and Engineering,* vol. 723, 2020.

[6] P. Salvetr, J. Vavřík, A. Školáková and P. Novák, "Formation of intermetallics layers in Ni-Ti

system prepared by direct energy deposition," *Manufacturing Technology,* vol. 19, no. 4, pp. 674-679, 2020.

[7] A. Yakovlev, E. Trunova, D. Grevey, M. Pilloz and I. Smurov, "Laser-assisted direct manufacturing of functionally graded 3D objects," *Surface and Coatings Technology,* vol. 190, no. 1, pp. 15-24, 2005.

[8] M. Brázda, P. Salvetr, J. Dlouhý and J. Vavřík, "Influence of laser power in direct laser deposition on the proportion of ferrite and austenite in duplex steel SAF2507," 2020.

[9] K. J. Terrassa, J. C. Haley, B. E. MacDonald and J. M. Schoenung, "Reuse of powder feedstock for directed energy deposition," vol. 338, pp. 819-829, 2018.

[10] K. Riener, N. Albrecht, S. Ziegelmeier, R. Ramakrishnan, L. Haferkamp, A. B. Spierings and G. J. Leichtfried, "Influence of particle size distribution and morphology on the properties of the powder feedstock as well as of AlSi10Mg parts produced by laser powder bed fusion (LPBF)," *Additive Manufacturing,* no. 34, 2020.

IOP Conf. Series: Materials Science and Engineering 1178 (2021) 012064 doi:10.1088/1757-899X/1178/1/012064

# Effect of inclusions on the fatigue strength of a bearing steel

**F Nový[1] O Bokůvka[1] M Vicen[1] M Činčala[2] R Nikolić[1]**

[1]University of Žilina, Univerzitná 1, 010 26 Žilina, Slovakia
[2]Volkswagen a. s., Priemyselná 1/9042, 036 01 Martin, Slovakia

E-mail: martin.vicen@fstroj.uniza.sk

**Abstract**. Here are presented results of experimental investigation about effect of inclusions on the fatigue strength of a bearing steel 100Cr6. Fatigue tests were carried out at sinusoidal cyclic tension-compression loading, f = 190 Hz, R = -1, room temperature, in the region ranged from $N \approx 3.5 \times 10^5$ to $N \approx 1.0 \times 10^8$ cycles of loading. Initiation of the fatigue cracks was observed in the bulk of the tested specimens and the so-called "fish eyes" were observed, mainly on the TiN inclusions (66 % Ti, 19 % N, 9 % Cr, remainder Fe, V, Zr). The TiN inclusions are hard, with sharp edges causing a significant stress concentration in a small localized area, resulting in an initiation of a fatigue crack. On the other hand, in standard for the chemical composition of this steel, the content of Ti was not prescribed.

## 1 Introduction

Fatigue has been a predominant fracture mode of load-bearing machine elements. Therefore, through the years, its prevention has become a fundamental design criterion, [1-4]. The microstructural heterogeneity, micro-impurity, holes, shrinkages, pores, grain size, grain boundaries and inclusions significantly affect the useful properties of construction materials, including the fatigue strength, [5-10]. In this paper are presented experimental results about influence of inclusions on the fatigue strength of the bearing steel 100Cr6.

## 2 Experiments and results

Experiments were carried out on the structural high-grade steel for rolling bearings 100Cr6, delivered in the soft annealing condition.
Verification of the qualitative and quantitative chemical composition was performed by the spark emission SPECTROMAXx device. The material was then heat treated according to the following procedure: austenitization (840 ± 5 °C, 20 min), cooling in the JS4oil at temperature T = 20 ± 2 °C and tempering (160 ± 5 °C, 90 min) with post-cooling in free air. The microstructure after the grinding, polishing and etching was observed with use of the electron microscope TESCAN Vega LMU. The Vickers hardness test HV10 was carried out on the Poldi-Vickers apparatus in the specimens' cross section in accordance with the STN EN ISO 6507-1 standard. The fatigue tests were performed at a low-frequency fatigue test (sinusoidal cyclic tension-compression loading, f = 190 Hz, R = -1, room temperature, T = 20 ± 2 °C) on the Roell – Amsler HFP 5100 device. Round bar specimens (21 pieces) of diameter of 3 mm, ground and polished by the metallographic procedures, were used for the fatigue tests. The investigation region of number of cycles was from $N \approx 3.5\times10^5$ to $N \approx 1.0\times10^8$ loading cycles. The basic chemical composition of the tested structural steel 100Cr6 (widespread, demonstrated by experiment vs. material standard) is shown in table 1. It was in accordance with the material standard [11]. The microstructure created after the heat treatment (quenching and low tempering)

contained the tempered martensite, retained austenite and uniformly distributed cementitious carbides, figure 1.

**Table 1.** Chemical composition (wt. %) of the tested structural steel 100Cr6.

| | C | Si | Mn | Cr | Ni | Cu | Ni+Cu | P | S | Ti | V | Co |
|---|---|---|---|---|---|---|---|---|---|---|---|---|
| Exper. | 1.0 | 0.28 | 0.34 | 1.48 | 0.1 | 0.12 | 0.2 | 0.01 | 0.01 | **0.163** | 0.033 | 0.021 |
| Mat. | 0.9 | 0.15 | 0.30 | 1.30 | max. | max. | max. | max. | max. | | | |
| Stand. | 1.0 | 0.35 | 0.50 | 1.65 | 0.30 | 0.25 | 0.50 | 0.027 | 0.030 | | | |

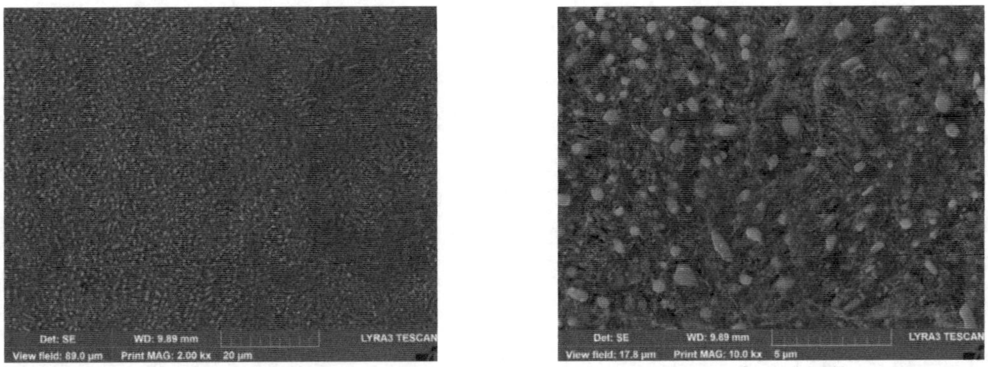


a) general view b) detail

**Figure 1.** Microstructure of the structural steel 100Cr6, etched by the picric acid; SEM

Measured hardness (average over five indentation points) was 713 HV10. Results of the fatigue tests are shown in figures 2 to 4 and in table 2. The fatigue tests were performed in the range from $3.5\times10^5$ to $1.0\times10^8$ loading cycles. The reason is the fact that fracture caused by fatigue occurs in structural steels far beyond the so-called "basic number" $N_c = 1.0\times10^6$ to $1.0\times10^7$ loading cycles, [1-2, 4].

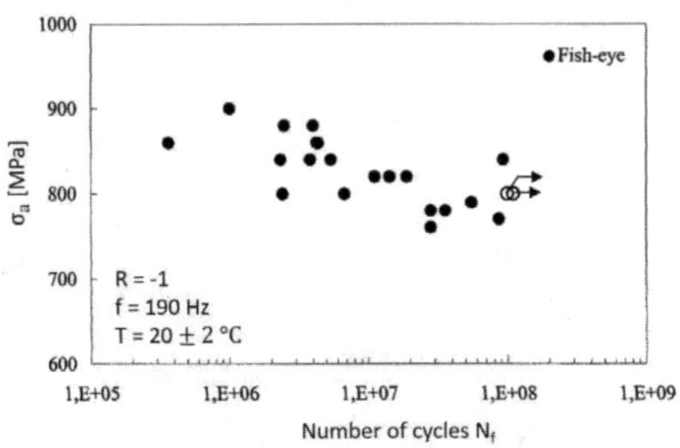


**Figure 2.** $\sigma_a = f(N_f)$, steel 100Cr6

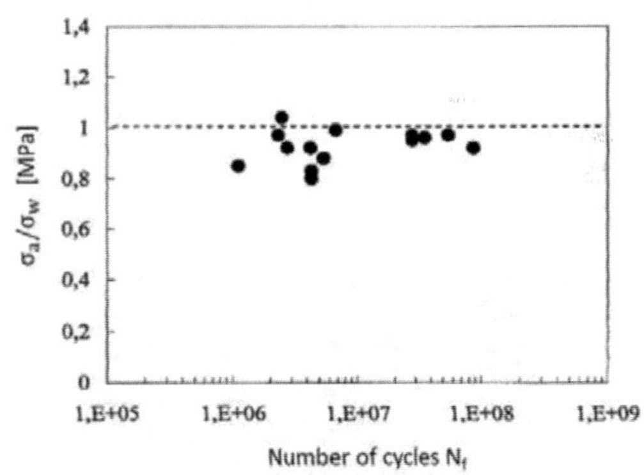


**Figure 3.** $\sigma_a/\sigma_w = f(N_f)$, steel 100Cr6

The fatigue limit $\sigma_w$ was calculated by the Murakami model for predicting the fatigue life from the non-metallic inclusions area, [6]. For the internal defect or inclusion, estimated fatigue limit is calculated as $\sigma = 1.56\times(HV+120)/\sqrt{area}^{1/6}\times[(1-R)/2]^{\alpha}$, with HV as the matrix hardness (100÷740), R as the cycle asymmetry coefficient, $\alpha = 0.266 + HV\times10^{-4}$ and $\sqrt{area}$ is the projection area of the maximum inclusion in a direction perpendicular to the highest tensile stress, figure 5.

**Table 2.** Selected results of the fatigue tests, steel 100Cr6

| $\sigma_a$ [MPa] | $N_f$ [cycles] | $\sqrt{area}$ [μm] | $\sigma_w$ [MPa] | $\sigma_a/\sigma_w$ |
|---|---|---|---|---|
| 880 | 2.48 x $10^6$ | 13.5 | 842 | 1.04 |
| 860 | 4.23 x $10^6$ | 7 | 931 | 0.92 |
| 860 | 4.34 x $10^6$ | 4 | 1014 | 0.85 |
| 840 | 2.33 x $10^6$ | 11.5 | 865 | 0.97 |
| 820 | 5.39 x $10^6$ | 6 | 965 | 0.85 |
| 840 | 4.34 x $10^6$ | 4 | 1014 | 0.80 |
| 820 | 1.11 x $10^7$ | 6 | 965 | 0.85 |
| 800 | 2.76 x $10^6$ | 11.5 | 865 | 0.92 |
| 800 | 6.75 x $10^6$ | 17 | 810 | 0.99 |
| 790 | 5.5 x $10^7$ | 17 | 810 | 0.97 |
| 780 | 2.80 x $10^7$ | 18 | 803 | 0.97 |
| 770 | 8.66 x $10^7$ | 14 | 837 | 0.92 |
| 760 | 2.80 x $10^7$ | 18 | 803 | 0.95 |

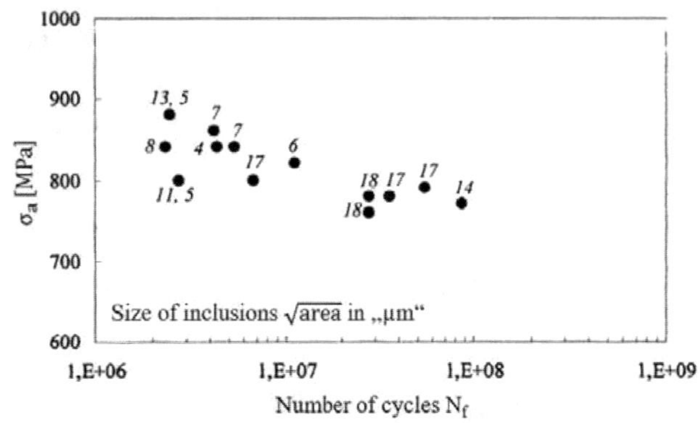


**Figure 4.** $\sigma_a = f(N_f)$, size of inclusions as a control factor of fatigue strength, steel 100Cr6

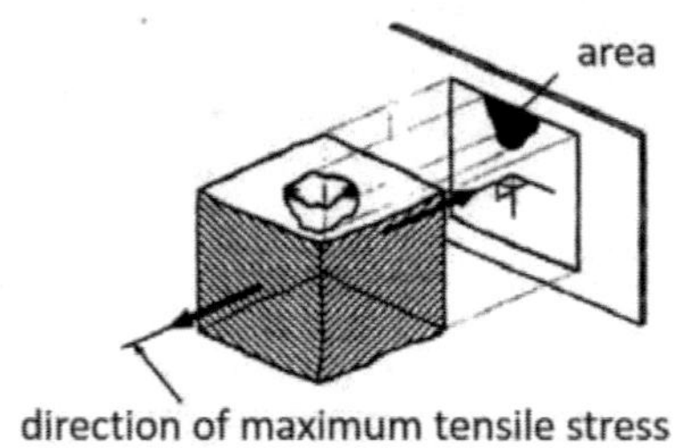


**Figure 5.** The $\sqrt{area}$ definition [6]

Initiation of the fatigue cracks was mainly from the inside of the tested specimens. The so-called "fish eyes" were observed on the TiN inclusions (66 % Ti, 19 % N, 9 % Cr, remainder Fe, V, Zr), figures 6 and 7.

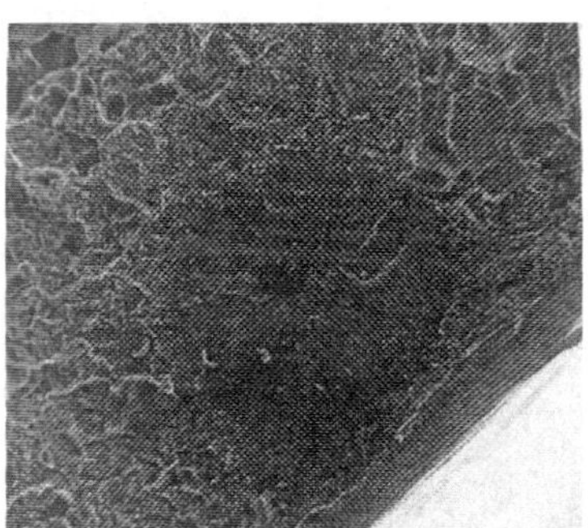

**Figure 6.** "Fish eye", TiN inclusion, steel 100Cr6

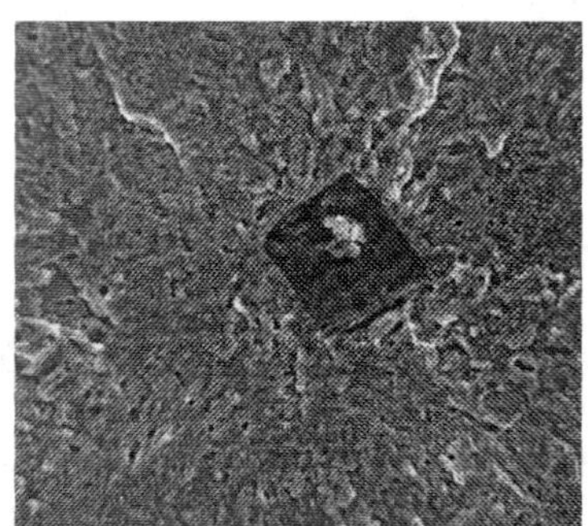

**Figure 7.** TiN inclusion, $\sigma_a$ = 800 MPa, $N_f$ = 2.76 x $10^6$ cycles, $\sqrt{area}$ = 11.5 μm

The TiN inclusions cause a significant stress concentration in a small localized area [2, 5-10]. There are two basic types of internal stresses around inclusions –deformation induced and mosaic stresses.

The deformation induced stresses arise as a result of different stress-deformation characteristics of the inclusions vs. matrix, while the mosaic stresses are caused by differences in the thermal expansion coefficients of the matrix and inclusions. The inclusion – matrix interface is incoherent and inclusions are the most efficient source of micro-cavities and micro-cracks [10].

The ratio $(\sigma_a / \sigma_w)$ ranged from 0.80 to 1.04; the required value must be $(\sigma_a / \sigma_w) > 1$. Despite this fact, all the specimens (figures 2-4) were in fact broken on the TiN inclusions. In standard for this steel chemical composition, no attention was paid to content of Ti, which is here emphasized as being very important [11].

## 3 Conclusions

Taking into account the performed experimental work, the following can be stated:

- Dependence $\sigma_a = f(N_f)$ shows a decreasing character after the so-called basic number $N_c = 1.0\times10^6$ to $1.0\times10^7$ of loading cycles.
- Fractography analysis showed the "fish eyes" on the fracture surface of the non-metallic TiN inclusions.
- The TiN inclusions can be classified as cracking due to their character, properties, quantity, distribution, shape (edge) and dimensions.
- The bearing steel 100Cr6 is very sensitive to the notch effects, not only on the surface, but on the non-metallic inclusions in the bulk material, as well.
- To ensure the high fatigue strength of the bearing steel 100Cr6, in the area of a high number of loading cycles, one must ensure adequate micro-purity by the precise production technology.

**Acknowledgements**

This research was partially financially supported by projects VEGA 1/0951/17, KEGA 012ŽU-4/2019 and ITMS2014+NFP313010T426 "Research and development activities of Žilina University in Žilina for the 21st century industry in the field of materials and nanotechnologies" financed by European regional development fund and Slovak state budget.

## References

[1] Bokůvka O, Nicoletto G, Guagliano M, Kunz L, Palček P, Nový F, Chalupová M, 2014 Fatigue of Materials at Low and High Frequency Loading EDIS ŽU Žilina
[2] Činčala M 2005 High cycle fatigue of bearing steel 14 109, PhD. thesis, UNIZA Žilina (in Slovak)
[3] Závodská D, Kuchariková L, Tillová E, Guagliano M, Chalupová M, Uhríčik M, Belan J, 2019 The effect of iron content on fatigue lifetime of AlZn10Si8Mg cast alloy, International Journal of Fatigue, vol 128, 105189
[4] Bokůvka O, Nicoletto G, Kunz L, Palček P, Chalupova M, 2002 Low and High Frequency Fatigue Testing, EDIS ŽU Žilina
[5] Murakami Y 1989 Efects of small defects and nonmetallic inclusions on the fatigue strength of metals, JSME Int. Journal, vol 32(2), p 167
[6] Murakami Y and Endo M 1994 Effects of defects, inclusions and inhomogeneities on fatigue strength, Fatigue, vol 16(3), p 163
[7] Murakami Y, Kodama S, Konuma S, 1989 Quantitative evaluation of effects of non-metallic inclusions on fatigue strength of high strength steels (part I.), Int. J. Fatigue, p 291
[8] Murakami Y and Usuki H 1989 Quantitative evaluation of effects of non-metallic inclusions on fatigue strength of high strength steels (part II), Int. J. Fatigue, p 299
[9] Mitura K and Landová S 1986 Inclusions in the steels and their influence on the usefull properties of steels, SNTL Praha (in Czech)
[10] Pokluda J, Kroupa L, Obdržálek L, 1994 Mechanical Properties and Structure of Solid Matter, PC-DIR s. r. o. Brno (in Czech)
[11] Skočovský P, Palček P, Konečná R, Varkoly L, 2000 Structural Materials, EDIS ŽU Žilina

IOP Conf. Series: Materials Science and Engineering 1178 (2021) 012065 doi:10.1088/1757-899X/1178/1/012065

# Possibilities of using thermally sprayed coatings for surface treatment of rail vehicles components

**M Vostřák[1], J Brodský[2], Š Houdková[1], P Šulcová[1]**

[1]Výzkumný a zkušební ústav Plzeň s.r.o., Tylova 1581/46, 301 00 Plzeň Česká republika

[2]Škoda Transportation a.s., Emila Škody 2922/1, 301 00 Plzeň, Česká republika

vostrak@vzuplzen.cz

**Abstract**. Thanks to their versatile properties and economical applicability, thermally sprayed coatings are being used in more and more areas of industry. Within the research project, suitable alternatives are being sought for the replacement of surface treatments of selected components of rail vehicles with thermal spraying. The aim is to increase the functionality and service life of selected parts and thus the entire vehicle. Based on the requirements for the functional properties of components, variants for the material and technology of thermal spraying are selected. Technologies considered include HVOF, TWAS, FS, and APS, materials include Fe, Ni, Zn, and Mo based alloys; WC and CrC based cermets. The basic functional properties of these coatings are analyzed and discussed. The further steps of the work to test suitability of these coatings for selected applications and are outlined.

## 1 Introduction

There are several methods and technologies for coatings formation, as well as an almost unlimited number of materials suitable for the application of surface layers. To corectly choose the suitable coating layer, it is necessary to know the application of the part, the method of loading and types of acting stresses, the working environment and expected lifespan of the part. [1, 2, 3].

Thermal spraying is a general term for forming metallic or non-metallic coatings with the thickness usually greater than 40 μm. The filler material (in the form of powder, wire, or rods) is partially melted using a heat source. It can be electric arc, plasma, or flame. The heated and molten particles are accelerated towards the surface either by process gases or by atomizing nozzles. Due to the high kinetic energy, the particles are anchored on a roughened surface. This creates a bond with the surface of the substrate and the particles form a so-called lamellar structure [1, 2, 3, 4, 5].

The main advantage of thermal spraying technologies is the wide range of materials that can be used to produce desired coatings. Practically any material that does not disintegrate below the melting point can be used. The second major advantage is the ability of most thermal sprays to apply coatings to the substrate without significantly thermally affecting the substrate surface. The third advantage is an ability to repair worn or damaged coatings in most cases without changing the properties or dimensions of the components [2].

The disadvantage of thermal spraying might be then the nature of the application technology. It is practically impossible to apply the coating to small or deep cavities and holes.

This work is a part of the project "Increasing of resistance of rail vehicle components by the means of modern thermal spraying technologies". The goals of the work are following: 1) to consider the types of loading of selected rail vehicle components, 2) choose suitable candidates for coating materials and technology for applying surface treatment, 3) optimize the thermal spraying application and in detail analyze the functional properties of the coatings and components with the new applied surface treatment.

## 2 Types of loading of selected parts

Within the research project, the surface treatment of several selected parts of rail vehicles is taken into consideration to be replaced by thermal spraying technologies. The parts are now used without surface treatment or with a surface treatment that can be significantly overcome by replacing by the new type. There are types of parts with similar loading.

The type of parts is located outside the vehicle, with a large surface area. The representative example would be "walkable footplates" which are located on the front of the vehicle. The main load of these parts is mainly external influences such as salt, water, rain, and aerodynamics action. The part will be further affected by stones because the area is walkable. The magnitude of the force will depend on the weight of the person when the weight of the person is calculated to be 200 kg. These parts require protection against external environment and basic protection against abrasion and erosion by stones, weather, and walking. For these parts, the state of the surface after the coating application has to be taken into account, as it is the final surface treatment and the parts would be used in the as-sprayed state.

Another type is parts of the vehicle chassis. These parts are loaded again with external influences as well as with the higher forces resulting from their function in the chassis. An example would be the "handle" which is a part that secures the wheelset when lifting the chassis or the whole vehicle. The force acting on the handle depends on the weight of the wheelset. The surface area requiring protection for these parts is not large, but the applied coating has to withstand significantly higher forces, moreover high resistance against abrasion and friction wear is required.

Another representative of loading would be a combination of pivot with high demand on hardness and abrasion resistance with a combination of its counterpart with a low friction coefficient.

## 3 Selected thermal spraying technologies and materials

Within the thermal spraying group, there are several technologies with differ in operating temperature, speed of molten particles, amount of additional material, etc. The technologies can be distinguished by the type of heat source. The first group utilizes electric energy – the examples are Twin Wire Arc Spray (TWAS), Atmospheric Plasma Spray (APS) or Vacuum Plasma Spray (VPS) The second group utilizes the combustion of mixed fuel and oxygen – examples are High Velocity Oxygen Fuel (HVOF), Detonation spraying (D-Gun) or Flame Spray (FS) [1, 2, 3].

In this work, the four technologies available at Research and testing institute in Pilsen are considered to be used for treating components of rail vehicles. These technologies are HVOF, TWAS, APS, and FS. The HVOF typical characteristics are high coating density and low porosity, low oxidation rate in a flight of particles and, compressive residual internal stress, while the drawbacks are high noise level and higher costs. Typical applications are coatings with high wear resistance, oxidation, or chemical resistance even at high temperatures. The materials commonly sprayed are superalloys based on Ni, Co, Fe, carbides based on WC or CrC and MCrAlY.

The TWAS typical characteristic is high deposition efficiency but high porosity, low cost, mobility, possibility of hand-spray and low noise. Typical applications are corrosion protection sliding surfaces, surface renovation (adding worn-out material), materials commonly used are metal alloys (Fe, Ni, Cu, Al, and Zn based).

The biggest advantage of APS is high temperature enabling the spraying of ceramic materials. Other typical characteristics are high deposition efficiency, high cost and high noise level. Typical

IOP Conf. Series: Materials Science and Engineering 1178 (2021) 012065 doi:10.1088/1757-899X/1178/1/012065

applications are thermal barrier coatings (TBC), wear-resistant ceramic base coatings, corrosion resistance. Typical coating materials are ceramics, carbides based hard-metals, abradable coatings, and alloys based on Ni, Co, and Fe.

The FS typical characteristics are high deposition efficiency, low adhesion, high porosity, low cost, low noise, and the possibility to hand-spray. Typical applications are wear and corrosion protection, surface renovation. Typical coating materials are Fe and Ni-based alloys, abradable coatings, self-fluxing alloys.

The selected materials considered for surface treatment are listed in Table 1. For a high demanding application for wear and corrosion resistance, the WC-CoCr and $Cr_3C_2$-NiCr sprayed by HVOF or flame sprayed NiCrBSi are selected. For corrosion and sufficient wear protection of large parts, Fe based stainless steel, Fe based alloy, or ZnAl sprayed by TWAS are chosen. For applications demanding good sliding properties along with sufficient wear resistance, Mo sprayed by TWAS and APS are selected.

**Table 1.** Materials used for coating application

| Coating material | Form | Technology | Commercial designation |
|---|---|---|---|
| Mo | Powder | APS | Amdry 313X |
| NiCrBSi | Powder | FS | FST 771.33 |
| WC-10Co4Cr | Powder | HVOF | Woka 3652 |
| $Cr_3C_2$-25NiCr | Powder | HVOF | Amperit 588.074 |
| Mo | Wire | TWAS | W400.1 |
| Fe13Cr | Wire | TWAS | W504.1 |
| Zn15Al | Wire | TWAS | Zinacor |
| Fe-based alloy | Wire | TWAS | Metco 8294 |

## 4 Coating microstructure and properties

The microstructure of HVOF sprayed WC-CoCr and $Cr_3C_2$-NiCr coatings are depicted in Figure 1. Both coatings represent typical microstructure for HVOF coating, high density with minimal porosity, vaguely distinguished individual splats, absence of cracks. These carbide-based coatings have the highest hardness and wear resistance of presented coatings. The surface hardness HRC is 59 ± 2, the microhardness on cross-section HV0.3 is 1170 ± 15 for WC-CoCr. The $Cr_3C_2$-NiCr has surface hardness HRC is 54 ± 4, the microhardness on cross-section HV0.3 is 948 ± 69.

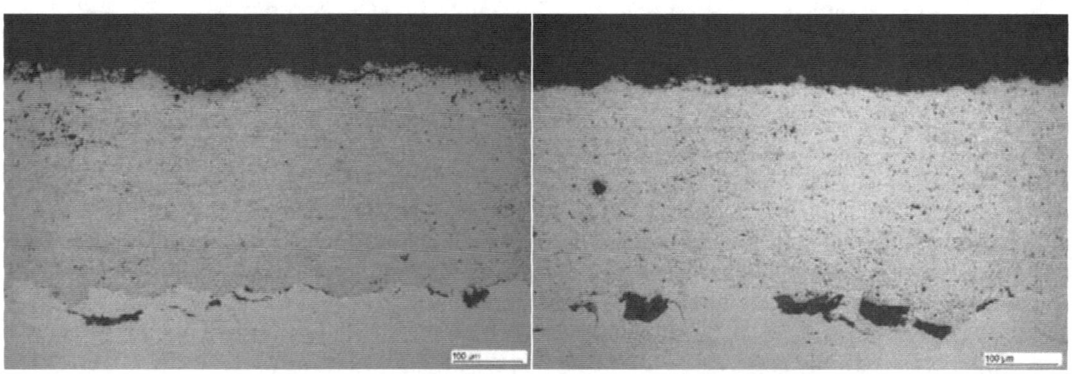

**Figure 1.** WC-CoCr (left) and $Cr_3C_2$-NiCr (right)

The microstructure of NiCrBSi coating sprayed by FS is depicted in Figure 2. It shows typical features of the microstructure of FS sprayed coatings, i.e. higher porosity and easily distinguishable individual splats, and the occasional occurrence of un-melted particles. The surface hardness HR15N is 78 ± 2, the microhardness on cross-section HV0.3 is 810 ± 57.

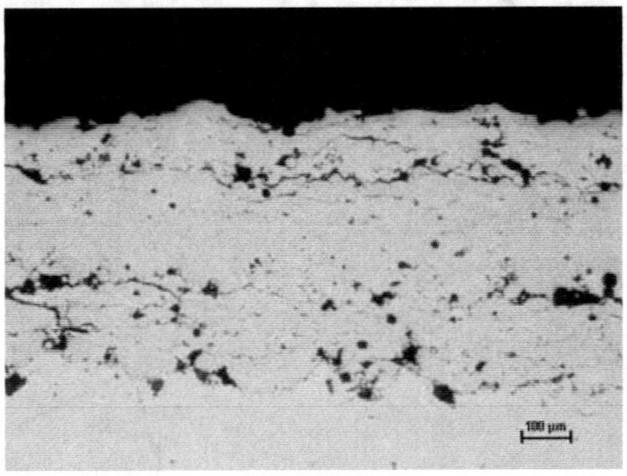


**Figure 2.** NiCrBSi FS sprayed

Zn15Al has very good corrosion resistance in the atmosphere and immersion in either fresh or saltwater. This coating offers better corrosion resistance than either pure zinc or aluminum, by combining advantages of both materials. ZnAl coating sprayed by TWAS, presented in Figure 3, has a very fine microstructure with almost zero porosity and without the occurrence of cracks or other defects. Differentiation of individual splats is possible only at higher magnification. The surface hardness HR15Y 85 ± 2. To comparison, the TWAS sprayed stainless steel Fe13Cr has a structure more typical for TWAS coatings, easily distinguished splats, and higher porosity. The surface hardness HR15N is 79 ± 1.

**Figure 3.** TWAS sprayed Zinacor (left) and Fe13Cr (right)

Molybdenum coatings are widely used as surface treatment due to their high wear and scuffing resistance and excellent sliding properties. Molybdenum coating deposited by the TWAS method (depicted in Figure 5) is without noticeable defects in the form of cracks. There is significant porosity on the boundary of individual splats. For comparison, the Mo coating sprayed by APS is depicted in Figure 5. There is also significant porosity, nevertheless, the bottom layer of the coating is more densified, the splats are smaller. The surface hardness HR15N of Mo coating sprayed by TWAS or APS is in the same range, is 67 ± 3.

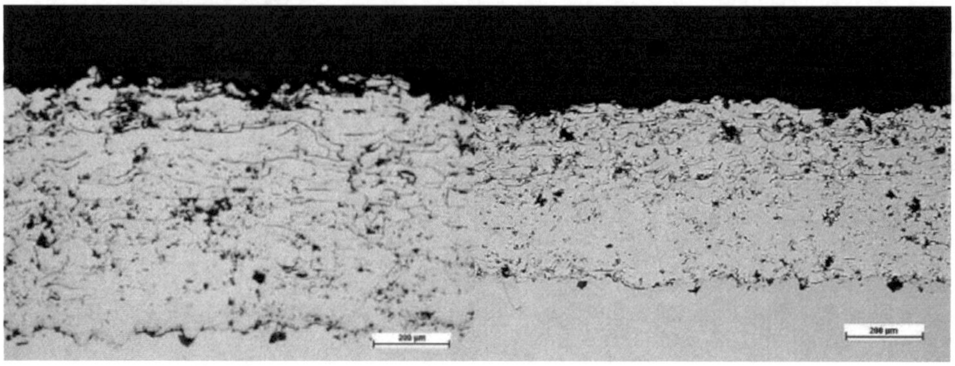

**Figure 5.** Mo sprayed by TWAS (left) and by APS (right)

## 5 Conclusion

The paper presents possible alternatives to the surface treatment of components for rail vehicles. The demands and expected types of loading and stresses acting on the selected parts are shortly discussed. Then, the four thermal spraying methods which are considered for these applications are presented, the technologies are HVOF, TWAS, APS, and FS. Based on the requirements of individual parts, the suitable coating materials were chosen, the materials include WC and CrC based carbide coatings for hard abrasive resistance coatings, Fe and Zn based alloys, and Mo for low friction applications. The microstructure of these selected coating sprayed above mentioned technologies was presented, along with surface hardness microhardness on cross-section, which represents some basic functional properties of the coatings. Based on preliminary results, the potentially suitable coatings for selected components were chosen. For each component, at least two coatings variants will be applied. The treatment of selected components is underway. For the components located outside the vehicle, with a large surface, the Fe and Zn based coatings are chosen. They provide sufficient protection and the spraying by TWAS is easily adjustable to spray these components. For the components requiring high wear and corrosion resistance, the HVOF sprayed cermet coatings or Flame sprayed NiCrBSi coating is considered. For applications demanding good sliding properties along with sufficient wear resistance, Mo coatings are selected. The sprayed components along with components treated with current methods will be tested for corrosion resistance in salt fog. Further, the abrasion and erosion resistance of the applied coatings will be tested. The most suitable surface treatment for selected parts will be considered for series production.

**Acknowledgements**
The paper has originated in the framework of the solution of project number FV40165 "Increasing of resistance of rail vehicles components by the means of modern thermal spraying technologies"

**References**

[1] Pawlowski L *The Science and Engineering of Thermal Spray Coatings*. Second edi. B.m.: John Wiley & Sons, Ltd., 2008. ISBN 978-0-471-49049-4
[2] Davis, J R 2004 ASM Handbook, Volume 5A: Thermal Spray Technology. Prepared under the direction of the ASM Thermal Spray Society Training Committee. ISBN: 978-1-61503-996-8
[3] Vardelle A M, Moreau CH, Vuoristo P, Akedo J 2016 The 2016 Thermal Spray Roadmap. Journal of Thermal Spray Technology. 25. 10.1007/s11666-016-0473-x
[4] CARTIER M 2003 Handbook of surface treatments and coatings. ASME Press, New York 464 str. ISBN: 978-1-86058-375-9
[5] Wank, A Basics of Thermal Spray Technology. GTV Verschleiss-Schutz GmbH, Luckenbach

# Corrosion behaviour of hydroxyapatite (HAP) coated AISI 316Ti austenitic biomaterial

**V Zatkalíková[1], L Markovičová[1], M Uhríčik[1], P Hanusová[1]**

[1] Department of Material Engineering, Faculty of Mechanical Engineering, University of Žilina, Univerzitná 8215/1, 01026 Žilina, Slovakia

viera.zatkalikova@fstroj.uniza.sk

**Abstract**. HAP bioactive coatings are widely applied to improve the bone response in orthopedic and dental stainless steels implants. Mechanical properties of the metallic implant are preserved and moreover, HAP improves the corrosion resistance. This article deals with the corrosion resistance of AISI 316Ti austenitic biomaterial with electrochemically deposited HAP coating. Corrosion behaviour is evaluated on the bases of two independent methods - the electrochemical potentiodynamic polarization test and exposure immersion tests, both carried out in physiological solution (0.9 % NaCl) at the temperature 37 ± 0.5 °C. The microstructure and chemical composition of HAP coating is evaluated by SEM and EDX analysis. According to the results of the potentiodynamic test, the HAP coated biomaterial shows the considerably higher corrosion resistance compared to that without the HAP coating. However, the corrosion losses and SEM analysis after the exposure test revealed severe disruption of the HAP coating and the stainless steel substrate started to corrode.

## 1 Introduction

Austenitic stainless steel implants are characterized by acceptable corrosion resistance, high strength and fracture toughness [1, 2], but their bonding ability to bone tissue is very low. In order to obtain bioactive and strong material, the formation of hydroxyapatite (HAP) ceramic coating on an austenitic implant with appropriate mechanical properties is considered a good approach [3].

Hydroxyapatite is a naturally occurring mineral form of calcium apatite with the formula $Ca_5(PO_4)_3(OH)$ (usually written as $Ca_{10}(PO_4)_6(OH)_2$ to express that the crystal unit cell comprises two entities). HAP is chemically, structurally and biologically similar to the calcium phosphate minerals in biological hard tissue and it has ability to form a strong chemical bond with bone [4]. However, the mechanical weakness of the HAP limits its practical applications to those requiring little or no load-bearing locations. If used in the form of surface coating, mechanical properties of the metallic implant are preserved and moreover HAP improves the corrosion resistance of the biomaterial by forming a barrier against the dissolution of metal ions from the substrate at the same time as promoting its bone bonding ability [3, 5, 6]. Therefore, HAP coating is widely applied to improve the bone response in orthopedic and dental stainless steels implants.

There are several techniques to obtain HAP surface coating including plasma spraying [7], sol-gel method [8], pulse laser deposition [9], biomimetic methods [10], and electrochemical deposition [11, 12].

Currently, plasma spraying is the most widely used commercially available method for preparation of the HAP coatings [7, 9]. However, the plasma-sprayed HAP coatings indicate some deficiencies in the

clinical applications related to the low adherence on the metal surface and to the absence of consistency in morphology and crystallinity [13].
The electrochemical deposition (based on the reaction: $10Ca^{2+} + 6PO_4^{3-} + 2OH^- \rightarrow Ca_{10}(PO4)_6(OH)_2$) is one of the simplest and most widely used methods of HAP coating. Recently, this HAP coating method became technologically important for various biomedical applications [12].
Presented article is focused on an assessment of the corrosion resistance of AISI 316Ti austenitic biomaterial with electrochemically deposited HAP coating. Corrosion behavior is evaluated on the bases of two independent methods - the electrochemical potentiodynamic polarization test and exposure immersion tests, both carried out in physiological solution (0.9 % NaCl) at the temperature 37 ± 0.5 °C. The microstructure and chemical composition of HAP coating is evaluated by SEM and EDX analysis.

## 2 Material and methods

The experimental material AISI 316Ti is Cr-Ni-Mo austenitic stainless steel stabilised by titanium, with the chemical composition shown in table 1.

**Table 1.** Chemical composition of AISI 316Ti stainless steel (wt. %)

| Cr | Ni | Mo | Mn | Ti | N | C | Si | P | S | Fe |
|---|---|---|---|---|---|---|---|---|---|---|
| 16.50 | 10.50 | 2.12 | 1.69 | 0.41 | 0.012 | 0.04 | 0.043 | 0.026 | 0.002 | balance |

HAP surface coating was obtained by electrochemical deposition (on the mechanically polished steel surface) in the electrolyte containing 0.042 M $Ca(NO_3)$ and 0.025 M $NH_4H_2PO_4$, pH=4.2. The deposition process proceeded at a stable cathode potential ranging from 2.25 to 3 V for 30 minutes at the room temperature. Microstructure of the obtained HAP coating was observed by SEM and its chemical composition was evaluated by EDX analysis.
0.9 % sodium chloride solution at the temperature of 37 ± 0.5 °C for simulation the internal environment of the human body was used as the corrosion environment for electrochemical potentiodynamic test and exposure immersion test.
The potentiodynamic polarization test was carried out in a three electrode cell of the measuring system (VoltaLab 10 with VSP unit). Polarization curves were obtained by the EC-LAB SOFT software. Potential between the sample and the electrolyte has been settled for 5 minutes before the polarization. Scan range was -0.3V – 0.9 V vs the open circuit potential and the scan rate was 1 mV/s. The measured surface of 1 $cm^2$ area was rinsed with ethanol before measurement. The saturated calomel electrode (SCE) was applied as the reference electrode and platinum foil as a counter electrode. At least five experiment repeats were carried out for both HAP coated surface and the for the steel surface without HAP coating (HAP coating was removed by grinding). For each of both types of surface, the most appropriate curve was chosen and according it, the electrochemical parameters were determined.
The exposure immersion test was carried out for three HAP coated rectangular-shaped specimens (it means steel substrate + Ca/P coating) with dimensions 75 mm x 30 mm x 4 mm. The specimens were degreased by ethanol and weighted out with accuracy ± 0.000 01 g. After 50-days exposure the specimens were washed by de-mineralized water and by ethanol, freely dried up and weighted out again [14]. The specimen surfaces were SEM observed after immersion test. Average corrosion rate ($g.m^{-2}.day^{-1}$) was calculated from corrosion losses ($g.m^{-2}$) of the three tested specimens.

## 3 Experiment results and discussion

The SEM analysis revealed a non-uniform HAP-like coating with microstructure created by flake-like rounds shapes (figure 1).

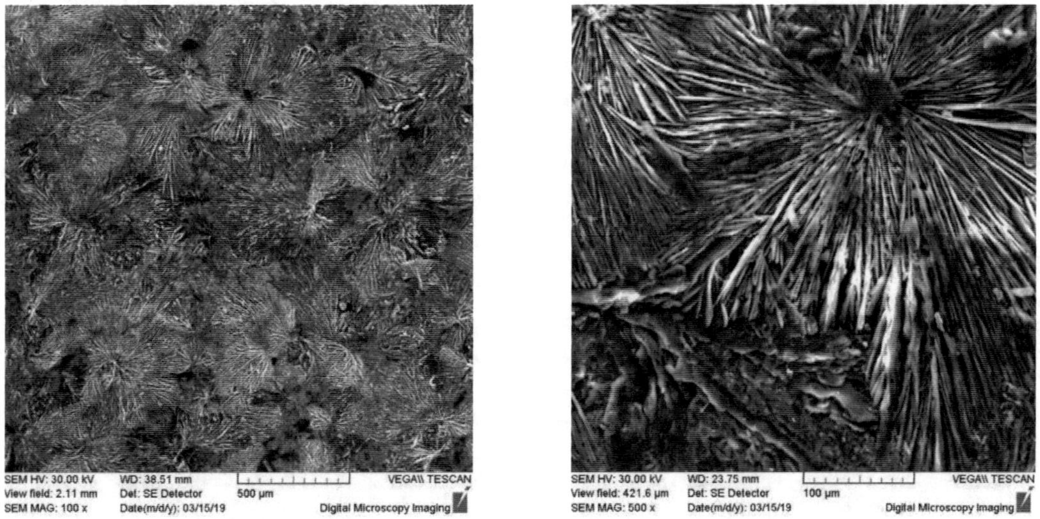


**Figure 1.** Microstructure of the HAP coating: left - general view, right - detail of flake-like shape

EDX point analysis (figure 2) confirmed the high content of Ca and P, Ca/P ratio 1.35. This ratio differs from the ideal Ca/P one (1.67) considered by the authors [15] and it points to the defected HAP coating obtained by used experimental method.

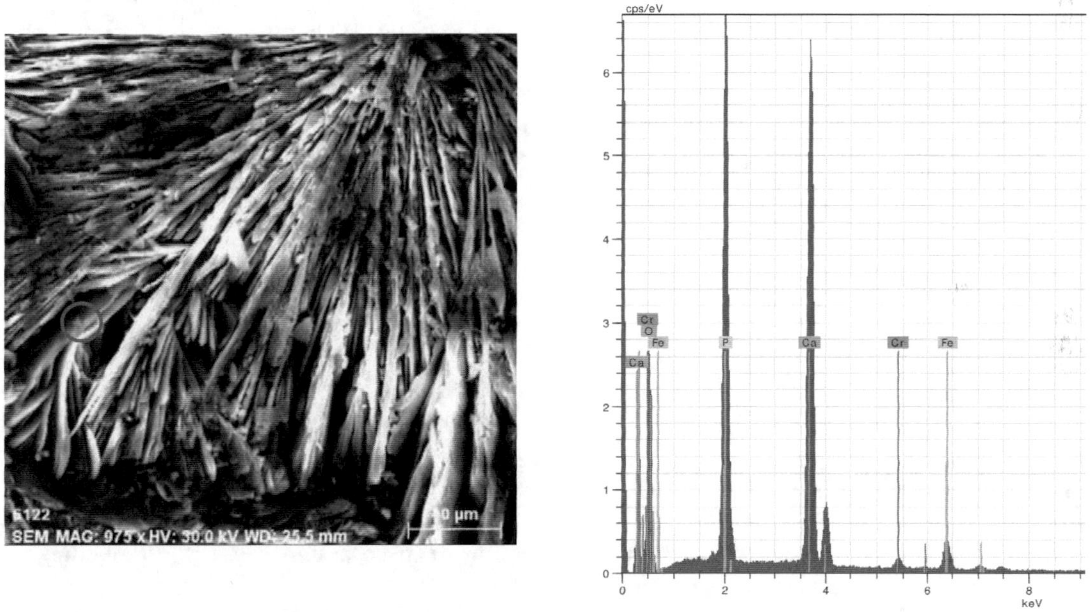


**Figure 2.** EDX analysis of HAP coating: left - the circled point at which the EDX analysis was carried out, right - diagram of the chosen chemical elements content

Potentiodynamic polarization curves of the HAP coated surface and of the steel ground surface without HAP coating are shown and can be compared in figure 3. The electrochemical characteristics determined from the curves are listed in table 2. The pitting potentials ($E_p$) of both tested curves were determined as the potentials of a sudden, strong increase of current density, which denotes the breakdown of the passive film and the beginning of stable pit growth. The shift of the HAP coated steel $E_p$ to the more noble value ($E_p$ position shown by the blue arrow) and the lower passive current density compared to the steel without HAP coating point to the rise of the resistance to the pitting [16-18]. The corrosion potential ($E_{corr}$) and the corrosion current density ($i_{corr}$) as the additional electrochemical characteristics were determined according to the Tafel analysis of the curves [19, 20]. The $E_{corr}$ values of the both curves are similar, but the thermodynamic stability of HAP coated material is slightly higher (higher $E_{corr}$ value). The $i_{corr}$ values, which have the relationship to the kinetics of the corrosion process, also point to the higher corrosion resistance of the HAP coated steel (lower $i_{corr}$ value).

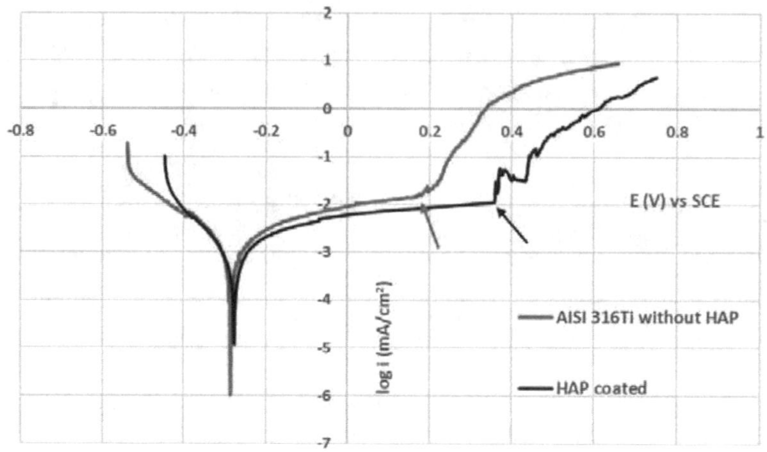


**Figure 3.** Potentiodynamic polarization curves of the AISI 316Ti working electrode with/without HAP coating. The arrows point to the $E_p$ positions.

**Table 2.** The electrochemical characteristics of the AISI 316Ti working electrode with/without HAP coating.

| | With HAP coating | Without HAP coating |
|---|---|---|
| $E_p$ (V) | 0.385 | 0.191 |
| $E_{corr}$ (V) | -0.277 | -0.288 |
| $i_{corr}$ (μA.cm$^{-2}$) | 0.776 | 1.402 |

As shown by SEM (figure 4), the HAP coating was markedly reduced during the long-term (50-days) exposure immersion test in 0.9 % NaCl solution at the temperature of 37 ± 0.5 °C. Moreover, the uncovered steel surface was locally attacked by the pitting. The observed phenomena were reflected in corrosion losses of the three tested samples and the average corrosion rate 0.128851 g.m$^{-2}$.day$^{-1}$ was calculated (standard deviation 0.0133). The authors [14] recorded the significantly lower average corrosion rate of the similar Cr-Ni-Mo stainless steel (without HAP) exposed to 0.9 % NaCl solution at the same temperature for the comparable time. Therefore, can be assumed that the reduction of HAP-like coating by the mechanical exfoliation during the immersion test (clearly visible in figure 4) contributed most to the higher corrosion loss of the tested AISI 316Ti biomaterial. It can be stated that the electrochemically deposited HAP-like coating did not show sufficient stability in the given long-term test conditions.

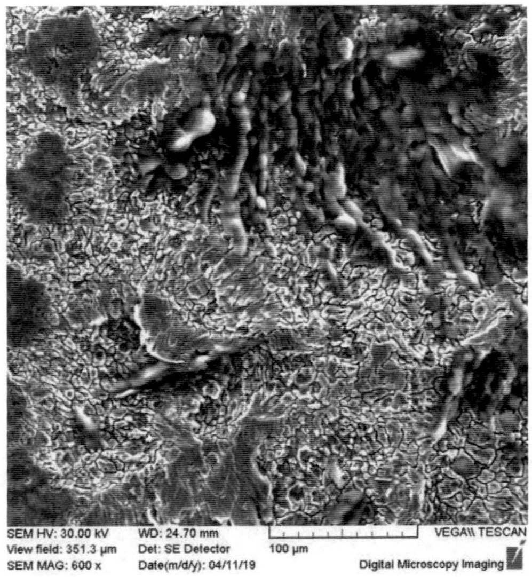


**Figure 4.** Reduced HAP coating after 50-days exposure immersion test. Stainless steel substrate is attacked by the pitting corrosion.

## 4 Conclusion

Based on the results of performed experiments can be concluded:

- The SEM and EDX analysis before the corrosion tests confirmed the non-uniform HAP-like coating with flake-like microstructure and with the Ca/P ratio 1.35.
- The corrosion losses and the SEM analysis after the exposure test showed that the HAP coating was severely disrupted upon exposure and the stainless steel substrate started to corrode locally by the pitting.
- According to the results of the electrochemical potentiodynamic test the HAP coated biomaterial had the considerably higher pitting potential, lower passive current density and also lower corrosion current density ($i_{corr}$) compared to that without the HAP coating. Thus, the stainless steel substrate (without the HAP coating) showed the lower corrosion resistance in this test.

It can be concluded that the experiments confirmed the HAP-like material properties close to commercially produced "young" hydroxyapatite-based materials. If this type of biomaterial comes into long-term contact with saline, it could lead to corrosion damage and thus to a reduction in the biocompatibility and the lifetime of the implant.

**Acknowledgement**

The research was supported partially by Scientific Grant Agency of Ministry of Education, Science and Sport of Slovak Republic and Slovak Academy of Sciences, grants KEGA No. 013ŽU-4/2019 and VEGA No. 1/0134/20.

## References

[1] Ratner B D, Hoffmann A S, Schoen F J and Lemons J E 2004 *Biomaterials Science* Elsevier San Diego USA pp 1337-1368
[2] Nagase M 1995 Host Reactions to Particulate Biomaterials. Encyclopedic handbook of biomaterials and bioengineering 1 Part A New York pp 95-97
[3] Babu N R, Manwatkar S, Rao K P and Kumar T S 2004 Bioactive Coatings on 316L Stainless Steel Implants *Trends Biomater Artif Organs* **17** 2 pp 43-47
[4] Shores E C and Holmes R E 1993 An Introduction to Bioceramics *World Scientific* pp 181–198

[5] Liu D, Savino K and Yates M Z, 2011 Coating of hydroxyapatite films on metal substrates by seeded hydrothermal deposition *Surface & Coatings Technology* **205** pp 3975-3986
[6] Kwok C, Wong P, Cheng F and Man H 2009 Characterization and corrosion behavior of hydroxyapatite coatings on Ti6-Al-4V fabricated by electrophoretic deposition *Applied Surface Science* **255** pp 6736-6744
[7] Bogya E S, Károly Z and Barabás R 2015 Atmospheric plasma sprayed silica–hydroxyapatite coatings on magnesium alloy substrates *Ceramics International* **41** 4 pp 6005–6012
[8] Masouleh M P and Asgharzadeh H 2013 Optimization of sol-gel technique for coating of metallic substrates by hydroxyapatite using the Taguchi method *Materials Science-Poland* **31** 3 pp 424-433.
[9] Gnanavel S, Ponnusamy S, Mohan L and Muthamizhchelvan C 2018 In Vitro Corrosion Behaviour of Ti–6Al–4V and 316L Stainless Steel Alloys for Biomedical Implant Applications *J Bio Tribo Corros* **4** 1 https://doi.org/10.1007/s40735-017-0118-8
[10] Ma J, Wong H, Kong L and Peng K 2003 Biomimetic processing of nanocrystallite bioactive apatite coating on titanium *Nanotechnology* **14** pp 1239-1324
[11] Paunovic M and Schlesinger M 2006 Fundamentals of electrochemical deposition New Jersey: John Wiley and Sons Inc. Hoboken 978-0471712213
[12] Nam P T, Lam T D, Huong H T, Phuong N T, Trang N T, Hoang T, Huong N T, Thang L B, Drouet C, Grossin D, Kergourlay E, Bertrand G, Devilliers D and Thanh D T 2015 Electrodeposition and Characterization of Hydroxyapatite on TiN/316LSS *Journal of Nanoscience and Nanotechnology* **15** pp 1–11
[13] Fazan F and Marquis P M 2000 Dissolution behavior of plasmasprayed hydroxyapatite coatings *J Mater Sci Mater Med* **11** pp 787–793
[14] Zatkalíková V, Markovičová L and Chalupová M 2016 Corrosion resistance of Cr-Ni-Mo stainless steel in chloride and fluoride containing environment *Manufacturing technology: journal for science, research and production* **16** 5 pp 1193-1198
[15] Miryousefi, H 2016 Effects of Substrate and Electrodeposition Parameters on the Microstructure of Hydroxyapatite Coating University of Tehran, Iran, [online.PDF] https://www.researchgate.net/publication/297615321_Effects_of_Substrate_and_Electrodeposition_Parameters_on_the_Microstructure_of_Hydroxyapatite_Coating
[16] Szklarska–Smialowska Z 2005 Pitting and crevice corrosion NACE International Houston Texas pp 15-138
[17] Liptáková T, Alaskari A and Halamová M 2012 Surface treatment of the AISI 316L after welding and study of corrosion behaviour *Periodica Polytechnica Transportation Engineering* **41** 2 pp 143-147
[18] Oravcová M, Palček P, Zatkalíková V, Tański T and Król M 2017 Surface treatment and corrosion behaviour of austenitic stainless steel biomaterial *IOP Conference Series: Materials Science and Engineering* **175** 1
[19] Hadzima B and Liptáková T 2008 Základy elektrochemickej korózie kovov (The bases of electrochemical corrosion of metals*)* EDIS Žilina pp 96-98
[20] Kuchariková L, Liptáková T, Tillová E, Kajánek D and Schmidová E 2018 Role of Chemical Composition in Corrosion of Aluminum Alloys *Metals* **8** 8 581 https://doi.org/10.3390/met8080581

# Chemical-mechanical polishing of TRIP steel surface

**O Ambrož[1], D Černík[1], D Rychlý[1], P Jozefovič[1] and Š Mikmeková[1]**

Institute of Scientific Instruments of the CAS, v. v. i., Královopolská 147, Brno 612 64, Czech Republic

E-mail: ondrej@isibrno.cz

**Abstract.** TRIP steels are mostly used due to their high mechanical properties in the automotive industry. The excellent mechanical properties of these steels are inseparably related to the retained austenite phase (γ) phase. Therefore, accurate knowledge of γ phase volume fraction, size, and distribution within a matrix is crucial for a correct understanding of the structure-properties relationship. This study aims to clarify the effect of the sample preparation technique on the results of phase analysis obtained by an electron back-scattered diffraction (EBSD) method. For EBSD analysis, the TRIP steel specimens were prepared by various techniques, such as mechanical polishing, chemical-mechanical polishing (CMP), and electropolishing. The CMP technique shows the most promising results. Therefore, this work focuses mainly on a methodology for optimizing this technique.

## 1 Introduction

Quality of specimen surface is crucial for an electron back-scatted diffraction (EBSD) analysis due to the extremely shallow escape depth of the diffracted electrons. The EBSD is a surface-sensitive technique and the information depth is only a few tens of nm. Hence, the surface top layers play an important role and are responsible for the quality of the EBSD analysis [1,2]. Various metallographic techniques are available for the EBSD characterization, such as mechanical polishing, chemical polishing, chemical-mechanical polishing, and electropolishing. For multiphase steels, such as TRIP steels, the most common technique used is the electropolishing technique. The electropolishing technique provides high surface quality but on the other hand, this technique results in selective etching of the phases, and martensite and austenite phases arise from the ferritic-bainitic matrix. Such surface relief creates errors in EBSD data collection. The chemical polishing (i.e. etching) is also problematic due to increasing of the surface roughness. The mechanical polishing is usually not sufficient for the EBSD phase analysis of the TRIP steels because of austenite to martensite transformation induced by deformation. Chemical-mechanical polishing seems to be a promising technique for obtaining a flat and deformation-free surface [3]. This paper focuses on the investigation of the effect of the chemical-mechanical polishing process parameters on the EBSD data collected from the TRIP steel samples.

## 2 Experimental part

The specimens were characterized using a dual-beam microscope Helios 4G (Thermofisher Scientific). The microscope is equipped with a sophisticated detection system enabling effective angular and energy separation of the signal electrons. The scanning electron microscope (SEM) images were collected by a mirror detector (MD) situated within a column. This detector collects mainly very slow secondary images. It enables us to obtain surface-sensitive images containing information about the surface state, such as contamination, oxidation, etc. The surface roughness was investigated with a concentric back-scattered detector (CBS). The CBS detector enables effective angular and energy filtering of the back-scattered electrons and collection of the low take-off angle elastic electrons provides the micrographs with high sensitivity on the surface topography. The SEM images were collected at 1 keV landing energy of the primary electrons which secures low information depth and high surface-sensitive imaging. The EBSD analysis was performed in an ultra-high-resolution SEM Magellan 400L. The Magellan is equipped with a Hikari high-speed EBSD Camera (EDAX) and the EBSD data were analyzed using OIM Analysis 7 (EDAX). The material under investigation was a transformation induced plasticity (TRIP) steel having a bainitic-ferritic matrix and containing 12% of retained austenite (based on the X-Ray Diffraction analysis). The steel had low carbon content of 0.2% and was further alloyed by manganese and silicon. Manganese content of the steel was 2%, and there was 1.5% of silicon. Heat treatment of the steel is not known (not provided). The specimens for the chemical-mechanical polishing experiments were prepared by standard metallographic techniques, i.e. using conventional grinding followed by a polishing process finished with 0.25 µm diamond paste polishing. Several types of colloidal silica were used in the experiment. The samples were finished with OP-A, OP-U, and OP-S suspensions. The OP-S is a colloidal silica suspension with a pH of 9.8 and a grain size of about 0.25 µm. This suspension is suited for very ductile metals, such as Al, Ti, Cu, and others. The OP-U is also a colloidal silica suspension, like OP-S, with a pH of 9.8 and 0.04 µm particle size, but with different composition, which makes it less abrasive. This suspension is suited for the final polishing of all materials. The OP-A is an acidic aluminum oxide suspension with a pH of 3 and is suited for the final polishing of ferrous metals, Ni-base alloys, and ceramics. The grain size of alumina particles is 0.02 µm.

## 3 Results and discussion

Figure 1 shows the TRIP steel specimens prepared by different metallographic techniques, namely by mechanical polishing (MP), chemical polishing using 5 ml HF in 95 ml $H_2O_2$ (CHP), chemical-mechanical polishing with colloidal silica suspension OP-U (CHMP), and conventional electropolishing (EP). The SEM images were obtained with the MD detector (upper row) and with the last segment of the CBS detector (bottom row). The MP process finished with 0.25 µm diamond paste polishing results in a low-quality surface containing various defects, such as scratches and the presence of abrasive particles embedded in the matrix. The CHP specimen was finished with 10 sec in fresh 5 ml HF in 95 ml $H_2O_2$ and as visible, the etching rate is not homogeneous the secondary phases are etched preferentially. The same phenomenon, i.e. preferential etching of selected phases, occurs in the case of the EP process. Based on the SEM results, the specimen surface finished with the CHMP process exhibits the highest quality of the surface. The CHPM specimen is flat and without any defects observed in the case of the MP specimen.

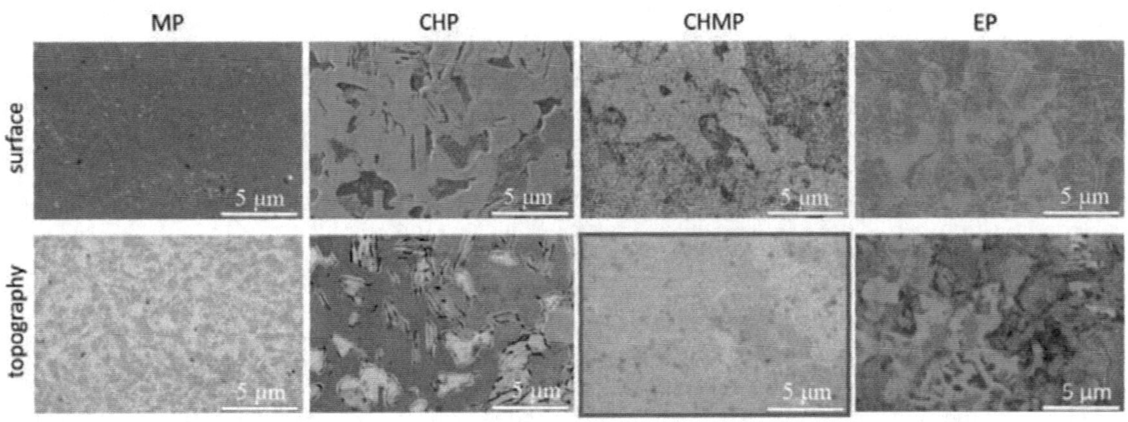

**Figure 1.** SEM micrographs of the TRIP steel specimens prepared by: mechanical polishing (MP), chemical polishing (CHP), chemical-mechanical polishing (CHMP), and electropolishing (EP). The upper row micrographs: low energy secondary electrons sensitive on surface state (MD detector); the bottom row micrographs: low take off angle back-scattered electrons sensitive on topography (CBS detector, mode D).

The SEM micrographs presented in figure 1 were collected at 1 keV landing energy of the primary electrons, 0.8 nA probe current and the working distance of 7 mm. Imaging at low landing energies is important due to the achievement of high surface sensitivity. Figure 2 show a Monte Carlo simulation of the primary electrons trajectories having 1 keV landing energy in Fe target (software Casino). The penetration depth is shallower than 10 nm.

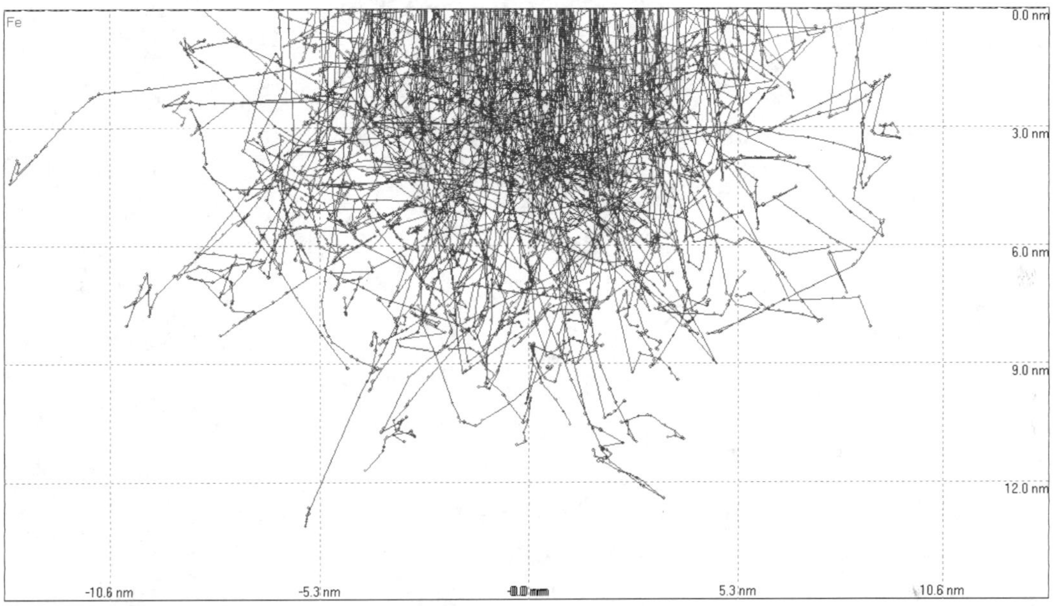


**Figure 2.** Trajectories of the primary electrons in Fe target: 1 keV landing energy of projectiles, number of projectiles in simulation 200 000, perpendicular angle of incidence. Red trajectories: elastically-scattered electrons; bleu trajectories: inelastically-scattered electrons.

The next part of the study is focused on the CHMP process. Figure 3 shows the results of the EBSD analysis of the specimens finished with various active oxide polishing suspensions. The EBSD analysis was performed with a step size of 100 nm. The TRIP steel specimens were finished using identical conditions, i.e. 0.25 μm diamond polishing followed by the CHMP process. The CHMP processes were performed for 30 min with loads of 5 N and the OP-A, the OP-S, and the OP-U suspensions were utilized. As is visible, the OP-A suspension with acidic pH of 3 is very aggressive and its utilization results in the destroying of the surface and secondary phases. The OP-A specimen EBSD results show the extremely low quality of secondary phases. Martensite and austenite phases are dark in the image quality (IQ) map and are not indexed in the phase map. The ferrite matrix is indexed in an orientation map but the quality of patterns is low. The surface becomes probably corroded inside the acidic environment created by the OP-A. In contrast, the secondary phases become indexed in the case of samples prepared with the OP-S and the OP-U solutions. The OP-S and the OP-U specimens show similar total fraction of the FCC phase, as mentioned in the table. The indexing quality parameters for the FCC phase, i.e. a confidence and an image quality index, are better in the case of the OP-U specimen.

For a more careful comparison between the OP-S and the OP-U samples, the high-resolution EBSD analysis was performed. Figure 4 shows the EBSD orientation, image quality, and phase maps of the OP-S and the OP-U specimens collected with a very fine step size of 20 nm.
The specimen finished with the OP-U suspension exhibits less artifacts than the specimen prepared with the OP-S. The OP-S specimen IQ map shows the presence of shadows near the grain boundaries (marked by yellow arrows) as a consequence of not optimal surface planeness close to the boundaries.

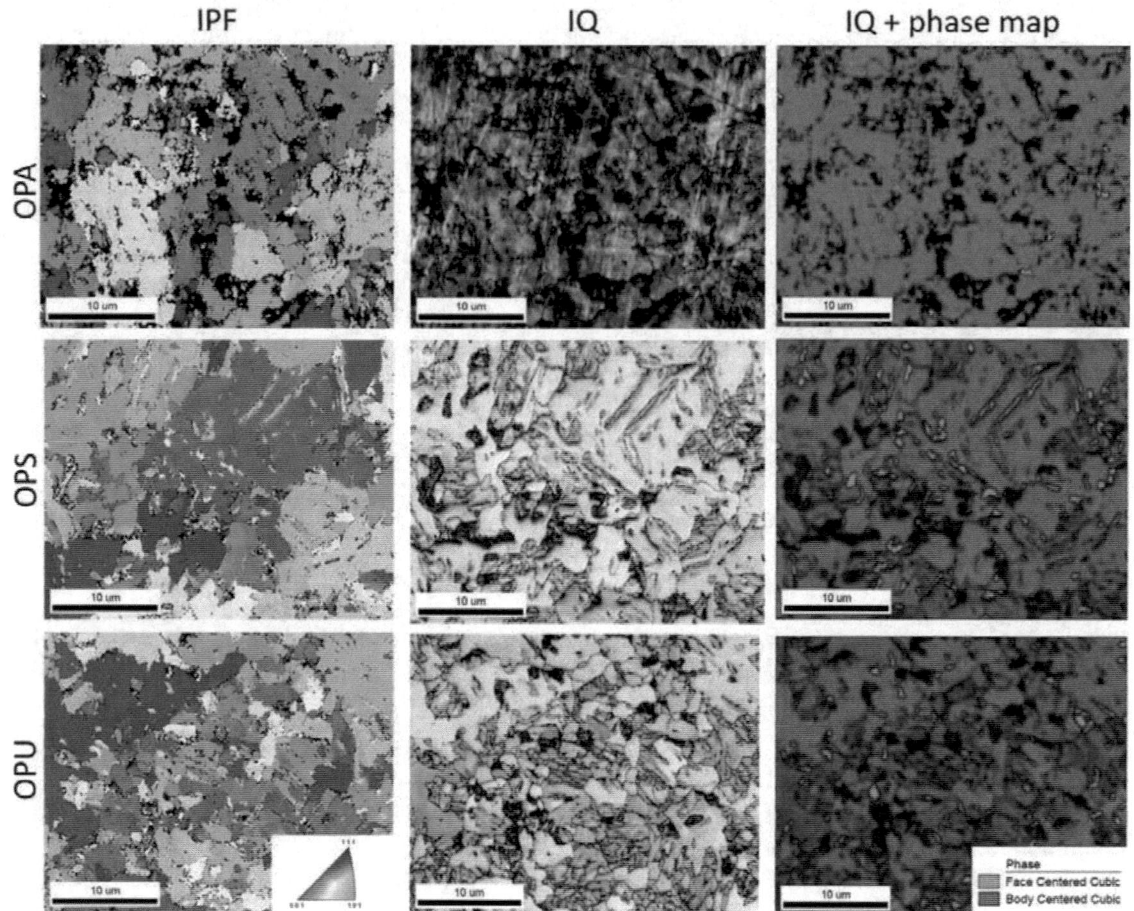


| specimen | total fraction BCC | total fraction FCC | FCC average confidence index | FCC avera image quality |
|---|---|---|---|---|
| OP-A | 0.767 | 0.006 | 0.07 | 1995.02 |
| OP-S | 0.868 | 0.075 | 0.50 | 4663.32 |
| OP-U | 0.861 | 0.064 | 0.62 | 5085.39 |

**Figure 3.** The EBSD analysis of OP-A, OP-S, and OP-U specimens: orientation map (IPF), image quality map (IQ), IQ + phase map. The EBSD analysis parameters: 20 keV primary beam energy, 3.2 nA probe current, step size 100 nm. The table situated below shows total fraction of the phases and average confidence index and image quality parameters of FCC lattice.

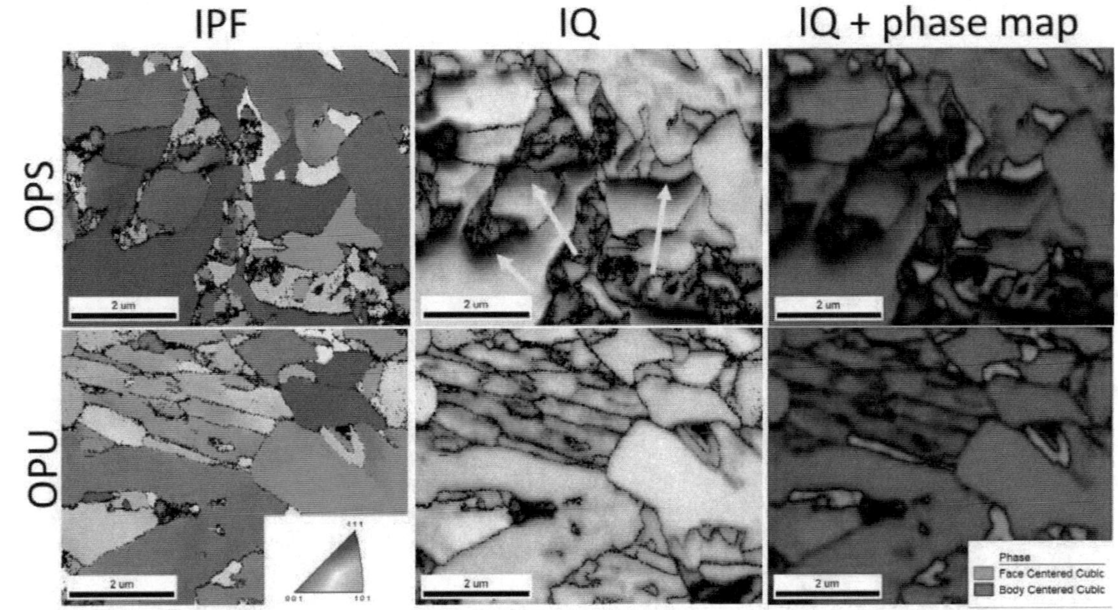


**Figure 4.** The EBSD analysis of OP-S and OP-U specimens: orientation map (IPF), image quality map (IQ), IQ + phase map. The EBSD analysis parameters: 20 keV primary beam energy, 3.2 nA probe current, step size 20 nm.

## 4 Conclusion

The TRIP steel surfaces were finished with various active oxide polishing suspensions, namely with the OP-S, the OP-U, and the OP-A. The specimens were prepared using the identical polishing process parameters, i.e. polishing time of 30 minutes and with loads of 5 N. Only suspension type was different for each process. The OP-A suspension is formed by the finest particles but its acidic pH is very aggressive which results in total destroying of the secondary phases (martensite, austenite). The OP-S and the OP-U show similar results to the EBSD analysis. Comparison of the EBSD indexing quality parameters for the FCC lattice, such as the average confidence index and the image quality index, show higher and thus better values in the case of the OP-U specimen. Furthermore, the high-resolution EBSD analysis of the OP-S and the OP-U specimens shows that the OP-U specimen has lower roughness and thus contains fewer artefacts close to the grain boundaries. Based on this study, the OP-U is the most suited suspension for the TRIP steels specimen preparation and secures high-quality EBSD data.

## Acknowledgement

The research was supported by Technology Agency of the Czech Republic (TN01000008).

## References

[1] Humphreys F J 2001 Grain and subgrain characterisation by electron backscatter diffraction *J. Mater. Sci.* **36** 3833-54

[2] Thomsen K, Schmidt N H, Bewick A, Larsen K and Goulden J 2013 Improving the accuracy of orientation measurements using EBSD *Microsc. Microanal.* **19** 724-5

[3] Girault E, Jacques P, Harlet Ph, Mols K, Van Humbeeck J, Aernoudt E and Delannay F 1998 Metallographic methods for revealing the multiphase microstructure of TRIP-assisted steels *Mater. Charact.* **40** 111-8

# AUTHOR INDEX